Pitman Research Notes in Ma
Series

Submission of proposals for consideration
Suggestions for publication, in the form of outlines and representative samples, are invited by the Editorial Board for assessment. Intending authors should approach one of the main editors or another member of the Editorial Board, citing the relevant AMS subject classifications. Alternatively, outlines may be sent directly to the publisher's offices. Refereeing is by members of the board and other mathematical authorities in the topic concerned, throughout the world.

Preparation of accepted manuscripts
On acceptance of a proposal, the publisher will supply full instructions for the preparation of manuscripts in a form suitable for direct photo-lithographic reproduction. Specially printed grid sheets can be provided and a contribution is offered by the publisher towards the cost of typing. Word processor output, subject to the publisher's approval, is also acceptable.

Illustrations should be prepared by the authors, ready for direct reproduction without further improvement. The use of hand-drawn symbols should be avoided wherever possible, in order to maintain maximum clarity of the text.

The publisher will be pleased to give any guidance necessary during the preparation of a typescript, and will be happy to answer any queries.

Important note
In order to avoid later retyping, intending authors are strongly urged not to begin final preparation of a typescript before receiving the publisher's guidelines. In this way it is hoped to preserve the uniform appearance of the series.

Addison Wesley Longman Ltd
Edinburgh Gate
Harlow, Essex, CM20 2JE
UK
(Telephone (0) 1279 623623)

Titles in this series. A full list is available from the publisher on request.

251 Stability of stochastic differential equations with respect to semimartingales
X Mao
252 Fixed point theory and applications
J Baillon and M Théra
253 Nonlinear hyperbolic equations and field theory
M K V Murthy and S Spagnolo
254 Ordinary and partial differential equations. Volume III
B D Sleeman and R J Jarvis
255 Harmonic maps into homogeneous spaces
M Black
256 Boundary value and initial value problems in complex analysis: studies in complex analysis and its applications to PDEs 1
R Kühnau and W Tutschke
257 Geometric function theory and applications of complex analysis in mechanics: studies in complex analysis and its applications to PDEs 2
R Kühnau and W Tutschke
258 The development of statistics: recent contributions from China
X R Chen, K T Fang and C C Yang
259 Multiplication of distributions and applications to partial differential equations
M Oberguggenberger
260 Numerical analysis 1991
D F Griffiths and G A Watson
261 Schur's algorithm and several applications
M Bakonyi and T Constantinescu
262 Partial differential equations with complex analysis
H Begehr and A Jeffrey
263 Partial differential equations with real analysis
H Begehr and A Jeffrey
264 Solvability and bifurcations of nonlinear equations
P Drábek
265 Orientational averaging in mechanics of solids
A Lagzdins, V Tamuzs, G Teters and A Kregers
266 Progress in partial differential equations: elliptic and parabolic problems
C Bandle, J Bemelmans, M Chipot, M Grüter and J Saint Jean Paulin
267 Progress in partial differential equations: calculus of variations, applications
C Bandle, J Bemelmans, M Chipot, M Grüter and J Saint Jean Paulin
268 Stochastic partial differential equations and applications
G Da Prato and L Tubaro
269 Partial differential equations and related subjects
M Miranda
270 Operator algebras and topology
W B Arveson, A S Mishchenko, M Putinar, M A Rieffel and S Stratila
271 Operator algebras and operator theory
W B Arveson, A S Mishchenko, M Putinar, M A Rieffel and S Stratila
272 Ordinary and delay differential equations
J Wiener and J K Hale
273 Partial differential equations
J Wiener and J K Hale
274 Mathematical topics in fluid mechanics
J F Rodrigues and A Sequeira
275 Green functions for second order parabolic integro-differential problems
M G Garroni and J F Menaldi
276 Riemann waves and their applications
M W Kalinowski
277 Banach C(K)-modules and operators preserving disjointness
Y A Abramovich, E L Arenson and A K Kitover
278 Limit algebras: an introduction to subalgebras of C*-algebras
S C Power
279 Abstract evolution equations, periodic problems and applications
D Daners and P Koch Medina
280 Emerging applications in free boundary problems
J Chadam and H Rasmussen
281 Free boundary problems involving solids
J Chadam and H Rasmussen
282 Free boundary problems in fluid flow with applications
J Chadam and H Rasmussen
283 Asymptotic problems in probability theory: stochastic models and diffusions on fractals
K D Elworthy and N Ikeda
284 Asymptotic problems in probability theory: Wiener functionals and asymptotics
K D Elworthy and N Ikeda
285 Dynamical systems
R Bamon, R Labarca, J Lewowicz and J Palis
286 Models of hysteresis
A Visintin
287 Moments in probability and approximation theory
G A Anastassiou
288 Mathematical aspects of penetrative convection
B Straughan
289 Ordinary and partial differential equations. Volume IV
B D Sleeman and R J Jarvis
290 *K*-theory for real *C**-algebras
H Schröder
291 Recent developments in theoretical fluid mechanics
G P Galdi and J Necas
292 Propagation of a curved shock and nonlinear ray theory
P Prasad
293 Non-classical elastic solids
M Ciarletta and D Ieșan
294 Multigrid methods
J Bramble
295 Entropy and partial differential equations
W A Day
296 Progress in partial differential equations: the Metz surveys 2
M Chipot
297 Nonstandard methods in the calculus of variations
C Tuckey
298 Barrelledness, Baire-like- and (LF)-spaces
M Kunzinger
299 Nonlinear partial differential equations and their applications. Collège de France Seminar. Volume XI
H Brezis and J L Lions

300 Introduction to operator theory
T Yoshino
301 Generalized fractional calculus and applications
V Kiryakova
302 Nonlinear partial differential equations and their applications. Collège de France Seminar Volume XII
H Brezis and J L Lions
303 Numerical analysis 1993
D F Griffiths and G A Watson
304 Topics in abstract differential equations
S Zaidman
305 Complex analysis and its applications
C C Yang, G C Wen, K Y Li and Y M Chiang
306 Computational methods for fluid-structure interaction
J M Crolet and R Ohayon
307 Random geometrically graph directed self-similar multifractals
L Olsen
308 Progress in theoretical and computational fluid mechanics
G P Galdi, J Málek and J Necas
309 Variational methods in Lorentzian geometry
A Masiello
310 Stochastic analysis on infinite dimensional spaces
H Kunita and H-H Kuo
311 Representations of Lie groups and quantum groups
V Baldoni and M Picardello
312 Common zeros of polynomials in several variables and higher dimensional quadrature
Y Xu
313 Extending modules
N V Dung, D van Huynh, P F Smith and R Wisbauer
314 Progress in partial differential equations: the Metz surveys 3
M Chipot, J Saint Jean Paulin and I Shafrir
315 Refined large deviation limit theorems
V Vinogradov
316 Topological vector spaces, algebras and related areas
A Lau and I Tweddle
317 Integral methods in science and engineering
C Constanda
318 A method for computing unsteady flows in porous media
R Raghavan and E Ozkan
319 Asymptotic theories for plates and shells
R P Gilbert and K Hackl
320 Nonlinear variational problems and partial differential equations
A Marino and M K V Murthy
321 Topics in abstract differential equations II
S Zaidman
322 Diffraction by wedges
B Budaev
323 Free boundary problems: theory and applications
J I Diaz, M A Herrero, A Liñan and J L Vazquez
324 Recent developments in evolution equations
A C McBride and G F Roach
325 Elliptic and parabolic problems: Pont-à-Mousson 1994
C Bandle, J Bemelmans, M Chipot, J Saint Jean Paulin and I Shafrir
326 Calculus of variations, applications and computations: Pont-à-Mousson 1994
C Bandle, J Bemelmans, M Chipot, J Saint Jean Paulin and I Shafrir
327 Conjugate gradient type methods for ill-posed problems
M Hanke
328 A survey of preconditioned iterative methods
A M Bruaset
329 A generalized Taylor's formula for functions of several variables and certain of its applications
J-A Riestra
330 Semigroups of operators and spectral theory
S Kantorovitz
331 Boundary-field equation methods for a class of nonlinear problems
G N Gatica and G C Hsiao
332 Metrizable barrelled spaces
J C Ferrando, M López Pellicer and L M Sánchez Ruiz
333 Real and complex singularities
W L Marar
334 Hyperbolic sets, shadowing and persistence for noninvertible mappings in Banach spaces
B Lani-Wayda
335 Nonlinear dynamics and pattern formation in the natural environment
A Doelman and A van Harten
336 Developments in nonstandard mathematics
N J Cutland, V Neves, F Oliveira and J Sousa-Pinto
337 Topological circle planes and topological quadrangles
A E Schroth
338 Graph dynamics
E Prisner
339 Localization and sheaves: a relative point of view
P Jara, A Verschoren and C Vidal
340 Mathematical problems in semiconductor physics
P Marcati, P A Markowich and R Natalini
341 Surveying a dynamical system: a study of the Gray–Scott reaction in a two-phase reactor
K Alhumaizi and R Aris
342 Solution sets of differential equations in abstract spaces
R Dragoni, J W Macki, P Nistri and P Zecca
343 Nonlinear partial differential equations
A Benkirane and J-P Gossez
344 Numerical analysis 1995
D F Griffiths and G A Watson
345 Progress in partial differential equations: the Metz surveys 4
M Chipot and I Shafrir
346 Rings and radicals
B J Gardner, Liu Shaoxue and R Wiegandt
347 Complex analysis, harmonic analysis and applications
R Deville, J Esterle, V Petkov, A Sebbar and A Yger
348 The theory of quantaloids
K I Rosenthal
349 General theory of partial differential equations and microlocal analysis
Qi Min-you and L Rodino

350 Progress in elliptic and parabolic partial differential equations
A Alvino, P Buonocore, V Ferone, E Giarrusso, S Matarasso, R Toscano and G Trombetti
351 Integral representations for spatial models of mathematical physics
V V Kravchenko and M V Shapiro
352 Dynamics of nonlinear waves in dissipative systems: reduction, bifurcation and stability
G Dangelmayr, B Fiedler, K Kirchgässner and A Mielke
353 Singularities of solutions of second order quasilinear equations
L Véron
354 Mathematical theory in fluid mechanics
G P Galdi, J Málek and J Necas
355 Eigenfunction expansions, operator algebras and symmetric spaces
R M Kauffman
356 Lectures on bifurcations, dynamics and symmetry
M Field
357 Noncoercive variational problems and related results
D Goeleven
358 Generalised optimal stopping problems and financial markets
D Wong
359 Topics in pseudo-differential operators
S Zaidman
360 The Dirichlet problem for the Laplacian in bounded and unbounded domains
C G Simader and H Sohr
361 Direct and inverse electromagnetic scattering
A H Serbest and S R Cloude
362 International conference on dynamical systems
F Ledrappier, J Lewowicz and S Newhouse
363 Free boundary problems, theory and applications
M Niezgódka and P Strzelecki
364 Backward stochastic differential equations
N El Karoui and L Mazliak
365 Topological and variational methods for nonlinear boundary value problems
P Drábek
366 Complex analysis and geometry
V Ancona, E Ballico, R M Mirò-Roig and A Silva
367 Integral expansions related to Mehler–Fock type transforms
B N Mandal and N Mandal
368 Elliptic boundary value problems with indefinite weights: variational formulations of the principal eigenvalue and applications
F Belgacem
369 Integral transforms, reproducing kernels and their applications
S Saitoh
370 Ordinary and partial differential equations. Volume V
P D Smith and R J Jarvis
371 Numerical methods in mechanics
C Conca and G N Gatica
372 Generalized manifolds
K-G Schlesinger
373 Independent axioms for Minkowski space-time
J W Schutz
374 Integral methods in science and engineering Volume one: analytic methods
C Constanda, J Saranen and S Seikkala
375 Integral methods in science and engineering Volume two: approximation methods
C Constanda, J Saranen and S Seikkala
376 Inner product spaces and applications
T M Rassias
377 Functional analysis with current applications in science, technology and industry
M Brokate and A H Siddiqi
378 Classical and quantic periodic motions of multiply polarized spin-particles
A Bahri
379 Analysis, numerics and applications of differential and integral equations
M Bach, C Constanda, G C Hsiao, A-M Sändig and P Werner

Abbas Bahri

Rutgers University

Classical and quantic periodic motions of multiply polarized spin-particles

LONGMAN

Addison Wesley Longman Limited
Edinburgh Gate, Harlow
Essex CM20 2JE, England
and Associated Companies throughout the world.

*Published in the United States of America
by Addison Wesley Longman Inc.*

First published 1998

AMS Subject Classifications (Main) 21, 37, 39
(Subsidiary) 45

ISSN 0269-3674

ISBN 0 582 32749 0

British Library Cataloguing in Publication Data

A catalogue record for this book is
available from the British Library

Printed in Great Britain by Henry Ling Ltd.,
at the Dorset Press, Dorchester, Dorset.

Table of contents

Preface

A word of warning to the adventurous reader: this work, as were several of my earlier works, is a huge construction, relying on many technical pieces. Each of these pieces needs to be sound for the ensemble to work.

While I have been careful in my proofs and I read them again and again, I would not dare claim that there is no major mistake in here (in crucial details), let alone minor ones or misprints: I am confident that my statements are sound, but there are limits to my absolute certitudes.

In earlier works, I also could not make such a claim. However, all these works turned out to be right, either because some independent proofs of some partial results confirmed them or because, with time, I have built a better understanding or because my students have pushed the related directions. I have also, countless times, repeated these proofs, to myself and to my students, under various forms.

For this present work, I claim that the phenomena which I describe in the introduction and those which I study in the second and third sections of this paper are qualitatively correct, I have no doubt about this. They had also been displayed, by a different method, in my earlier work "Pseudo-Orbits of contact forms". The general aim and scope of the first section is also certainly sound, since this is only a new proof, with very few hypotheses, of the technical results of "Pseudo-Orbits of contact forms". Hence, the results of this section also certainly hold. Have I indeed attained the full generality which I claim? Only time, and teaching will tell.

This is the end of a long journey, started seventeen years ago. Thanks are due to several persons who were my companions along this road: my parents, my late father and my mother; in their home, I always found the quietness and the calm to think and study almost without noticing, because I was so happy, I have been lucky to be born in a family of scholars, of the quiet and discrete type; Eva and Felix Browder, Haim Brezis. Eva and Felix, I have known since I undertook this work, our relationship has become wonderful. While I was not producing any tangible result during two years at the University of Chicago, Felix never doubted that I was working, I never heard a word from him about my complete absence of productivity. This is again why I have been lucky: I found myself in the hands of scholars.

Haim Brezis supported this work and was keen to have it published at the most difficult times, during these seventeen years. I am deeply grateful to him for this: he has very sound intellectual choices, he knows the true value of several things and he is a friend I am lucky to have had on my way.

My sister and my brothers Akiçà, Seoud, Annes, my wife Diana, all of the same blend, this combination of love, affection and respect for scholarship and intellectual achievement. Diana managed to make something of my hectic life, Annes has become

Typeset by $\mathcal{AMS}$-TEX

the friend of the conversations I used to have with my father, Seoud ... I am lucky to have them.

Yvan Charmion, Emmanuel Hebey, Jallal Shatah, Sagun Chanillo and Belhassen Dehman, we are tied by a deep friendship and a mutual respect, all my thanks and my affection to my friends, Raouf Bennaceur with whom I had several conversations.

My children, the twins Thouraya and Kahena, Salima and Mohamed El Hedi, they give to my life its true dimension. As they mature, Thouraya and Kahena are becoming my friends, the two little ones give me the impression of being something and someone.

To all of them, these words of thanks and of gratitude. May I add to them, because of her outstanding work, Ms. Robin Campbell and Mrs. Barbara Mastrian who typed this manuscript. I am very, very much indebted to them. Mrs. Daisy Calderon also helped in the typing of this manuscript. My thanks extend also to her. Finally, it is my great pleasure to thank the National Science Foundation for its constant support since 1984.

Montgomery, January 6th 1997
Abbas Bahri

Introduction

The following work is concerned by classical and quantic periodic motions of particles. Our framework is purely mathematical, our motivation started with the study of periodic motions of a natural generalization, pointed at by A. Weinstein [12], of the Hamiltonian-vector-fields of classical mechanics, a classical generalization which is called, in the mathematical framework, contact vector-fields. However, our approach of the Weinstein-conjecture, which came after the work of P. H. Rabinowitz [10] and A. Weinstein [11], had led us very far away of our initial goal. Studying this conjecture, we discovered [1] new motions of non-classical type, but which we could not explain physically. We have called these new motions critical points at infinity, have described them as "pseudo-orbits of contact forms" in [1], have generated then a mathematical concept in variational theory naturally associated to those critical points at infinity ([2], [3]) and, then, have applied it to other problems in geometry (Yamabe problem, scalar-curvature problem, C. R. Yamabe problem [4], [5], [6], [7], [8]) and in Newtonian mechanics ([9]).

We had abandoned our earlier effort on contact forms, for two reasons: one was mathematical and could be summed up in one serious, major difficulty which we described in the introduction of [1], namely our variational problem had no Fredholm structure; in particular, the periodic orbits, as critical points, did not induce any difference of topology in the level sets of the functional, even after completing Galerkin approximations. We were left with variations without any hope for finding the critical points of these variations. The other reason was that we were unable to find, despite various attempts as well as conversations with physicists, an appropriate physical interpretation for these non classical periodic motions, besides some unclear resemblance with WKB techniques in some regard—our problem is highly non linear and, as we will see, has no linear version — and tunnel effect.

Physics, which claims to describe the laws of Nature, cannot accept a complication without motivation, we had to keep away from any interpretation. With this paper, we remove the two reasons, that is we establish all the results and concepts of [1] on a Fredholm ground on one hand; on the other hand, we provide a physical, very natural, interpretation for these critical points at infinity, which fits completely the quantized motions of the spin of an elementary particle and provides an explanation for sudden changes in the polarization of their spins.

We prove that these quantized motions are very essential to the classical motions of the spin, that they are in some way, as we tried to explain throughout all our work ([1], [2], [3]), more important than the classical motions. We also prove that there are periodic forms of these motions.

We provide natural manifolds (Γ_{2k} and Σ_k), for each integer k, of curves with k

jumps and a variational problem on them, corresponding to the least action principle, and whose critical points are these quantized motions (for Γ_{2k}).

Furthermore, our framework goes beyond the framework of S^1 bundles which is the one used in the quantization theory, we work on a general three-dimensional compact manifold M^3, equipped with a contact form α, M^3 is not necessarily an S^1 bundle over the space of positions of spins, usually S^2.

Finally, with respect to the mathematical framework of [1], we remove almost all the hypotheses about the dynamics of α along v (the vector-field of the fibers in quantization theory) and give to our results a vast generality.

Let us now describe, in this introduction, more precisely our framework and discuss its physical interpretation. M^3 is a three-dimensional compact manifold equipped with a form α such that

$$\alpha \wedge d\alpha$$

is never zero (is a volume form).

ξ is the Reeb vector-field of α i.e., ξ satisfies:

$$\alpha(\xi) \equiv 1 \quad d\alpha(\xi, \cdot) \equiv 0$$

ξ is the natural generalization of the Hamiltonian vector-field of classical mechanics. The framework of (M^3, α, ξ) can be encountered in classical mechanics: given a free particle which lives in a two-dimensional manifold N^2, equipped with a Riemannian metric g, its motions are along the geodesics of the metric, which are the projections on N^2 of the orbits of the Hamiltonian vector-field on the unit S^1 (sphere) bundle over N^2.

However, a fact which is not usually emphasized is that it is *very natural* in quantum mechanics. Indeed, when the classical motions of a particle with spin are quantized, an S^1 bundle is usually built (see [13] where this presentation is very clear) over the direct product of the space-time space E with S^2. The position in S^2 modelizes the self-orientation of the particle. There is a natural contact form on this S^1 bundle, which couples the usual classical action (the form $\Sigma\, p_i\, dq_i$) with the spin action. Embedded in it, there is an S^1 bundle over S^2 equipped with a contact form. This S^1 bundle varies depending on the value of the spin number. For example, for spin 1, we obtain $P\mathbb{R}^3$, the projective S^1 bundle (built out of the unit sphere tangent bundle) over S^2 it modelizes the motion of the spin of a photon. For spin 1/2, we obtain S^3, this modelizes for example the electron-quantic spin motion, etc.

In an overwhelming number of regards, we share the framework of quantization theory for the spin. However, our *projection* is different from the standard one, although it allows to track perfectly the spin position on S^2. We claim that it is equivalent for several reasons explained below.

In the models discussed in the literature, the contact form part over S^2 is independent of the position in space-time. It depends only on the spin and the position above the spin in the S^1 bundle (see [13] for example). The model examples are symmetric in the S^1 direction. However, this symmetry can be broken. This phenomenon is usually called, in physics, *polarization.* This is a geometrical phenomenon which can be linear in some directions, circular in other directions, elliptic in third directions. The circular polarization of the particles is divided into two classes: the clock-wise and counterclockwise polarizations. From what we can understand the mathematical presentation of the notion of polarization is ambiguous in quantum mechanics, because it is unclear whether distinct polarizations correspond to different particles or to one particle having possibly different polarizations. One can feel this ambiguity when confronted to the two ways of producing a spin equal to $\frac{n}{2}$: either "fusion", a framework which fits more the one particle scheme, or by "assembly of identical particles", which seems to be more commonly used. Polarization, for us, is related to a notion of rotation of the contact form along a vector-field v in its kernel. It is based on the notions of coincidence and conjugate points described below. Our point of view, while different, is closer to the quantization by fusion. We keep one "spin direction", i.e. one base point on S^2 with various possible polarizations, tracked by the projection of the characteristic direction of α, ξ. This construction is not identical to the usual ones. But it is easy to reconstruct from it the usual spin direction, which corresponds to another Hopf projection map. A particle can therefore have multiplicity. Our scheme, is completely compatible with the various polarizations which are possible for spin 1/2, for example. These polarizations are of two types: those bearing an inversion of time; and finer ones, corresponding to two consecutive inversions of time. Indeed, for example for the electron, there are four "equivalent" positions in the S^1-bundle. We can prove the existence of periodic motions up to quantic jumps and inversions of time between them. This is completed through the study of another similar variational problem, briefly described below, see the variatinal problem on $\mathcal{L}_\beta$.

Two of these polarizations relate to the spins 1/2 and - 1/2 with time inversion. They correspond to the two dimensions of the Planck space. The two other ones (the ones which we will study in this work) are finer: they distinguish between a wave function and its opposite. One knows that, for spin 1/2, two consecutive time inversions bring the wave function to its opposite i.e. do not bring back to the initial situation. For the electron, this corresponds to the fact that, corresponding to a spin position in S^2, together with an orientation of its speed at this position, there are two possible (symmetric) positions in the S^1 bundle:

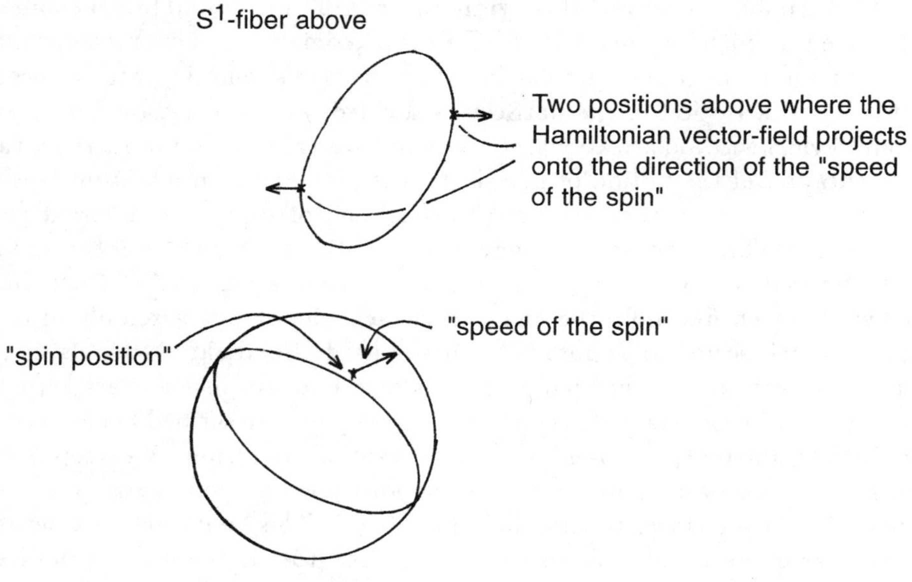

FIGURE A

The direction of projection of the Hopf-bundle which tracks down the spin position is usually transverse to the direction of the polarizations i.e. spin 1/2 and spin -1/2, for example, project onto opposite points on S^2. What we are doing is to take a Hopf fibration which contains the two polarizations in the same fiber and project on S^2 along this direction.

Our result is not the usual spin direction on S^2; but, if we add to this point on S^2 the direction of the projection of the characteristic vector-field ξ of α on S^2, we can come back to S^3, but for exceptional points, namely the ones on the conjugate points hypersurface. We can then reproject along another Hopf fibration (the "standard" one), and find the usual spin direction. Thus, our framework fits the physical framework and, we think, is more natural when we try to analyze the polarizations in a given direction. In the symmetric case, these two positions in the S^1 bundle, the *polarizations*, cannot be distinguished.

However, some materials, crystals, etc. such as the quartz (dextrogyre and levogyre) have the effect of polarizing the particles. This also can be obtained through the application of a magnetic field (Zeeman effect) or electric field (Stark effect). In the Zeeman effect, an important feature of the polarization effect is that it is created *independently* of the physical motion of the particle, that is the experimental device is such that the polarization effect is clearly seen to be related to the self-movement of the particle on itself, not to its physical movement in the space-time space.

Therefore, this polarization is obtained through a pure action on the S^1 bundle over S^2, not through an action on the S^1 bundle over the direct product of the space-time with S^2. In other words, there are experiments where a dissymmetry in the polarization is created which is not due to the movement in the space-time but rather to the quantic movement of the spin of the particle.

In mathematical terms, this corresponds to a gauge change in the form α, that is to the multiplication of α by a function $\lambda : M^3 \longrightarrow \mathbb{R}^+ - \{0\}$: the application of a magnetic field in the Zeeman effect could be thought of as changing the symmetric α into $\lambda\alpha$, where λ depends on the magnetic field. Using λ, and depending on the spin position over S^2, the particle might choose the + or − polarization. Indeed, the effect of λ on the Hamiltonian vector-field is the following: at two symmetric points, this vector-field still projects onto the same speed direction over S^2, but with possibly different lengths. It is natural then to think that the particle which we are tracking will follow the shortest length, once the speed orientation is given, until it hits some hypersurface in S^3 where there is an exchange of relative lengths between the two positions. The particle polarized positively, for example, should then jump to the symmetric position, the negative polarization, which in the future, is preferred. Since this jump is *not in space-time*, but in the quantic space, it can happen with infinite speed without contradicting the fundamental laws of physics.

The next questions, which brings us back to the mathematical side of this work, are the following: How do these jumps occur? Are there periodic motions of this type? etc. These questions are also tied to a highly nonlinear form of the WKB, or tunnel, effects and techniques on the phase of the spin (its magnetic moment).

Let us come back, for a moment, to the mathematical framework and discuss the periodic orbit variational problem for ξ. In the abstract, general framework of a contact manifold, there is one natural variational problem to seek the periodic motions of the quantic spin on the loop space of M, $H^1(S^1, M)$. If x designates such a loop, the functional is:

$$J(x) = \int_0^1 \alpha_x(\dot{x})^2 \, dt$$

It could also be :
$J(x) = \int_0^1 \alpha_x(\dot{x})\, dt$ (a form completely equivalent to the Maupertius least-action principle). In the work of Floer-Hofer-Viterbo [14] [15] and other related works, other functionals are used, but they involve more "structure" than the contact form α only. H. Hofer [26] has settled the three dimensional Weinstein conjecture, using pseudo-holomorphic curves. However, for higher winding numbers, he seems also to encounter problems of non-compactness. We therefore proceed with our work.

As we previously explained in [1] (see, in particular, the introduction of [1]), J is a very bad variational problem on $H^1(S^1, M)$, because the variational flows do not seem to be Fredholm, at least the natural (and even not so natural) ones and the periodic motions, as critical points of J, have infinite Morse index.

J on $H^1(S^2, M)$, appears to be useless. In view of the fact that the Morse index is infinite, it is natural to try to restrict the variations of the curve. There are natural mathematical candidates to play the role of new spaces of variations: there are the Legendre manifolds, which are submanifolds of $H^1(S^1, M)$ defined by the equation

$$\mathcal{L}_\beta = \left\{x \in H^1(S^1, M) \text{ s.t } \beta(\dot{x}) \equiv 0\right\}$$

β is a one-dimensional, non singular form.

When β is a contact form, the space of $\mathcal{L}_\beta$ have been studied by Smale [16], Boothby [17], ... They have established that these spaces have the topology of the loop space.

In our framework, β can be picked up under the form

$$\beta = d\alpha(v, \cdot)$$

where v is a vector-field in the kernel of α.

The fact that β is non-singular translates into a hypothesis on α, maybe removable, namely that there is a non-zero vector-field in its kernel; in mathematical language, the bundle over M^3 whose fiber at each point x is the hyperplane $\ker \alpha_x$ is trivializable. The fact that β can be taken to be itself a contact form, in the $S^3 \longrightarrow S^2$ framework for example, which corresponds to the spin 1/2-particles, is related to the convexity of the Hamiltonian (or the gauge function), see [1] Proposition 1 of Chapter 1.

We need some physical motivation for this restriction, and for the choice of v, and there is again a natural one; if we come back to the basic conceptual construction of the S^1 bundle over S^2, for a given spin, see for example [13], we find that the vector-field ξ is tangent to the S^1 fiber at each point, a fact that misleads in several directions, because, eventually, it leads to assume that, in the quantization process, the fiber should be transverse to the kernel of α. The motivation for this does not exist physically; the only possible motivation that we could possibly see relies in the fact that ξ is the tangent to the fiber and its trajectories are nice and easy.

However, as soon as the $SO(3)$-symmetry of the polarization is broken, i.e., as soon as we move away from the standard contact form on S^3, ξ is not anymore tangent to the fibers and its trajectories are, besides a few, quite far from being closed. In fact, the dynamics of ξ becomes an extremely complicated dynamics of a Hamiltonian vector-field, with recurrent orbits, possibly symbolic dynamics, etc. There is no reason, anymore, to prefer this S^1 bundle with the particular projection it involves. We could very well seek for other, more convenient projection maps $S^3 \longrightarrow S^2$; and there is a very natural family of these, which is related to α and does not change when the gauge changes. By the same token, it allows a precise tracking of the quantic position of the spin in relation to its speed and position on S^2 and also a precise tracking, in function of these data on the basis, of the polarization of the particle. Such projections come in a 1-parameter (parametrized by S^1) family in the $S^3 \longrightarrow S^2$ framework, i.e., the framework of the spin 1/2-particles. The choice of the projection , in this S^1 family, can be either thought as an additional mathematical construction which helps the physical observation, or more probably, the experimental apparatus might discriminate between all these projections and choose one among them. This projectin should be directly related, if not equal, to the direction along which we are studying the polarization of the particle.

Let us describe this set of projections:

In the kernel of the standard contact form α_0, over S^3, $\alpha_0 = x_1dx_2 - x_2dx_1 + x_3dx_4 - x_4dx_3$, $S^3 = \{x = (x_1, x_2, x_3, x_4)$ such that $x_1^2 + x_2^2 + x_3^2 + x_4^2 = 1\}$, there is a one-parameter family of vector-fields, very easy to write, each of them will be denoted v, whose orbits are all of them closed. Thus, each v define an S^1 action on S^3 and the quotient by this S^1 action (identifying points along the same v-orbit) is S^2. We thus have a projection map:

$$S^3 \longrightarrow S^2.$$

The only intrinsic notion is in S^3: it represents the quantic spin. The projection on S^2 represents the spin of the particle, in its space-time, i.e., an orientation of the particle as well as the speed of this orientation, which depends on the projection, hence on the way with which we are trying to detect the spin and polarization of the particle. Here, we encounter the usual *only quantic* rigorous definition of the spin, which, as a kinetic moment, has an indeterminacy built in it, following Heisenberg's uncertainty principle. And we could embed this measurement problem in the relatively arbitrary choice of v in this family. However, v should be partly determined by this direction, along which we are studying the polarization of the particle, as explained above. (Some other criteria related to the relation between v, $[\xi, v]$, $[v, [\xi, v]]$ might lead to better v's than others, see Proposition 2 of [1], for formulae relating all these quantities. A quantity $\bar{\mu}$, a function on the manifold, plays a key role in the polarization effects. Some $\bar{\mu}$'s may therefore be better than others, leading to restrictions on v).

The choice of v allows to find completely the quantic position of the spin of the free particle in function of measurements in S^2, of the orientation and its speed because of a phenomenon described in the introduction of [1] and which we hinted earlier. Namely, given a point z on S^2 and a direction w_z tangent to S^2 at z, there are on the S^1-orbit (the v-orbit) above z in S^3, only two positions $\hat{z}_1$, $\hat{z}_2$ such that the Hamiltonian vector-field $\xi_{\hat{z}_1}$ and $\xi_{\hat{z}_2}$ projects (parallel to v) on w_z (with orientation and without length). In another language which is ours in [1], the plane ker α_x along the v-fiber rotates twice on itself, in the v transport, when we complete a full v-orbit. β, which has also v in its kernel by construction ($\beta = d\alpha(v, \cdot)$), and which is transverse to α, has also to rotate twice and ξ is forced to follow the movement, thereby projecting twice on the same direction.

The only parameter left is the length at $\hat{z}_1$ and $\hat{z}_2$ measured in term of α or in terms of the gauge function, which induces the polarization effect of the ambient space (crystal or magnetic field, etc.). $\hat{z}_1$ may be preferred to $\hat{z}_2$, or the converse, depending on these relative lengths. Furthermore, the jump from $\hat{z}_1$ to $\hat{z}_2$ along the

v-orbit is costless, since $\alpha(v) = 0$, hence the contribution of such a jump is zero in the computation of the Maupertius action along the curve.

The quantic trajectory will therefore follow a classical pattern along the shortest length polarization among the two polarizations until a moment when the two lengths become equal and there is a switch in their relative order (along the classical trajectory). Then, the quantic spin prefers to jump to the other polarization at no cost and continue its classical trajectory thereafter, following the new pattern. The jumps occur on a special two-dimensional submanifold of S^3, called the submanifold of conjugate points (there where $\lambda(u) = \lambda(-u)$).

As we explain below, the jump can be analyzed through extremely nonlinear WKB technique, involving "tunnel effects" between the two polarizations. The reason for this is to be found in the fact *the motion is not anymore the motion of a free particle*, which is linear and induces linear phenomena. This is the motion of *the spin of this particle*, for which we are trying to build, as we will see—and this is related to v—a phase (the quantic position in S^1 of the spin), (hence probably a Schrödinger equation in some sense) and we are trying to find the limit of this motion when the viscosity along v tends to zero (see below for more explanations).

From the mathematical point of view, our phenomena is much more involved: v is an arbitrary non singular vector field in the kernel of α. Its dynamics can be arbitrary, very far from being periodic. Still, α rotates along v (see [1] Proposition 0.3), there is a whole family of polarizations, sometimes an infinite number on a given orbit, sometimes none. There is a whole family of lengths, for each of these polarizations and a least length one and jumps occur, from one to the other, when they switch.

Furthermore, we can ask that these global motions are closed, or with prescribed positions on S^2.

From a mathematical point of view, we consider the same functional on $\mathcal{L}_\beta$. In fact, we bring in one more restriction, which is not really important, but makes life easier, we could work on $\mathcal{L}_\beta$ and the results would be marginally modified (maybe not physically, we would then allow switches from ξ to $-\xi$, i.e., switches, from time to time, of polarizations and reversibility of the classical system; this is compatible with the introduction of helicity.)

The new manifold is obtained by requiring that $\alpha(\dot{x})$ is a positive constant (of arbitrary value) on the curve. This corresponds (exactly) to the apparently less restrictive requirement that $\alpha(\dot{x})$ is non negative, a choice of direction of time. $\alpha(\dot{x})$ equal to a constant corresponds to the energy conservation in classical mechanics.

Let

$$\begin{cases} C_\beta = \mathcal{L}_\beta \cap \{\alpha_x(\dot{x}) \equiv \text{ non-negative constant } \} \\ J(x) = \displaystyle\int_0^1 \alpha_x(\dot{x})\, dt = \alpha_x(\dot{x}) \end{cases}$$

where

$$\beta = d\alpha(v, \cdot)$$

The curve x of C_β have their tangent vector equal to $\dot{x} = a\xi + bv$, where a is a positive constant.

The function is

$$J(x) = a$$

It controls the a-component. But, the control, through the variations, i.e., through several possible gradient flows, of the b-component, which, here, plays the role of a phase (it is the component along the S^1 fiber in the $S^3 \longrightarrow S^2$ model) is extremely weak. In this new space of variations, the periodic orbits of ξ are of finite Morse index. However, the second derivative is not Fredholm at such a critical point. In Chapter 10 of [1], we have explained how this phenomenon yielded a very serious problem here, producing periodic orbits with finite Morse index, but which did not induce any difference of topology in the level sets of the functional J. We briefly recall here, for the sake of the qualitative understanding, the underlying phenomenon: We consider a piece of ξ-orbit, hence a local extremal of the classical mechanics problem

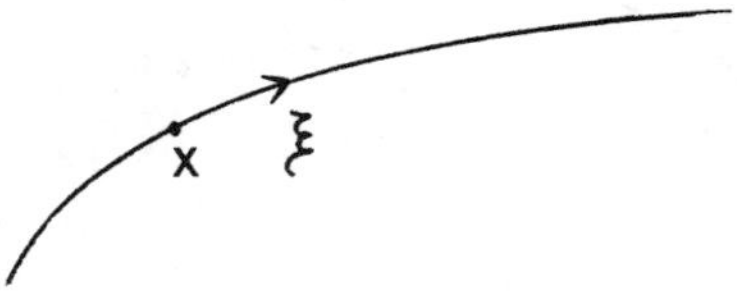

FIGURE B

At each x of this curve, it costs zero, since $\alpha_x(v) \equiv 0$ to describe a piece of v-orbit back and forth in time zero.

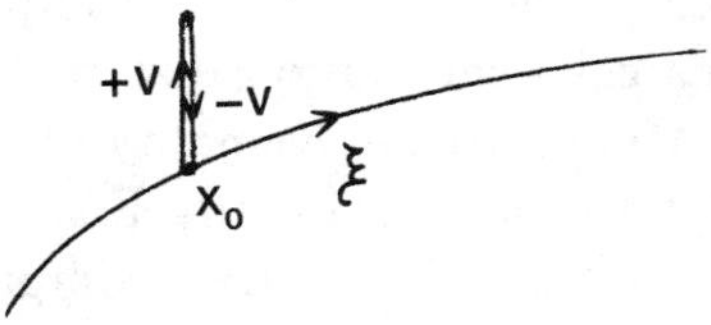

FIGURE C

Describing such curves, in fact as long as $\dot{x}$ can be split on ξ and v, with a non-negative component on ξ, we stay among our admissible variations. Let ϕ_s be the one-parameter group generated by v. Let $\phi_s(x_0)$ the orbit through x_0 and $(\phi_s^*\,\alpha)_{x_0}$ the pull-back of α from x_0 to $\phi_s(x_0)$ along this one parameter group. We can compare at each point $(\phi_s^*\,\alpha)_{x_0}(\xi)$ with $\alpha_{\phi_s(x_0)}(\xi) = 1$, i.e., $\alpha_{x_0}\left(D_{\phi_{-s}}(\xi_{\phi_s(x_0)})\right)$ and $\alpha_{\phi_s(x_0)}(\xi_{\phi_s(x_0)})$, comparing thus the length of $\xi_{\phi_s(x_0)}$ along the form α (which is one) and its length along the pull back of α from x_0 to $\phi_s(x_0)$. The trajectory of the spin component, to stay classical, has to have

$$\alpha_{x_0}\left(D_{\phi_{-s}}(\xi_{\phi_s(x_0)})\right) \leq 1$$

that is the length along the pull-back has to be less than 1. Otherwise, we can complete the following construction:

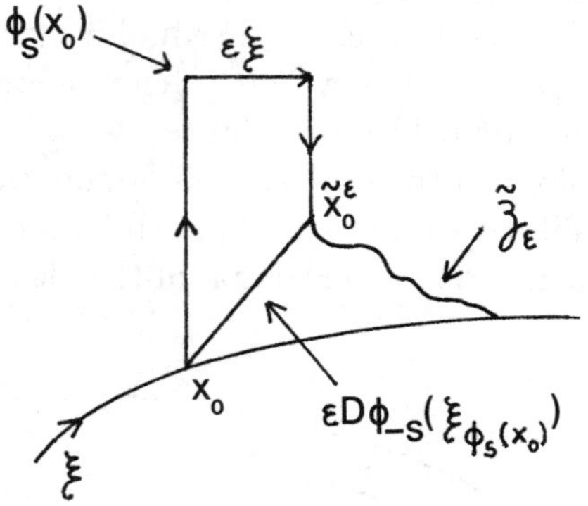

FIGURE D

i.e., follow the classical trajectory until x_0, then jump in time zero, with cost zero, from x_0 to $\phi_s(x_0)$, follow again a new classical trajectory along ξ during the time ε, then move back to $\tilde{x}_0^\varepsilon$ by ϕ_{-s} and close back to the initial classical trajectory, using a piece of curve in C_β or $C_\beta^+ = \{x \in \mathcal{L}_\beta \text{ s.t } \alpha(\dot{x}) \geq 0\}$, $\tilde{z}^\varepsilon$. If $1 < \alpha_{x_0}\left(D_{\phi_{-s}}(\xi_{\phi_s(x_0)})\right)$, the action of the new curve is less than the one of the initial ξ-orbit. The variation is not local, it involves huge motions along v, but the deformation can be fit into a continuous deformation (see section II of the present work) which deforms a whole neighborhood of this classical motion downwards, thereby contradicting—at the topological level—the Maupertius extremality of this curve. The deformation is not Fredholm and cannot be produced by any Fredholm deformation. Somewhat, at each time of a classical motion, the particle compares all the other classical positions where it could run along the v-orbit and chooses the best one; when we try to find the extrema of the function $1 - \alpha_{x_0}\left(D_{\phi_{-s}(\xi)}\right)$, we find (see [1]) that these occur at other points x_i along the v-orbit $\phi_s(x_0)$ through x_0 such that the kernel of α, which is independent of the gauge function λ and is a hyperplane, has turned in the v-transport, an even number of times and has come back onto itself. x_i is then called a coincidence point of x_0. In the spin 1/2-particle framework, for example, a coincidence point of x_0 is $-x_0$ or x_0 itself. It is the particle with the opposite polarization.

Thus, the particle, in order to extremize its motions, has to go into non Fredholm variations and compare its action with the one of its polarized twin. If the one of the polarized twin is better, it moves to it, along v. Otherwise, it continues its quiet classical path. At each point x_0 of a classical trajectory, or equivalently of M^3, we can associate an eigenvalue problem along the v-orbit through x_0, which is linear (and is probably the form that the Schrödinger equation for the phase takes here): this eigenvalue problem modelizes the transport equation of the form α along the one-parameter group general by v. It has been extensively described in [1], Proposition 0.9. It takes the form:

$$\begin{cases} \dfrac{d^2\varphi}{ds^2} + \bar{\mu}\,\dfrac{d\varphi}{ds} + \varphi = 0 \\ \varphi(s) = 1 \\ \dot{\varphi}(0) = 0 \end{cases} \tag{1}$$

$\bar{\mu}$ is a function on the v-orbit, s is the time along the v-orbit. We are seeking the points s_i such that $\dot{\varphi}(s_i)$ is zero, $\varphi(s_i) = \lambda_i > 0$. Then, $\phi_{s_i}(x_0)$ is a coincidence point of x_0, λ_i its eigenvalue. At all coincidence points of x_0, λ_i has to be less than or equal to 1; when there is a switch, hence when we cross a new sink with $\lambda_i = 1$, the particle moves to this new position. (1) is linear, but changes as the particle moves along ξ. Hence, we do not have one eigenvalue problem (with a countable number of eigenvalues, or quantic-spin states), but a continuous family of linear eigenvalue problems over the continuous motion. As described in [1] , (1) is a curious transformation of the critical equation of

$$J_\varepsilon(x) = J(x) + \varepsilon\, I(x)$$

with

$$I(x) = \int_0^1 b^2\, dt \quad x \in C_\beta.$$

i.e., if we try to bring in some viscosity, or some compactness to J with the addition of εI (plus lots of other technical devices, including a tricky deformation lemma, see [1]), we find critical points x_ε of J_ε, which, when ε tends to zero, have the following property: the critical equation is an equation on the component b along v, which takes the form of the second order non-linear (damped) pendulum equation of the type (2)$\varepsilon\,\ddot{b} + b^3 + — = 0$. When ε tends to zero, along the b-uncontrolled pieces of time, i.e., the time where $\dot{x}_\varepsilon$ is rather parallel to v, the critical equation transforms, through a curious change of variables, into (1) (see [1], the introduction, in particular). Hence, b or rather $(1 - \varepsilon \int_0^1 b^2,\ \dot{b}/\omega b)$ becomes some kind of phase which is parallel to $\left(\varphi,\ \frac{d\varphi}{ds}\right)$ and measures the rotation of α and its length along the v-orbit.

(1) and (2) have strong—at least at the interpretation level—similarities with the WKB limit ($\varepsilon = \bar{h} \to 0$) and tunnel effects. However, they are here extremely nonlinear, because they are involved in the motion of the spin of the particle. The transformation of (2) into (1) is not at all linear phenomenon. (2), furthermore, is nonlinear.

From a mathematical point of view, the convergence proof in [1] to the critical point at infinity was extremely long, technical and required several hypotheses on the dynamics of α along v and on ϕ_s. The flow was still not Fredholm.

We have removed here all these hypotheses and we have built a Fredholm flow, a very geometric one, built out of a—quite complicated—diffusion equation on b, which spouses closely the oscillations of b. The flow loses some regularity, since we have to start with H^2-data, the existence is proven in H^2, but the continuity is in H^1. It is tricky to prove that the continuity holds in H^2 a.e. in s. Furthermore, there is globally loss of H^1 and H^2 control. The only control left, paid with a high price of technical involvement, is a control on $\int |b|$ (globally on the flow-line) as well as a monotonicity of b outside of sudden oscillations, involving, such as in Yamabe type problems, at least a δ_0—amount in $\int |b|$, δ_0 being a given fixed positive number. The main ingredient in this construction is a flow which cancels the small oscillations of b (those of energy less than δ_0). The large oscillations cannot be touched, i.e., cannot be cancelled, but they can be regularized and brought progressively into Dirac-jumps along v. They are in finite number. This flow has the marvelous properties of the normal curvature flow, in dimension 1: it decreases only the zeros of b (does not increase them) and is locally Fredholm. Hence, these non-classical oscillations, which destroyed the criticality of the classical motions, cannot occur along its flow-lines. Although we lack for the moment a general existence mechanism—we can produce several partial ones, see section III.2 of this work—the deformations along those flow-lines can reasonably lead to detect the periodic motions , which induce, along these deformations, differences of topology.

However, this flow has still critical points at infinity, i.e., the Palais-Smale condition is not satisfied and the curves which are responsible for this failure are those quantic motions which we described above. The new fact, with respect to [1], which we establish here is the existence of a full variational problem at infinity which is, in some sense, a retract (locally) of the variational problem in C_β. In physical terms, we could claim that there exist spaces of quantic motions with a variational problem whose extremals are the quantic periodic motions. These spaces are of two types: Γ_{2k} and Σ_k. Each of them is associated to an integer k. Γ_{2k} has been described above as the space of closed curves alternatively tangent to $\pm v$ or to ξ, with $2k$-jumps. The functional is the sum of the lengths along ξ. For Σ_k, the $\pm v$ jumps occur only between coincidence points (associated to eigenvalues of (1)). The functional is again the sum of the ξ-jumps. Key properties of these manifolds are, first, that some of the critical points of the variational problems on them (which are the natural extensions of our variational problem on C_β) are the critical points at infinity of our initial variational problem (i.e., those quantic period motions); [on Γ_{2k}, and on Σ_k, there are some other critical points which are also very natural.] Second, the difference of the Morse indices of the critical points at infinity in restriction to Γ_{2k} is very much tied to the same differences of the full space (i.e., in $C_\beta^+ = \{x \in \mathcal{L}_\beta \text{ s.t } \alpha(\dot{x}) \geq 0\}$), if there is a path between these two critical points at infinity in Γ_{2k}.

The difference in Γ_{2k} is, in some sense, only half of the difference in C_β^+ and represents the index difference due to the jumps. The phenomenon on Σ_k is even more interesting dynamically: we can define what it means, for such a quantic period motion, to be hyperbolic on elliptic, such as in the classical case. (The definition involves as usual the eigenvalues of the Poincarè-return map, which is not the classical one, but the natural one for these quantic motions, see section II). The difference of the Morse indices in Σ_k is then the same than in a canonical (with respect to α) subbundle of TC_β^+ near Σ_k of co-dimension k, if the two orbits are both elliptic or both hyperbolic. There is a jump of 1, otherwise. This opens a whole line of interesting problems on the dynamics of this "piecewise-vector-field" made of combination of ξ-pieces and of $\pm v$-jumps between coincidence points and their relation to their Morse indices, such as for the classical motions in dynamical systems. There is, then, a natural construction of a product of $\Sigma_k \times [-1,1]$ where the difference of the Morse indices is exactly equal to the same difference for this canonical subbundle of TC_β^+. $\Sigma_k \times [-1,1]$ is contained into $\Gamma_{2k} \times [-1,1]$.

Besides Γ_{2k} and Σ_k, we also define another submanifold of Σ_k, denoted $\widetilde{\Sigma}_k$, of dimension $k-1$, defined by a canonical relation: The form α, after being transported along any curve of $\widetilde{\Sigma}_k$, should come back exactly on itself.

We prove that the difference of Morse indices at two critical points at infinity for J and the same difference for $J|_{\widetilde{\Sigma}_k}$ are the same. (All critical points at infinity belong to $\widetilde{\Sigma}_k$). Thus, $\widetilde{\Sigma}_k$ and J, up to the surprising fact that there might be other critical points besides the critical points at infinity, are really a "retract" of J on C_β.

These phenomena deserve further analysis, we believe that we are only at the early stages of this work. We also point out, in various other sections of this paper, the close relationship between these quantic periodic motions and the classical ones. We suggest a scheme which ties them. We finally, extend the framework of the present paper to include cases where the spin-component of the Hamiltonian depends on the spatial/physical position of the particle. It seems to us that some extensions suggested by the Stark effect contain such a feature. Future work should address this natural framework in full generality, englobing the spin and spatial components in a single variational problem, which could be of "classical" type. We discuss this fact below, i.e. we discuss briefly in the next section, an addition to this introduction, the notion of spin and spinors in quantum mechanics, the related notions of polarization and helicity, and their relation to Schrödinger's equation.

Additional remarks about the notion of spin and spinors in quantum mechanics

The notion of spin is central in quantum mechanics even though, for several founders of the field, it "competes" with Schrödinger's equation. Schrödinger himself [18] tried repeatedly to show that all the phenomena which led to quantum mechanics, including the phenomena of polarization and helicity in the Zeeman and Stark effects, the Balmer's rays, Lande's formula, everything was contained in his equation.

That might be the case, but the equation had to be severely twisted for this, the Hamiltonian changed with new terms which did not have any physical explanation, and, as far as selection and polarization are concerned, the argument seems to us to be weak (see [18], Chapter III) and certainly does not account for the spinorial interpretation of the notion of polarization and helicity. In fact, we feel that Schrödinger's equation provides only a probabilistic approach to particle motions and, as far as spin is concerned, can only account for very symmetric phenomena, where polarization and helicity are somewhat found, not acquired. This is why the introduction of the spin and the spinors in quantum mechanics has not been led to its full end and has rather remained confined to what we feel to be an unrealistic framework.

The Schrödinger's equation has been considered to be accurate because it has allowed to compute the intensities in the Stark effect. It has also allowed to account for some of the Zeeman effect, but only after the introduction of the spin. Schrödinger [18] also explains why his equation, coupled with some probabilistic point of view, allows to derive the rules of selection and polarization. For example, for the Stark effect, Schrödinger [18] computed the new eigenvalues of the perturbed operator after addition of Fz due to the electric field. The preliminary conceptual work of Courant-Hilbert paved the way to such a theory.

However, an argument due to Ehrenfest reported by Heisenberg [19] shows that the center of mass of the wave packet follows more or less a classical pattern. Therefore, any phenomenon explained by Schrödinger's equation with a *classical Hamiltonian* has a classical "shadow".

It is probable that our modern more sophisticated tools in Morse Theory allow to improve the classical understanding of the Stark effect and of Lande's formula. The concept of connecting orbit between critical points (classical periodic orbits) of Morse index difference equal to 1 is probably suited to account for such transitions. The fact that, in the Stark effect, the initial situation is radially symmetric, later perturbed by Fz, a perturbation in the z-direction which destroys the symmetry, probably yields several of these connecting orbits. The underlying symmetries, which are preserved (some of them) even after perturbation should naturally lead to the selection and polarization rules.

Of course, some probability of transition related to the flow-lines connecting one periodic orbit to the other has to be chosen, fitting the physical known constraints.

This indicates that genuinely quantic effects are rather related to the notion of spin, which had to be introduced in order to explain the Zeeman effect.

From the spinorial point of view, the geometric quantization of quantum mechanics, see [13] for examples, is so symmetric that it looks artificial. The understanding of the polarization and helicity, for which it has been built, bears the trace of this feature and seems also to be artificial. We already pointed out earlier how the contact form in the spinorial component was entirely symmetric, so symmetric that its characteristic vector-field has all its orbits closed. What was needed was an artificial construction to account for the Zeeman effect, a symmetric effect. Beyond this effect, the solidity of the construction is most probably wrong, this symmetry seems very unnatural.

This is also tied to the understanding of the polarization of the light. Since all is so symmetric, there is no actual difference between a particle before it entered quartz and after. The only effect of the quartz is to have miraculously found the direction where the wave function of the particle was polarized. For example, the creation of polarized light amounts to finding the various directions of polarizations of the various photons and to bring them to be a single direction. The problem, then, is that, in case of multiple polarizations, i.e. when there is helicity, and there must other be several directions or several orientations of the spin, there should still be, for such a model, a perfect symmetry. There is no preferred helicity, no preferred polarization, since this property is a property of the wave function of the particle, which is only adjusted by the quartz, or the magnetic field, etc.

It seems to us more natural to reverse the problem, taking as variable the spinorial representation of the wave function, if we want to work with Schrödinger's equation, or taking as variable the quantic spin in S^3 if we want to work with the geometric quantization, and determine through a variational problem the evolution of this variable. The basic symmetry , isotropy, which is a companion to quantum mechanics is then dropped, even if the key interpretations are kept and we can understand how a particle acquires a polarization in one direction, how and why it might change helicity, etc. An interesting additional remark, here, is that the spin, from the theoretical point of view, is very much related to the underlying symmetries of the Hamiltonian which imply the existence of a multiple eigenfunction, hence an indeterminacy for the value of its wave function in this eigenvalue space. These symmetric models correspond to central configuration forces. When the symmetry is broken—and it is difficult to imagine that all fields of forces are symmetric—the indeterminacy disappears, hence the spin should disappear. This also does not seem to be very reasonable.

The ideas which we described above, seem to us to be more reasonable, although such ideas have to be taken with caution, until the related interpretations are seen to be sound and are confirmed by other directions and phenomena.

Critical points at infinity and the Einstein-Podolsky-Rosen paradox

After having written this physical introduction to our work, we have learnt of the existence of the so-called E.P.R. paradox, which had been extensively explained in various technical and less technical books ([20], [21], [22]) and which relates to the foundation of quantum mechanics. The similarity of the notions involved in this paradox and the notions involved in our physical interpretation of the critical points at infinity is striking. It strongly suggests to us that the particles of spin 1/2 and spin -1/2 involved in this paradox are in fact *one particle* jumping from one polarization to another. The particle is, in our point of view, able to switch, in a time equal to zero, its polarization, keeping the same spatial position if an admissible polarization is better than the one it has. The signals which it sends are related to the several possible polarizations, in case of symmetry (hence, we then can expect several signals). When this symmetry is broken (is it the case when a detector is introduced in one of the slits of the two-slits-experiment (see [20])?), one polarization prevails and the signals reduce to one signal. The particle loses its "multiplicity". Also, the presence of the vector-field v, which was partly arbitrary in our construction, seems to possibly fit with the introduction of the experimental apparatus, as we did suggest above.

However, these interpretations are only conceptual at this stage. It is too early for any serious claim, which would have to be checked with scientists involved in the experiments and other involved in the building of the conceptual framework. Furthermore, we do not believe that our construction, as it stands, can account for any continuous passage, from the quantic to the classical world. We provide in Appendix 3, some hints about the ways such a transition could occur: Again, we are only claiming an unexpected contribution and we have been solely led by our interest for a fascinating mathematical problem to us.

Optical polarization of molecules

Molecules are intermediate between particles and bodies of classical type. Therefore, the spin makes sense for them as self-movement of rotation, to which nearly classical laws of Physics apply.

This is another area where the present work might give some contribution.

Pseudo-Orbits of contact forms

We start this work reproducing the first few pages of Pseudo-Orbits of contact forms ([1]). This allows us to set the geometric framework, describe the variational problems and the evolution which we will be studying in the next several sections. These pages have their own numbering of formulae and Propositions, *independent of the numbering in the remainder of the paper.* There are a simple reproduction of a tiny part of [1], which is useful for our present purpose.

P1. Geometric Data

α is a contact form on a three dimensional compact and orientable manifold M. ξ will denote its Reeb vector-field, thus satisfying:

$$\alpha(\xi) \equiv 1; \ d\alpha(\xi, \cdot) \equiv 0. \tag{1.1}$$

We assume that the α-fiber bundle over M to be trivial; hence we can choose a non-vanishing vector-field v in the kernel of α.

Once v is fixed, we get another vector-field $\widetilde{w}$, in the kernel of α, such that $d\alpha(v, \widetilde{w}) = 1$

$$\alpha(v) \equiv \alpha(\widetilde{w}) \equiv 0;\ d\alpha(v, \widetilde{w}) \equiv 1. \tag{1.2}$$

We have

$$\alpha \wedge d\alpha(\xi, v, \widetilde{w}) \equiv 1. \tag{1.3}$$

We introduce the non singular one differential form $\beta = d\alpha(v, \cdot)$ and we denote:

$$p = \beta \wedge d\beta(\xi, v, \widetilde{w}) = d\beta(\xi, v) = d\alpha(v, [\xi, v]). \tag{1.4}$$

p might vanish on M if β is not a contact form. We will assume at a certain point of this paper that this is not the case. However, for the moment, we introduce $\Sigma = \{x \in M | p(x) = 0\}$, which we may assume by a general position argument to be a hypersurface of M. In case Σ is not empty, we will need to study the trace of the plane-field β on Σ, i.e.,

$$q = \beta \cap T\Sigma; T\Sigma \quad \text{being the tangent space to } \Sigma\ . \tag{1.5}$$

Again, by a general position argument, q can be assumed to have non degenerate singular points as well as non degenerate periodic orbits.

An interesting question arises about the geometrical significance of β being a contact form. There is an example which shows this significance: let $N \overset{i}{\hookrightarrow} \mathbb{R}^4$ be a compact hypersurface in $\mathbb{R}^4$ and $\alpha = i^*\alpha_0$, where $\alpha_0 = x_1 dx_2 - x_2 dx_1 + x_3 dx_4 - x_4 dx_3$.

We have the following proposition, whose proof is straightforward computation:

Proposition P1. *If N bounds a strictly convex open set in $\mathbb{R}^4$ and if v is a vector-field in the kernel of α defining a Hopf fibration of N, then $\beta = d\alpha(v, \cdot)$ is a contact form on N.*

Hence the hypothesis β is a contact form means some convexity of (M, α) in the v-direction.

For the three sections (P1, P2, P3) to come, we will assume nothing on β but the fact that it lies in general position.

However, later, we will assume that β is a contact form. In this situation, we normalize v by multiplication by a positive function λ so that

$$\beta \wedge d\beta(\xi,\ v,\ \widetilde{w}) \equiv 1. \tag{1.6}$$

As p is positive when β is a contact form (by transversality to α), the function λ is $\frac{1}{\sqrt{p}}$. We denote then:

$$\bar{\mu} = \alpha(w);\ \bar{\mu}_\xi = d\bar{\mu}(\xi);\ \bar{\mu}_{\xi\xi} = d\bar{\mu}_\xi(\xi) \tag{1.7}$$

where w is the Reeb vector-field of β, thus satisfying:

$$\beta(w) \equiv 1;\ d\beta(w) \equiv 0. \tag{1.8}$$

We then have the following proposition:

Proposition P2. *$w = -[\xi, v] + \bar{\mu}\xi$,; $[\xi, [\xi, v]] = -\tau v; \tau \in C^\infty(M, \mathbb{R}); \bar{\mu}_\xi = d\alpha([\xi, v], [v, [\xi, v]])$: $\bar{\mu}_{\xi\xi} + \tau\bar{\mu} = -\tau_v$ where $\tau_v = d\tau(v)$. All these relations hold under the hypothesis $\beta \wedge d\beta(\xi, v, w) \equiv 1$.*

The proof of Proposition P2 is also straightforward computations. (see Appendix 2)

We will consider submanifolds of $H^1(S^1; M)$, where $H^1(S^1; M)$ is the space of H^1 loops on M. If δ is a C^∞ one-differential on M, we can compute, along a curve $x(\cdot)$ belonging to $H^1(S^1; M)$ the function $\delta_x(\dot{x})$ which belongs to $L^2(S^1)$. We will need the expression of the first variation of $\delta_x(\dot{x})$ along a tangent vector z to $H^1(S^1; M)$ at $x(\cdot)$ which is given by the following formula, one can obtain in local coordinates.

Proposition P3. *Let z be the tangent vector to $H^1(S^1; M)$ along the curve $x(\cdot)$. The first variation of $\delta_x(\dot{x})$ with respect to z is $\frac{d}{dt}\delta(z) - d\delta(\dot{x}, z)$.*

P2. The Spaces of Variations $\mathcal{L}_\beta$ and C_β

We consider in this section the two following subsets of $H^1(S^1; M)$:

$$\mathcal{L}_\beta = \{x \in H^1(S^1; M) | \beta(\dot{x}) \equiv 0\} \tag{2.1}$$

$$C_\beta = \{x \in \mathcal{L}_\beta | \alpha_x(\dot{x}) \equiv \text{ strictly positive constant}\}. \tag{2.2}$$

We want to know when $\mathcal{L}_\beta$ and C_β are submanifolds of $H^1(S^1; M)$; and more generally, if this is not the case, which type of singularities we can expect.

This will be carried out, for $\mathcal{L}_\beta$, through the study of its tangent space, which is given, by Proposition P3, by the following equation:

$$x \in \mathcal{L}_\beta; \quad z \text{ is tangent to } \mathcal{L}_\beta \text{ along } x \text{ if and only if} \tag{2.3}$$
$$\frac{d}{dt}\beta_x(z) = d\beta_x(\dot{x}, z)$$

while, for C_β, we get:

$$x \in C_\beta; \quad z \text{ is tangent to } C_\beta \text{ along } x \text{ if and only if} \tag{2.4}$$

$$\begin{cases} \dfrac{d}{dt}\beta_x(z) = d\beta_x(\dot{x}, z) \\ \dfrac{d}{dt}\beta_x(z) - d\beta_x(\dot{x}, z) = \text{ constant }. \end{cases}$$

As $(\xi, v, \widetilde{w})$ provides with a basis to the tangent space to M, $\dot{x}$ can be decomposed on this basis, thus giving:

$$\dot{x} = a\xi + bv + c\widetilde{w}; \quad a, b, c, \in L^2(S^1) \tag{2.5}$$

$x \in \mathcal{L}_\beta$ if and only if $\beta_x(\dot{x}) = d\alpha_x(v, \dot{x}) = c \equiv 0$; while $x \in C_\beta$ if $c \equiv 0$ and if a is a positive constant.

Let z be a tangent vector to $\mathcal{L}_\beta$ (respectively C_β) at x. We decompose z in $z = \lambda\xi + \mu v + \eta\widetilde{w}$, where $\lambda, \mu, \eta \in H^1(S^1; \mathbb{R})$.

Equations (2.3) and (2.4) then provide:

$$z = \lambda\xi + \mu v + \eta\widetilde{w} \quad \text{is tangent at } x \text{ to } \mathcal{L}_\beta \text{ if and only if} \tag{2.6}$$

$$\dot{\eta} = d\beta(\dot{x}, \widetilde{w})\,\eta + (\mu a - \lambda b)\,p$$

$$z = \lambda\xi + \mu v + \eta\widetilde{w} \quad \text{is tangent at } x \text{ to } C_\beta \text{ if and only if} \tag{2.7}$$

$$\dot{\eta} = d\beta(\dot{x}, \widetilde{w})\,\eta + (\mu a - \lambda b)\,p; \ \dot{\lambda} = b\eta + \text{ constant}$$

We then have the following theorem:

Theorem P1. *If β is a contact form $\mathcal{L}_\beta - M$ and $C_\beta - M$ are submanifolds of $H^1(S^1; M)$. In the general case, when β is in generic position, the singularities of $\mathcal{L}_\beta - M$ are the contractible loops (with any parametrization) lying on the trajectories of q on Σ; while the singularities of $C_\beta - M$ lie on the α-positively oriented periodic of q.*

Proof of theorem P1. We will discuss the general case.

From equation (2.6), we derive, given μ and λ, the function η:

$$\eta = e^{\int_0^t d\beta(\dot{x}, \widetilde{w})\,ds}\left(\int_0^t e^{-\int_0^s d\beta(\dot{x}, \widetilde{w})\,d\tau}(\mu a - \lambda b)\,p\,ds + C\right). \tag{2.8}$$

Given μ and λ in $H^1(S^1; \mathbb{R})$, η has to be periodic. Hence:

$$C = e^{\int_0 d\beta(\dot{x}, wtdw)\,ds}\left(\int_0^t e^{-\int_0^s d\beta(\dot{x}, \widetilde{w})\,d\tau}(\mu a - \lambda b)\,p\,ds + C\right). \tag{2.9}$$

Equation (2.9) allows to compute C if $\int_0^1 d\beta(\dot{x}, \widetilde{w})\,ds$ is not zero.

Hence, on those curves of $\mathcal{L}_\beta$, the tangent space is given by the choice of μ and λ arbitrary in $H^1(S^1; \mathbb{R})$ and is thus $H^1(S^1; \mathbb{R}) \times H^1(S^1; \mathbb{R})$.

Otherwise, if $\int_0^1 d\beta(\dot{x}, \widetilde{w})\,ds = 0$, we necessarily have:

$$\int_0^1 e^{-\int_0^s d\beta(\dot{x}, \widetilde{w})\,d\tau}(\mu a - \lambda b)\,p\,ds = 0 \tag{2.10}$$

while C is arbitrary.

Hence, if $p(x(\cdot))$ is not identically zero, (2.10) provides a constraint on μ and λ which then belong to $H^1(S^1; \mathbb{R}) \times H^1(S^1; \mathbb{R})/\mathbb{R}$. But C is an arbitrary parameter; and the tangent space to $\mathcal{L}_\beta$ is thus $H^1(S^1;\mathbb{R}) \times H^1(S^1; \mathbb{R})/\mathbb{R} \times \mathbb{R}$, providing no singularity.

However, if $p(x(\cdot))$ is zero, hence if the curve lies on the trajectories of $q = B \cap T\Sigma$, then the constraint given by (2.10) disappears, providing a singularity.

It follows that singularities of $\mathcal{L}_\beta - M$ are drawn on the trajectories of q and must satisfy

$$\int_0^1 d\beta(\dot{x}, \widetilde{w})\, dt = \int_c d\beta(\cdot, \widetilde{w}) = 0; \quad c \text{ being the curve }. \tag{2.11}$$

In general position of β, (2.11) implies we are dealing with contractible loops on the trajectories of q. Hence the result on $\mathcal{L}_\beta - M$.

The result about $C_\beta - M$ follows from the analysis of the differential system:

$$\begin{pmatrix} \dot{\eta} \\ \dot{\lambda} \end{pmatrix} = a \begin{bmatrix} d\beta(\xi, \widetilde{w}) & 0 \\ 0 & 0 \end{bmatrix} \begin{bmatrix} \eta \\ \lambda \end{bmatrix} + b \begin{bmatrix} 0 & -p \\ 0 & 0 \end{bmatrix} \begin{bmatrix} \eta \\ \lambda \end{bmatrix} + \begin{bmatrix} \mu a p \\ C \end{bmatrix} = [a\,A + b\,B] \begin{bmatrix} \eta \\ \lambda \end{bmatrix} + \begin{bmatrix} \mu a p \\ C \end{bmatrix}. \tag{2.12}$$

Here again, we want periodic H^1 functions.

We know that a is constant, while b is L^2.

Hence, the matrix differential equation:

$$\left| \begin{array}{l} \dot{R} = (a\,A + b\,B)\,R \\ R(0) = Id \end{array} \right. \tag{2.13}$$

has a unique solution, which is continuous and invertible. Its inverse G is continuous and satisfies:

$$\left| \begin{array}{l} \dot{G} = -G(a\,A + b\,B) \\ G(0) = Id. \end{array} \right. \tag{2.14}$$

The general solution of (2.12) can be written:

$$\begin{pmatrix} \eta \\ \lambda \end{pmatrix}(t) = z(t) = R(t) \left[\int_0^1 R(s)^{-1} \begin{bmatrix} \mu a p \\ C \end{bmatrix} ds + K \right] \tag{2.15}$$

where $K = \begin{bmatrix} K_1 \\ K_2 \end{bmatrix}$ is a constant.

Since we want z to be periodic, we need:

$$K = R(1) \left[\int_0^1 R(s)^{-1} \begin{bmatrix} \mu a p \\ C \end{bmatrix} ds + K \right] \tag{2.16}$$

which can be written into:

$$[Id - R(1)]\,K = \int_0^1 R(s)^{-1} \begin{bmatrix} \mu a p \\ C \end{bmatrix} ds. \tag{2.17}$$

A first case is easy to solve:

1. $\det[Id - R(1)] \neq 0$ Then $\begin{bmatrix} \mu a p \\ C \end{bmatrix}$ is arbitrary and K can be computed through (2.17). The tangent space to $C_\beta - M$ on such curves is thus $\mathbb{R} \times H^1(S^1; \mathbb{R})$.
2. $R(1) = Id$

Eq. (2.17) becomes then

$$\int_0^1 R(s)^{-1} \begin{bmatrix} \mu ap \\ C \end{bmatrix} ds = 0; \quad \text{while } K \text{ is arbitrary.} \tag{2.18}$$

Eq. (2.18) gives in general two constraints on C and μ.

We have to check that these two constraints do not degenerate and are independent.

Let us denote:

$$R(s)^{-1} = \begin{bmatrix} \widehat{A}(s) & \widehat{B}(s) \\ \widehat{C}(s) & \widehat{D}(s) \end{bmatrix} \quad \text{where } \widehat{A},\ \widehat{B},\ \widehat{C},\ \widehat{D} \text{ are continuous functions.} \tag{2.19}$$

Eq. (2.18) amounts then to:

$$\begin{aligned} a\left[\int_0^1 \widehat{A}(s)\,\mu p\,ds\right] + C\int_0^1 \widehat{B}(s)\,ds = 0 \\ a\left[\int_0^1 \widehat{C}(s)\,\mu p\,ds\right] + C\int_0^1 \widehat{D}(s)\,ds = 0. \end{aligned} \tag{2.20}$$

If these two constraints degenerate, we have:

$$\left(\int_0^1 \widehat{D}(s)\,ds\right)\widehat{A}p = \left(\int_0^1 \widehat{B}(s)\,ds\right)\widehat{C}p. \tag{2.21}$$

We are assuming that $p(x(\cdot))$ is not identically zero. Otherwise, such a curve would lie on an α-positively oriented periodic orbit of q.

By the obvious S^1 invariance of the expression of the tangent space, we may also assume that $p(x(0))$ is not zero.

As $\widehat{A}(0) = 1$, while $\widehat{C}(0) = 0$, we deduce from (2.21) that $\int_0^1 D(s)\,ds = 0$; which implies, if the constraints degenerate, that $\int_0^1 \widehat{B}(s)\,ds = 0$ and $\widehat{C}p = \lambda\widehat{A}p$. But $C(0) = 0$ while $\widehat{A}(0) = 1$; hence $\lambda = 0$ and we derive:

$$\widehat{C}p = 0; \quad \int_0^1 \widehat{D}(s)\,ds = \int_0^1 \widehat{B}(s)\,ds = 0. \tag{2.22}$$

On the other hand, $R(s)^{-1} = G$ satisfies (2.14); hence:

$$\dot{\widehat{D}} = -bp\,\widehat{C}. \tag{2.23}$$

Thus, by (2.22), $D(s) = 1$; which contradicts $\int_0^1 \widehat{D}(s)\,ds = 0$.

It follows that the two constrains do not degenerate; the tangent space to $C_\beta - M$ on such curves is $[H^1(S^1;\mathbb{R}) \times \mathbb{R}/\mathbb{R} \times \mathbb{R}] \times \mathbb{R} \times \mathbb{R} \simeq H^1(S^1;\mathbb{R}) \times \mathbb{R}$. Hence no singularity in this case again.

The remaining case to study is:

3. $\dim Ker(Id - R(1)) = 1$.

Let $E_1 = Im(Id - R(1))$; E_2 a supplement in $\mathbb{R}^2$ of E_1 and P the projection on E_2.

Eq. (2.17) can only be satisfied if $\int_0^1 R(s)^{-1} \left[\begin{smallmatrix}\mu ap \\ C\end{smallmatrix}\right] ds$ belongs to E_1; thus if:

$$P\left(\int_0^1 R(s)^{-1} \begin{bmatrix}\mu ap \\ C\end{bmatrix} ds\right) = 0 \tag{2.24}$$

which is equivalent to:

$$\int_0^1 P\, R(s)^{-1} \begin{bmatrix}\mu ap \\ C\end{bmatrix} ds = 0. \tag{2.25}$$

These constraints can be seen not to degenerate by the same technique as in case 2. Under (2.25), (2.17) has a unique solution in K up to translations in $Ker(Id - R(1))$, whose dimensions is one.

The tangent space is then $[\mathbb{R} \times H^1(S^1; \mathbb{R})/\mathbb{R}] \times \mathbb{R} \times H^1(S^1; \mathbb{R})$ and the proof of theorem 1 is thereby complete. □

Remark P1. We point out here that α-positively oriented periodic orbits q are likely not to exist. Indeed, these periodic orbits are due to elliptic singularities of q on Σ; they appear then as limit cycles. But, on such a singularity, $dp(\xi) = dp(v) = 0$. Hence, there is a curve $dp(v) = 0$ passing through this point. This curve is likely to intersect the limit cycle which is tangent to $q = (-dp(v), dp(\xi))$. But, in this case, $dp(v)$ changes sign on the periodic orbit, forbidding α-positiveness (or either negativeness).

Remark P2. When β is a contact form, we normalize v such that $\beta \wedge d\beta = \alpha \wedge d\alpha$ (this is equivalent to (1.6)). Then $\tilde{w}$ can be replaced by w in a given basis of the tangent space to M and the equation of the tangent space to C_β becomes

$$\begin{cases} \overset{\cdot}{\overbrace{\lambda + \bar{\mu} n}} = bn + Cte \\ \dot{\eta} = \mu a = \lambda b\ , \end{cases}$$

where $\bar{\mu}$ has been defined in proposition P2.

Further information on the topology of the spaces of variations $\mathcal{L}_\beta$ and C_β is provided later on in this paper.

P3. The Functional

We assume from now on that β is a contact form.

We are considering a normalized vector-field v, such that $\beta \wedge d\beta(\xi, v, w) = 1$; or else $\beta \wedge d\beta = \alpha \wedge d\alpha$. w is the Reeb vector-field of β; and $w = -[\xi, v] + (\bar{\mu} = \alpha(w))$. We also have $[\xi, [\xi, v]] = -\tau v$ and $\bar{\mu}_{\xi\xi} + \tau v = -\tau_v$ (see proposition A2).

On C_β, we consider the functional:

$$J(x) = \alpha_x(\dot{x}) = \int_0^1 \alpha_x(\dot{x})\, dt. \tag{3.1}$$

We first compute the first variation of J. The gradient of J with respect to some fixed metric will be denoted ∂J:

Proposition P4. *J is a C^2 functional on C_β whose critical points are periodic orbits to ξ. If z is a tangent vector to C_3 along the curve $x(\cdot)$, then:*

$$\partial J(x) \cdot z = -\int_0^1 d\alpha_x(\dot{x},\, z)\, dt.$$

Proof. The first variation of $\alpha_x(\dot{x})$ is $\frac{d}{dt}\,\alpha(z) - d\alpha(\dot{x},\, z)$ (see proposition P3). Hence, the first variational of $J(x)$ is $\partial J(x) \cdot z = \int_0^1 \left(\frac{d}{dt}\,\alpha(z) - d\alpha(\dot{x},\, z)\right)\, dt = -\int_0^1 d\alpha(\dot{x},\, z)\, dt$.

Now if $z = \lambda\xi + \mu v + \eta w$ and $\dot{x} = a\xi + bv$, we have:

$$\partial J(x) \cdot z = -\int_0^1 b\eta\, dt. \tag{3.2}$$

Eq. (3.2) shows that J is C^1, as b is L^2 and η is H^1.

To see that J is C^2, we notice that:

$$b = d\alpha_x(\dot{x},\, w) = Y_x(\dot{x})$$

where

$$Y_x = d\alpha_x(\cdot,\, w) \quad \text{is a } C^\infty\text{-one differential form .} \tag{3.3}$$

The first variation of $Y_x(\dot{x})$ along z is

$$\frac{d}{dt}\, Y(z) - dY(\dot{x},\, z).$$

Hence, this first variation can be rewritten:

$$\begin{aligned} \frac{d}{dt}\, Y(z) - dY(\dot{x},\, z) &= \dot{\mu} - dY(a\xi + bv,\, \lambda\xi + \mu v + \eta w) = \\ &= \dot{\mu} - (\mu a - \lambda b)\, dY(\xi,\, v) - a\eta\, dY(\xi,\, w) - b\eta\, dY(v,\, w) \end{aligned} \tag{3.4}$$

which is again an L^2-function.

As $\partial J(x) \cdot z = -\int_0^1 b\eta\, dt = \int_0^1 Y_x(\dot{x})\eta\, dt$ and as the first variation of $Y_x(\dot{x})$ is L^2, J is at least C^2 on C_β.

It remains to show that critical points of J are periodic orbits to ξ. Indeed, $\partial J(x) \cdot z = 0$ is equivalent to $\int_0^1 b\eta\, dt = 0$ for any variation η such that:

$$\left\{ \begin{aligned} &\frac{d\alpha(z)}{dt} - d\alpha(\dot{x},\, z) = Cte \\ &\frac{d\beta(z)}{dt} - d\beta(\dot{x},\, z) = 0 \end{aligned} \right. \iff \left\{ \begin{aligned} \overset{\cdot}{\overbrace{\lambda + \bar{\mu}\eta}} &= b\eta + C \\ \dot{\eta} &= \mu a - \lambda b \end{aligned} \right. \quad \lambda,\, \mu,\, \eta \text{ periodic of period one,} \tag{3.5}$$

(see remark P2).

In (3.5) by periodicity of $\lambda+\bar{\mu}\eta$, C has to be equal to $\int_0^1 b\eta\, dt$. Hence, x is critical to J if and only if (3.5) has no periodic solution with a non vanishing C. This implies $b=0$ by similar arguments than the ones used in the previous sections to study the tangent space to C_β.

There is also a heuristic way to see this fact, if we assume b to be C^1.

Then, if b is not zero, we can compute $\lambda+\bar{\mu}\eta=\int_0^t b\eta\, ds - t\int_0^1 b\eta\, ds$ from the first equation of (3.5) choosing η in H^2 such that $\int_0^1 b\eta\, ds$ is not zero. μ is then easily derived as a $H^1(S^1, \mathbb{R})$ function equal to $\frac{\dot{\eta}+\lambda b}{a}$ from the second equation of (3.5). Hence the result with this heuristic argument (which can be made rigorous).

But if $b=0$, $\dot{x}=a\xi$, with a being a strickly positive constant.

Hence x is a periodic orbit to ξ.

The result given by proposition P4 extends to the more general case where β is no more a contact form, but is in general position:

Proposition P5. *Assume β is in general position. Then the critical points of J on C_β of finite Morse index are periodic orbits to ξ.*

The proof is somewhat technical and hence will be published elsewhere since this situation is not our purpose here.

Coming back to (3.4), (3.5), we see that the variation of $\alpha(x)$, the a-component along $z=\lambda\xi+\mu v+\eta w$, z tangent to C_β at x is:

$$z\cdot a=\frac{\partial a}{\partial s}=-\int_0^1 b\eta \quad \text{(after integration of the first equation of (3.5) between 0 and 1)}$$

while the variation of b is $(d\gamma(\xi,v)=\gamma([\xi,v])=0, \quad -d\gamma(\xi,w)=z, \ -d\gamma(v,w)=\bar{\mu}\xi$ as some easy computation show):

$$z\cdot b=\frac{\partial b}{\partial s}=\dot{\mu}+a\eta\tau-b\eta\bar{\mu}\xi \quad \text{(follows from (3.4) and the above identities).}$$

This couple of equations

$$\begin{cases} \dfrac{\partial a}{\partial s}=-\displaystyle\int_0^1 b\eta \\ \dfrac{\partial b}{\partial s}=\dot{\mu}+a\eta\tau-b\eta\bar{\mu}\xi \quad \left(\dot{\eta}=\mu a-\lambda b,\ \dot{\overbrace{\lambda+\bar{\mu}\eta}}=b\eta-\displaystyle\int_0^1 b\eta\right)\end{cases}$$

will be the focus of all the forthcoming work.

I. Technical construction of a diffusion flow and convergence theorems

I.1. Technical construction of a diffusion flow

We take $b \in H^1$. Hence, b is continuous. We assume that b has an oscillation, i.e. we have an interval $[t_1, t_2]$ such that

$$\tilde{\bar{\mu}} = \underset{[t_1,t_2]}{\mathrm{Max}} |b| > \mathrm{Sup}(|b|(t_1), |b|(t_2)) = \underline{\tilde{\mu}} \qquad (*)$$

and **we assume that b does not take the value zero on** $[t_1, t_2]$.

Let $|b|_\infty$ be the L_∞ -norm of b.

$\varepsilon_0 = c_0 > 0$ is a small fixed constant (positive), upon which conditions will be imposed later. We will distinguish, in the following construction two cases: C is a universal constant larger than 1 related to the data of the problem

Case 1) $\quad C(|I| + \int_I b^2)|I| < \varepsilon_0$ where $I = [t_1, t_2]$

Case 2) $\quad C(|I| + \int_I b^2)|I| \geq \varepsilon_0/2$ where $I = [t_1, t_2]$

We subdivide the case 2) into two other cases.

We first introduce the following definition, which we will use in case 2) holds:

Let $\varepsilon_{15}, \varepsilon_{16} > 0$ be two small positive constants, which ultimately will tend to zero.

Definition 1. *An oscillation is* **simple** *if*
$I(t_1, t_2) = \{t \in [t_1, t_2] s.t \quad |b|(t) \geq (1 - \varepsilon_{15})\tilde{\bar{\mu}}\}$ *is connected.* ε_{15} *is a small positive constant, which we can make as small as we wish.*
M is, in the sequel, a large constant to be chosen later.

If an oscillation is simple and if the set $I(t_1, t_2)$ is connected, the set

$$J(t_1, t_2) = \{t \in [t_1, t_2] s.t \quad |b|\,(t) < (1 - \varepsilon_{15})\tilde{\bar{\mu}}\}$$

is made of two connected components $J^+(t_1, t_2)$ and $J^-(t_1, t_2)$, having lengths $\Delta t^+, \Delta t^-$ (functions of (t_1, t_2)).

Case 2) will be subdivided, for simple oscillations, into two subcases:

Case	2)a)	$\Delta t^+(\tilde{\bar{\mu}} + M) \geq \varepsilon_{16}$	or	$\Delta t^-(\tilde{\bar{\mu}} + M) \geq \varepsilon_{16}$
	2)b)	$\Delta t^+(\tilde{\bar{\mu}} + M) < \varepsilon_{16}$	or	$\Delta t^-(\tilde{\bar{\mu}} + M) < \varepsilon_{16}$

ε_{16} is again a small positive constant, which we can make as small as we wish. Given an oscillation satisfying $(*)$, we take two values

$$\mu_1 \text{ and } \mu_2 \text{ such that}$$

$$\underline{\tilde{\mu}} < \mu_1, \mu_2 < \frac{\bar{\tilde{\mu}} + \underline{\tilde{\mu}}}{2}. \tag{1}$$

Let

$$\bar{t} \text{ be a point in } [t_1, t_2] \text{ such that } |b(\bar{t})| = \bar{\tilde{\mu}}. \tag{2}$$

Let

$$\begin{cases} t_1^+ = \text{Sup}\,\{t_1 \le t \le \bar{t} \text{ such that } |b(t)| = \mu_1\} \\ t_2^+ = \text{Inf}\,\{\bar{t} \le t \le t_2 \text{ such that } |b(t)| = \mu_2\}. \end{cases} \tag{3}$$

All our arguments below will be applied to intervals of the type $[t_1^+, t_2^+]$, associated to oscillations of the type $(*)$ on intervals $[t_1, t_2]$. In particular, the subdivisions 1), 2), 2)a), 2)b) will be applied to such intervals $[t_1^+, t_2^+]$ with one further specification. We claim that, given a oscillation satisfying $(*)$, we have:

Lemma 1. *μ_1 and μ_2, satisfying (1), can be chosen so that: There exists $+\infty > \gamma > 0$ such that, for any $\bar{\bar{\delta}} > 0$, there exists $t_1^- \in [t_1^+ - \bar{\bar{\delta}}, t_1^+]$ and $t_2^- \in [t_2^+, t_2^+ + \bar{\bar{\delta}}]$ such that:*

(i) $\mu_1^- = |b|(t_1^-) < \mu_1 = |b|(t_1^+), \mu_2^- = |b|(t_2^-) < \mu_2 = |b|(t_2^+)$

(ii) $\left|\frac{b(t_i^+) - b(t_i^-)}{t_i^+ - t_i^-}\right| \ge \gamma$.

Proof of Lemma 1. Assume b is positive or $[t_1, t_2]$ (the other case is symmetric).

Let, for $t < \bar{t}$ (a similar construction is completed for $t > \bar{t}$).

$$\tilde{b}(t) = \inf_{\tau \in [t, \bar{t}]} b(\tau). \tag{4}$$

Clearly, $\tilde{b}$ is a continuous function, which increases with t and

$$\tilde{b}(t) \le b(t).$$

Clearly, also, if $\tilde{b}(t_0) < b(t_0)$ then $\tilde{b}$ is constant near t_0, thus

$$\dot{\tilde{b}}(t_0) = 0. \tag{5}$$

We claim that if Lemma 1 does not hold, then, $\forall t$, $\exists \delta > 0$ such that

$$\tilde{b}(\tau) \ge \tilde{b}(t) \qquad \forall \tau \in [t - \delta, t]. \tag{6}$$

We will prove this claim later. It follows that

$$\tilde{b}(\tau) \geq \tilde{b}(t) \qquad \forall \tau \leq t. \tag{7}$$

($\tau \geq t'_0$, a given lowerbound on the t's).

Thus, we derive that $\tilde{b}(t)$ decreases or is non increasing for $\tilde{b}(t) \in \left(\underline{\tilde{\mu}}, \frac{\tilde{\mu} + \underline{\tilde{\mu}}}{2}\right)$, a contradiction.

We now prove (6):

Taking μ_1 a variable in $\left(\underline{\tilde{\mu}}, \frac{\tilde{\mu} + \underline{\tilde{\mu}}}{2}\right)$, if $\tilde{b}(t_0) = b(t_0) = \mu_1$, then either $t_1^+(\mu_1) > t_0$ and $\tilde{b}(t)$ is constant on $[t_0, t_1^+(\mu_1)]$, equal to μ_1, hence $\dot{\tilde{b}}(t)$ is zero. Or $t_1^+(\mu_1) = t_0$. Arguing by contradiction, if lemma 1 does not hold at $t_0 = t_1^+(\mu_1)$, then for any $\gamma_k = \varepsilon_k = \frac{1}{k}$, there exists $\delta_k > 0$ such that for any $t \in [t_0 - \delta_k, t_0]$

$$\frac{b(t) - b(t_0)}{t - t_0} < \frac{1}{k}.$$

Thus, $b(t) \geq b(t_0) + \frac{1}{k}(t - t_0)$ for $t \in [t_0 - \delta_k, t_0]$. This yields: $\tilde{b}(t) \geq \tilde{b}(t_o) + \frac{1}{k}(t - t_0)$ for $t \in [t_0 - \delta_k, t_0]$. If $\tilde{b}(t_0) < b(t_0)$, this inequality also holds. Thus, we derive:

$$\tilde{b}(t) \geq \tilde{b}(t_0) + \frac{1}{k}(t - t_0) \text{ for } t'_0 \leq t \leq t_0 \tag{8}$$

where t'_0 is a lowerbound on the t's involved in this argument. Letting k tend to $+\infty$, we derive:

$$\tilde{b}(t) \geq \tilde{b}(t_0) \, \forall t'_0 \leq t \leq t_0 \tag{9}$$

as claimed. The result follows.

Observe that, given an interval with an oscillation satisfying $(*)$ and constructing on associated interval $[t_1^+, t_2^+]$, we can make it satisfy Lemma 1 with only a small perturbation of μ_1 and μ_2; indeed the values of $\underline{\tilde{\mu}}$ and $\frac{\underline{\tilde{\mu}} + \tilde{\mu}}{2}$ did not play any role in the above argument besides the fact that $\underline{\tilde{\mu}}$ was less than μ_1, μ_2 and $\tilde{\mu}$ larger than these same values.

We now consider oscillations satisfying Lemma 1. and 1). We introduce on such an interval $[t_1^-, t_2^-]$ the function η_0 solution of:

$$\begin{cases} \frac{d^2}{dx^2}\left(e^{-\int_{t_1^-}^t \bar{\mu} b} \eta_0\right) + C(1 + b^2)\eta_0 e^{-\int_{t_1^-}^t \bar{\mu} b} = \frac{-2}{|x_1^- - x_2^-|^2}, \\ \eta_0(x_1^-) = \eta_0(x_2^-) = 0 \end{cases}$$

$$\text{where} \begin{cases} x = x(t) = \int_{t_1^-}^t e^{-\int_{t_1^-}^z \bar{\mu} b d\tau} dz \\ x_1^- = 0, x_2^- = \int_{t_1^-}^{t_2^-} e^{-\int_{t_1^-}^z \bar{\mu} b d\tau}. \end{cases} \tag{10}$$

Lemma 2. *In case 1), (10) has a unique positive solution.*

Proof. If we are in case 1), we can assume, taking t_1^-, t_2^- to be very close to t_1 and t_2 that

$$C\left(1+\int_{t_1^-}^{t_2^-} b^2\right)|t_1^- - t_2^-| < \varepsilon_0.$$

Since C is larger than 1

$$\int_{t_1^-}^{t_2^-} |b| < 1.$$

Thus, $x(t)$ and t is a change of variables with bounded derivative and inverse derivative. Thus,

$$C\left(1+\int_{x_1^-}^{x_2^-} b^2\right)|x_1^- - x_2^-| < C_1\varepsilon_0$$

where C_1 is a universal constant.

We then observe that, for a function $f \in H_0^1[x_1^-, x_2^-]$, we have:

$$\int_{x_1^-}^{x_2^-} \dot{f}^2 dx \geq \frac{1}{|x_1^- - x_2^-|}|f|_\infty^2 \geq \frac{C}{C_1\varepsilon_0}\left(|x_2^- - x_1^-| + \int_{x_1^-}^{x_2^-} b^2\right)|f|_\infty^2 \geq$$

$$\geq \frac{1}{C_1\varepsilon_0}\int_{x_1^-}^{x_2^-} C(1+b^2)f^2.$$

Thus, if

$$C_1\varepsilon_0 < 1$$

(10) has a unique solution.

We, furthermore, have:

Lemma 2′. *Given a bound on $|b|_\infty$, we have*

$$|\eta_0|_\infty + |x_1^- - x_2^-|\left|\frac{d}{dx}\eta_0\right|_\infty + |x_1^- - x_2^-|^2\left|\frac{d^2}{dx^2}\eta_0\right|_\infty \leq C(|b|_\infty).$$

Proof. We need only to prove this result when $C|x_1^- - x_2^-|(1+|b|_\infty)$ is small. Otherwise, we have a bound depending on $|b|_\infty$. When $C|x_1^- - x_2^-|(1+|b|_\infty)$ tends to zero, this bound is immediate. □

Given $\theta > 0$, we define

$$\psi_{\theta,\varepsilon}(b) \tag{11}$$

to be a $C^\infty\varepsilon$-regularization of the function $(|b|-\theta)^+$, which behaves as follows $\left(\frac{\partial\psi_{\theta,\varepsilon}}{\partial b}(b)\cdot b \geq 0\right)$

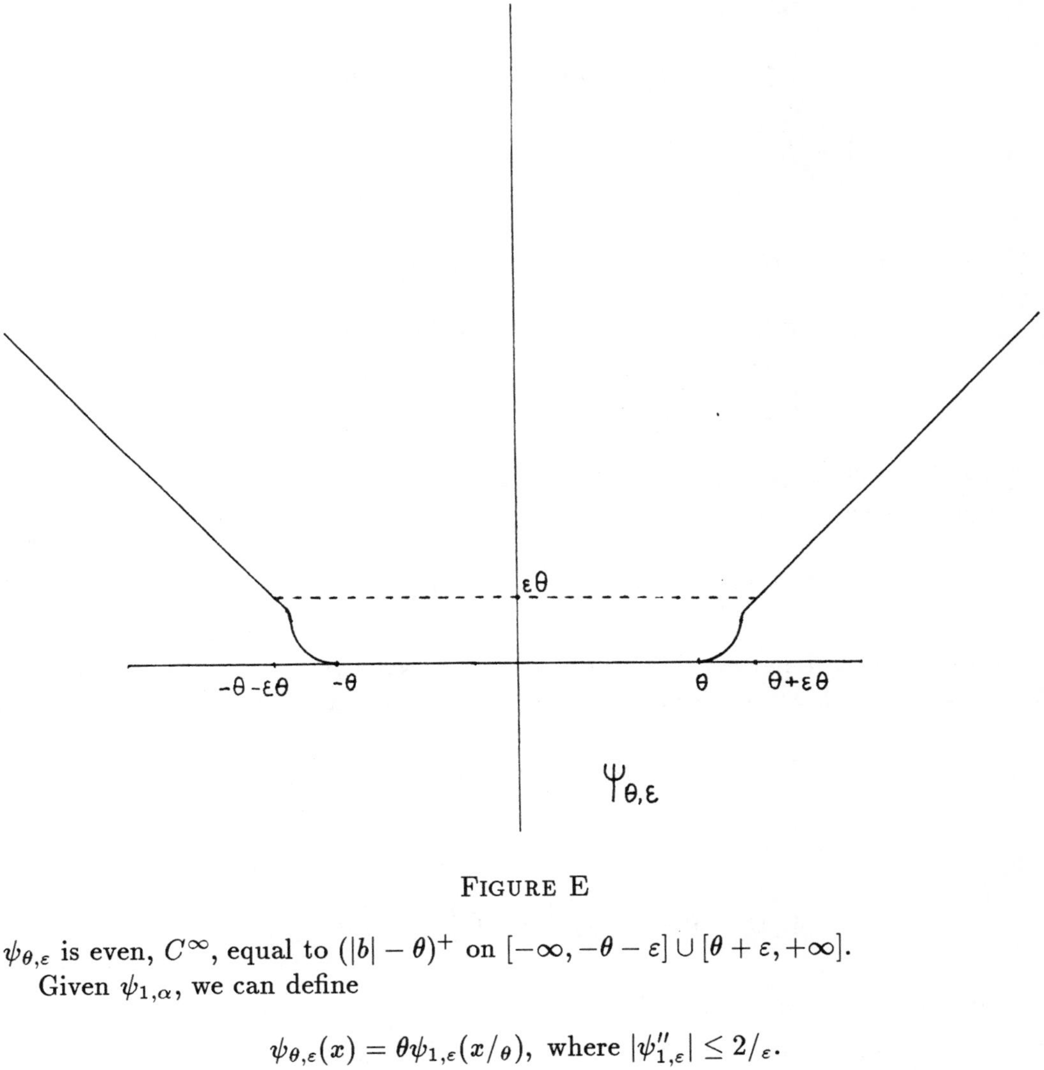

FIGURE E

$\psi_{\theta,\varepsilon}$ is even, C^∞, equal to $(|b|-\theta)^+$ on $[-\infty,-\theta-\varepsilon]\cup[\theta+\varepsilon,+\infty]$.

Given $\psi_{1,\alpha}$, we can define

$$\psi_{\theta,\varepsilon}(x)=\theta\psi_{1,\varepsilon}(x/_\theta),\ \text{where } |\psi''_{1,\varepsilon}|\le 2/_\varepsilon.$$

Observe that

$$\frac{\partial}{\partial\theta}\psi_{\theta,\varepsilon}(x)=\psi_{1,\varepsilon}(x/_\theta)-\psi'_{1,\varepsilon}(x/_\theta)x/_\theta$$

is bounded independently of θ,ε if $|\psi'_{1,\varepsilon}|\le 2$, which we can assume. Indeed, for $x\ge\theta+\varepsilon\theta$,

$$\frac{\partial}{\partial\theta}\psi_{\theta,\varepsilon}(x)=x/_\theta-1-x/_\theta=-1.$$

For $x\le\theta$, $\frac{\partial}{\partial\theta}\psi_{\theta,\varepsilon}(x)=0$. For $x\in[\theta,\theta+\varepsilon\theta]$, $\frac{\partial}{\partial\theta}\psi_{\theta,\varepsilon}(x)$ is obviously bounded.

Lastly, $\frac{\partial^2 \psi_{\theta,\varepsilon}}{\partial x \theta \varepsilon} = \frac{\partial}{\partial \varepsilon}\left((\frac{\partial}{\partial y}\psi_{1,\varepsilon})(x/_\theta)\right) = 0(1/_\varepsilon)$.

We will take a $\psi_{\theta,\varepsilon}$ of this type below.

Observe that, taking $0 < \varepsilon < 1$:

$$(\psi)\begin{cases} \left|\frac{\partial^2 \psi_{\theta,\varepsilon}}{\partial \theta^2}\right| = |(\psi''_{1,\varepsilon}(x/_\theta) \cdot x/_\theta \cdot x/_\theta)\frac{1}{\theta}| \leq 4/_{\varepsilon\theta}(1+\varepsilon)^2 \leq 8/_{\varepsilon\theta} \\ \left|\frac{\partial^2 \psi_{\theta,\varepsilon}}{\partial \theta \partial x}\right| = |-\psi''_{1,\varepsilon}(x/_\theta) \cdot x/_\theta \cdot 1/_\theta| \leq 8/_{\varepsilon\theta} \\ \left|\frac{\partial^2 \psi_{\theta,\varepsilon}}{\partial x^2}\right| = \frac{1}{\theta}|\psi''_{1,\varepsilon}(x/_\theta)| \leq 8/_{\varepsilon\theta}, \quad \left|\frac{\partial^2 \psi_{\theta,\varepsilon}}{\partial x \partial \varepsilon}\right| \leq C/_\varepsilon. \end{cases}$$

Construction of $\theta(t), \varepsilon(t)$ and of $\psi_{\theta(t),\varepsilon(t)}(b), \psi_{\tilde{\theta}(t),\tilde{\varepsilon}(t)}(\tilde{b})$

Let

$$\begin{cases} \theta(t) \text{ be a } C^\infty \text{ positive function such that} \\ \theta(t_1^-) = |b|(t_1^-) = \mu_1^- \quad \theta(t_2^-) = |b|(t_2^-) = \mu_2^-, \\ \theta(t) \leq \text{ Sup } (\mu_1^-, \mu_2^-) \end{cases} \tag{12}$$

μ_1 and μ_2 will be chosen extremely close, $\tilde{\tilde{\delta}}$ in Lemma 1 very small, and we can assume that we constructed θ so that

$$|\theta - 1|_{C^3} \underset{\mu_2 \to \mu_1}{\longrightarrow} 0, \text{ with a lower bound on } t_2^- - t_1^-. \tag{13}$$

Observe that, in our construction of oscillations satisfying Lemma 1, we can take $|\mu_2 - \mu_1|$ as small as we wish and have still a lower bound on $t_2^+ - t_1^+$. Indeed, since

$$\underline{\tilde{\mu}} < \mu_1, \mu_2 < \frac{\tilde{\tilde{\mu}} + \underline{\tilde{\mu}}}{2}$$

we have:

$$t_2^+(\mu_2) - t_1^+(\mu_1) \geq t_2^+\left(\frac{\tilde{\tilde{\mu}} + \underline{\tilde{\mu}}}{2}\right) - t_1^+\left(\frac{\tilde{\tilde{\mu}} + \underline{\tilde{\mu}}}{2}\right) > 0. \tag{14}$$

We further specify $\psi_{\theta,\varepsilon}$ as follows:

We set

$$\begin{cases} \varepsilon(t) \equiv \frac{\mu_1 - \mu_1^-}{2\mu_1^-} \text{ on } [t_1^-, t_1^+], \varepsilon(t) \equiv \frac{\mu_2 - \mu_2^-}{2\mu_2^-} \text{ on } [t_2^+, t_2^-] \\ \theta(t) \equiv \mu_1^- \text{ on } [t_1^-, t_1^+] \quad \theta(t) \equiv \mu_2^- \text{ on } [t_2^+, t_2^-]. \end{cases} \tag{14'}$$

If we extend $\varepsilon(t)$ on $[t_1^+, \bar{t}]$ by $\frac{\mu_1 - \mu_1^-}{2\mu_1^-} = \varepsilon_1$, then

$$\psi_{\mu_1^-,\varepsilon_1}(b) = |b| - \mu_1^- \text{ on } [t_1^+, \bar{t}], \text{ since}$$

$|b(t)|$ is larger than $\mu_1 > \mu_1^-(1+\varepsilon_1)$ on $[t^+, \bar{t}]$.

We choose $\mu_2 > \mu_1$. There exists then since $\bar{\bar{\mu}} > \mu_2$, a point $\tilde{t}_2^+$ in $(t_1^+, \bar{t})$ such that $|b(t)| \geq \mu_2 + \bar{\bar{\delta}}$ on $[\tilde{t}_2^+, \bar{t}]$, $|b(\tilde{t}_2^+)| = \mu_2 + \bar{\bar{\delta}}$. $\bar{\bar{\delta}}$ is small positive parameter which satisfies $\mu_2 + \bar{\bar{\delta}} < \frac{\mu_2 + \bar{\bar{\mu}}}{2}, \mu_2 + \bar{\bar{\delta}} < \nu$ when $\nu > \mu_2$ and which is used in the globalization process.

Since $\mu_2^-(1+\varepsilon) = \mu_2^-\left(1 + \frac{\mu_2 - \mu_2^-}{2\mu_2^-}\right) < \mu_2$,

$$\psi_{\mu_2^-, \varepsilon_2}(b) = |b| - \mu_2^- \text{ on } [\tilde{t}_2^+, \bar{t}].$$

We can then glue up $\psi_{\mu_2^-, \varepsilon_2}$ and $\psi_{\mu_1^-, \varepsilon_1}$ on $[\tilde{t}_2^+, \bar{t}]$. Indeed, using the freedom given to us by Lemma 1, we can take

$$0 < \varepsilon_2 = \frac{\mu_2 - \mu_2^-}{2\mu_2^-} < \varepsilon = \frac{\mu_1 - \mu_1^-}{2\mu_1^-}. \tag{14''}$$

Then

$$\psi_{\mu_1^-, \varepsilon_1}(b) = \psi_{\mu_1^-, \varepsilon_2}(b) \text{ on } [\tilde{t}_2^+, \bar{t}].$$

since

$$\mu_1^-(1+\varepsilon_2) < \mu_1^-(1+\varepsilon_1) < \mu_1 < \mu_2.$$

Thus, $\psi_{\mu_2^-, \varepsilon_2}$ and $\psi_{\mu_1^-, \varepsilon} = \psi_{\mu_1^-, \varepsilon_2}$ can be glued up by taking a function $\theta(t)$ which will evolve from μ_1^- to μ_2^- in $[\tilde{t}_2^+, \bar{t}]$ rather near $\bar{\bar{\mu}}$, hence there where the variations have not effect; the corresponding precise requirement is provided in (γ) below. The glueing function in ε (from ε_1 to ε_2) has no effect on the function $\psi_{\theta,\varepsilon}$. Again $|\theta - 1|_{C^3}$ will tend to zero as μ_2 tends to μ_1 and $\mu_2^- - \mu_1^-$ to zero because $\tilde{t}_2^+$ is upperbounded, away from $\bar{t}$, using arguments similar to the ones used for (14). $\dot{\theta}$ has support in $[\tilde{t}_2^+, \bar{t}]$.

Then, $\psi_{\theta(t), \varepsilon_2}(b) = |b| - \theta(t)$ on $[\tilde{t}_2^+, \bar{t}]$, because $\theta(t)(1+\varepsilon_2) \leq \mu_2^-(1+\varepsilon_2) < \mu_2 \leq |b|(t) \forall t \in [\tilde{t}_2^+, \bar{t}]$ if we are careful to choose $\mu_2^- > \mu_1^-$ and $\theta \in [\mu_1^-, \mu_2^-]$. We then have glued up the two functions.

Observe that the resulting function $\psi_{\theta(t), \varepsilon(t)}$ is such that:

$$(\gamma)\begin{cases} |b|(t) \geq \mu_2 + \bar{\bar{\delta}} \text{ on } [\tilde{t}_2^+, t_2^+], \psi_{\theta(t),\varepsilon(t)}(b) = |b| - \theta(t) \text{ on } [t_1^+, t_2^+] \\ \theta(t)(1+\varepsilon(t) < \operatorname{Sup}\left(\frac{\mu_1 + \mu_1^-}{2}, \frac{\mu_2 + \mu_2^-}{2}\right) \\ \frac{\partial \psi_{\theta,\varepsilon}}{\partial \theta}(x)\dot{\theta} + \frac{\partial \psi_{\theta,\varepsilon}}{\partial \varepsilon}(x)\dot{\varepsilon} = -\dot{\theta}\chi_{[\tilde{t}_2^+, \bar{t}]} = -\dot{\theta}, \forall x \text{ if } t \notin [\tilde{t}_2^+, \bar{t}] \\ (\text{ then}, \dot{\theta} = \dot{\varepsilon} = 0), \forall x \text{ such that } |x| \geq \mu_2 \text{ otherwise} \end{cases}$$

Since

$$(\gamma)\begin{cases} (\theta,\varepsilon)\equiv(\mu_1^-,\varepsilon_1) \text{ on } [t_1^-,\tilde{t}_2^+], \psi_{\theta(t),\varepsilon(t)}(x)=\psi_{\theta(t),\varepsilon_2}(x)= \\ |x|-\theta(t) \text{ on } [\tilde{t}_2^+,\bar{t}] \text{ if } |x|\geq\mu_2 \quad (|b(t)|\geq\mu_2 \text{ on } [\tilde{t}_2^+,\bar{t}]) \\ \text{Observe that if we are given } \nu>\mu_2, \quad \nu\leq\bar{\tilde{\mu}}, \text{ we can introduce } \bar{\bar{t}}, \\ \text{the first value after } \tilde{t}_2^+, \text{ before } \bar{t}, \text{ such that } |b|(\bar{\bar{t}})=\nu \\ \text{and ask that } \dot{\theta} \text{ has support in } (\tilde{t}_2^+,\bar{\bar{t}}). \text{ Again }, \bar{\bar{t}} \text{ is bounded away from} \\ \tilde{t}_2^+ \text{ as } \mu_2\to\mu_1. \text{ Thus, } |\theta-1|_{C^3}\underset{\mu_2\to\mu_1}{\longrightarrow}0\,. \end{cases}$$

If $\tilde{b}$ is close to b in H^1, in fact in L^∞ we can build using t_1^-,t_1^+,t_2^-,t_2^+ a related $\psi_{\tilde{\theta},\tilde{\varepsilon}}(\tilde{b})$ which will satisfy (14′) - (14″) and all of (γ) including $\psi_{\tilde{\theta}(t),\tilde{\varepsilon}(t)}(\tilde{b})=|\tilde{b}|-\tilde{\theta}$ on $[t_1^+,t_2^+]$ and $|\tilde{b}(t)|\leq\tilde{\mu}_2+\delta_{20}$ on $[\tilde{t}_2^+,\bar{t}]$, since we can make sure that $|\tilde{b}|\geq\mu_1-\delta>\mu_1^-(1+\varepsilon_1)=\underline{\frac{\mu_1^-+\mu_1}{2}}$ on $[t_1^+,\bar{t}]$ and than $\mu_2-\delta>\mu_2^-(1+\varepsilon_2)=\underline{\frac{\mu_2+\mu_2^-}{2}}$ on $[\bar{t},t_2^+]$. Choosing $\tilde{b}$ close enough to b will keep these inequalities, hence the claim. Furthermore, with the same $\tilde{t}_2^+$, we can ask that $|\tilde{b}(t)|\geq\tilde{\mu}_2+\frac{\bar{\bar{\delta}}}{2}>\mu_2+\frac{\bar{\bar{\delta}}}{4}$ on $[\tilde{t}_2,\bar{t}]$ and that $|\tilde{b}(\tilde{t}_2^+)|<\nu$ if $\nu>\mu_2$. We can also ask, then, that $\dot{\tilde{\theta}}$ has support in $\{|\tilde{b}|(t)|<\nu\}$, using the continuity of this construction process and the fact that $\dot{\theta}$ has support in $\{|b(t)|<\nu\}$, if $\nu>\mu_2$. Observe that the fact that $\frac{\partial\psi_{\tilde{\theta},\tilde{\varepsilon}}}{\partial\theta}(x)\dot{\tilde{\theta}}+\frac{\partial\psi_{\tilde{\theta},\tilde{\varepsilon}}}{\partial\tilde{\varepsilon}}\dot{\tilde{\varepsilon}}=-\dot{\tilde{\theta}}\chi_{[\tilde{t}_2^+,\bar{t}]}=-\dot{\tilde{\theta}}\forall x$ if $\notin[\tilde{t}_2^+,\bar{t}],\forall x,|x|\geq\tilde{\mu}_2$ otherwise, can be preserved because $(\tilde{\theta},\tilde{\varepsilon})$ is kept equal to $(\tilde{\mu}_1^-,\tilde{\varepsilon})$ on $[t_1^-,\tilde{t}_2^+]$ and $|\tilde{b}(t)|\geq\tilde{\mu}_2$ on $[\tilde{t}_2^+,\bar{t}]$. We also have kept (14′) and (14″). Thus, the previous construction for θ,ε can be repeated step by step. Observe that, then, $\tilde{\theta},\tilde{\varepsilon}$ depends only on $|b-\tilde{b}|_{L^\infty}$ and can be taken to be Lipschitz as functions of $\tilde{b}$ in $W^{1,1}$ for example. This concludes on construction of $\psi_{\tilde{\theta},\tilde{\varepsilon}}(b)$.

Let c be a positive constant, on which quite stringent requirements will be imposed later. c will be less than the variation of a along a flow-line of the pseudo-gradient which we are building, namely we require:

$$0<c\leq\inf\left(-\frac{\tilde{C}a}{1+|b|_\infty^3}\frac{\partial a}{\partial s},1\right). \tag{15}$$

$\tilde{C}$ is a fixed, universal constant. We will analyze the meaning of (15) and the choice of c later.

Let $\nu>0$ be a fixed positive constant

$$\nu\ll M. \tag{16}$$

The precise value of ν will be given later. Given an oscillation, we will always require:

$$\mu_1,\mu_2>\nu \quad \text{or} \quad \mu_1,\mu_2<\nu. \tag{16'}$$

If $\mu_1, \mu_2 < \nu$, we choose $\tilde{\tilde{\delta}}$ so small in Lemma 1 so that

$$\inf(\tilde{\bar{\mu}}, \nu) > \operatorname{Sup}\left(\mu_1, \mu_2, \mu_1 + \frac{\mu_2 - \mu_2^-}{2}, \mu_2 + \frac{\mu_1 - \mu_1^-}{2}\right). \tag{16''}$$

Assume now that we are given a single oscillation, with:

$$\mu_1, \mu_2 > \nu. \tag{17}$$

In the case where $J^+(t_1, t_2)$ or $J^-(t_1, t_2)$ satisfies 2)a), we introduce a function w^+ or w^- on J^+ or J^- with the following properties:

$$\begin{aligned} & w^\pm \in C_0^\infty(J^\pm), 1 \geq w^\pm \geq 0, |\dot{w}|^{\pm}_{C^2} \leq 10000 c_0^{100}(|b|_\infty + M)^3 \\ & \int w^\pm \geq \frac{1}{100}\Delta t^\pm. \end{aligned} \tag{18}$$

We then have:

Definition 2. *We define $\eta = \eta_{12}$ of an oscillation to be:*
(i) *in case 1: We assume the oscillation to satisfy lemma 1 and we set $\eta_{12} = (\psi_{\theta(t),\varepsilon}(b)\eta_0) sgn b + cb : \theta, \varepsilon, c$ will be determined appropriately below.*
(ii) *The oscillation is assumed to be simple. In case 2), 2)a): $\eta_{12} = w^\pm sgn b$ under (17), which we can assume if $\tilde{\bar{\mu}} > \nu$. In case 2), 2)b): $\eta_{12} = 0$.*
(iii) *If $|b| < (1 - \varepsilon_{15})|b|_\infty$ and $\big||b| - \nu\big| \geq \varepsilon_{15}$ on an interval $[t_1, t_2]$ such that $|t_1 - t_1| \geq \varepsilon_{16}$ we set $\tilde{\eta} = \tilde{\eta}_{12}$ to be a C^∞-function, positive on $[t_1, t_2]$ typically a regularization of $\sin \frac{\pi(t - t_1)}{t_2 - t_1}$, which satisfies $|\tilde{\eta}|_{C^2} \leq C/(t_2 - t_1)^2, C$ a universal constant, $\tilde{\eta} \in [0, 1]$.*

Any such η_{12} creates an H^1-vector-field Z_{12}. Observe that we are working in H^2. Therefore, the existence for Z_{12} will be established using some monotonicity properties. We now claim that:

Lemma 3. *let s be the time along a flow-line generated by Z_{12}. Then,provided ε, θ are chosen appropriately and provided we are careful in the construction of η_0, we have:*
(i) $\frac{\partial}{\partial s}\left(\frac{|b|_\infty}{a}\right) \leq -\frac{C_1}{a}\frac{\partial a}{\partial s}$, when C_1 is a universal constant depending only on $\tilde{C}$ in (15) and on the contact form α.

(ii) There exists a universal constant $\bar{C}$, which depends only on ν and γ_0 not on the choice of $\tilde{C}$ in (15) such that:

$$\frac{\partial}{\partial s}\left(\int (|b| - \nu)^+ + \bar{C}(1 + a)\right) \leq 0 \text{ if } a \geq \gamma_0$$

(iii)

$$\frac{1}{2}\frac{\partial}{\partial s}\int_0^1 \dot{b}^2 \leq -\frac{9c}{10}\int_0^1 \ddot{b}^2 + \frac{C_3(|b|_\infty(0),\varepsilon_{16})}{c}(1+\frac{1}{\Gamma^2}+\frac{1}{\Gamma}(C_4(|b|_\infty(0)) - \frac{\partial(\bar{\tilde{\mu}}/a)}{\partial s} - $$
$$- C(\frac{\partial a}{\partial s})) \times \left(\int_{t_1^-}^{t_2^-} \dot{b}^2 + t_2^- - t_1^-\right) - C_4(|b|_\infty(0))(1+\int_0^1 \dot{b}^2)\frac{\partial a}{\partial s},$$

where Γ is a lower bound on all oscillations satisfying $(*)$ of type 1) of $\bar{\tilde{\mu}} - \mu_2$ and where C_3, C_4 are appropriate functions of $|b|_\infty(0)$. $-\frac{\partial}{\partial s}\left(\frac{\bar{\tilde{\mu}}}{a}\right) - Ca$ can be replaced by $\frac{1}{|x_1^- - x_2^-|^2}$ in the formula above.

Proof of Lemma 3. We have

$$\begin{aligned}\frac{\partial b}{\partial s} &= \dot{\mu} + a\eta\tau - b\eta\bar{\mu}_\xi = \frac{\ddot{\eta} + \lambda\dot{b} + \dot{\lambda}b}{a} + a\eta\tau - b\eta\bar{\mu}_\xi = \\ &= \frac{\ddot{\eta} + b^2\eta - \dot{\bar{\mu}}\eta b}{a} + \frac{\lambda\dot{b}}{a} + a\eta\tau - b\eta\bar{\mu}_\xi - \frac{b\int_0^1 b\eta}{a}\end{aligned} \tag{19}$$

A property such as (i) can be seen on $b \in C^1$, by density since we will be proving the H^1-continuity in some sense (H^1 on b) of the flow defined by Z_{12}. We thus need only to prove that:

$$(sgnb)\frac{\partial}{\partial s}\left(\frac{b}{a}\right) \leq -\frac{C_1}{a}\frac{\partial a}{\partial s} \text{ at any } t \text{ such that } |b|(t) = |b|_\infty. \tag{20}$$

This is obvious in the case of 2)a) because η_{12} has support in $J^\pm(t_1, t_2)$, far from any point where $|b|(t) = |b|_\infty$. Hence, at any such point, these η_{12} vanish C^2. (19) implies that $(sgnb)\frac{\partial b}{\partial s} = -\frac{|b|\int_0^1 b\eta_{12}}{a}$ at any t such that $|b|(t) = |b|_\infty$. The same argument holds in case (iii) of Definition 2. Since $\tilde{C}$ is less than $|b|_\infty, \tilde{\eta} = \tilde{\eta}_{12}$ also vanishes at any t such that $|b|(t) = |b|_\infty$.

If $\eta_{12} = cb$, then, using (19) and the fact that $(sgnb)\ddot{b}$ is always negative at any such point, we have:

$$(sgnb)\frac{\partial}{\partial s}(b/a) \leq c(|b|_\infty^3 + C_0) \tag{21}$$

where C_0 is an appropriate constant.

Using (15), (i) follows also in this case.

We now consider the case where $\eta_{12} = \psi_{\theta(t),\varepsilon}(b)\eta_0 sgnb$.

Observe that near t_1^-, t_2^-, $|b|$ is strictly below $\bar{\tilde{\mu}} \leq |b|_\infty$. Hence the possible discontinuity of $\dot{\eta}_0$ does not matter there. This will allow to keep (i) after C^∞-regularization of η_{12}. The case where $\bar{\tilde{\mu}} < |b|_\infty$ is immediate: $\frac{\partial}{\partial s}|b|_\infty$ is zero then. We therefore assume that $(1 + a^2 + \bar{\bar{\mu}}^2) < \varepsilon_0/(t_1 - t_2)^2$ with $\bar{\bar{\mu}} \geq |b|_\infty/2$. We compute far from t_1^-, t_2^-, in the interval $[t_1^-, t_2^-]$, at a point where $\dot{b}(t) = 0, \ddot{b}(t) sgn b(t) \leq 0$, the value of

$$\overline{\dot{\eta}_{12} - \bar{\mu} b \eta_{12}} = \left(e^{\int_{t_1^-}^t \bar{\mu} b d\tau} \overline{e^{-\int_{t_1^-}^t \bar{\mu} b d\tau} \eta_{12}}^{\bullet} \right) = e^{-\int_{t_1^-}^t \bar{\mu} b d\tau} \frac{d}{dx^z} \left(\eta_{12} e^{-\int_{t_1^-}^t \bar{\mu} b d\tau} \right).$$

Assume, for simplicity, that $b(t)$ is positive. We use (10), the fact that $\ddot{b}(t) \leq 0$. Observe that since $[t_1, t_2]$ did satisfy 1), so does $[t_1^-, t_2^-]$, thus $\int_{t_1^-}^{t_2^-} |b| d\tau \leq 1$ if ε_0 is small enough, hence:

$$\frac{1}{\bar{c}} |t_1^- - t_2^-| \leq |x_1^- - x_2^-| \leq \bar{c} |t_1^- - t_2^-| \tag{22}$$

where $\bar{c}$ depends only on α.

Also, (22′) $0 \leq \eta_0 \leq C(|b|_\infty), \dot{\eta}_0 = 0(C(|b|_\infty)/_{|x_1^- - x_2^-|})$.

We then have, for any such $t \in [t_1^-, t_2^-]$, using also (1), (γ), and the bound on $\int_{t_1^-}^t |\bar{\mu} b| d\tau$:

$$\overline{\dot{\eta}_{12} - \bar{\mu} b \eta_{12}}^{\bullet} = e^{-\int_{t_1^-}^t \bar{\mu} b d\tau} \frac{d^2}{dx^2} \left(\eta_{12} e^{-\int_{t_1^-}^t \bar{\mu} b d\tau} \right) = e^{-\int_{t_1^-}^t \bar{\mu} b d\tau} \frac{d^2}{dx^2} \left(\psi_{\theta,\varepsilon} \eta_0 e^{-\int_{t_1^-}^t \bar{\mu} b} \right) = \tag{23}$$

$$= \left[\frac{-2}{(x_1^- - x_2^-)^2} \psi_{\theta,\varepsilon} + 0 \left(\frac{C(|b|_\infty)}{|x_1^- - x_2^-|} \frac{\partial \psi_{\theta,\varepsilon}}{\partial \theta} \dot{\theta} \right) - (C(1 + \bar{\bar{\mu}})^2 \psi_{\theta,\varepsilon} + \right.$$

$$\left. + 0 \left((a^2 + \bar{\bar{\mu}}^2) \dot{\theta} \right) \right) \eta_0 + \left(-\ddot{\theta} \chi_{[\bar{t}_2^+, t]} + \frac{\partial \psi_{\theta,\varepsilon}}{\partial b} \ddot{b} \right) \eta_0 e^{-\int_{t_1^-}^t \bar{\mu} b} \Bigg] e^{-\int_{t_1^-}^t \bar{\mu} b dx} = (1).$$

Observe that $\ddot{b}(t) \leq 0, \eta_0 \geq 0$. $\int_0^1 |b|$ is bounded as we will see. Thus,

$$(1) \leq \frac{-1}{2(x_1^- - x_2^-)^2} \psi_{\theta,\varepsilon} e^{-\int_{t_1^-}^t \bar{\mu} b d\tau} + \frac{0(|\dot{\theta}|_{C^1} \times C(|b|_\infty))}{|x_1^- - x_2^-|} - \frac{C}{2} (1 + \bar{\mu}^2) \psi_{\theta,\varepsilon} \eta_0 e^{-\int_{t_1^-}^t \bar{\mu} b dx}, \tag{23′}$$

where $0\left(|\dot{\theta}|_{C^1}\right)$ depends only on $|x_1^- - x_2^-|, \bar{\bar{\mu}}$ and the first derivative of $\psi_{\theta,\varepsilon}$ with respect to θ, which is bounded independently of ε, θ.

We now keep $\mu_1, \mu_2 \leq \gamma < \bar{\bar{\mu}}$. This allows us to lower bound $|x_2^- - x_1^-|$ and $|t_2^- - t_1^-|$ and $\int b \psi_{\theta,\varepsilon} \eta_0$ in function of γ only, as soon as $|\theta - 1|_{C^3} \leq \bar{\bar{\delta}}$ (i.e μ_2 close enough to $\mu_1, \mu_1, \mu_2 \leq \gamma$).

We then make μ_2 tend to $\mu_1, \bar{\bar{\delta}}$ of Lemma 1 to tend to zero, thus θ tends to 1 in C^3.

Hence, any $0(|\dot{\theta}|_{C^1})$ can be absorbed in the lower bound of $\int b \psi_{\theta,\varepsilon} \eta_0$, using also the lower bound on $|x_2^- - x_1^-|$.

Thus, at any such $b(t)$, after absorbing $\frac{b^2\psi_{\theta,\varepsilon}\eta_0}{a} + (a\tau - b\bar{\mu}_\xi)\psi_{\theta,\varepsilon}\eta_0$ in $-\frac{C}{2}(1+\bar{\bar{\mu}}^2)e^{\int_{t_1^-}^t \bar{\mu}b}\psi_{\theta,\varepsilon}\eta_0$ by taking C large enough in (10), we have:

$$\frac{\partial(b/_a)}{\partial s} \leq \frac{1}{a}\left(\frac{-1}{2(x_1^- - x_2^-)^2}\psi_{\theta,\varepsilon}e^{-\int_{t_1^-}^t \bar{\mu}bd\tau}\right) - \frac{1}{a}\frac{\partial a}{\partial s}. \tag{24}$$

Since $\int_{t_1^-}^{t_2^-} |b|d\tau \leq \bar{\bar{\mu}}|t_1^- - t_2^-| < \sqrt{\varepsilon_0}$, since (22) holds, we easily derive, using (1), at any such $b(t)$, for example at $b(t) = \bar{\bar{\mu}}$

$$\frac{\partial(b/_a)}{\partial s} + \frac{1}{a}\frac{\partial a}{\partial s} \leq \frac{-\bar{\bar{c}}}{a(t_1^- - t_2^-)^2}\inf_{\tau\in[t_1^-,t_2^-]}\psi_{\theta(\tau),\varepsilon(\tau)}(\bar{\bar{\mu}}) = \frac{-\bar{\bar{c}}}{a}\frac{\bar{\bar{\mu}} - \mu_2^-}{(t_1^- - t_2^-)^2} \leq -\frac{\bar{\bar{c}}}{2a}\frac{\bar{\bar{\mu}} - \mu_2}{(t_1^- - t_2^-)^2} \tag{25}$$

if ε_0 is chosen small enough, where $\bar{\bar{c}}$ is a universal constant. (i) follows. Observe that, in case of (ii) or (iii) of Definition 2, we have

$$\frac{\partial}{\partial s}\bar{\bar{\mu}} + \frac{\partial a}{\partial s} \leq C(\varepsilon_{15}, \varepsilon_{16})(1 + |b|_\infty)^{1000}. \tag{25'}$$

We now prove (ii). We provide the argument when $|b| - \nu$ has a finite number of zeros. This argument certainly extends to any $b \in H^1$ using Proposition A1 of Appendix 1.

Assume that

$$\text{on } [t_j, \tilde{t}_{j+1}], |b| - \nu > 0, \text{ with } |b| = \nu \text{ at } t_j, \tilde{t}_{j+1}. \tag{26}$$

Then:

$$\dot{b}(t_j)sgnb(t_j) \geq 0, \dot{b}(\tilde{t}_{j+1})sgnb(\tilde{t}_{j+1}) = \dot{b}(\tilde{t}_{j+1})sgnb(t_j) \leq 0. \tag{27}$$

Observe that, since we required (16'), η_0 is either identically zero near t_j and $\tilde{t}_{j+1}$, if $\mu_1, \mu_2 > \nu$ and $\bar{\bar{\delta}}$ is small enough so that $\mu_1^-, \mu_2^- > \nu$.

If $\mu_1, \mu_2 < \nu$, we then have

$$t_1^- < t_j < t_2^- \Rightarrow \begin{cases} \nu \leq \bar{\bar{\mu}} \\ t_1^- < t_j < \tilde{t}_{j+1} < t_2^-. \end{cases}$$

($|b|$ is larger than ν, hence than μ_2^- on $[t_j, \tilde{t}_{j+1}]$.) (28)

After regularization, if needed, if (28) does not hold, we have, of course: $(sgnb(t_j) = sgnb(\tilde{t}_{j+1}))$

$$\begin{aligned} &sgnb(t_j)(\dot{\eta}(t_j) - \bar{\mu}(x(t_j))\nu sgnb(t_j)\eta(t_j)) = 0 \geq \\ &\geq sgnb(\tilde{t}_{j+1})(\dot{\eta}(\tilde{t}_{j+1}) - \bar{\mu}(x(\tilde{t}_{j+1}))\nu sgnb(\tilde{t}_{j+1})\eta(\tilde{t}_{j+1})) = 0 \end{aligned} \tag{29}$$

since $\eta, \dot{\eta}$ are then zero at $t_j, \tilde{t}_{j+1}$.

The same argument works in case (iii) of Definition 2, since b is constant, distinct from ν on the support of $\tilde{\eta}$.

We claim that the inequality of (29), slightly modified, still holds under (28), with $\eta = \eta_{12} = (\psi_{\theta,\varepsilon}(b)\eta_0)sgnb + cb$.

Assume, for example, that b is positive, on $[t_j, t_{j+1}]$, for sake of simplicity. Then, using (27) and (12) (16″) and (18) (in particular, $\frac{\partial \psi_{\theta,\varepsilon}}{\partial \theta}\dot{\theta} + \frac{\partial \psi_{\theta,\varepsilon}}{\partial \varepsilon}\dot{\varepsilon} = -\dot{\theta}$ on $[t_j, \tilde{t}_{j+1}]$, since $\nu > \theta + \varepsilon\theta \quad \forall t$ and $\psi_{\theta(t_j),\varepsilon(t_j)}(\nu) - \psi_{\theta(\tilde{t}_{j+1}),\varepsilon(\tilde{t}_{j+1})}(\nu) = \theta(\tilde{t}_{j+1}) - \theta(t_j)$)

$$\dot{\eta}(t) - \bar{\mu}\nu\eta(t)\Big|_{t_j}^{\tilde{t}_{j+1}} \leq \eta_0 \left(\frac{\partial \psi_{\theta,\varepsilon}}{\partial \theta}\dot{\theta}(t) + \frac{\partial \psi}{\partial \varepsilon}\dot{\varepsilon}\right)\Big|_{t_j}^{\tilde{t}_{j+1}} + \psi_{\theta,\varepsilon}\dot{\eta}_0\Big|_{t_j}^{\tilde{t}_{j+1}} - \tag{30}$$

$$-\bar{\mu}\nu\psi_{\theta,\varepsilon}(\nu)\eta_0\Big|_{t_j}^{\tilde{t}_{j+1}} - c\bar{\mu}\nu^2\Big|_{t_j}^{\tilde{t}_{j+1}} = -\eta_0\dot{\theta}(t)\Big|_{t_j}^{\tilde{t}_{j+1}} +$$

$$+\psi_{\theta,\varepsilon}\frac{d}{dx}\left(e^{-\int_{t_1^-}^{t} \bar{\mu} b d\tau}\eta_0\right)\Big|_{t_j}^{\tilde{t}_{j+1}} - c\bar{\mu}\nu^2\Big|_{t_j}^{\tilde{t}_{j+1}} \leq$$

$$\leq \underbrace{\psi_{\theta(t_j),\varepsilon}(\nu)\left(\frac{d}{dx}(e^{-\int_{t_1^-}^{t} \bar{\mu} b d\tau}\eta_0)\Big|_{t_j}^{\tilde{t}_{j+1}}\right)}_{(I)} +$$

$$+\bar{\bar{C}}(|b|_\infty)\left(\frac{1}{|x_1^- - x_2^-|}\int_{t_j}^{\tilde{t}_{j+1}} |\dot{\theta}|\right) + c\nu^2\int_{t_j}^{\tilde{t}_{j+1}} |a\bar{\mu}_\xi + b\bar{\mu}_v|dt - \eta_0\dot{\theta}(t)\Bigg|_{t_j}^{\tilde{t}_{j+1}}$$

(I) is negative because the function $e^{-\int_{t_1^-}^{t} -\bar{\mu}b}\eta_0$ is concave. Observe that $\nu \leq \bar{\bar{\mu}}$. Since $|b| \geq \nu$ on $[t_j, \tilde{t}_{j+1}]$, and since $\dot{\theta}$ has support in $[\tilde{t}_2^+, \bar{\bar{t}}]$ (see (γ)), with $|b|(t) < \nu$ on $[\tilde{t}_2^+, \bar{\bar{t}}]$, the term $-\eta_0\dot{\theta}\Big|_{t_j}^{\tilde{t}_{j+1}}$ drops. This term can be taken very small, anyway, as we show below.

As we pointed out earlier, $e^{-\int_{t_1^-}^{t} \bar{\mu} b d\tau}\eta_0$ is concave in x; thus $\frac{d}{dx}\left(e^{-\int_{t_1^-}^{t} \bar{\mu} b d\tau}\eta_0\right)\Big|_{t_j}^{\tilde{t}_{j+1}}$ is non positive. As we also pointed out above, we can keep $\mu_1, \mu_2 \leq \gamma < \bar{\bar{\mu}}$ and let μ_2 tend to μ_1 so that θ tends to 1 in C^3. η changes then. However $\int_{t_1^+}^{t_2^+} b\eta$ remains lower bounded by a constant $\delta > 0$, depending only on γ.

Thus, we can take μ_2 so close to μ_1 so that:

$$\sum_j 0\left(\frac{\bar{\bar{C}}(|b|_\infty)}{|x_2^- - x_1^-|}\int_{t_j}^{\tilde{t}_{j+1}} |\dot{\theta}|\right) \leq \int_{t_1^+}^{t_2^+} b\eta \leq -\frac{\partial a}{\partial s}. \tag{31}$$

Furthermore, since $|b| > \nu$ on $[t_j, t_{j+1}]$, we have

$$c\nu^2 \sum_j \int_{t_j}^{\tilde{t}_j+1} |a\bar{\mu}_\xi + b\bar{\mu}_v| dt \leq C_1 \sum_j \left(a \int_{t_j}^{\tilde{t}_j+1} cb^2 + \nu \int_{t_j}^{\tilde{t}_j+1} cb \right)$$
$$\leq \tilde{C}_1(1+a) \sum_j c \int_{t_j}^{\tilde{t}_j+1} b^2 \tag{32}$$
$$\leq \tilde{C}_1(1+a) \int_{t_1^-}^{t_2^-} b\eta \leq -\tilde{C}_1(1+a)\frac{\partial a}{\partial s}.$$

Similarly

$$\sum_j \left| \int_{t_j}^{\tilde{t}_j+1} a\eta\tau - b\eta\bar{\mu}_\xi \right| \leq -\tilde{C}_1 \left(\frac{a}{\nu} + 1 \right) \int_{t_1^-}^{t_2^-} b\eta \leq -\tilde{C}_1 \left(\frac{a}{\nu} + 1 \right) \frac{\partial a}{\partial s}. \tag{33}$$

Using then (29), (30), (31), (32), (33), the fact that $\frac{d}{dx}\left(e^{-\int_{t_1^-}^{t} \bar{\mu} b d\tau} \eta_0 \right) \Big|_{t_j}^{\tilde{t}_j+1}$ is non positive and (19), we derive (ii).

Observe that the above argument extends to any $\tilde{b}$ close enough in H^1 to b which is also such that $|\tilde{b}| - \nu$ has a finite number of zeros if we modify $\psi_{\theta,\varepsilon}$ in $\psi_{\tilde{\theta},\tilde{\varepsilon}}$ satisying (γ) for $\tilde{b}$ and if we are careful in our extension. Indeed, since $\mu_1, \mu_2 > \nu$ or $\mu_1, \mu_2 < \nu$, we can choose the neighborhood of b so small that the same property holds for $\tilde{b}$. We can, in fact, keep the same $[t_1^-, t_2^-]$ and the same η_0, modify $\psi_{\theta,\varepsilon}(b)$ into $\psi_{\tilde{\theta},\tilde{\varepsilon}}(\tilde{b})$ where $\tilde{\theta}, \tilde{\varepsilon}$ are very slight modifications of θ, ε built for $\tilde{b}$ as follows: (We recall the construction of $\psi_{\tilde{\theta},\tilde{\varepsilon}}(b)$) we keep $t_1^-, t_1^+, t_2^-, t_2^+$ unchanged, but use the values of $|\tilde{b}|$ at these points to define $\tilde{\theta}, \tilde{\varepsilon}$ as in $(14)'$. (γ) still holds including the fact that $\psi_{\theta(t),\varepsilon(t)} = |b| - \theta(t)$ on $[t_1^+, t_2^+]$, which we can make hold because $\mu_1^-(1+\varepsilon_1) < \mu_1$ and $\mu_2^-(1+\varepsilon_2) < \mu_2$, inequalities which we can keep for $\tilde{b} \cdot \tilde{t}_2$ is changed into a $\tilde{\tilde{t}}_2$. We keep the notation $\tilde{t}_2$ for the initial construction. $\dot{\theta}$ can be assumed to have support in $(\tilde{t}_2^+, \bar{\bar{t}})$ and, in fact in $\{|\tilde{b}| < \nu\}$, by continuity. Indeed, $\dot{\theta}$ has support in $(\tilde{t}_2^+, \bar{\bar{t}})$, thus in $\{|b| < \nu\}$. The construction of $\tilde{\theta}, \tilde{\varepsilon}$ is very parallel to the one of θ, ε but with modified values of $\tilde{\theta}(t_i^-), \tilde{\theta}(t_i^+), \tilde{\varepsilon}(t_i^-), \tilde{\varepsilon}(t_i^+)$due to the fact that b is changed into $\tilde{b}$. Therefore, by continuity, if $\tilde{b}$ is close enough to b in $H^1, \tilde{\theta}$ can be assumed to have support in $\{|\tilde{b}| < \nu\}$. This allows to extend the above argument.

This fact will be used later on when we globalize our flow. The same remark holds for (i), because Lemma 1 $ii)$ extends to a neighborhood of b in H^1, if we allow μ_1, μ_2 to vary slightly with $\tilde{b}$ and keep t_1^+, t_2^+ unchanged. This other fact will also be used when we globalize our flow.

We now prove (iii):

We first focus on case i) of Definition 2. The other cases are very easy.

We use (19) and we compute, in the formal sense:

$$\frac{1}{2}\frac{\partial}{\partial s}\int_0^1 \ddot{b}^2 = -\int_0^1 \ddot{b}\frac{\partial b}{\partial s} = -\int_0^1 \ddot{b}(\dot{\mu} + a\eta\tau - b\eta\bar{\mu}_\xi) =$$
$$= -\int_0^1 \ddot{b}(\frac{\overline{\dot{\eta} + \lambda b}}{a} + a\eta\tau - b\eta\bar{\mu}_\xi) = \tag{34}$$
$$= -\int_0^1 \ddot{b}\frac{\overline{\dot{\eta} + (\lambda + \bar{\mu}\eta)b}}{a} + \int_0^1 \ddot{b}\frac{\overline{\mu\eta\dot{b}}}{a} - \int_0^1 \ddot{b}(a\eta\tau - b\eta\bar{\mu}_\xi).$$

We first consider $-\int_0^1 \ddot{b}\ddot{\eta} = -c\int_0^1 \ddot{b}^2 - \int_0^1 \ddot{b}sgnb\ddot{\overline{\psi_{\theta,\varepsilon}\eta_0}}$.

We have (in the formal sense):

$$-\int_0^1 \ddot{b}\ddot{\eta} = -c\int_0^1 \ddot{b}^2 - \int_0^1 \ddot{b}sgnb\left(\ddot{\overline{\psi_{\theta,\varepsilon}}}\eta_0 + 2\dot{\psi}_{\theta,\varepsilon}\dot{\eta}_0 + \psi_{\theta,\varepsilon}\ddot{\eta}_0\right) = \tag{35}$$
$$= -\int_0^1 \ddot{b}sgnb\left(\left[-\ddot{\theta} + \frac{\partial^2\psi_{\theta,\varepsilon}}{\partial b^2}\dot{b}^2 + \frac{\partial\psi_{\theta,\varepsilon}}{\partial b}\ddot{b}\right]\eta_0 + \right.$$
$$\left. +2\left(-\dot{\theta} + \frac{\partial\psi_{\theta,\varepsilon}}{\partial b}\dot{b}\right)\dot{\eta}_0 + \psi_{\theta,\varepsilon}\ddot{\eta}_0\right) - c\int_0^1 \ddot{b}^2.$$

Let us first observe that, since η_0 is positive and since $\frac{\partial\psi_{\theta,\varepsilon}}{\partial b} = sgnb$ on $[\tilde{t}_2^+, \bar{t}]$

$$-\int_0^1 \ddot{b}^2 sgnb\frac{\partial\psi_{\theta,\varepsilon}}{\partial b}\eta_0 \le -\int_{\tilde{t}_2^+}^{\bar{t}} \ddot{b}^2\eta_0. \tag{36}$$

As μ_2 tends to $\mu_1, [\tilde{t}_2^+, \bar{t}]$ does not collapse. Thus, there exists $d > 0, \underline{t}$ depending on b and on the oscillation, not on $|\mu_1 - \mu_2|$, such that:

$$-\int_{\tilde{t}_2^+}^{\bar{t}} \ddot{b}^2\eta_0 \le -d\int_{\underline{t}}^{\bar{t}} \ddot{b}^2. \tag{37}$$

If we come back to the construction of θ, we can ask that $\dot{\theta}$ has support in $[\underline{t}, \bar{t}]$ and still have $|\theta - 1|_{C^3}$ tend to zero when μ_2 tends to μ_1. This is coherent with (γ), including its requirements about$\nu, \dot{\theta}$ would then have support in $[\underline{t}, \bar{\bar{t}}]$, and $\underline{t}$ could easily be taken fixed, strictly less than $\bar{\bar{t}}$. Since $|\eta_0|_{C^1}$ remains bounded because the intervals do not collapse, (see Lemma 2′) we can ask that ($\dot{\theta}$ has support in $[\underline{t}, \bar{t}]$):

$$|\int_0^1 \ddot{b}sgnb(\ddot{\theta}\eta_0 + 2\theta\dot{\eta}_0)| \le \bar{t} - \underline{t} + \frac{d}{2}\int_{\underline{t}}^{\bar{t}} \ddot{b}^2. \tag{38}$$

Thus

$$- \int_0^1 \ddot{b}^2 sgnb \frac{\partial \psi_{\theta,\varepsilon}}{\partial b} \eta_0 - \int_0^1 \ddot{b} sgnb (-\ddot{\theta}\eta_0 - 2\dot{\theta}\dot{\eta}_0) \leq \tag{39}$$

$$\leq -\frac{1}{2} \int_0^1 \ddot{b}^2 sgnb \frac{\partial \psi_{\theta,\varepsilon}}{\partial b} \eta_0 + \bar{t} - \underline{t} \leq \bar{t} - \underline{t}.$$

Considering now $-2\int_0^1 \ddot{b}\dot{b} sgnb \frac{\partial \psi_{\theta,\varepsilon}}{\partial b} \dot{\eta}_0$, we have:

$$| -2 \int_0^1 \ddot{b}\dot{b} sgnb \frac{\partial \psi_{\theta,\varepsilon}}{\partial b} \dot{\eta}_0 | \leq C \int_{t_1^-}^{t_2^-} |\ddot{b}||\dot{b}| e^{\tilde{c}\int_{t_1^-}^{t_2^-} |b|} \left(\frac{1}{|x_2^- - x_1^-|} + \bar{\tilde{\mu}} \right) dt \tag{40}$$

when C depends on $|b|_\infty$.

Clearly:

$$|x_2^- - x_1^-| \geq e^{-\tilde{c}\int_{t_1^-}^{\bar{t}_2} |b|} |t_2^- - t_1^-|. \tag{41}$$

Thus

$$| \int_0^1 \ddot{b}\dot{b} sgnb \frac{\partial \psi_{\theta,\varepsilon}}{\partial b} \dot{\eta}_0 | \leq C e^{\tilde{c}(t_2^- - t_1^-)\bar{\tilde{\mu}}} \left(\bar{\tilde{\mu}} + \frac{1}{|t_2^- - t_1^-|} \right) \times \tag{42}$$

$$\times \int_{t_1^-}^{t_2^-} |\ddot{b}\dot{b}| \leq \frac{c}{10} \int_{t_1^-}^{t_2^-} \ddot{b}^2 + \frac{4C^2 e^{2\tilde{c}\bar{\tilde{\mu}}}}{c} \left(\bar{\tilde{\mu}}^2 + \frac{1}{|t_2^- - t_1^-|^2} \right) \int_{t_1^-}^{t_2^-} \dot{b}^2.$$

On a given flow-line, we know that $\bar{\tilde{\mu}}$ is bounded above, since $|b|_\infty$ is bounded above by i) of Lemma 3.

Thus, using (25), we derive: ($|b|_\infty(0)$ is the L^∞-norm of b at the time zero on the flow-line).

$$\frac{\bar{\tilde{\mu}} - \mu}{(t_1^- - t_2^-)^2} \leq C_1(|b|_\infty(0)) \left(-\frac{\partial \bar{\tilde{\mu}}}{\partial s} - \frac{\partial a}{\partial s} + C_2(|b|_\infty(0)) \right). \tag{43}$$

Thus, denoting Γ any positive lowerbound of $\bar{\tilde{\mu}} - \mu_2$:

$$-2 \int_0^1 \ddot{b} sgnb \frac{\partial \psi_{\theta,\varepsilon}}{\partial b} \dot{b}\dot{\eta}_0 - c \int_0^1 \ddot{b}^2 \leq \tag{44}$$

$$\leq -9\frac{c}{10} \int_0^1 \ddot{b}^2 + \frac{C_3(|b|_\infty(0))}{c} \left(1 + \frac{1}{\Gamma^2} + \frac{1}{\Gamma}(C_4(|b|_\infty(0)) - \frac{\partial \bar{\tilde{\mu}}}{\partial s} - \frac{\partial a}{\partial s}) \right) \times$$

$$\times \left(\int_{t_1^-}^{t_2^-} \dot{b}^2 + t_2^- - t_1^- \right) - C_4(|b|_\infty(0))(1 + \int_0^1 \dot{b}^2) \frac{\partial a}{\partial s}.$$

We now consider what is left in (34), after $-\int \ddot{b}\bar{\eta}$. We are trying to upperbound in a similar way. For $\psi_{\theta,\varepsilon}\eta_0$, we have:

$$\int_0^1 \left|\ddot{b}(a\psi_{\theta,\varepsilon}\eta_0\tau - b\psi_{\theta,\varepsilon}\eta_0\bar{\mu}_\xi)\right| \leq \frac{c}{100}\int_{t_1^-}^{t_2^-} \ddot{b}^2 + \frac{C_s(|b|_\infty(0))}{c}|t_1^- - t_2^-|. \tag{45}$$

Similarly

$$\left|\int_0^1 \frac{\ddot{b}}{a}\dot{\overline{\bar{\mu}\psi_{\theta,\varepsilon}\eta_0 b}}\right| \leq \frac{C(|b|_\infty(0))}{a}\int_{t_1^-}^{t_2^-} |\ddot{b}|(|\dot{\theta}| + \frac{\psi_{\theta,\varepsilon}}{|t_1^- - t_2^-|} + |\dot{b}| + 1) \leq \tag{46}$$

$$\leq \frac{C(|b|_\infty(0))}{a}\left(\left(\int_{t_1^-}^{t_2^-} \ddot{b}^2\right)^{1/2}\left(O(|t_1^- - t_2^-|^{1/2} + \left(\int_{t_1^-}^{t_2^-} \dot{b}^2\right)^{1/2} + \frac{1}{|t_1^- - t_2^-|^{1/2}}\right)\right) \leq$$

$$\leq \frac{c}{100}\int_{t_1^-}^{t_2^-} \ddot{b}^2 + \frac{C'(|b|_\infty(0))}{c}\left(|t_1^- - t_2^-| + \int_{t_1^-}^{t_2^-} \dot{b}^2 + \frac{1}{|t_1^- - t_2^-|}\right).$$

Observe that

$$\Gamma \leq \int_{t_1^-}^{t_2^-} |\dot{b}| \leq \sqrt{|t_2^- - t_1^-|}\left(\int_{t_1^-}^{t_2^-} \dot{b}^2\right)^{1/2}.$$

Thus

$$\frac{1}{|t_2^- - t_1^-|} \leq \frac{\int_{t_1^-}^{t_2^-} \dot{b}^2}{\Gamma^2}$$

and we can replace $\frac{1}{|t_1^- - t_2^-|}$ by $\frac{\int_{t_1^-}^{t_2^-} \dot{b}^2}{\Gamma^2}$ in (46). This allows to keep the same upperbound than the one of (44).

Finally, designating by $\widetilde{\lambda + \bar{\mu}\eta}$ the part of $\lambda + \bar{\mu}\eta$ related to $\psi_{\theta,\varepsilon}\eta_0$:

$$\left|\int_0^1 \ddot{b}\dot{\overline{\widetilde{\lambda + \bar{\mu}\eta}b}}\right| = \left|\int_0^1 \left(\ddot{b}b - \frac{\dot{b}^2}{2}\right)\left(b\psi_{\theta,\varepsilon}\eta_0 - \int_0^1 b\psi_{\theta,\varepsilon}\eta_0\right)\right| = \tag{47}$$

$$= \left|\frac{1}{2}\int_0^1 \dot{b}^2 \int_0^1 b\psi_{\theta,\varepsilon}\eta_0 + \frac{c}{20}\int_{t_1^-}^{t_2^-} \ddot{b}^2 + \frac{C(|b|_\infty(0))}{c}|t_1^- - t_2^-| + C(|b|_\infty(0))\int_{t_1^-}^{t_2^-} \dot{b}^2\right.$$

$$\leq -\frac{1}{2}\int_0^1 \dot{b}^2\frac{\partial a}{\partial s} + \frac{c}{200}\int_{t_1^-}^{t_2^-} \ddot{b}^2 + \frac{C(|b|_\infty(0))}{c}\left(|t_1^- - t_2^-| + \int_{t_1^-}^{t_2^-} \dot{b}^2\right).$$

There are similar terms for cb:($\underline{\lambda + \bar{\mu}\eta}$, the part of $\lambda + \bar{\mu}\eta$ due to cb)

$$c\left|\int_0^1 \ddot{b}(ab\tau - b^2\bar{\mu}_\xi)\right| \leq \frac{c}{400}\int_0^1 \ddot{b}^2 + cC(|b|_\infty(0)) \tag{45'}$$

$$\left|c\int_0^1 \frac{\ddot{b}}{a}\dot{\overline{\bar{\mu}b^2}}\right| \leq \frac{c}{400}\int_0^1 \ddot{b}^2 + cC(|b|_\infty(0)(1+\int_0^1 \dot{b}^2)) \tag{46'}$$

$$\left|\int_0^1 \ddot{b}\,\dot{\overline{\underline{(\lambda+\bar{\mu}\eta)}b}}\right| \leq 10c\int_0^1 (b^2+\dot{b}^2)(b^2+\int_0^1 b^2) \leq -C(|b|_\infty(0)\frac{\partial a}{\partial s}\left(\int_0^1 \dot{b}^2+1\right) \tag{47'}$$

(45′), (46′), (47′) are the only estimates where the integrals are taken on $[0,1]$ (besides $-\frac{9c}{10}\int_0^1 \ddot{b}^2$). They are all due to cb.

We are thus left with $-\int_0^1 \ddot{b}sgnb\left(\frac{\partial^2\psi_{\theta,\varepsilon}}{\partial b^2}\dot{b}^2\eta_0 + \psi_{\theta,\varepsilon}\ddot{\eta}_0\right)$.

As for $-\int_0^1 \ddot{b}\frac{\partial^2\psi_{\theta,\varepsilon}}{\partial b^2}\dot{b}^2\eta_0$, observe that, by construction, see (γ), $\frac{\partial^2\psi_{\theta,\varepsilon}}{\partial b^2}$ is zero on $[t_1^+, t_2^+]$ and is otherwise bounded by $8/_{\varepsilon\theta}$.

On $[t_1^-, t_1^+], \varepsilon = \varepsilon_1, \theta = \theta(t_1^-) = \mu_1^-$, thus $\theta\varepsilon = \underline{\mu_1 - \mu_1^-}_{\,2}$. Similarly, on $[t_1^+, t_2^-], \varepsilon\theta = \underline{\mu_2 - \mu_2^-}_{\,2}$.

Thus: (Observe that $\dot{b}$ takes the value zero on $[t_1^-, t_2^-]$)

$$\left|\int_0^1 \ddot{b}sgnb\frac{\partial^2\psi_{\theta,\varepsilon}}{\partial b^2}\dot{b}^2\eta_0\right| \leq \frac{8}{\mu_1-\mu_1^-}\int_{t_1^-}^{t_1^+}|\ddot{b}||\dot{b}|^2\eta_0 + \frac{8}{\mu_2-\mu_2^-}\int_{t_2^+}^{t_2^-}|\ddot{b}|\dot{b}^2\eta_0 \leq \tag{47''}$$

$$\leq 8\left(\int_{t_1^-}^{t_2^-}\ddot{b}^2\dot{b}^2\right)^{1/2}\left(\left(\int_{t_1^-}^{t_1^+}\eta_0^2\dot{b}^2\right)^{1/2}\times\frac{1}{\mu_1-\mu_1^-}+\left(\int_{t_2^+}^{t_2^-}\eta_0^2\dot{b}^2\right)^{1/2}\times\frac{1}{\mu_2-\mu_2^-}\right) \leq$$

$$\leq 8\int_{t_1^-}^{t_2^-}\ddot{b}^2\left(\left(\int_{t_1^-}^{t_1^+}\frac{\eta_0^2\dot{b}^2}{(\mu_1-\mu_1^-)^2}\right)^{1/2}+\left(\int_{t_2^+}^{t_2^-}\frac{\eta_0^2}{(\mu_2-\mu_2^-)^2}\dot{b}^2\right)^{1/2}\right).$$

Observe, now, that on $[t_i^-, t_i^+]$, we have:

$$|\eta_0(t)| \leq C(b)|t_i^- - t_i^+| \tag{48}$$

where C depends on b, but not on how close μ_i^- is to μ_i, i.e. on how small $\bar{\bar{\delta}}$ is taken in Lemma 1. Indeed, $\eta_0(t_i^-)$ is zero, the interval $[t_1^-, t_2^-]$ does not collapse and η_0 is of a prescribed type, hence (48) with a finite constant $C(b)$ which depends only on b via η_0, that is on b in a weak sense (less than H^1) as we will see later.

Thus:

$$\frac{|\eta_0(t)|}{\mu_i-\mu_i^-}\le C(b)\frac{|t_i^- - t_i^+|}{\mu_i-\mu_i^-}\le\frac{C(b)}{\gamma}\quad\forall t\in[t_i^-,t_i^+]. \tag{49}$$

Thus

$$\Big|\int_0^1 \ddot{b}sgnb\frac{\partial\psi_{\theta,\varepsilon}}{\partial b^2}\dot{b}^2\eta_0\Big|\le\frac{8C(b)}{\gamma}\int_{t_1^-}^{t_2^-}\ddot{b}^2\left(\left(\int_{t_1^-}^{t_1^+}\dot{b}^2\right)^{1/2}+\left(\int_{t_2^+}^{t_2^-}\dot{b}^2\right)^{1/2}\right). \tag{50}$$

Since $\dot{b}$ is L^2, we can choose $\bar{\bar{\delta}}$ so small in Lemma 1 so that

$$\Big|\int_0^1 \ddot{b}sgnb\frac{\partial^2\psi_{\theta,\varepsilon}}{\partial b^2}\dot{b}^2\eta_0\Big|\le\frac{c}{1000}\int_{t_1^-}^{t_2^-}\ddot{b}^2. \tag{51}$$

Hence, $|\int_0^1 \ddot{b}sgnb\frac{\partial^2\psi_{\theta,\varepsilon}}{\partial b^2}\dot{b}^2\eta_0|$ can be incorporated in the left hand side of (44) also. (We will modify this argument below in order to use the H^1-weak topology.)

Finally, we consider $-\int_0^1 \ddot{b}sgnb\psi_{\theta,\varepsilon}\ddot{\eta}_0\cdot\ddot{\eta}_0$ presents Diracs at t_1^-,t_2^-. But $\psi_{\theta,\varepsilon}$ vanishes there.

No other term presented this problem. Before upperbounding $-\int_0^1 \ddot{b}sgnb\psi_{\theta,\varepsilon}\ddot{\eta}_0$, we show what meaning it should be given. For this purpose, we multiply η_0 by $\omega_{\bar{\bar{\varepsilon}}}$ a function with compact support in (t_1^-,t_2^-) and we show that, as $\omega_{\bar{\bar{\varepsilon}}}$ tends to 1 on (t_1^-,t_2^-), $-\int_0^1 \ddot{b}sgnb\psi_{\theta,\varepsilon}\ddot{\overline{\omega_{\bar{\bar{\varepsilon}}}\eta_0}}$ converges to $-\int_{t_1^-}^{t_2^-}\ddot{b}sgnb\psi_{\theta,\varepsilon}\ddot{\eta}_0$ which is defined without any ambiguity. Since all the other terms converge, with $\omega_{\bar{\bar{\varepsilon}}}\eta_0$ instead of η_0, without any problem, we will have given a meaning to our computations after (34), up to the integration by parts $\frac{\partial}{\partial s}\int_0^1\dot{b}^2=-\int_0^1\ddot{b}\frac{\partial b}{\partial s}$, which is to be understood in the distributional sense. We also need to upperbound$-\int_{t_1^-}^{t_2^-}\ddot{b}sgnb\psi_{\theta,\varepsilon}\ddot{\eta}_0$.

Observe that, since $\psi_{\theta,\varepsilon}(t_i^-)=0$ and $|\dot{\overline{\psi_{\theta,\varepsilon}}}|\le|\dot{\theta}|+2|\dot{b}|=2|\dot{b}|$ on $[t_i^-,t_i^+]$:

$$\begin{aligned}\psi_{\theta,\varepsilon}(t)&\le 2\Big|\int_{t_i^-}^t|\dot{b}|\Big|\le 2|t-t_i^-|\operatorname*{Sup}_{[t_1^-,t_2^-]}|\dot{b}|\le\quad\forall t\in[t_i^-,t_i^+]\\&\le 2|t-t_i^-|\left(\int_{t_1^-}^{t_2^-}\ddot{b}^2\right)^{1/2}\qquad(\dot{b}\text{ vanishes somewhere on }[t_1^-,t_2^-]).\end{aligned} \tag{52}$$

Thus, assuming that $\omega_{\bar{\bar{\varepsilon}}}$ is 1 on $[t_1^+,t_2^+]$ and using standard upperbounds on η_0:

$$\begin{aligned}&\Big|\int_{t_1^-}^{t_2^-}\ddot{b}sgnb\psi_{\theta,\varepsilon}\left(\ddot{\overline{\omega_{\bar{\bar{\varepsilon}}}\eta_0}}-\ddot{\eta}_0\right)\Big|\le\\&\le C\left(\int_{t_1^-}^{t_2^-}\ddot{b}^2\right)^{1/2}\left[\sum_{i=1}^2\int_{t_i^-}^{t_i^+}|\ddot{b}|\left(|\ddot{\omega}_{\bar{\bar{\varepsilon}}}|\,|t-t_i^-|^2+|\omega_{\bar{\bar{\varepsilon}}}-1|+|\dot{\omega}_{\bar{\bar{\varepsilon}}}|\,|t-t_i^-|\right)\right]\end{aligned} \tag{53}$$

where C depends on t_1^-, t_2^-, η_0, but not on $\tilde{\tilde{\varepsilon}}$.

As $\tilde{\tilde{\varepsilon}}$ tends to zero, we can assume that

$$|\omega_{\tilde{\tilde{\varepsilon}}} - 1| + |\ddot{\omega}_{\tilde{\tilde{\varepsilon}}}|\,|t - t_i^-|^2 + |\dot{\omega}_{\tilde{\tilde{\varepsilon}}}|\ |t - t_i^-| \le 10\chi_{\{|\omega_{\tilde{\tilde{\varepsilon}}} - 1| > 0\}}. \tag{54}$$

Thus:

$$\left| \int_0^1 \ddot{b} sgn b \psi_{\theta,\varepsilon} \overline{\omega_{\tilde{\tilde{\varepsilon}}} \ddot{\eta}_0} - \int_{t_1^-}^{t_2^-} \ddot{b} sgn b \psi_{\theta,\varepsilon} \ddot{\eta}_0 \right| \le$$

$$\le 10C \left(\int_{t_1^-}^{t_2^-} \ddot{b}^2 \right)^{1/2} \int_{\mathrm{Sup}|\omega_{\tilde{\tilde{\varepsilon}}} - 1| \cap [t_1^-, t_2^-]} |\ddot{b}| \le 10C \int_{t_1^-}^{t_2^-} \ddot{b}^2 \times |\ \mathrm{Sup}\ |\omega_{\tilde{\tilde{\varepsilon}}} - 1| \cap [t_1^-, t_2^-]|^{1/2}. \tag{54''}$$

Thus, when $\tilde{\tilde{\varepsilon}}$ tends to zero, such a term has no weight against $-\frac{c}{1000} \int_0^1 \ddot{b}^2$ and we can make sense of $\int_0^1 \ddot{b} sgn b \psi_{\theta,\varepsilon} \overline{\omega_{\tilde{\tilde{\varepsilon}}} \ddot{\eta}_0}$ if we can make sense of $\int_{t_1^-}^{t_2^-} \ddot{b} sgn b \psi_{\theta,\varepsilon} \ddot{\eta}_0$. We then see how we can make sense of all our previous computations. As to $\int_{t_1^-}^{t_2^-} \ddot{b} sgn b \psi_{\theta,\varepsilon} \ddot{\eta}_0$, we have, using (10):

$$\left| \int_{t_1^-}^{t_2^-} \ddot{b} sgn b \psi_{\theta,\varepsilon} \ddot{\eta}_0 \right| \le \left| \int_{t_1^-}^{t_2^-} \ddot{b} sgn b \psi_{\theta,\varepsilon} \left(\ddot{\eta}_0 - \overline{\bar{\mu} b \eta_0}^{\cdot} \right) \right| + \left| \int_{t_1^-}^{t_2^-} \ddot{b} sgn b \psi_{\theta,\varepsilon} \overline{\bar{\mu} b \eta_0}^{\cdot} \right| \tag{55}$$

$$\le \left| \int_{t_1^-}^{t_2^-} |\ddot{b} sgn b \psi_{\theta,\varepsilon}| C(|b|_\infty(0)) \left(0(|b|_\infty^2(0) + 1)(1 + a^2)) + |\dot{b}| + \frac{0(|b|_\infty(0))}{|t_1^- - t_2^-|} \right) \right| +$$

$$+ \left| \int_{t_1^-}^{t_2^-} \ddot{b} sgn b \psi_{\theta,\varepsilon} \frac{2}{|x_1 - x_2|^2} \times e^{-\int_{t_1^-}^{t} \bar{\mu} b d\tau} \right) \Bigg| \le$$

$$\le \frac{c}{1000} \int_{t_1^-}^{t_2^-} \ddot{b}^2 + \frac{C_5(|b|_\infty(0))(1 + a^2)}{c} \left(\int_{t_1^-}^{t_2^-} (\dot{b}^2 + \frac{\psi_{\theta,\varepsilon}}{|t_1^- - t_2^-|^2}) \right) +$$

$$+ \frac{2}{|x_1^- - x_2^-|^2} \left| \int_{t_1^-}^{t_2^-} \dot{b} \left(sgn b \psi_{\theta,\varepsilon} e^{-\int_{t_1^+}^{t} \bar{\mu} b d\tau} \right)^{\cdot} \right|$$

In (55), we integrate by parts the last term and we used the fact that $\psi_{\theta,\varepsilon}(|b|(t_i^-))$ is zero. This integration by parts is justified if $\int_0^1 \ddot{b}^2$ is finite. Otherwise, all these expressions, in any approximation (regularization), are of smaller order with respect to $c \int_0^1 \ddot{b}^2$. Observe now that (we use (γ))

$$\begin{aligned} &\frac{2}{|x_1^- - x_2^-|^2} \left| \int_{t_1^-}^{t_2^1} \dot{b} \left(sgn b \psi_{\theta,\varepsilon} e^{-\int_{t_1^-}^{t} \bar{\mu} b d\tau} \right)^{\cdot} \right| \\ &\le \frac{2}{|t_1^- - t_2^-|^2} C_6(|b|_\infty(0)) \int_{t_1^-}^{t_2^-} |\dot{b}|(0(|b|_\infty(0) + |\dot{\theta}| + |\dot{b}|). \end{aligned} \tag{56}$$

Using our previous arguments on $\dot{\theta}$, it is again easy, from (55) and (56), to estimate $\int_{t_1^-}^{t_2^-} \ddot{b} sgnb \psi_{\theta,\varepsilon} \ddot{\eta}_0$, thus $\int_0^1 \ddot{b} sgnb \psi_{\theta,\varepsilon} \overline{\omega_{\tilde{\tilde{\varepsilon}}} \eta_0}^{\cdot\cdot}$, when $\tilde{\tilde{\varepsilon}}$ tends to zero, which stands for $\int_0^1 \ddot{b} sgnb \psi_{\theta,\varepsilon} \ddot{\eta}_0$ in the left hand side of (44). (iii) follows in the case 1) of Definition 2.

In cases (ii) and (iii) of Definition 2, we can require:

$$|\eta_{12}|_{C^3} \leq C(|b|_\infty(0), \varepsilon_{16}) \tag{57}$$

(57) is easily built in the construction of η_{12}. (iii) follows then easily. the proof of Lemma 3 is thereby complete. The integration by parts in (34) clearly holds if $b \in H^2$. Otherwise, (iii) of Lemma 3 means that the inequality holds for any $\tilde{b}$, as an H^1_- neighborhood of b, which is in H^2. The definition of $\eta_{12}(\tilde{b})$ can follow closely the definition of $\eta_{12}(b)$ since to any oscillation of b, we can associate an oscillation of $\tilde{b}$.

We observe here the following important fact: denote

$$\eta(\tilde{b}) = \psi_{\tilde{\theta},\tilde{\varepsilon}}(\tilde{b}) \tilde{\eta}_0 sgn\tilde{b} + c\tilde{b} \tag{58}$$

where $\psi_{\tilde{\theta},\tilde{\varepsilon}}$ is constructed for $\tilde{b}$ using $t_1^-, t_1^+, t_2^-, t_2^+$ such as $\psi_{\theta,\varepsilon}$ has been constructed for b and where $\tilde{\eta}_0 = \eta_0 e^{\int_{t_1^-}^t \bar{\mu}\tilde{b} - \int_{t_1^-}^t \bar{\mu} b}$ ((58)$'$). η_0 has been constructed for b.

Observe that

$$\dot{\overline{\tilde{\eta}_0 - \bar{\mu}\tilde{b}\tilde{\eta}_0}} = e^{\int_{t_1^-}^t \bar{\mu}\tilde{b}} \dot{\overline{e^{-\int_{t_1^-}^t \bar{\mu}\tilde{b}} \tilde{\eta}_0}} = \tag{59}$$

$$= \dot{\overline{e^{\int_{t_1^-}^t \bar{\mu}\tilde{b}} \dot{\overline{e^{-\int_{t_1^-}^t \bar{\mu} b}}}}} \eta_0 = \dot{\overline{e^{\int_{t_1^-}^t \bar{\mu} b} \dot{\overline{e^{-\int_{t_1^-}^t \bar{\mu} b}}}}} \eta_0 +$$

$$+ \dot{\overline{\left[\left(e^{\int_{t_1^-}^t \bar{\mu}\tilde{b}} - e^{\int_{t_1^-}^t \bar{\mu} b}\right) e^{-\int_{t_1^-}^t \bar{\mu} b}\right] e^{\int_{t_1^-}^t \bar{\mu} b} \dot{\overline{e^{-\int_{t_1^-}^t \bar{\mu} b}}}}} \eta_0.$$

Thus, using (10):

$$\dot{\overline{\tilde{\eta}_0 - \bar{\mu}\tilde{b}\tilde{\eta}_0}} = \frac{-2}{(x_2^- - x_1^-)^2} e^{-\int_{t_1^-}^t \bar{\mu} b d\tau} - C(1+b^2)\eta_0 e^{-2\int_{t_1^-}^t \bar{\mu} b d\tau} + \tag{60}$$

$$+0\left(\left[\left|\bar{\mu}\tilde{b} - \bar{\mu} b\right| + \left|e^{\int_{t_1^-}^t \bar{\mu}\tilde{b}} - e^{\int_{t_1^-}^t \bar{\mu} b}\right|\right] \times\right.$$

$$\left.\times \left(|\eta_0| + |\dot{\eta}_0| + \frac{1}{|x_1^- - x_2^-|^2} + C(1+b^2)\eta_0\right)\right).$$

Thus, if $\tilde{b}$ is close enough to b in H^1, we have:

$$\dot{\overline{\tilde{\eta}_0 - \bar{\mu}\tilde{b}\tilde{\eta}_0}} + C(1+\tilde{b}^2)\tilde{\eta}_0 e^{-2\int_{t_1^-}^t \bar{\mu}\tilde{b} d\tau} \leq -\frac{3}{2(x_1^- - x_2^-)^2} e^{-\int_{t_1^-}^t \bar{\mu}\tilde{b} d\tau}. \tag{61}$$

(60)-(61) will replace (10) in our arguments below. All the proofs will be unchanged. We now observe:

Lemma C. *If* $|b-\tilde{b}|_{H^1}$ *is small enough, all the computations of Lemma 3 hold for* $\tilde{b}$.

Proof. (i) is easily seen to hold using (γ) which holds for $\tilde{b}$ (ii) has been discussed earlier. We are left with (iii).

The only points which might be a little bit delicate relate to (36)-(37), (47)-(48) and (52)-(56); observe that $\frac{\partial\psi_{\tilde{\theta},\tilde{\varepsilon}}}{\partial\tilde{b}} sgn\tilde{b}$ is always poitive. Choosing $\underline{t}$ for b such that $|b(\underline{t})| > \text{Sup}(\mu_1,\mu_2)$, we can make use of $\underline{t}$ for $\tilde{b}$ if $|b-\tilde{b}|_\infty$ is small enough.

The other steps in the proof of (iii) are related to (47)-(48) and (52)-(56); in (48), $C(b)$ depends on b. Since $\tilde{\eta_0}$ is equal to $\eta_0 e^{\int_{t_1^-}^t \bar{\mu}\tilde{b} - \int_{t_1^-}^t \bar{\mu}b}$, (48) still holds.

On the other hand, $|\frac{\partial^2\psi_{\tilde{\theta},\tilde{\varepsilon}}}{\partial\tilde{b}^2}|$ has again support in $[t_1^-,t_1^+]\cup[t_2^+,t_2^-]$ and is bounded by $\frac{1}{\tilde{\varepsilon}\tilde{\theta}}$, which we can assume to be less than $\frac{2}{\varepsilon\theta}$, if $\tilde{b}$ is close enough to b. Thus, this arguments of (47)-(51) extend. The computations of (52)-(56) are unchanged. qed

Our goal is now to globalize the flow (semi-flow) defined in Definition 2 and to prove existence and continuity for this flow.

We need, for this purpose, to improve (iii) of Lemma 3; namely, we will prove that there is a time-dependent control on $\int(\dot{b})^+$, for each flow-line and we will derive from it a time-dependent control on $\int\dot{b}^2$. In order to establish existence and continuity, we will need a complicated smoothing process, in order to take into account the fact that the curves are closed.

We start with the following lemma:

Lemma D. *Let* $\eta=\eta_{12}+cb$ *as in Definition 2, case i). Assume that* b_0 *is* C^∞ *and* $\dot{b}_0$ *has simple zeros. Then,* $C\frac{\partial a}{\partial s}+\frac{\partial}{\partial s}\int\dot{b}^+|_{s=0}\leq-\frac{1}{|x_1^- - x_2^-|^2}e^{-C|b|_\infty(0)}(\bar{\bar{\mu}}-\mu_2)+C_{22}(|b|_\infty(0))\left(\frac{1}{\bar{\bar{\mu}}-\mu_2}\right)-C\frac{\partial a}{\partial s}\int_0^1\dot{b}^+$ *where* C, C_{22} *do not depend on* b_0.

Proof. As we will see later, we can assume, using a generecity argument, that on a given flow-line, if we perturb it slightly, then at a given time and around this given time, $b(s)$ is C^1 in t and $\dot{b}$ has simple zeros. Furthermore, the map $s\mapsto\dot{b}(s)$ will be continuous in a suitable sense, so that the zeros of $\dot{b}$ are differentiable functions of s. (Proposition A1 of the Appendix). The computation of $\frac{\partial}{\partial s}\int\dot{b}^+$, at such values, is easier, since there is no boundary contribution due to the zeros of $\dot{b}$ and

$$\frac{\partial}{\partial s}\int\dot{b}^+=\sum_j\overline{\frac{\dot{\overline{\eta+\lambda b}}}{a}}+a\eta\tau-b\eta\bar{\mu}_\xi\Big|_{z_j^-}^{z_j^+}$$

where z_j^- is a minimum of b and z_j^+ is the consecutive maximum of b. Observe that $\ddot{b}(z_j^+)\leq0,\ddot{b}(z_j^-)\geq0,\dot{b}(z_j^+)=\dot{b}(z_j^-)=0$.

We estimate:

$$\Delta_j = \frac{\overline{\dot{\eta + \lambda b}}}{a} + a\eta\tau - b\eta\bar{\mu}_\xi\Big|_{z_j^-}^{z_j^+}, \eta = sgnb\psi_{\theta,\varepsilon}(b)\eta_0 + cb.$$

From (γ), we observe that, at b:

$$\frac{\partial\psi_{\theta,\varepsilon}}{\partial\theta}(b)\dot{\theta} + \frac{\partial\psi_{\theta,\varepsilon}}{\partial\varepsilon}\dot{\varepsilon} = -\dot{\theta}$$

$\psi_{\theta,\varepsilon}(b)$ vanishes at t_1^-, t_2^-. We extend it by zero outside of this interval.
Let

$$\bar{\psi}_j = \psi_{\theta,\varepsilon}(b(z_j^+)) \quad \underline{\psi}_j = \psi_{\theta,\varepsilon}(b(z_j^-)).$$

Let $\tilde{\psi}_{\theta,\varepsilon}(b) = \psi_{\theta,\varepsilon}(b)sgnb$.
We then have:

$$\Delta_j = \frac{\overline{\ddot{\tilde{\psi}_{\theta,\varepsilon}\eta_0 + cb}} + \dot{\lambda b}}{a} + (\tilde{\psi}_{\theta,\varepsilon}\eta_0 + cb)(a\tau - b\bar{\mu}_\xi)\Big|_{z_j^-}^{z_j^+}$$

$$= \frac{\overline{\ddot{\tilde{\psi}_{\theta,\varepsilon}\eta_0 + cb}} + b(b(\tilde{\psi}_{\theta,\varepsilon}\eta_0 + cb) - \int_0^1 b(\tilde{\psi}_{\theta,\varepsilon}\eta_0 + cb))}{a} -$$

$$-\frac{\overline{\dot{\bar{\mu}(\tilde{\psi}_{\theta,\varepsilon}\eta_0 + cb)b}}}{a} + (\tilde{\psi}_{\theta,\varepsilon}\eta_0 + cb)(a\tau - b\bar{\mu}_\xi)\Big|_{z_j^-}^{z_j^+}.$$

Observe that:

$$\dot{\tilde{\psi}}_{\theta,\varepsilon} = -\dot{\theta} + \frac{\partial\psi_{\theta,\varepsilon}}{\partial b}\dot{b} = -\dot{\theta} \text{ at } z_j^-, z_j^+$$

$$\ddot{\tilde{\psi}}_{\theta,\varepsilon} = -\ddot{\theta} + \frac{\partial\psi_{\theta,\varepsilon}}{\partial b}\ddot{b} + \frac{\partial^2\psi_{\theta,\varepsilon}}{\partial b^2}\dot{b}^2 = -\ddot{\theta} + \frac{\partial\psi_{\theta,\varepsilon}}{\partial b}\ddot{b} \text{ at } z_j^-, z_j^+.$$

Since $\ddot{b}(z_j^+) \leq 0, \ddot{b}(z_j^-) \geq 0, \frac{\partial\psi_{\theta,\varepsilon}}{\partial b}sgnb \geq 0$, we have:

$$\Delta_j \leq \left[\frac{-\ddot{\theta}\eta_0 sgnb - 2\dot{\theta}\dot{\eta}_0 sgnb}{a} + \dot{\theta}\eta_0 sgnb\frac{\bar{\mu}b}{a}\right]\Bigg|_{z_j^-}^{z_j^+} \tag{63}$$

$$+\left[\frac{b(b(\tilde{\psi}_{\theta,\varepsilon}\eta_0 + cb) - \int_0^1 b(\tilde{\psi}_{\theta,\varepsilon}\eta_0 + cb))}{a} + \right.$$

$$\left. +(cb + \tilde{\psi}_{\theta,\varepsilon}\eta_0)(a\tau - b\bar{\mu}_\xi)\right]_{z_j^-}^{z_j^+} + \left[\tilde{\psi}_{\theta,\varepsilon}\frac{\overline{\dot{\eta}_0 - \bar{\mu}\eta_0 b}}{a}\right]_{z_j^-}^{z_j^+}.$$

In view of (10), we have:

$$\dot{\eta}_0 - \bar{\mu}\eta_0 b = 0\left(\frac{1}{t_1^- - t_2^-}\right) \times C(|b|_\infty(0)) \tag{64}$$

and

$$\overline{\dot{\eta_0 - \bar{\mu}\eta_0 b}} = -\frac{2}{(x_2^- - x_1^-)^2} e^{-\int_{t_1^-}^t \bar{\mu} b d\tau} - C(1+b^2)\eta_0 e^{-2\int_{t_1^-}^t \bar{\mu} b d\tau} \tag{65}$$

in $[t_1^-, t_2^-]$. Since $\overline{\dot{\eta_0 - \bar{\mu}\eta_0 b}}$ is multiplied by $\tilde{\psi}_{\theta,\varepsilon/_a}$, which vanishes outside of $[t_1^-, t_2^-]$, we can use (65) in (63). (slightly changed into (60) for $\tilde{\eta}_0$).

We can use the above formula (C is large enough) in order to upperbound Δ_j since we have extended $\tilde{\psi}_{\theta,\varepsilon}$ to all of $[0,1]$. The formula holds even if $z_j^\pm = t_1^-, t_2^-$ because, then $\tilde{\psi}_{\theta,\varepsilon}$ and $\dot{\widetilde{\psi_{\theta,\varepsilon}}} = -\dot{\theta} + \frac{\partial \tilde{\psi}_{\theta,\varepsilon}}{\partial b}\dot{b}$ vanish at those points ($\dot{b}$ is zero at $z_j^\pm, \tilde{\psi}_{\theta,\varepsilon}$ and $\dot{\theta}$ are zero at t_1^-, t_2^-). Thus, with obvious notations,

$$\Delta_j \leq \frac{-2}{(x_2^- - x_1^-)^2}\left(\bar{\tilde{\psi}}_j e^{-\int_{t_1^-}^{z_j^+} \bar{\mu} b} - \tilde{\underline{\psi}}_j e^{-\int_{t_1^-}^{z_j^-} \bar{\mu} b}\right) + \Gamma_j.$$

Observe that all intervals I where η_0 is non zero have $\int_I |b| < 1$, if η_0 is small enough and C large enough, because they are of type (1). In fact, $\int_{t_1^-}^{t_2^-} |b|$ is bounded by 1.

We need to estimate Γ_j, which is made of differences. Besides the terms multiplied by c, which can be upperbounded by $\underline{C}c\left[(1+|b|_\infty^2)(b(z_j^+) - b(z_j^-)) + \right.$ $\left.(1+|b|_\infty^4)|z_j^+ - z_j^-|\right], \underline{C}$ a universal constant, each difference can be upperbounded by the integral of the derivative between z_j^- and z_j^+. $-C(1+b^2)\eta_0\tilde{\psi}_{\theta,\varepsilon} e^{-2\int_{t_1^-}^t \bar{\mu} b d\tau}\Big|_{z_j^-}^{z_j^+}$ can be split as:

$$-C\tilde{\psi}_{\theta,\varepsilon}(z_j^+)\eta_0(z_j^+) e^{-2\int_{t_1^-}^{z_j^+} \bar{\mu} b d\tau}(b^2(z_j^+) - b^2(z_j^-))(=(1)) -$$

$$-C(1+b^2(z_j^-))(\eta_0(z_j^+) e^{-2\int_{t_1^-}^{z_j^+} \bar{\mu} b d\tau}\tilde{\psi}_{\theta,\varepsilon}(z_j^+) - \eta_0(z_j^-) e^{-2\int_{t_1^-}^{z_j^-} \bar{\mu} b d\tau}\tilde{\psi}_{\theta,\varepsilon}(z_j^-))$$

$$= (1) - C(1+b^2(z_j^-))\eta_0(z_j^+) e^{-2\int_{t_1^-}^{z_j^+} \bar{\mu} b}\left(\bar{\tilde{\psi}}_j - \tilde{\underline{\psi}}_j\right)(=(2)) + 3$$

Using (γ), we know that $\tilde{\psi}_{\theta,\varepsilon} = (|x| - \theta)^+ sgnx$ on $[t_1^+, t_2^+]$, that θ, ε are constants on $[t_1^-, t_1^+]$ and $[t_2^+, t_2^-]$, that $\tilde{\psi}_{\theta,\varepsilon}$ increases with x and that $|\tilde{\psi}_{\theta,\varepsilon}(x) - (|x| - \theta)^+ sgnx|$ is less than $\varepsilon\theta = 0(\mu_1 - \mu_1^- + \mu_2 - \mu_2^-)$.

Furthermore, there are at most two intervals $[z_j^-, z_j^+]$ having t_1^+ or t_2^+ in their interior.

Thus,

$$\bar{\tilde{\psi}}_j - \underline{\tilde{\psi}}_j \geq - \int_{z_j^-}^{z_j^+} |\dot{\theta}| \tag{66}$$

but for possibly two terms where

$$\bar{\tilde{\psi}}_j - \underline{\tilde{\psi}}_j \geq - \int_{z_j^-}^{z_j^+} |\dot{\theta}| - \tilde{C}(\mu_1 - \mu_1^- + \mu_2 - \mu_2^-). \tag{67}$$

Thus, the sum of the terms (2) is $0((1+|b|_\infty^2)\left(\int_0^1 |\dot{\theta}| + \tilde{C}(\mu_1 - \mu_1^- + \mu_2 - \mu_2^-)\right)$.

(2) can be used, because $-C\eta_0(1+b^2)e^{-2\int_{t_1^-}^{z_j^+} \bar{\mu} b}$ is negative, C is very large and $\int_{t_1^-}^{z_j^+} |b|$ is bounded by 1 if $\eta_0(z_j^+)$ is non zero, to absorb in the same lowerbound several other similar terms coming from $b^2(\tilde{\psi}_{\theta,\varepsilon}\eta_0)$ etc.

(1) is negative and can be used also to absorb all the other similar terms, of the same type. Thus all the difference terms due to b^2 or $\tilde{\psi}_{\theta,\varepsilon}$ can be thrown away. Differences involving $b(z_j^+) - b(z_j^-)$ can be upperbounded by $\int \dot{b}^+$.

The zeros of b are in finite number, controlled on a given flow-line. This number depends only on the initial condition. Therefore, there are at most k_0 of terms involving change of sign of b, independently of the time along the flow-line. We also have, if an interval $[z_j^-, z_j^+]$ contains t_1^- or t_2^-, the possiblility that the difference reduces to one term. However, this only happens at most twice. (Observe that these are boundary terms, which yield a 2 in the formula below. We will get rid of $k_0 + 2$ later.)

We thus have:

$$\frac{\partial}{\partial s} \int \dot{b}^+ = \sum \Delta_j \leq C(|b|_\infty(0))\Big[(1 + \frac{1}{|t_1^- - t_2^-|})\left(\frac{1}{a}+1\right)\int_0^1 |\dot{\theta}|_{C^2} + \tag{68}$$

$$+\left(1+|\dot{\theta}|_{C^2}\right)\left(\int_0^1 |\dot{\eta}_0| + \int_{t_1^-}^{t_2^-} |\ddot{\eta}_0| + 1\right)$$

$$+(k_0+2)\left(1+|\dot{\theta}|_{C^2}\left(1+\frac{1}{|t_1^- - t_2^-|}\right)\right)\Big] + C(1+c(1+|b|_\infty^2))\int \dot{b}^+ -$$

$$-\frac{2}{(x_2^- - x_1^-)^2}\sum_j \left[\tilde{\psi}_{\theta,\varepsilon} e^{-\int_{t_1^-}^{t} \bar{\mu} b d\tau}\right]_{z_j^-}^{z_j^+}.$$

We rewrite

$$\tilde{\psi}_{\theta,\varepsilon} e^{-\int_{t_1^-}^{t} \bar{\mu} b d\tau}\Big|_{z_j^-}^{z_j^+} \quad \text{as}$$

$$\left(\bar{\tilde{\psi}}_j - \underline{\tilde{\psi}}_j\right) e^{-\int_{t_1^-}^{z_j^+} \bar{\mu} b d\tau} + e^{-\int_{t_1^-}^{z_j^+} \bar{\mu} b d\tau} \underline{\tilde{\psi}}_j \left(-e^{\int_{z_j^-}^{z_j^+} \bar{\mu} b d\tau} + 1\right).$$

$\int_0^1 |b|$ will remain bounded along all our deformation. We could use this stronger bound. But, we need only

$$\int_0^1 |b| \leq \underline{C} |b|_\infty(0). \tag{69}$$

Therefore,

$$\left| e^{-\int_{z_j^-}^{z_j^+} \bar{\mu} b d\tau} - 1 \right| \leq C(|b|_\infty(0)) \left| \int_{z_j^-}^{z_j^+} |b| d\tau \right|.$$

Thus:

$$\tilde{\psi}_{\theta,\varepsilon} e^{-\int_{t_1^-}^{t} \bar{\mu} b d\tau}\Big|_{z_j^-}^{z_j^+} \geq \left(\bar{\tilde{\psi}}_j - \underline{\tilde{\psi}}_j\right) e^{-\int_{t_1^-}^{z_j^-} \bar{\mu} b d\tau} - C_1\left(|b|_\infty(0)\right) \int_{z_j^-}^{z_j^+} |b| d\tau \underline{\tilde{\psi}}_j.$$

Assuming, for example, that b is positive on the support of the oscillation (the other case is similar) we are studying, we know, by construction of t_1^+ and t_2^+ that $\dot{b}(t_1^+)$ is nonnegative and $\dot{b}(t_2^+)$ is nonpositive. Therefore, t_1^+ has to belong to an interval $[z_{j_0}^-, z_{j_0}^+]$ while t_2^+ has to belong to an interval $[z_{j_1}^+, z_{j_1+1}^-]$.

$\tilde{\psi}_{\theta,\varepsilon}$ has support in $[t_1^-, t_2^-]$, which is slightly larger than $[t_1^+, t_2^+]$. The two intervals can be chosen however to be as close as we wish.

Therefore, we can ask that, since the interval $[t_1^+, t_2^+]$ does not collapse:

$$\frac{|\underline{\tilde{\psi}}_{j_1+1}|}{|x_1^- - x_2^-|^2} + \frac{|\underline{\tilde{\psi}}_{j_0}|}{|x_1^- - x_2^-|^2} \leq \frac{1}{\underline{C}(1+|b|_\infty^3)} \int_{t_1^-}^{t_2^-} b\tilde{\psi}_{\theta,\varepsilon}\eta_0 \leq \inf\left(\frac{-\frac{\partial a}{\partial s}}{\underline{C}(1+|b|_\infty^3)}, 1\right) \tag{70}$$

and this can be built in the construction of our vector-field, it is a condition which will hold in an L^∞ neighborhood of b, if it holds at b, since $|x_1^- - x_2^-|^2$ remains lowerbounded away from zero when μ_1 tends to μ_2 and μ_1^- tends to μ_2^-.

The right way to think about this is to start with an arbitrary function $b_1 \in H^1$ and build $\psi_{\theta,\varepsilon}$ and η_0 as before, with the following precisions: considering t_1^+ (or t_2^+), there is, as close as we wish to t_1^+, on the right of t_1^+, a point $t_{0,1}^+$ where b is slightly larger than b at t_1^+ (respectively, on the left and also slightly larger). We can then build $\psi_{\theta,\varepsilon}$ so that (70) holds on $[t_1^+, t_{0,1}^+]$ and $[t_{0,2}^+, t_2^+]$. This only requires making μ_1 tend to μ_2, μ_1^- to μ_1, μ_2^- to μ_2 and taking then, if necessary, $t_{0,1}^+$ closer to t_1^+ (respectively $t_{0,2}^+$ closer to t_2^+). Indeed, $|x_1^- - x_2^-|^2$ remains lowerbounded away from zero all over this process. The inequality, with 1 replaced by 2, will hold on an L^∞-neighborhood of b_1. We can ask that this neighborhood is so small so that $b(t_{0,i}^+) > b(t_i^+)$ for $i = 1,2$ on all this neighborhood. Then, $z_{j_0}^-$ has to be less than $t_{0,1}^+$ and $z_{j_1+1}^-$ larger than $t_{0,2}^+$. (70) thus holds with 1 replaced by 2.

The difference of signs between $\bar{\tilde{\psi}}_j$ and $\underline{\tilde{\psi}}_j$ can only occur for $j = j_0$ or $j = j_1 + 1$, since b is positive (respectively negative) on the support of the oscillation we are singling out. We have already stated in (70) these contributions. We also notice that, in the differences $\bar{\tilde{\psi}}_j - \underline{\tilde{\psi}}_j$, there is a slight dependence on θ, ε, since these are functions of t.

However, in view of (γ), of the fact that θ and ε are constants on $[t_1^-, t_1^+] \cup [t_2^+, t_2^-]$, that $|\psi_{\theta,\varepsilon}(x) - (|x| - \theta)^+| \le \varepsilon\theta = 0(\mu_1 - \mu_1^- + \mu_2 - \mu_2^-)$ and that there are at most two intervals $[z_j^-, z_j^+]$ having t_1^+ or t_2^+ in their interior, we have:

$$\begin{aligned}
\sum_j \left(\bar{\tilde{\psi}}_j - \underline{\tilde{\psi}}_j\right) e^{-\int_{t_1^-}^{z_j^-} \bar{\mu} b d\tau} &\ge -\left(\int_0^1 |\dot{\theta}|\right) e^{C|b|_\infty(0)} + \\
&+ \left(\sum_j \operatorname{Sup}_{t\in[z_j^-, z_j^+]} \left[\tilde{\psi}_{\theta(t),\varepsilon(t)}(b(z_j^+)) - \tilde{\psi}_{\theta(t),\varepsilon(t)}(b(z_j^-))\right]\right) e^{-C|b|_\infty(0)} \\
&- C(\mu_1 - \mu_1^- + \mu_2 - \mu_2^-).
\end{aligned} \tag{71}$$

Using then the the same facts, in particular (γ) and our remark about $j_0, j_1 + 1$, we have: ($\tilde{\psi}_{\theta,\varepsilon}(x)$ is increasing function of x)

$$\sum_j \operatorname{Sup}_{t\in[z_j^-, z_j^+]} \left(\tilde{\psi}_{\theta(t),\varepsilon(t)}(b(z_j^+)) - \tilde{\psi}_{\theta(t),\varepsilon(t)}(b(z_j^-))\right) \ge$$

$$\ge \sum_{\substack{j s.t \\ z_j^+ \in [t_1^+, t_2^+]}} \left||b(z_j^+)| - |b(z_j^-)|\right| - \int_0^1 |\dot{\theta}| - 4|x_1^- - x_2^-|^2 - C(\mu_1 - \mu_1^-) - C(\mu_2 - \mu_2^-) \ge$$

$$\ge \frac{1}{2}(\bar{\tilde{\mu}} - \mu_2) - \int_0^1 |\dot{\theta}| - 4|x_1^- - x_2^-|^2.$$

(If b is replaced by $\tilde{b}, \theta, \varepsilon$ by $\tilde{\theta}, \tilde{\varepsilon}$, the above argument extends readily. The extra-term in (60) globally yield a contribution which is $0(\int_0^1 |a - \tilde{a}| + |b - \tilde{b}| + |\dot{b} - \dot{\tilde{b}}|)$ which we can assume to be so small that the same estimates still hold. $\tilde{a}$ is the a component of the curve corresponding to $\tilde{b}$).

On the other hand, taking into account our arguments above about $z^-_{j_0}$ and $z^-_{j_1+1}$, we have :

$$\sum_j C_1\left(\int_{z_j^-}^{z_j^+}|b|d\tau\right)\underline{\tilde{\psi}}_j \leq C(|b|_\infty(0))(|t_1^- - t_2^-| + |\underline{\tilde{\psi}}_{j_1+1}| + |\tilde{\psi}_{j_0}|) \leq$$

$$\leq 2C_1(|b|_\infty(0))(|t_1^- - t_2^-|)$$

if our flow is built appropriately so that (70) is satisfied.

Summing up, we have: $\left(|\ddot{\eta}_0| \leq \frac{C}{|t_1^- - t_2^-|}\chi_{[t_1^-, t_2^-]}\right)$

$$\frac{\partial}{\partial s}\int \dot{b}^+$$

$$\leq C(|b|_\infty(0))\Bigg[\left(1+\frac{1}{|t_1^- - t_2^-|}(1+\frac{1}{a})\int_0^1|\dot{\theta}|_{C^2} + (1+|\dot{\theta}|_{C^2})\left(\int_0^1|\dot{\eta}_0| + \int_{t_1^-}^{t_2^-}|\ddot{\eta}_0| + 1\right)\right.$$

$$\left.+(k_0+2)\left(1+|\dot{\theta}|_{C^2}(1+\frac{1}{|t_1^- - t_2^-|}\right)\right)\Bigg] -$$

$$-\frac{2}{|x_1^- - x_2^-|^2}\left(-2e^{C|b|_\infty(0)}\int_0^1|\dot{\theta}| + e^{-C|b|_\infty(0)}\frac{(\tilde{\bar{\mu}} - \mu_2)}{2} - 2C_1(|b|_\infty(0))|t_1^- - t_2^-|\right) +$$

$$(C + Cc(1+|b|^2_\infty))\int \dot{b}^+ \leq$$

$$\leq \tilde{C}(|b|_\infty(0))\left(1+\frac{1}{|t_1^- - t_2^-|}(1+\frac{1}{a})\int_0^1|\dot{\theta}|_{C^2} + (1+|\dot{\theta}|_{C^2})\left(1+\frac{1}{|t_1^- - t_2^-|}\right)\right. +$$

$$\left.+(k_0+2)\left(1+|\dot{\theta}|_{C^2}\left(\frac{1}{|t_1^- - t_2^-|}\right)\right)\right) - \frac{2}{|x_1^- - x_2^-|^2}\left(-\tilde{C}_1(|b|_\infty(0))\int_0^1|\dot{\theta}| +\right.$$

$$\left.+e^{-C|b|_\infty(0)}\frac{(\tilde{\bar{\mu}} - \mu_2)}{2}\right) + C\left(1+c(1+|b|^2_\infty)\right)\int \dot{b}^+$$

$|\dot{\theta}|_{C^2}$ can be made as small as we wish, keeping $|t_1^- - t_2^-|$ lowerbounded away from zero, when $\mu_1 \to \mu_2, \mu_1^-$ to μ_1 and μ_2^- to μ_2. Thus

$$\frac{\partial}{\partial s}\int \dot{b}^+ \leq -\frac{2e^{-C|b|_\infty(0)}(\tilde{\bar{\mu}} - \mu_2)}{|x_1^- - x_2^-|^2} + \tilde{\tilde{C}}(|b|_\infty(0))\left(\frac{1}{|x_1^- - x_2^-|} + k_0 + 2\right) + \tag{72}$$

$$(C + Cc(1+|b|^2_\infty))\int \dot{b}^+ \leq \frac{-1}{|x_1^- - x_2^-|^2}e^{-C|b|_\infty(0)}\frac{(\tilde{\bar{\mu}} - \mu_2)}{2} + \frac{C_2(|b|_\infty(0))}{\tilde{\bar{\mu}} - \mu_2} +$$

$$+C_{21}(|b|_\infty(0)) \times (k_0+2) + C\left(1+c(1+|b|^2_\infty)\right)\int \dot{b}^+.$$

We claim now that we can remove k_0+2 from the above estimate, because we have been considering a single oscillation, hence by construction of $[t_1^+, t_2^+]$, there is no change of sign in this interval (it suffices to take $\bar{\bar{\mu}} > 0$), hence on $[t_1^-, t_2^-]$. Thus, all the $\Delta_j' s$ such that $[z_j^-, z_j^+]$ is included in $[t_1^-, t_2^-]$ are not concerned by these sign changes. We are left with those which do not meet this interval, which do not count since $\bar{\psi}_j, \underline{\psi}_j$ are zero there, and the two intervals, which might exist with $z_j^- < t_1^- < z_j^+$ or $z_j^- < t_2^- < z_j^+$. Thus, we can bring back k_0+2 to 4.

However, because t_2^+ belongs to $[z_{j_1}^+, z_{j_1+1}^-]$, hence z_j^- belongs to $[t_2^+, t_2^-]$, these boundary terms can be better tackled using (70). In (68), k_0+2 was multiplied by $(1+|\dot{\theta}|_{C^2}(1+\frac{1}{|t_1^- - t_2^-|}))$.Using (70), k_0 is replaced by 2 and 1 in $1+|\dot{\theta}|_{C^2}(1+\frac{1}{|t_1^- t_2^-|})$ was due to $\frac{b^2\tilde{\psi}_{\theta,\varepsilon}\eta_0}{a}\Big|_{z_j^-}^{z_j^+}$ or $\frac{Cb^2\tilde{\psi}_{\theta,\varepsilon}\eta_0}{a}\Big|_{z_j^-}^{z_j^+}$ or $\tilde{\psi}_{\theta,\varepsilon}\eta_0(a\tau - b\bar{\mu}\xi)\Big|_{z_j^-}^{z_j^+}$. It, therefore, since these are boundary terms, i.e. arise when t_1^- or t_2^- is in $[z_j^-, z_j^+]$, can be replaced using (70) by$-\frac{\partial a}{\partial s}$. The same bound holds for $|\dot{\theta}|_{C^2}(1+\frac{1}{|t_1^- - t_2^-|})$. Thus

$$C\frac{\partial a}{\partial s} + \frac{\partial}{\partial s}\int \dot{b}^+ \leq -\frac{1}{|x_1^- - x_2^-|^2}e^{-C|b|_\infty(0)}(\bar{\bar{\mu}} - \mu_2) + \tag{73}$$

$$+C_{22}(|b|_\infty(0))\left(\frac{1}{\bar{\bar{\mu}} - \mu_2} + 1\right) + C((1+c(1+|b|^2_\infty))\int \dot{b}^+.$$

As pointed out in this proof, the result extends to $\psi_{\tilde{\theta},\tilde{\varepsilon}}(\tilde{b})\tilde{\eta}_0 sgn\tilde{b}$, for $\tilde{b}$ close to b in H^1, where $\psi_{\tilde{\theta},\tilde{\varepsilon}}$ is the extension of $\psi_{\theta,\varepsilon}$ indicated after (γ).

The result extends further as follows:

In the above computations, we did not really use the fact that we were considering one single oscillation. Namely, if we consider

$$\eta = \sum_{i=1}^{m} \eta_{12,i} + cb$$

where

$$\eta_{12,i} = \tilde{\psi}_{\theta_i,\varepsilon_i}\eta_{0,i} \tag{74}$$

with

(75) The supports of the $\eta_{o,i}$ are disjoint.

(76) On the support of each $\eta_{o,i}, b$ does not change sign.

(77) (70) is required for the sum of the related expressions for i running from 1 to m i.e.

$$\sum_{i=1}^{m} \frac{|\tilde{\psi}_{j_1,i+1}|}{|x^-_{1,i} - x^-_{2,i}|^2} + \frac{|\tilde{\psi}_{j_0,i}|}{|x^-_{1,i} - x^-_{2,i}|^2} \leq \frac{1}{C(1+|b|^3_\infty)} \sum_{i=1}^{m} \int_{t^-_{1,i}}^{t^-_{2,i}} \tilde{\psi}_{\theta_i,\varepsilon_i} \eta_{0,i}$$

we will derive, using the same techniques:

Lemma D'. *Let $\eta = \sum_{i=1}^{m} \eta_{12,i} + cb$. Assume b_0 is C^∞ and $\dot{b}_0$ has simple zeros. Then,*

$$C\frac{\partial a}{\partial s} + \frac{\partial}{\partial s} \int \dot{b}^+ \Big|_{s=0} \leq -e^{-C|b|_\infty(0)} \left(\sum_{i=1}^{m} \frac{\bar{\bar{\mu}}_i - \mu_{2,i}}{|x^-_{1,i} - x^-_{2,i}|^2} \right) +$$

$$+C_{22}(|b|_\infty(0)|) \sum_{i=1}^{m} \frac{1}{\bar{\bar{\mu}}_i - \mu_{2,i}} + C_{23}(|b|_\infty(0))(1+c) \int \dot{b}^+.$$

Lemma D' is easily derived from the argument of Lemma D, from (72) in particular, simply noticing furthermore that we were able to drop, using (70) and (68), k_0+2 from the right hand side of (72). Also, we do not find $m \times (1+c)C_{23}(|b|_\infty(0)) \int \dot{b}^+$, i.e. we do not find m times the last term in the right hand side of our upperbound because we are not summing (73) on all the oscillations, but rather rerwriting the proof of Lemma with $\tilde{\psi}_{\theta,\varepsilon}\eta_0$ replaced by $\sum_{i=1}^{m} \tilde{\psi}_{\theta_i,\varepsilon_i}\eta_{0,i}$, the $\eta_{0,i}$'s having disjoint supports. Observe that the bound on $|\dot{\theta}|_{C^2}$ is easily replaced by similar bounds on $\sum_{i=1}^{m} |\dot{\theta}_i|_{C^2}$. With this, (77) and the same remark about k_0, the proof extends easily q.e.d.

Corollary E. *Assume that we, in addition, require that $\forall i, |\bar{\bar{\mu}}_i - \mu_{2,i}| \geq$ Sup $\left(\varepsilon_{11}, \frac{\delta_{k,q}}{40}\right) = \zeta > 0$. ($\varepsilon_{11}$ and $\delta_{k,q}$ are constants to come in the next sections). Then:*

$$C\frac{\partial a}{\partial s} + \frac{\partial}{\partial s} \int \dot{b}^+ \Big|_{s=0} \leq -\tilde{C}_1(|b|_\infty(0))\zeta \sum_{i=1}^{m} \frac{1}{|x^-_{1,i} - x^-_{2,i}|^2} + \tilde{C}_2(|b|_\infty(0), \zeta)$$
$$- C\left(Ca + \int \dot{b}^+\right)\frac{\partial a}{\partial s} + C\int \dot{b}^+.$$

Proof. Observe that there are at most $n(\zeta, |b|_\infty(0))$ indices i such that $|x^-_{1,i} - x^-_{2,i}| > \frac{1}{2}\frac{e^{-C/2|b|_\infty(0)}}{\sqrt{C_{22}|b_\infty(0)|}}\zeta$ since these intervals are disjoint.

The result follows.

We then have:

Lemma 3′. *Let us consider a convex combination $cb+\sum\alpha_i\eta_{12,i}$ of $\eta_{12,i}$ made each of them of several $\psi_{\tilde\theta,\tilde\varepsilon}\eta_0$ with disjoint support, and with amplitude lower bounded by a uniform $\Gamma > 0$.*

We then have:

(a) i) and ii) of Lemma 3 hold

(b) Let s be the time along this new flow. Then

$$\frac{1}{2}\frac{\partial}{\partial s}\int_0^1 \dot{b}^2 + \frac{9c}{10}\int_0^1 \ddot{b}^2 \leq \frac{\tilde{\tilde{C}}_3(|b|_\infty(0),\varepsilon_{16})}{c}\Big(-\frac{1}{\Gamma}\frac{\partial}{\partial s}(Ca+\int \dot{b}^+) + C\int \dot{b}^+ +$$

$$+\tilde{\tilde{C}}_4(|b|_\infty(0),\Gamma) + \tilde{\tilde{C}}(|b|_\infty(0))(Ca+\int \dot{b}^+)(-\frac{\partial a}{\partial s})\Big)(1+\int_0^1 \dot{b}^2)$$

$$-\tilde{\tilde{C}}_4(|b|_\infty(0))(1+\int_0^1 \dot{b}^2)\frac{\partial a}{\partial s}.$$

Remark. We could write η as $\sum\alpha_i(\eta_{12,i}+cb)$ or more generally take $\eta = \sum\alpha_i(\eta_{12,i}+c_ib)$. Letting $c=\sum\alpha_i c_i$, we are back to our framework.

Corollary F. *When s tends to the explosion time or to $+\infty$, $c\int_1^1 b^2$ has to tend to zero on a subsequence. If there is blow-up in finite time, c has to tend to zero on a subsequence.*

Proof. When s tends to $+\infty$, we use $\int_0 c\int_0^1 b^2 < a(0) < +\infty$. Thus, $c\int_0^1 b^2$ has to tend to zero, as claimed.

If we have blow-up in finite time, we use b).

It is easy to see that $-\frac{1}{\Gamma}\frac{\partial}{\partial s}((a+\int\dot{b}^+)+\tilde{\tilde{C}}_4+\tilde{\tilde{C}}_5(Ca+\int\dot{b}^+)\times(-\frac{\partial a}{\partial s})+C\int\dot{b}^+$ is positive (it upper bounds $\sum\frac{1}{|x_{1,i}^- - x_{2,i}^-|^2}$, up to a multiplicative coefficient).

Thus, arguing by contradiction, we can lower bound c by $\theta > 0$ and we have:

$$\frac{\theta(\frac{1}{2}\frac{\partial}{\partial s}\int_0^1\dot{b}^2+\frac{90}{10}\int_0^1\ddot{b}^2)}{1+\int_0^1\dot{b}^2} \leq C\left(-\frac{1}{\Gamma}\frac{\partial}{\partial s}\left(Ca+\int\dot{b}^+\right)+1+\int\dot{b}^+ +\left(Ca+\int\dot{b}^+\right)\frac{\partial}{\partial s}(-a)-\frac{\partial a}{\partial s}\right). \tag{78}$$

Using Corollary E (which holds for a convex combination such as the one of Lemma 3′ because the inequality of Corollary E behaves well under convex combination) and the fact that the time is finite, we derive that $Ca+\int\dot{b}^+$ is bounded. $\frac{\partial}{\partial s}(-a)$ is positive. Therefore, the integral of the right hand side of (78) is finite, thus $\int_0^1\dot{b}^2$ is bounded. There cannot be explosion q.e.d.

Proof of Lemma 3′. a) is obvious. For b), we first notice that (iii) of Lemma 3 yields almost immediately, for a disjoint sum of oscillations $\sum_{i=1}^m\psi_{\tilde\theta_i,\tilde\varepsilon_i}\eta_{0,i}+cb$

$$\frac{1}{2}\frac{\partial}{\partial s}\int_0^1 \dot{b}^2 + \frac{9c}{10}\int_0^1 \ddot{b}^2$$
$$\leq \frac{C_3}{c}\left(1+\frac{1}{\Gamma^2}+\frac{1}{\Gamma}\sum_{i=1}^{m}\frac{1}{|x_{1,i}^- - x_{2,i}^-|^2}\right)\left(\int_0^1 \dot{b}^2+1\right) - C_4\left(1+\int_0^1 \dot{b}^2\right)\frac{\partial a}{\partial s}. \qquad (iii)$$

Using Corollary E, we can replace $\sum_{i=1}^{m}\frac{1}{|x_{1,i}^- - x_{2,i}^-|^2}$ by

$$\frac{1}{\Gamma\tilde{C}_1}\left(-C\frac{\partial a}{\partial s} - \frac{\partial}{\partial s}\int \dot{b}^+ + C\int \dot{b}^+ + \tilde{C}_2 + \tilde{C}_3(Ca+\int \dot{b}^+)\left(-\frac{\partial a}{\partial s}\right)\right).$$

This yields b) for $\sum_{i=1}^{m}\psi_{\bar{\theta}_i,\bar{\varepsilon}_i}\eta_{0,i}+cb$.
However, b) behaves very well under consvex combination.
The result follows.

We now establish the local existence for our flow, after globalizing it. We build a partition of unity $(\alpha_i)_{i\in I}$ of H^1 (on b) so that Lemmas 3 and 3′ hold, i.e for each $b\in H^1$, we single out an oscillation, if possible, satisfying (1) or time intervals where we can plug functions η_{12} of Definition 2. We then consider $\sum_{j=1}^{m_i}\eta_{12,j}+cb=\eta_{12}+cb$, c subject to (15) and, using Lemma 1, we can modify η_{12} slightly so that Lemmas 3 and 3′ hold. Considering then the $\psi_{\theta,\varepsilon}$ built as in (γ) and the same time interval $[t_1^-,t_2^-]$, as well as the same c, there is a small neighborhood of $b, U(b)$, in H^1 such that $\sum_{j=1}^{m} sgn\varphi\psi_{\tilde{\theta}_j,\tilde{\varepsilon}_j}(\varphi)\tilde{\eta}_{0,j}+c\varphi$ satisfies Lemmas 3 and 3′, with φ as initial data if φ is in $U(b)$, where $\tilde{\theta}_j=\tilde{\theta}^j_\varphi(t)$ and $\tilde{\varepsilon}_j=\tilde{\varepsilon}^j_\varphi(t)$ are associated to φ in the way that $\psi_{\tilde{\theta},\tilde{\varepsilon}}(\tilde{b})sgn\tilde{b}\eta_0$ has been built for $\tilde{b}$ close to b.

We then construct, subject to $(U(b))_{b\in H^1}$ a locally finite partition of unity $(\alpha_i)_{i\in I}$.

If α_i has its support in $U_i(b)$, we denote Z_i the corresponding vector-field defined by $\eta_i=\sum_{j=1}^{m_i} sgn\varphi\psi_{\tilde{\theta}_i^j,\tilde{\varepsilon}_i^j}(\varphi)\tilde{\eta}_{0,j}+c\varphi\cdot\sum_{i\in I}\alpha_i Z_i$ is the global flow for which we wish to establish a statement of existence, uniqueness and continuous dependence, in a suitable sense, on the initial conditions - α_i is Lipschitz in the H^1- sense on b.

The above construction has to be refined, in fact: There is, here, an important remark which needs to be made in order to transform a bound on $\int_0^1\dot{b}^2$ into a strong convergence in H^1 along a flow-line, for s finite. This remark allows to reduce accidents along a flow-line, in finite time, to explosions. It goes as follows:

We complete the construction of the oscillatory parts $\sum\psi^i_{\tilde{\theta},\tilde{\varepsilon}}\tilde{\eta}^i_0$ and of their related domains of definition $U(b_i)$ so that, on a given ball of H^1, there are only a finite number of functions $\eta_0^i\psi^i_{\theta,\varepsilon}$ involved *for all of this ball.* Of course, the $\tilde{\eta}_0^i\psi^i_{\tilde{\theta},\tilde{\varepsilon}}$ depend on $\tilde{b}$ hence are in infinite number. But, because they originate in a finite number of $\eta_0^i\psi^i_{\theta,\varepsilon}$, they are bounded C^4, which is what we need. Then, computing $\frac{\partial}{\partial s}\int_0^1\ddot{b}^2$, we find that, if there is no explosion and $\int_0^1 \dot{b}^2$ remains bounded, $\int_0^1 \ddot{b}^2$ also remains bounded if the $\psi_{\theta,\varepsilon}$ have been chosen appropriately (we strongly use here the fact that the functions $\eta_0^i\psi_{\theta,\varepsilon}^i$ are in finite number); thus, b converges strongly in H^1 to its weak limit when t tends to T. There is no explosion of any kind, the solution can be continued thereafter. This argument requires some work. Alternatively, we can bypass this argument and work with the H^1-weak topology on balls, proving, in this weaker topology, uniqueness and continuity. This is the way which we use below.

The way we have completed our construction, the $\eta_0^i\psi_{\theta,\varepsilon}^i$ need not to be in finite number on a given ball of H^1. Indeed, we required that $\int_{t_i^-}^{t_i^+} \dot{\tilde{b}}^2$ should be small with respect to c.

This can be overcome as follows: the bothering term is of the type

$$\int_{t_1^-}^{t_1^+} |\ddot{\tilde{b}}| \left|\frac{\partial^2 \psi_{\tilde{\theta},\tilde{\varepsilon}}}{\partial x^2}\right| \dot{\tilde{b}}^2 \tilde{\eta}_0$$

$\left|\frac{\partial^2 \psi_{\tilde{\theta},\tilde{\varepsilon}}}{\partial x^2}\tilde{\eta}_0\right|$ is bounded by a constant $\frac{C}{\gamma}$ which is independent of how close t_1^- is chosen to t_1^+, as long as Lemma 1 holds.

Indeed, $|\tilde{\eta}_0|$ is bounded by $\frac{C}{|t-t_1^-|}$, on $[t_1^-, t_1^+]$ where C is upperbounded in function of $|t_2^+ - t_1^+|$. γ is provided by Lemma 1. This estimate is largely independent of $\tilde{b}$, as long as this $\tilde{b}$ stays in a weak H^1-neighborhood of b. Thus, on such a weak-neighborhood, we have:

$$\int_{t_1^-}^{t_1^+} |\ddot{\tilde{b}}| \left|\frac{\partial^2 \psi_{\tilde{\theta},\tilde{\varepsilon}}}{\partial x^2}\right| \dot{\tilde{b}}^2 \tilde{\eta}_0 \leq \frac{c}{\gamma} \int_{t_1^-}^{t_1^+} |\ddot{\tilde{b}}| \dot{\tilde{b}}^2 .$$

Assume that $\tilde{b}$ is in the ball $B(o,R)$ of H^1. If $\tilde{b}$ is close enough, in the weak topology to b, we have (t_1^+, t_2^+ are given by b)

$$\int_{t_1^-}^{t_1^+} |\ddot{\bar{b}}|\,|\dot{\bar{b}}|^2 \leq \int_{t_1^-}^{t_1^+} \Bigg(|\ddot{\bar{b}}|\,|\dot{\bar{b}}|\,|\dot{b}| + |\ddot{\bar{b}}| \left| \dot{b} - \frac{1}{t_1^+ - t_1^-} \int_{t_1^-}^{t_1^+} \dot{b} \right| |\dot{\bar{b}}| +$$

$$+ |\ddot{\bar{b}}|\,|\dot{\bar{b}}| \left| \frac{1}{t_1^+ - t_1^-} \int_{t_1^-}^{t_1^+} (\dot{b} - \dot{\bar{b}}) \right| + |\ddot{\bar{b}}|\,|\dot{\bar{b}}| \left| \dot{\bar{b}} - \frac{1}{|t_1^+ - t_1^-|} \int_{t_1^-}^{t_1^+} \dot{\bar{b}} \right| \Bigg) \leq$$

$$\leq \int_{t_1^-}^{t_2^-} \ddot{\bar{b}}^2 \left(\left(\int_{t_1^-}^{t_1^+} \dot{b}^2 \right)^{1/2} + \left(\int_{t_1^-}^{t_1^+} \left| \dot{b} - \frac{1}{t_1^+ - t_1^-} \int_{t_1^-}^{t_1^+} \dot{b} \right|^2 \right)^{1/2} + \right.$$

$$\left. + \frac{1}{|t_1^+ - t_1^-|} \left| \int_{t_1^-}^{t_1^+} \dot{b} - \dot{\bar{b}} \right| \right) + \left(\int_{t_1^-}^{t_1^+} |\ddot{\bar{b}}| \right) \times \int_{t_1^-}^{t_1^+} |\ddot{\bar{b}}|\,|\dot{\bar{b}}|$$

$$\leq \int_{t_1^-}^{t_2^-} \ddot{\bar{b}}^2 \left(\left(\int_{t_1^-}^{t_1^+} \dot{b}^2 \right)^{1/2} + \left(\int_{t_1^-}^{t_1^+} \left| \dot{b} - \frac{1}{t_1^+ - t_1^-} \int_{t_1^-}^{t_1^+} \dot{b} \right|^2 \right)^{1/2} + \right.$$

$$\left. + \frac{1}{|t_1^+ - t_1^-|} \left| \int_{t_1^-}^{t_1^+} \dot{b} - \dot{\bar{b}} \right| + \sqrt{t_1^+ - t_1^-} \times R \right).$$

We then observe that, on a weak H^1-neighborhood, c is lowerbounded in function only of the oscillations of b: $\bar{b}$ has to have oscillations close to those of b as soon as $|b - \bar{b}|_{L^\infty}$ is small, a weak statement in H^1. This will be made precise in Lemmas 5-8, below. We thus can choose, using Lemma 1, t_1^- close enough to t_1^+ so that

$$\frac{C}{\gamma} \sum_i \left(\int_{t_{1,i}^-}^{t_{1,i}^+} \dot{b}^2 \right)^{1/2} + \left(\int_{t_{1,i}^-}^{t_{1,i}^+} \left| \dot{b} - \frac{1}{t_{1,i}^+ - t_{1,i}^-} \int_{t_{1,i}^-}^{t_{1,i}^+} \dot{b} \right|^2 \right)^{1/2} + R\sqrt{t_{1,i}^+ - t_{1,i}^-} \leq \frac{c}{10000}$$

$t_{1,i}^-, t_{2,i}^-$ being found in this way, we can ask that

$$\frac{C}{\gamma} \sum_i \frac{1}{|t_{1,i}^+ - t_{1,i}^-|} \left| \int_{t_{1,i}^-}^{t_{1,i}^+} \dot{b} - \dot{\bar{b}} \right| < \frac{c}{1000}$$

and a similar statement should hold for $t_{2,i}^-, t_{2,i}^+$. These are again weak statements.

Our construction is completed on each $B(o,R)$: since these balls are weakly compact, we can extract from the related weak coverings finite ones. We then have a finite number of functions $\psi_{\theta,\varepsilon}^i \eta_0^i$ on the ball $B(o,n)$. Convex-combining on annuli, we globalize our construction. The $U(b_i)$'s should be of a related type. In this way, if the solution does not converge in finite time $\int_0^1 \dot{b}^2$ has to tend to $+\infty$. This remark does not destroy the properties of the α_i - in particular, the fact that they are locally finite and Lipschitz with respect to b- which we described earlier. We resume now our reasoning.

Let x_0 be a curve in C_β such that $b_0 \in H^1$.

Let $\varepsilon > 0$ be a small parameter. For ε small enough, $0 < \varepsilon < \varepsilon_0$, we denote

$$\phi_\varepsilon(b)$$

an ε regularization of b, which we can obtain solving the equation - $\varepsilon\overline{\ddot{\phi_\varepsilon(b)}}+\phi_\varepsilon(b) = b$. ε is not the ε of $\psi_{\theta,\varepsilon}$. We denote also ε this ε but keep in mind that that they are different. This is for the sake of keeping uncomplicated notations. Thus, $\phi_\varepsilon(b)$ can be assumed to be in a given weak neighborhood V, where the α_i's are almost everywhere zero, i.e. only a finite number of them is non zero. At any rate, V will be bounded. Thus, by the construction above, there are only a finite number of $\psi_{\theta,\varepsilon}\eta_0$ involved in V. We can therefore rewrite the η of $\sum \alpha_i Z_i$ as $\sum \beta_i sgn\varphi\psi_{\bar{\theta}_i,\bar{\varepsilon}_i}\bar{\eta}_{o,i}$. Introducing a convenient abuse which has no importance in this section, we come back to the notation $\sum \alpha_i\eta_i$ which we will keep here.

The map:

$$L^2 \longrightarrow H^2$$

$$b \longrightarrow \sum \alpha_i(\phi_\varepsilon(b))\eta_i(\phi_\varepsilon(b)) = N_\varepsilon(b)$$

is obviously Lipschitz.

Let, for any curve x close to x_0 in the full space $H^1(S^1, M), Z(x)$ be the H^1-vector-field tangent to $H^1(S^1, M)$ at x defined by:

$$\begin{cases} Z_\varepsilon(x) & = \frac{\overline{\dot{N_\varepsilon(b)}+\lambda_\varepsilon\phi_\varepsilon(b)}}{\int_0^1 a} v(x) + \lambda_\varepsilon\xi(x) + N_\varepsilon(b)w \\ \text{where } \dot{x} & = a\xi + bv + cw, \quad w = -[\xi, v] + \bar{\mu}\xi, \\ \lambda_\varepsilon & = \left(\int_0^t \left(bN_\varepsilon(b) - c\frac{\overline{\dot{N_\varepsilon(b)}}}{\int_0^1 a}\right) - t\int_0^1 \left(bN_\varepsilon(b) - c\frac{\overline{\dot{N_\varepsilon(b)}}}{\int_0^1 a}\right)\right) - \bar{\mu}N_\varepsilon(b) \end{cases}$$

Z_ε is locally Lipschitz, even in this weaker topology, because we have smoothened b in $\phi_\varepsilon(b)$, hence the differential equation:

$$\left| \begin{array}{ll} \frac{\partial x}{\partial x} & = Z_\varepsilon(x) \\ x(0) & = \tilde{x}(0) \end{array} \right.$$

has locally a unique solution.

Let $a_\varepsilon(s), b_\varepsilon(s), c_\varepsilon(s)$ be the coordinates of $\dot{x}_\varepsilon(s)$ on ξ, v, w. We have:

$$\begin{cases} \frac{\partial a_\varepsilon}{\partial s} & = \overline{\dot{\lambda_\varepsilon + \bar{\mu}N_\varepsilon}} - b_\varepsilon N_\varepsilon + c_\varepsilon\frac{\dot{N}_\varepsilon+\lambda_\varepsilon\phi_\varepsilon}{\int_0^1 a_\varepsilon} = \\ & = \frac{c_\varepsilon}{\int_0^1 a_\varepsilon}\lambda_\varepsilon\phi_\varepsilon - \int_0^1 \left(b_\varepsilon N_\varepsilon - \frac{c_\varepsilon\dot{N}_\varepsilon}{\int_0^1 a_\varepsilon}\right) \\ \frac{\partial c_\varepsilon}{\partial s} & = \dot{N}_\varepsilon - \frac{\dot{N}_\varepsilon+\lambda_\varepsilon\phi_\varepsilon}{\int_0^1 a_\varepsilon} \times a_\varepsilon + \lambda_\varepsilon b_\varepsilon = \dot{N}_\varepsilon\left(1 - \frac{a_\varepsilon}{\int_0^1 a_\varepsilon}\right) + \lambda_\varepsilon\left(b_\varepsilon - \frac{\phi_\varepsilon a_\varepsilon}{\int_0^1 a_\varepsilon}\right) \\ \frac{\partial b_\varepsilon}{\partial s} & = \frac{\overline{\dot{\dot{N}+\lambda_\varepsilon\phi_\varepsilon}}}{\int_0^1 a_\varepsilon} + a_\varepsilon N_\varepsilon\tau - b_\varepsilon N_\varepsilon\bar{\mu}_\xi + c_\varepsilon d\gamma\left(w, \lambda_\varepsilon\xi + \frac{\dot{N}_\varepsilon+\lambda_\varepsilon\phi_\varepsilon v}{\int_0^1 a_\varepsilon}\right). \end{cases} \tag{78}$$

Clearly, the equation on $\frac{\partial b_\varepsilon}{\partial s}$ drives the other ones, because any estimate on $\int b_\varepsilon^2 + \int \overline{\dot{\phi}_\varepsilon(b_\varepsilon)}^2$ would yield estimates on a_ε and c_ε at least in an L^2 - sense and would also allow to continue the flow, as we will see.

Let us assume that the initial condition $x(0)$ is such that $c(0) = 0, a(0)$ is constant and $b(0)$ belongs to a small ball around b_0 in $H^1, B(b_0, \rho)$, included in V. There is then, for any $\varepsilon > 0$ small enough, a continuous family $x_\varepsilon(s)$, solution of our differential equation at least for a small time $s, s \in [0, s_0(\varepsilon, b(0))], s_0 > 0$. $\phi_\varepsilon(b_\varepsilon)(s)$ will belong to V for a small time $[0, s_1(\varepsilon, b(0))]$ because $N_\varepsilon, \phi_\varepsilon$ etc. are bounded C^∞ if b_ε is bounded L^2. ($\varepsilon > 0$). We are going to derive a lower bound on s_1, independent of ε and $b(0)$ in $B(b_0, \rho)$.

Lemma G. $\exists \gamma > 0, \exists \bar{\varepsilon} > 0, \exists \bar{\rho} > 0$, *such that if* $\rho < \bar{\rho}, \varepsilon < \bar{\varepsilon}$ *then* $s_1(\varepsilon, b(0)) \geq \gamma$ *for any* $b(0) \in B(b_{0,\rho})$.

Proof of Lemma G. We establish, as long as $\phi_\varepsilon(b_\varepsilon)$ is in V and that $\int_0^1 b_\varepsilon^2 \leq C'$, differential equations and inequalities that $\int_0^1 a_\varepsilon^2, \int_0^1 c_\varepsilon^2$ should satisfy:

Observe that ϕ_ε is then bounded in H^1, hence the α_i's are all zero, but a finite number of them and N_ε is therefore bounded in H^1, with a uniform bound C.

Thus,

$$\frac{1}{2}\frac{\partial}{\partial s}\int_0^1 a_\varepsilon^2 = \int_0^1 \frac{a_\varepsilon c_\varepsilon \lambda_\varepsilon \phi_\varepsilon}{\int_0^1 a_\varepsilon} - \left(\int_0^1 a_\varepsilon\right)\int_0^1\left(b_\varepsilon N_\varepsilon - \frac{c_\varepsilon \dot{N}_\varepsilon}{\int_0^1 a_\varepsilon}\right)$$

$$\text{where } |\lambda_\varepsilon|_{L^\infty} \leq C\left(1 + \left(\int_0^1 c_\varepsilon^2\right)^{1/2} \Big/ \left(\textstyle\int_0^1 a_\varepsilon\right)\right)$$

$$\left|\frac{1}{2}\frac{\partial}{\partial s}\left(\int_0^1 a_\varepsilon\right)^2\right| \leq C\left(1 + \int_0^1 c_\varepsilon^2 + \left(\int_0^1 a_\varepsilon\right)^2\right)$$

$$\frac{1}{2}\frac{\partial}{\partial s}(c_\varepsilon^2) = c_\varepsilon\frac{\partial c_\varepsilon}{\partial s} = \left(\int_0^s \dot{N}_\varepsilon\left(1 - \frac{a_\varepsilon}{\int_0^1 a_\varepsilon}\right) + \lambda_\varepsilon\left(b_\varepsilon - \phi_\varepsilon\frac{a_\varepsilon}{\int_0^1 a_\varepsilon}\right)\right) \times$$

$$\times\left(\dot{N}_\varepsilon\left(1 - a_\varepsilon \Big/ \textstyle\int_0^1 a_\varepsilon\right) + \lambda_\varepsilon\left(b_\varepsilon - \frac{\phi_\varepsilon a_\varepsilon}{\int_0^1 a_\varepsilon}\right)\right).$$

Observe that there is only a finite number of α_i non zero on V.

Thus, $|\dot{N}_\varepsilon|_\infty \leq C(V)\left(|\phi_\varepsilon(\dot{b})|_\infty + 1\right) \leq C(V)\left(+\left(\int_0^1\left(\overline{\ddot{\phi}_\varepsilon(b)}\right)^2\right)^{1/2}\right)$, since the $\dot{\eta}_0$ are globally bounded above (they are in finite number) and since (γ) holds.

Thus, since $c_\varepsilon(0) = 0$,

$$\frac{1}{2}\int_0^1 c_\varepsilon^2(s) = \frac{1}{2}\int_0^1 \left(\int_0^s \dot{N}_\varepsilon \left(1 - a_\varepsilon/_{\int_0^1 a_\varepsilon}\right) + \lambda_\varepsilon \left(\frac{b_\varepsilon - \phi_\varepsilon a_\varepsilon}{\int_0^1 a_\varepsilon}\right)\right)^2 \leq$$

$$\leq C\frac{s}{2}\left[\int_0^s \int_0^1 \left(\dot{N}_\varepsilon \left(1 - a_\varepsilon/_{\int_0^1 a_\varepsilon}\right)\right)^2 + \int_0^1 \lambda_\varepsilon^2 \left(b_\varepsilon - \frac{\phi_\varepsilon a_\varepsilon}{\int_0^1 a_\varepsilon}\right)^2\right]$$

$$\leq \frac{C's}{2}\left[\int_0^s \left\{\left(\int_0^1 \ddot{\phi}_\varepsilon^2\right)\left(1 + \int_0^1 a_\varepsilon^2/_{(\int_0^1 a_\varepsilon)^2}\right) + \left(1 + \frac{\int_0^1 c_\varepsilon^2}{\left(\int_0^1 a_\varepsilon\right)^2}\right)\right\}\right].$$

On the other hand,

$$\frac{1}{2}\frac{\partial}{\partial s}\int_0^1 a_\varepsilon^2 \leq C\left(\frac{\int_0^1 a_\varepsilon^2}{\left(\int_0^1 a_\varepsilon\right)^2} + 1 + \frac{\left(\int_0^1 c_\varepsilon^2\right)^2}{\left(\int_0^1 a_\varepsilon\right)^4} + \left(\int_0^1 a_\varepsilon\right)^2 + \int_0^1 c_\varepsilon^2\right).$$

And as noticed before,

$$\frac{\partial}{\partial s}\left(\int_0^1 a_\varepsilon^2\right)^2 \leq c\left(1 + \int_0^1 c_\varepsilon^2 + \left(\int_0^1 a_\varepsilon\right)^2\right).$$

We then take the equation on b_ε and multiply it by $\ddot{\phi}_\varepsilon$ and integrate between 0 and 1. Since our regularization is symmetric, we have:

$$\frac{\partial}{\partial s}\int \dot{b}_\varepsilon \overline{\dot{\phi_\varepsilon(b_\varepsilon)}} = 2\int \frac{\partial \dot{b}_\varepsilon}{\partial s}\overline{\dot{\phi_\varepsilon(b_\varepsilon)}} = -2\int \frac{\partial b_\varepsilon}{\partial s}\overline{\ddot{\phi_\varepsilon(b_\varepsilon)}}.$$

Thus

$$\frac{1}{2}\frac{\partial}{\partial s}\int \dot{b}_\varepsilon \overline{\dot{\phi_\varepsilon(b_\varepsilon)}} = \frac{1}{\int_0^1 a_\varepsilon}\int_0^1 \left(\overline{\dot{N_\varepsilon + \lambda_\varepsilon \phi_\varepsilon}}\right)\ddot{\phi}_\varepsilon +$$

$$+\int_0^1 (a_\varepsilon N_\varepsilon \tau - b_\varepsilon N_\varepsilon \bar{\mu}_\xi)\ddot{\phi}_\varepsilon + \int_0^1 c_\varepsilon d\gamma\left(w, \lambda_\varepsilon \xi + \frac{\dot{N}_\varepsilon + \lambda_\varepsilon \phi_\varepsilon}{\int_0^1 a_\varepsilon}\right)\ddot{\phi}_\varepsilon.$$

The expression $\int_0^1 \ddot{N}_\varepsilon \ddot{\phi}_\varepsilon$ can be estimated just as in Lemma 3, iii) for $\int \overline{\ddot{\psi_{\theta,\varepsilon}\eta_0}} sgnb\ddot{b}$. The only difference is that we have now a finite member of such $\psi_{\theta,\varepsilon}\eta_0$ combined along the partition of unity. Observe that, since there is only a finite number of such $\psi_{\theta,\varepsilon}\eta_0$, the related $\frac{1}{(t_1^- - t_2^-)^2}$ are upper bounded. Rather than using the statement of

iii) of Lemma 3, where $-\frac{\partial \bar{\mu}}{\partial s}$ is involved, we will use the other upper bounds derived when providing the proof of iii) of Lemmma 3, where $-\frac{\partial \bar{\mu}}{\partial s}$ is replaced by $\frac{1}{(t_1^- - t_2^-)^2}$.

We thus have: ($\int_0^1 \overline{\dot{\phi_\varepsilon(b)}}^2$ is bounded)

$$\frac{1}{\int_0^1 a_\varepsilon} \int_0^1 \ddot{N}_\varepsilon \ddot{\phi}_\varepsilon \leq -\frac{9c}{10} \int_0^1 \overline{\ddot{\phi_\varepsilon(b)}}^2 + \frac{C_3}{c}\left(1 + \frac{1}{\inf \Gamma_i^2} + \frac{\tilde{\tilde{C}}}{\inf \Gamma_i}\right)$$

$\inf \Gamma_i$ is taken on all oscillations which have a non zero α_i on V, hence it is a constant $\theta_V > 0$.

We also have, for any $\delta > 0$:

$$\left| \int_0^1 (a_\varepsilon N_\varepsilon \tau - b_\varepsilon N_\varepsilon \bar{\mu}_\xi)\, \ddot{\phi}_\varepsilon \right| \leq \delta \int \ddot{\phi}_\varepsilon^2 + C(\delta).$$

and, upper bounding $|\dot{N}_\varepsilon|_\infty$ by $\tilde{C}\left(\int \phi_\varepsilon(\ddot{b})^2\right)^{1/2}$

$$\left| \int_0^1 c_\varepsilon d\gamma\left(w, \lambda_\varepsilon \xi + \frac{\dot{N}_\varepsilon + \lambda_\varepsilon \phi_\varepsilon}{\int_0^1 a_\varepsilon} v\right) \ddot{\phi}_\varepsilon \right| \leq$$

$$\leq \delta \int \ddot{\phi}_\varepsilon^2 + C(\delta)\left(1 + \int_0^1 c_\varepsilon^2 / _{(\int_0^1 a_\varepsilon)^2}\right) \int_0^1 c_\varepsilon^2 \left(1 + \frac{1}{\left(\int_0^1 a_\varepsilon\right)^2}\right) +$$

$$+ C(\delta) \int_0^1 c_\varepsilon^2 / \left(\int_0^1 a_\varepsilon\right)^2 \int_0^1 \ddot{\phi}_\varepsilon(b)^2.$$

Finally, we have ($\int_0^1 \dot{\phi}_\varepsilon^2$ is bounded, $\int_0^1 \overline{\dot{N_\varepsilon(b)}}^2$ is also bounded on V since the functions η_0 are in finite number)

$$\frac{1}{\int_0^1 a_\varepsilon} \int_0^1 \overline{\dot{\lambda_\varepsilon \phi_\varepsilon}} \ddot{\phi}_\varepsilon \leq \frac{C(\delta)}{\left(\int_0^1 a_\varepsilon\right)^2}\left[\left(1 + \int_0^1 c_\varepsilon^2 / _{(\int_0^1 a_\varepsilon)^2}\right) + \frac{1}{\left(\int_0^1 a_\varepsilon\right)^2} \int_0^1 c_\varepsilon^2 \overline{\dot{N}_\varepsilon(b)}^2\right] +$$

$$+ \delta \int_0^1 \ddot{\phi}_\varepsilon^2 \leq \frac{C(\delta)}{\left(\int_0^1 a_\varepsilon\right)^2}\left[\left(1 + \int_0^1 c_\varepsilon^2 / _{(\int_0^1 a_\varepsilon)^2}\right) + \frac{\tilde{C}}{\left(\int_0^1 a_\varepsilon\right)^2} \int \ddot{\phi}_\varepsilon(b)^2 \int_0^1 c_\varepsilon^2\right] + \delta \int_0^1 \ddot{\phi}_\varepsilon^2.$$

Thus, assuming that c, for $\phi_\varepsilon(b)$, is lowerbounded by $c_0 > 0$ on the trajectory:

$$\frac{\partial}{\partial s} \int \dot{b}_\varepsilon \overline{\dot{\phi_\varepsilon(b_\varepsilon)}} \leq \frac{9c_0}{10} \int_0^1 \overline{\ddot{\phi_\varepsilon(b)}}^2 + 2\delta \int_0^1 \ddot{\phi}_\varepsilon^2 +$$

$$+C(\delta)\left(1+\frac{1}{\left(\int_0^1 a_\varepsilon\right)^2}\right)\cdot\frac{\int_0^1 c_\varepsilon^2}{\left(\int_0^1 a_\varepsilon\right)^2}\cdot\int_0^1 \ddot{\phi}_\varepsilon(b)^2+$$

$$+C(\delta)\left(1+\int_0^1 c_\varepsilon^2/_{(\int_0^1 a_\varepsilon^2)^2}\right)\left(1+\int_0^1 c_\varepsilon^2\right)\left(1+\frac{1}{\left(\int_0^1 a_\varepsilon\right)^2}\right).$$

Thus, if δ is small enough:

$$\begin{cases} \frac{\partial}{\partial s}\int \dot{b}_\varepsilon\overline{\dot{\phi}_\varepsilon(b_\varepsilon)} & \leq -\frac{4}{5}c_0\int_0^1 \overline{\ddot{\phi}_\varepsilon(b)}^2 + C(\delta)\left(1+\frac{1}{(\int_0^1 a_\varepsilon)^2}\right)\frac{\int_0^1 c_\varepsilon^2}{(\int_0^1 a_\varepsilon)^2}\int_0^1 \ddot{\phi}_\varepsilon(b)^2+ \\ & +C(\delta)\left(1+\int_0^1 c_\varepsilon^2/_{(\int_0^1 a_\varepsilon)^2}\right)\left(1+\int_0^1 c_\varepsilon^2\right)\left(1+\frac{1}{(\int_0^1 a_\varepsilon)^2}\right) \\ \left|\frac{\partial}{\partial s}\left(\int_0^1 a_\varepsilon\right)^2\right| & \leq C\left(1+\int_0^1 c_\varepsilon^2+\left(\int_0^1 a_\varepsilon\right)^2\right) \\ \frac{1}{2}\frac{\partial}{\partial s}\int_0^1 a_\varepsilon^2 & \leq C\left(\frac{\int_0^1 a_\varepsilon^2}{(\int_0^1 a_\varepsilon)^2}+1+\frac{(\int_0^1 c_\varepsilon^2)^2}{(\int_0^1 a_\varepsilon)^4}+\left(\int_0^1 a_\varepsilon\right)^2+\int_0^1 c_\varepsilon^2\right) \\ \frac{1}{2}\int_0^1 c_\varepsilon^2(s) & \leq C''s\left[\int_0^s\left\{\left(\int_0^1 \ddot{\phi}_\varepsilon^2\right)\left(1+\int_0^1 a_\varepsilon^2/_{(\int_0^1 a_\varepsilon)^2}\right)+\right.\right. \\ & \left.\left.+\left(1+\int_0^1 c_\varepsilon^2/_{(\int_0^1 a_\varepsilon)^2}\right)\left(1+\frac{\int_0^1 a_\varepsilon^2}{(\int_0^1 a_\varepsilon)^2}\right)\right\}\right]. \end{cases}$$

Observe that $\int_0^1 \dot{b}_\varepsilon\dot{\phi}_\varepsilon(b) = \int \dot{\phi}_\varepsilon(b)^2 + \varepsilon\int\ddot{\phi}_\varepsilon(b)^2 \geq \int\dot{\phi}_\varepsilon(b)^2$.
Let $\tilde{s}$ be the maximal positive time such that:

$$\begin{cases} \int_0^1 c_\varepsilon^2(s) \leq \tilde{\delta} < 1, \int_0^1 a_\varepsilon^2(s) \leq 1 + a_0^2 \\ \left(\int_0^1 a_\varepsilon\right)^2 \geq \frac{1}{2}a_0^2, \phi_\varepsilon(b_\varepsilon)(s)\in V, \int_0^1 b_\varepsilon^2 \leq C' \qquad \forall s\in[0,\tilde{s}]. \end{cases}$$

Using the above differential equations, we have:

$$\left.\begin{array}{ll} \left|\left(\int_0^1 a_\varepsilon\right)^2(s)\right. & \left.-\left(\int_0^1 a_\varepsilon(0)\right)^2\right| \leq Cs \\ \int_0^1 a_\varepsilon^2(s) & \leq C_1 s + \int_0^1 a_\varepsilon^2(0) \\ \int_0^1 c_\varepsilon^2(s) & \leq C_2 s + c_3 s\int_0^s\int_0^1 \ddot{\phi}_\varepsilon^2 \end{array}\right\}$$

where C, C_1, C_2, C_3 does not depend on $\tilde{\delta}$ as long as $\tilde{\delta} < 1$. Also $C_4(\delta), C_5(\delta), C_6(\delta)$, below

$$\left(\frac{4c_0}{5}-C_4(\delta)\tilde{\delta}\right)\int_0^s\int_0^1 \overline{\ddot{\phi}_\varepsilon(b)^2}+\frac{1}{2}\int\dot{\phi}_\varepsilon(b_\varepsilon)^2 \leq \int \dot{b}_\varepsilon\dot{\phi}_\varepsilon(b_\varepsilon)(0)+C_5(\delta)s \leq C_6(\delta).$$

Thus, taking $\tilde{\delta} < 4c_0/_{10C_4(\delta)}$, since $a_\varepsilon(0)$ is close to a_0: ($\varepsilon_2 > 0$ measures how close)

$$\begin{cases} \left|\left(\int_0^1 a_\varepsilon\right)^2(s) - a_0^2\right| & \leq Cs + \varepsilon_2 \\ \int_0^1 a_\varepsilon^2(s) & \leq C_1 s + a_0^2 + a\varepsilon_2 a_0 \qquad \forall s\in[0,\tilde{s}] \\ \int_0^1 c_\varepsilon^2(s) & \leq C_7(\delta)s \end{cases} \tag{79}$$

where $C_7(\delta)$ does not depend on $\tilde{\gamma}$.

We, in addition have:

$$\frac{2}{5}c_0\int_0^s\int_0^1 \overline{\ddot{\phi_\varepsilon(b)}}^2 + \int_0^1 \dot{b}_\varepsilon\dot{\phi}_\varepsilon(\dot{b}_\varepsilon)(s) \leq \int_0^1 \dot{b}_\varepsilon\dot{\phi}_\varepsilon(\dot{b})(0) + c_s(\delta)s. \tag{80}$$

Using the fact that $\int_0^1 \dot{b}_\varepsilon\dot{\phi}_\varepsilon(b)$ is lowerbounded by $\int_0^1 \dot{\phi}_\varepsilon(b)^2$, upperbounding also $\int_0^1 \dot{\phi}_\varepsilon(b)^4$ by $C\int_0^1|\ddot{\phi}_\varepsilon(b)|^2$, where C is upperbound on $\int_0^1 \dot{\phi}_\varepsilon(b)^2$ derived from (80), we derive that

$$\int_0^s \left|\frac{\partial b_\varepsilon}{\partial s}\right|^2_{L^2} + \left|\frac{\partial a_\varepsilon}{\partial s}\right|^2_{L^2} + \left|\frac{\partial c_\varepsilon}{\partial s}\right|^2_{L^2}$$

is finite for s finite. Thus, there is no explosion as long as we keep our bounds, in finite time.

We also have, as long as $\int_0^1 b_\varepsilon^2 \leq C'$ and $\phi_\varepsilon(b_\varepsilon)\in V$:

$$\frac{\partial}{\partial s}\int_0^1 b_\varepsilon^2 = \int_0^1 b_\varepsilon\left(\frac{\overline{\dot{N}_\varepsilon+\lambda_\varepsilon\phi_\varepsilon}}{\int_0^1 a_\varepsilon} + a_\varepsilon N_\varepsilon\tau - b_\varepsilon N_\varepsilon\bar{\mu}_\xi + c_\varepsilon d\gamma(w,\lambda_\varepsilon\xi + \frac{\dot{N}_\varepsilon+\lambda_\varepsilon\phi_\varepsilon}{\int_0^1 a_\varepsilon}v)\right) \leq$$

$$\leq C'''\left(\int_0^1 b_\varepsilon^2 + \int_0^1 \ddot{\phi}_\varepsilon^2 + \int_0^1 \dot{\phi}_\varepsilon^4\right) + \tilde{C}_4 \leq C''''\left(\int_0^1 b_\varepsilon^2 + \int_0^1 \ddot{\phi}_\varepsilon^2\right) + \tilde{C}_4.$$

Combining (80) and (81), and the fact that $\int_0^1 \dot{b}_\varepsilon\dot{\phi}_\varepsilon(b_\varepsilon)$ is larger than $\int_0^1 \dot{\phi}_\varepsilon(b)^2$, we derive that if $\int_0^1 b_\varepsilon^2(o) \leq \frac{C'}{2}$, taking V to be a ball $B(o,R)$ and assuming that $b_\varepsilon(o)$ and ε satisfy:

$$\int_0^1 \dot{b}\dot{\phi}_\varepsilon(\dot{b})(o) \leq \frac{R}{2}.$$

(This is tantamount, taking ε very small, to assume that $\int_0^1 \dot{b}^2(o) \leq \frac{R}{2}$) $\phi_\varepsilon(b)$ will be still in $B(o,R)$, by (80), and $\int_0^1 b_\varepsilon^2$ will be less than C' for some sizable time Δs, independent of ε, provided ε is small enough, Δs depending only on C_1', R and c. Also, (79) will hold. The Lemma follows as long as c, for $\phi_\varepsilon(b)$ does not approach zero on the trajectory. Putting a cut-off function in fron of z_ε, we can make it zero as soon as c is small. Then, our lemma holds without restriction.

We remove this restriction, below.

Lemma H. *As $\varepsilon\to 0, c_\varepsilon$ tends to zero in $L^1(S^1)$ everywhere in $s, a_\varepsilon(s) - \int_0^1 a_\varepsilon(s)$ tends to zero in $L^1(S^1)$ everywhere in s.*

Proof of Lemma H. We have, using our bounds:

$$\frac{\partial}{\partial s}\int_0^1 |c_\varepsilon| \leq \int_0^1 |\dot{N}_\varepsilon|\left|1-\frac{a_\varepsilon}{\int_0^1 a_\varepsilon}\right| + C\left(\int_0^1 |b_\varepsilon-\phi_\varepsilon| + \int_0^1\left|1-\frac{a_\varepsilon}{\int_0^1 a_\varepsilon}\right|\right)$$

$$\leq \left(\left(\int_0^1 |\ddot{\phi}_\varepsilon|^2\right)^{\frac{1}{2}} + C'\right)\int_0^1\left|a_\varepsilon - \int_0^1 a_\varepsilon\right| + C\int_0^1 |b_\varepsilon-\phi_\varepsilon|.$$

On the other hand:

$$\frac{\partial}{\partial s}\int_0^1 |a_\varepsilon - \int_0^1 a_\varepsilon| \le \int_0^1 \left| \frac{c_\varepsilon \lambda_\varepsilon \phi_\varepsilon}{\int_0^1 a_\varepsilon} - \int_0^1 \frac{c_\varepsilon \lambda_\varepsilon \phi_\varepsilon}{\int_0^1 a_\varepsilon} \right| \le$$
$$\le C \int_0^1 |c_\varepsilon|.$$

Thus, denoting $y_\varepsilon(s) = \int_0^s \int_0^1 |c_\varepsilon|$, we have, since $a_\varepsilon(0) = \int_0^1 a_\varepsilon(0)$:

$$y_\varepsilon''(s) \le C\left(\left(\int_0^1 |\ddot{\phi}_\varepsilon|^2\right)^{\frac{1}{2}} + C'\right) y_\varepsilon(s) + C\int_0^1 |b_\varepsilon - \phi_\varepsilon|.$$

Since $y_\varepsilon'(0) = \int_0^1 |c_\varepsilon|(0) = 0$, we have:

$$y_\varepsilon'(s) \le C_{30} \operatorname*{Max}_{\tau\in[0,s]} y_\varepsilon(\tau)\left(\int_0^s \left(\int_0^1 |\ddot{\phi}_\varepsilon|^2\right)^{\frac{1}{2}} + 1\right) + \left(C\int_0^s \int_0^1 |b_\varepsilon - \phi_\varepsilon|(\tau)\right)$$

$\int_0^s \int_0^1 |\ddot{\phi}_\varepsilon|^2$ is bounded above independently of s, for $s \in [0,1]$. Observe that $b_\varepsilon - \phi_\varepsilon = -\varepsilon\ddot{\phi}_\varepsilon$.

Thus

$$y_\varepsilon'(s) \le C_{31} \operatorname*{Max}_{\tau\in[0,s]} y_\varepsilon(\tau) + C\varepsilon \int_0^s \int_0^1 |\ddot{\phi}_\varepsilon|.$$

Thus

$$y_\varepsilon'(s) \le C_{31} \operatorname*{Max}_{\tau\in[0,s]} y_\varepsilon(\tau) + C_{31}\varepsilon \qquad \forall\tau \in [0,s].$$

Integrating again:

$$y_\varepsilon(\tau) \le \left(C_{31} \operatorname*{Max}_{\tau_1\in(0,s]} y_\varepsilon(\tau_1)\right)\tau + C_{31}\varepsilon\tau \qquad \forall\tau \le s.$$

Taking $C_{31}s < \frac{1}{2}$, we derive:

$$\operatorname*{Max}_{\tau\in[0,s]} y_\varepsilon(\tau) \le C\frac{1}{2}\varepsilon s.$$

When ε tends to zero, $\operatorname*{Max}_{\tau\in[0,s]} y_\varepsilon(\tau)$ as well as $y_\varepsilon'(s)$ tends to zero. Hence, $\int_0^1 |c_\varepsilon|$ tends to zero. Also, $(a_\varepsilon - \int_0^1 a_\varepsilon)(s)$ tends to zero in $L^1(S^1)$ everywhere in s. It is then easy to see that:

$$\int_0^s \int_0^1 \left| \frac{\partial}{\partial s} a_\varepsilon + \int_0^1 b_\varepsilon N_\varepsilon \right| \leq C \int_0^s \int_0^1 |c_\varepsilon| \left(1 + |\dot{N}_\varepsilon|_\infty\right)$$

$$\leq C \int_0^s \int_0^1 |c_\varepsilon| \left(1 + \left(\int_0^1 |\ddot{\phi}_\varepsilon|^2\right)^{\frac{1}{2}}\right)$$

$$\leq C_1 \operatorname*{Sup}_{(0,s)} \int_0^1 |c_\varepsilon|$$

tends to zero.

Also

$$\int_0^s \int_0^1 |c_\varepsilon| \left| d\gamma \left(w, \lambda_\varepsilon + \frac{\dot{N}_\varepsilon + \lambda_\varepsilon \phi_\varepsilon}{\int_0^1 a_\varepsilon} v \right) \right|$$

tends to zero by the same argument.

The same argument implies that:

$$\int_0^s \left| \lambda_\varepsilon + \overline{\mu} N_\varepsilon(b_\varepsilon) - \left(\int_0^t b N_\varepsilon(b_\varepsilon) - t \int_0^1 b_\varepsilon N_\varepsilon(b_\varepsilon) \right) \right|_{L^\infty(S^1)} \tag{82}$$

tends to zero.

We in fact have the following better result:

Lemma K. *(i) There exist a constant $C > 0$ and $\overline{\varepsilon} > 0$ such that for any $b(0) \in V$ and $s \leq s_1$, we have:*

$$\int_0^s \left| \int_0^1 \left| \frac{\partial b_\varepsilon}{\partial s} \right|^2 + \int_0^1 \left(|\ddot{N}_\varepsilon|^2 + |\ddot{\phi}_\varepsilon|^2 \right) + \int_0^1 \dot{\phi}_\varepsilon(b_\varepsilon)^2 + \phi_\varepsilon(b_\varepsilon)^2 \right| ds dt \leq C$$

$$\int_0^s \left(\int_0^1 \left(\left| \frac{\partial x_\varepsilon}{\partial s} \right|^2 + |\dot{x}_\varepsilon|^2 \right) \right) ds dt \leq C$$

$$\forall s \leq \inf(s_1, 1)$$

(ii) $b_\varepsilon(s,t)$ converges therefore weakly in $L^2([0, s_1] \times S^1]$ to a function $b(s,t)$ and $\phi_\varepsilon(b_\varepsilon)$ converges strongly to $b(s,t)$ in $H^1(S^1)$, almost everywhere in s and in $L^2([o, s_1] \times S^1)$.

(iii) $x_\varepsilon(s,t)$ converges weakly in $H^1([0, s_1) \times S^1)$ to $x(s,t)$ and $x_\varepsilon(s,t)$ converges strongly to $x(s,t)$ in $L^\infty(S^1)$ for any s. $\dot{x}(s,t) = a\xi + b(s,t)v$ where $a(s)$ is the limit of $\int_0^1 a_\varepsilon(s)$ when ε tends to zero.

(iv) $N_\varepsilon(\phi_\varepsilon(b_\varepsilon))$ *converges strongly to* $N_0(b)$ *in* $H^1(S^1)$ *a.e. in* s *and is bounded in* $L^2([0,s_1],H^2(S^1))$. *It is also bounded in* $H_1(S^1)$ *for every* s. *Thus,* $N_\varepsilon(\phi_\varepsilon(b_\varepsilon))$ *converges to* $N_0(b)$ *in* $L^2([0,s_1],H^1(S^1))$

(v) The limit function $b(s,t)$ *satisfies:*

a) $\int_0^1 |\dot{b}(s,t)|^2 dt \le C \qquad \forall s \in [0,s_1]$

b) $\int_0^{s_1}\int_0^1 |\ddot{b}(s,t)|^2 ds dt \le C$

The identity holds a.e. in (s,t), *in the distributional sense and also holds in* $L^2(S^1)$ *a.e. in* s.

This Lemma settles the existence function. The next lemma settles the uniqueness and continuity:

Lemma L. *i) The solution defined in the previous lemma is locally unique and continuous from* $H^1(S^1)$ *into* $L^\infty(S^1)$ *in* b. *We, in fact, have the more precise statement:*

$$\operatorname*{Sup}_{t\in[0,s]} |b(\tau,t)-\tilde{b}(\tau,t)|^2_\infty + \int_0^s\int_0^1 |\dot{b}(\tau,t)-\dot{\tilde{b}}(\tau,t)|^2 dt d\tau$$

tends to zero when $|a_0) - \tilde{a}(0)| + \int_0^1 |b(0,t)-\tilde{b}(0,t)|^2 + d(x(0),\tilde{x}(0))dt$ *tends to zero, assuming that* $\int_0^1 |\dot{b}(0,t)|^2 + |\dot{\tilde{b}}(0,t)|^2$ *is bounded by a given constant* A, *i.e., in some sense, the map*

$$b(0,t') \mapsto b(s,t)$$

is, almost everywhere in s, *continuous from* $H^1(S^1)$ *into* $H^1(S^1)$.

ii) The same statements hold for the curves $x(s,t)$, *with* $H^2(S^1)$ *and* $W^{1,\infty}(S^1)$.

Proof of Lemma K. Proof of i): We know that $\int_0^1 \dot{\phi}_\varepsilon(b_\varepsilon)^2 + \phi_\varepsilon^2(b_\varepsilon)$ is bounded for $0 \le s \le s_1(b_\varepsilon \in V)$ and that $\int_0^s\int_0^1 |\ddot{\phi}_\varepsilon|^2$ is bounded. When we are in V, the α_i are all zero but a finite number. Observe that: (we use (γ))

$$|\overline{\psi_{\tilde{\theta},\tilde{\varepsilon}}(\phi_\varepsilon)\ddot{\mu}_0}| \le \frac{C}{|t_1^- - t_2^-|^2} + |\dot{\phi}_\varepsilon|^2\left(1 + |\frac{\partial^2\psi_{\theta,\varepsilon}}{\partial b^2}|\right) + |\ddot{\phi}_{\theta,\varepsilon}| + |\ddot{\theta}|$$

(the Diracs of $\ddot{\eta}_0$ at t_1^-, t_2^- do not matter because $\psi_{\tilde{\theta},\tilde{\varepsilon}}(\phi_\varepsilon)$ vanishes there, see the proof of Lemma 3 (*iii*)).

Therefore,

$$\int_0^s\int_0^1 |\ddot{N}_\varepsilon|^2 \le C(V)\int_0^s\int_0^1 |\ddot{\phi}_\varepsilon|^2.$$

Observe also that we can upper bound $\int_0^1 c_\varepsilon^2 \dot{N}_\varepsilon^2$ by $|\dot{N}_\varepsilon|_\infty^2 \int_0^1 c_\varepsilon^2$. $\int_0^1 c_\varepsilon^2$ is upper-bounded by (19) and $|\dot{N}_\varepsilon|_\infty^2$ can be also upper bounded by $\int_0^1 |\ddot{N}_\varepsilon|^2$. Thus

$$\int_0^s \int_0^1 c_\varepsilon^2 \dot{N}_\varepsilon^2 \leq C.$$

Coming back to (78), we then derive that

$$\int_0^s \int_0^1 \left(\frac{\partial b_\varepsilon}{\partial s}\right)^2$$

is bounded as claimed.

Observe that

$$\left|\frac{\partial x_\varepsilon}{\partial s}\right| \leq C \left(\frac{|\dot{N}_\varepsilon| + |\lambda_\varepsilon||\phi_\varepsilon|}{\int_0^1 a_\varepsilon} + |N_\varepsilon| + |\lambda_\varepsilon|\right).$$

The estimate on $\int_0^s \int_0^1 |\frac{\partial x_\varepsilon}{\partial s}|^2$ follows. $\int_0^1 |\dot{x}_\varepsilon|^2$ is upperbounded by $C\left(\int_0^1 a_\varepsilon^2 + \int_0^1 b_\varepsilon^2 + \int_0^1 c_\varepsilon^2\right)$ which is upperbounded by (79)

i) follows.
ii) follows immediately using the fact that $\phi_\varepsilon(b_\varepsilon)$ is bounded in $H^1([o, s_1] \times S^1)$ and that $\phi_\varepsilon(b_\varepsilon) - b_\varepsilon$ tends to zero weakly.

The first statement of *iii)* is immediate. Since $a_\varepsilon - \int_0^1 a_\varepsilon$ tends to zero in L^1 as well as c_ε, it is clear that $\dot{x}(s,t) = a\xi + b(s,t)v$ where $a = a(s)$ is independent of it.

We need therefore only to prove that $x_\varepsilon(s,t)$ converges to $x(s,t)$ strongly in $L^\infty(S^1)$ for any s. We observe, for this, that b_ε is bounded L^2. The curve $\hat{x}_\varepsilon$ defined by the equation:

$$\begin{cases} \dot{\hat{x}}_\varepsilon = \left(\int_0^1 a_\varepsilon\right)(s)\xi(\hat{x}_\varepsilon) + b_\varepsilon v(\hat{x}_\varepsilon) \\ \hat{x}_\varepsilon(s,0) = x_\varepsilon(s,0) \end{cases}$$

is bounded in H^1, almost closed because $\hat{x}_\varepsilon$ and x_ε start at the same point and the L^1-distance (defined locally) of $\dot{\hat{x}}_\varepsilon$ and $\dot{x}_\varepsilon$ in their components (as we ignore that ξ, v are taken at different points for $\dot{\hat{x}}_\varepsilon$ and $\dot{x}_\varepsilon$) tends to zero, since $\int_0^1 |c_\varepsilon|$ tends to zero as well as $\int_0^1 |a_\varepsilon - \int_0^1 a_\varepsilon|$. Thus, when ε tends to zero, $\hat{x}_\varepsilon$, for any s, converges strongly, for any s in $H^1(S^1)$, hence in $L^\infty(S^1)$. Hence x_ε also converges strongly, for any s, in $L^\infty(S^1)$. We have to prove the uniqueness of the sets of limits for each s. For this, we observe that

$$\left|\frac{\partial x_\varepsilon}{\partial s}\right|_\infty \leq C(1 + |\dot{N}|_\infty) \leq C' \left(1 + \left(\int_0^1 |\ddot{N}_\varepsilon|^2\right)^{\frac{1}{2}}\right).$$

Thus

$$\left| \int_{s'}^{s} |\frac{\partial x_\varepsilon}{\partial s}|_\infty \right| \leq C'(|s-s'| + (\int_{s'}^{s} (\int_0^1 (\ddot{N}_\varepsilon)^2)^{\frac{1}{2}})) \tag{83}$$

$$\leq C'' \left(|s-s'| + \sqrt{|s-s'|} \right).$$

Thus

Indeed, $x_\varepsilon(s')$ tends in $L^\infty(S^1)$ to $x_\varepsilon(s)$, when s' tends to s.

Of course, x_ε converges to x in $L^\infty(S^1)$, almost everywhere in s. Combining this convergence a.e. in s with (83), we can construct, by diagonal extraction, a sequence converging uniformly in s, in $L^\infty(S^1)$. It suffices for this, to pick up for $\delta_k = \frac{1}{2^k}$, a finite number of points spread on $[0, s_1]$, spread at a distance at most $\frac{1}{2^{k+1}}$ one from the next one, such that the sequence (x_{ε_j}) converges at these values of s. Combining with (83), (x_{ε_j}) will, on $[0, s_1]$, satisfy the Cauchy sequence requirement with a value of $\varepsilon = 0\left(\frac{1}{2^k}\right)$. Iterating with k and constructing a diagonal sequence, we derive the result.

Observe that the limit curve $a(s,t)$ satisfies (an easy consequence of (83)).

$$\operatorname*{Sup}_{t\in S^1} |x(s,t) - x(s',t)| \leq C'' \left(|s-s'| + \sqrt{|s-s'|} \right).$$

We now prove (iv) and (v) a) and b):

(v)a) is immediate. $\ddot{\phi}_\varepsilon$ converges weakly in $L^2([0,s_1] \times S^1)$. Its limit has to be $\ddot{b}(s,t)$. Using (i), we derive v)b).

(iv) follows from i) if we take into account the fact that, since we have proved that b_ε remains in V, there are only a finite number of α_i which are now zero. For each of them:

$$\left| \alpha_i(b_\varepsilon) - \alpha_i(b_{\varepsilon'}) \right| \leq C(V) \left| b_\varepsilon - b_{\varepsilon'} \right|_{H^1(S^1)}.$$

Since b_ε converges to b, almost everywhere in s, in $H^1(S^1)$ and since $|b_\varepsilon|_{H^1}$ is bounded, $\alpha_i(b_\varepsilon)$ converges to $\alpha_i(b)$ a.e. in s. Since α_i is bounded above by 1, (iv) follows.

Multiplying by a C^∞ function ω of (s,t) the equation on $\frac{\partial b_\varepsilon}{\partial s}$ and integrating on $[0, s_1] \times S^1$, we can use our various convergences after integration by parts. The only slightly bothering term is

$$\int_0^{s_1} \int_0^1 \frac{c_\varepsilon \dot{N}_\varepsilon d\gamma(w,v)\omega}{\int_0^1 a_\varepsilon}$$

which tends to zero because $\int_0^1 |c_\varepsilon|$ tends to zero and $|\dot{N}_\varepsilon|_{L^\infty(S^1)}$ is upperbounded by $|\ddot{N}_\varepsilon|_{L^2(S^1)}$ which is $L^1([0,s_1])$. Thus, we have:

$$\frac{\partial b}{\partial s} = \frac{\overline{\dot{\overline{N}_0 + \lambda_0 b}}}{a} + aN_0\tau - bN_0\overline{\mu}_\xi \qquad \text{a.e. in } (s,t) \text{ and in the distributional sense.}$$

By i) and iii), $\frac{\partial b}{\partial s}$ and $\ddot{\overline{N}}_0$ are in $L^2(S^1)$ a.e. in s. Thus, the identity holds in $L^2(S^1)$ a.e. in s. The proof of Lemma K is thereby complete.

Removal of the restriction on c, which depends, in fact, only on the limit behavior; control on $|b|_\infty, \int_0^1 \dot{b}^2, \int_0^1 \dot{b}^+$ along a flow-line

The evolution equation satisfied by b has, formally, the properties described in Lemmas $3, 3'$ etc.

We prove below that they actually hold.

They imply that c, along the flow-line, should remain lowerbounded for b, not for $\phi_\varepsilon(b)$.

Indeed, since $\phi_\varepsilon(b_\varepsilon)$ tends to b in H^1, a.e. in $s, \phi_\varepsilon(b_\varepsilon)$ will also be controlled in L^∞ a.e. in s, for ε tending to zero and the c related to $\phi_\varepsilon(b_\varepsilon)$ cannot tend to zero if the c related to its limit b does not. Furthermore, $b_\varepsilon = \phi_\varepsilon - \varepsilon\ddot{\phi}_\varepsilon$. $\varepsilon\ddot{\phi}_\varepsilon$ tends to zero in L^2, a.e. in s. Thus, $\int_0^1 b_\varepsilon^2$ is controlled a.e in s, with a control which depends only on the limit b, i.e on $|b|_\infty$. (We also use the fact that both evolution equations, the one on b_ε and the one on b behave well under weak limit. This allows to derive, out of almost everywhere bounds on c etc., uniform estimates). Thus, all the bounds needed for the proofs of Lemmas G,H,K now follow from the limit behavior, plus equations (79)-(81), including the lowerbound on c. This shows that the existence will continue as long as c, in the limit equation, does not tend to zero. The use of the word explosion is thereby justified.

We need now to establish Lemmas $3, 3'$ etc., not formally as we did before, but show that these estimates actually hold. We start with the one on $|b|_\infty$:

We first observe that, by (80):

$$\int_0^1 \dot{\phi}_\varepsilon(b_\varepsilon)^2(s) \le \int_0^1 \dot{b}_\varepsilon\dot{\phi}_\varepsilon(b_\varepsilon)(s) \le -c\int_0^s\int_0^1 \ddot{\phi}_\varepsilon(b_\varepsilon)^2 + \int_0^1 \dot{b}_\varepsilon\dot{\phi}_\varepsilon(b_\varepsilon)(o) + Cs.$$

When ε tends to zero, assuming $b_\varepsilon(o) = b(o)$ is H^1, $\int_0^1 \dot{b}_\varepsilon\dot{\phi}_\varepsilon(b_\varepsilon)(o)$ tends to $\int_0^1 \dot{b}^2(o)$.

Thus, since $\dot{\phi}_\varepsilon(b_\varepsilon)$ tends weakly to $\dot{b}$:

$$\int_0^1 \dot{b}^2(s) \le \int_0^1 \dot{b}^2(o) + Cs.$$

This shows that $b(s)$ tends to $b(o)$ in H^1 when s tends to $o, s > o$.

Let, now,

$$\rho = |b|_\infty(o).$$

Since we only build $\psi_{\theta,\varepsilon}\eta_0$ on oscillations of amplitude Γ to the least, and since $|b|_\infty(s) < |b|_\infty(o) + \frac{\Gamma}{10}$ for $s > o$, s small, all θ involved in the construction of $\psi_{\theta,\varepsilon}\eta_0$ for such b's are less than $|b|_\infty(o) - \frac{\Gamma}{2}$. We compute, for $s > o$ small, $\frac{\partial}{\partial s}\int(|b|-\rho)^{+2}$.

For the moment, we assume that there is no component ω in $\frac{\partial}{\partial s}$. (Observe that $\int(|b|-\rho)^{+2}$ is a continuous function of s).

We then have:

$$\frac{\partial}{\partial s}\int(|b|-\rho)^{+2} = 2\int_0^1(|b|-\rho)^+\left(\frac{\ddot{\overline{N(b)}}+\dot{\overline{\lambda b}}}{a} + aN\tau - bN\bar{\mu}_\xi\right) =$$

$$= -\frac{2c}{a}\int_{|b|>\rho}\dot{b}^2 - \frac{2}{a}\int_{|b|>\rho}\psi'_{\theta,\varepsilon}\dot{b}^2\eta_0 - \frac{2}{a}\int_{|b|>\rho}\dot{b}\dot{\theta}\eta_0 - 2\int_{|b|>\rho}\dot{b}\psi\dot{\eta}_0 +$$

$$-\frac{2}{a}\int_{|b|>\rho}\dot{b}b\lambda + 2\int(|b|-\rho)^+\left(a\psi_{\theta,\varepsilon}\eta_0\tau - b\psi_{\theta,\varepsilon}\eta_0\bar{\mu}_\xi + 2c(ab\tau - b^2\bar{\mu}_\xi).\right)$$

We can assume, for s small, that $\dot{\theta} = 0$ for $|b| > \rho$. Thus, $\psi_{\theta,\varepsilon}$ is then, for $|b| > \rho$, equal to $(|b| - \theta_o)^+$.

On the other hand,

$$-2\int_{|b|>\rho}b\dot{b}\lambda = \int_{|b|>\rho}(b^2-\rho^2)\left(b\eta - \int_0^1 b\eta - \dot{\overline{\bar{\mu}\eta}}\right)) =$$

$$= \int_{|b|>\rho}(b^2-\rho^2)\left(b\eta - \int_0^1 b\eta\right) + 2\int_{|b|>\rho}b\dot{b}\bar{\mu}\eta =$$

$$= \int_{|b|>\rho}(b^2-\rho^2)\left(b\eta - \int_0^1 b\eta\right) - 2\int_{|b|>\rho}\hat{\psi}_{\theta,\varepsilon}\dot{\overline{b\bar{\mu}\eta_0}}$$

$$-2c\int_{|b|>\rho}(a\bar{\mu}_\xi + b\bar{\mu}_v)\left(\frac{b^3-\rho^3}{3}\right).$$

Let

$$\hat{\psi}_{\theta,\varepsilon} = \int_\rho^x \psi_{\theta,\varepsilon}.$$

We then have:

$$-2\int_{|b|>\rho}\dot{b}\psi_{\theta,\varepsilon}\dot{\eta}_o = 2\int_{|b|>\rho}\hat{\psi}_{\theta,\varepsilon}\ddot{\eta}_0.$$

We thus have to estimate:

$$2\int_{|b|>\rho}\frac{\hat{\psi}_{\theta,\varepsilon}\ddot{\eta}_0}{a} + 2\int_{|b|>\rho}\psi_{\theta,\varepsilon}(|b|-\rho)^+\left(\frac{b+\rho}{2a}b\eta_0\right) + a\eta_0\tau - b\eta_0\bar{\mu}_\xi - \frac{2}{a}\int_{|b|>\rho}\hat{\psi}_{\theta,\varepsilon}\dot{\overline{b\bar{\mu}\eta_0}}.$$

Observe that, for C in (10) large enough, we have:

$$C\hat{\psi}_{\theta,\varepsilon} > \psi_{\theta,\varepsilon}(|b|-\rho)^+.$$

Indeed,

$$(C-1)(|b|-\theta_0)^+ > (|b|-(\rho))^+, \text{ since } \theta_0 < \rho.$$

Using, then (10), the quantity which we have to etimate is negative.

We are left with

$$2c\int(|b|-\rho)^+\left(ab\tau - b^2\bar{\mu}_\xi + \frac{(b+\rho)}{2}(b^2-\int_0^1 b^2)-\right.$$
$$\left.-\frac{1}{3}(a\bar{\mu}_\xi + b\bar{\mu}_v)(b^2+b\rho+\rho^2)\right) - \left(\int_0^1(|b|-\rho)^+(b+\rho)\right)\int_0^1 b\eta.$$

Observe that c is less than $-\frac{1}{1+|b|^3_\infty}\int_0^1 b\eta$. We can ask that c be less than $\frac{-\delta}{1+|b|^3_\infty}\int_0^1$, where δ is a small positive constant. The above expression becomes then negative.

Finally, if we introduce ω, we will have a contribution of the type

$$O(\varepsilon_{19})\int(|b|-\rho)^+.$$

If the support of ω is away from the set where $|b|=|b|_\infty$, this contribution is zero. Otherwise, we observe that such a contribution could be absorbed in $2\int_{|b|>\rho}\hat{\psi}_{\theta,\varepsilon}(\ddot{\eta}_0-\overline{b\bar{\mu}\eta_0}^{\cdot})$, if we multiply $\psi_{\theta,\varepsilon}\eta_0$ by M large, depending only on Γ. Indeed, $\hat{\psi}_{\theta,\varepsilon}$ is larger for $|x|>\rho$ than $\frac{\Gamma}{10}(|x|-\rho)^+$.

Thus, $\int_0^1(|b|-\rho)^{+2}$ is, for $s>0$ small, a continuous function with a negative right-derivative, i.e a negative distributional derivative, which is L^1.

The control on $|b|_\infty$ follows.

Let now φ be a C^∞, positive, convex function, with φ'' bounded.

We compute, in the distributional sense

$$\frac{\partial}{\partial s}\int_0^1 \varphi(\tilde{\phi}_\varepsilon(\dot{b}))$$

where $\phi_\varepsilon(f)$ is the solution of

$$-\varepsilon\ddot{\phi}_\varepsilon(f)+\phi_\varepsilon(f)=f \qquad f\in L^2(S^1)$$

and $\tilde{\phi}_\varepsilon = \phi_\varepsilon\circ\phi_\varepsilon\circ\phi_\varepsilon\circ\phi_\varepsilon$. It is easy to prove that $\tilde{\phi}_\varepsilon(\dot{b})$ is C^1 in s and t.

We have

$$\frac{\partial}{\partial s}\int_0^1 \varphi(\tilde{\phi}_\varepsilon(\dot{b})) = \int_0^1 \varphi'(\tilde{\phi}_\varepsilon(\dot{b}))\frac{\partial}{\partial s}\dot{\tilde{\phi}}_\varepsilon(b) =$$

$$= \int_0^1 \varphi'(\tilde{\phi}_\varepsilon(\dot{b}))\dot{\tilde{\phi}}_\varepsilon\left(\frac{\partial b}{\partial s}\right) =$$

$$= \int_0^1 \varphi'(\tilde{\phi}_\varepsilon(\dot{b}))\tilde{\phi}_\varepsilon\overline{\left(\frac{\ddot{N}+\dot{\overline{\lambda b}}}{a} + aN\tau - bN\bar{\mu}_\xi\right)}^{\,\cdot} =$$

$$= -\frac{1}{a}\int_0^1 \varphi''(\tilde{\phi}_\varepsilon(\dot{b}))\ddot{\tilde{\phi}}_\varepsilon(b)\left(c\ddot{\tilde{\phi}}_\varepsilon(b) + \tilde{\phi}_\varepsilon\left(\left(\frac{\partial\psi_{\theta,\varepsilon}}{\partial x}\ddot{b} + \frac{\partial^2\psi_{\theta,\varepsilon}}{\partial x^2}\dot{b}^2\right)\eta_0 + 2\frac{\partial\psi_{\theta,\varepsilon}}{\partial x}\dot{b}\dot{\eta}_0\right)\right)$$

$$+ \int_0^1 \varphi'\left(\tilde{\phi}_\varepsilon(\dot{b})\right)\tilde{\phi}_\varepsilon\overline{\left(\frac{2\dot{\theta}\dot{\eta}_0 + \ddot{\theta}\eta_0}{a} + \frac{\psi_{\theta,\varepsilon}\ddot{\eta}_0}{a} + \frac{\dot{\overline{\lambda b}}}{a} + aN\tau - bN\bar{\mu}_\xi\right)}^{\,\cdot}.$$

Observe that:

$$-c\int_0^1 \varphi''(\tilde{\phi}_\varepsilon(\dot{b}))\ddot{\tilde{\phi}}_\varepsilon^{\,2}(b) < o$$

$$-\int_0^1 \varphi''(\tilde{\phi}_\varepsilon(\dot{b}))\ddot{\tilde{\phi}}_\varepsilon(b)\tilde{\phi}_\varepsilon\left(\left(\frac{\partial\psi_{\theta,\varepsilon}}{\partial x}\ddot{b} + \frac{\partial^2\psi_{\theta,\varepsilon}}{\partial x^2}\dot{b}^2\right)\eta_0 + \right.$$

$$\left. +2\frac{\partial\psi_{\theta,\varepsilon}}{\partial x}\dot{b}\dot{\eta}_0\right) \text{ converges in } L^1([o,s_1]) \text{ to}$$

$$-\int_0^1 \varphi''(\dot{b})\ddot{b}\left(\left(\frac{\partial\psi_{\theta,\varepsilon}}{\partial x}\ddot{b} + \frac{\partial^2\psi_{\theta,\varepsilon}}{\partial x^2}\dot{b}^2\right)\eta_0 + 2\frac{\partial\psi_{\theta,\varepsilon}}{\partial x}\dot{b}\dot{\eta}_0\right).$$

Indeed, $\ddot{\tilde{\phi}}_\varepsilon(b)$ converges to $\ddot{b}$ weakly in $L^2([o,s_1]\times S^1)$ and $\int_0^1 \ddot{\tilde{\phi}}_\varepsilon^2(b)$ is less than $\int_0^1 \ddot{b}^2$. Thus, $\ddot{\tilde{\phi}}_\varepsilon(b)$ converges strongly to $\ddot{b}$ in $L^2([o,s_1]\times S^1)$. Also $\tilde{\phi}_\varepsilon\left(\left(\frac{\partial\psi_{\theta,\varepsilon}}{\partial x}\ddot{b} + \frac{\partial^2\psi_{\theta,\varepsilon}}{\partial x^2}\dot{b}^2\right)\eta_0 + 2\frac{\partial\psi_{\theta,\varepsilon}}{\partial x}\dot{b}\dot{\eta}_0\right)$ converges strongly to $\left(\frac{\partial\psi_{\theta,\varepsilon}}{\partial x}\ddot{b} + \frac{\partial^2\psi_{\theta,\varepsilon}}{\partial x^2}\dot{b}^2\right)\eta_0 + 2\frac{\partial\psi_{\theta,\varepsilon}}{\partial x}\dot{b}\dot{\eta}_0$.

Finally, $\tilde{\phi}_\varepsilon(\dot{b})$ converges to $\dot{b}$ a.e and φ'' is bounded above.

It is also clear that

$$\int_0^1 \varphi'(\tilde{\phi}_\varepsilon(\dot{b}))\,\tilde{\phi}_\varepsilon\overline{\left(\frac{2\dot{\theta}\dot{\eta}_0 + \ddot{\theta}\eta_0}{a} + \frac{\psi_{\theta,\varepsilon}\ddot{\eta}_0}{a} + \frac{\dot{\overline{\lambda b}}}{a} + aN\tau - bN\bar{\mu}_\xi\right)}^{\,\cdot}$$

converges.

We thus have:

$$\lim_{\varepsilon\to 0}\int_0^1 \varphi(\dot{\tilde{\phi}}_\varepsilon(b))(s) = \int_0^1 \varphi(\dot{b})(o) -$$

$$-\lim_{\varepsilon\to 0}\left(\int_0^s \frac{c}{a}\int_0^1 \varphi''(\dot{b})\ddot{\tilde{\phi}}_\varepsilon(b)^2 + \int_0^s\int_0^1 \varphi''(\dot{b})\frac{\partial\psi_{\theta,\varepsilon}}{\partial x}\ddot{\tilde{\phi}}_\varepsilon(b)^2\right)$$

$$+\int_0^s\int_0^1 \varphi'(\dot{b})\left(\frac{1}{a}\left(\overline{\frac{\partial^2\psi_{\theta,\varepsilon}}{\partial x^2}\dot{b}^2\eta_0}^{\,\cdot} + 2\overline{\frac{\partial\psi_{\theta,\varepsilon}}{\partial x}\dot{b}\dot{\eta}_0}^{\,\cdot} + \overline{2\dot{\theta}\dot{\eta}+\psi\ddot{\eta}_0}^{\,\cdot}\right)+\right.$$

$$\left.+\frac{\ddot{\overline{\lambda b}}}{a} + \overline{aN\tau - bN\bar{\mu}_\xi}^{\,\cdot}\right).$$

Using $\varphi = |x|^2$, we find the inequality of Lemma 3 after the usual tricks.

Using φ equal to a regularization of x^+, we derive (observe that $\dot{\tilde{\phi}}_\varepsilon(b)$ converges to $\dot{b}$ in L^2, for every s):

$$\int_0^1 \dot{b}^+(s) \le \int_0^1 \dot{b}^+(o) + \int_0^s\int_0^1 1_{\dot{b}>0}\dot{\Gamma}$$

where Γ is the natural expression.

Now, the computations of Corollary E hold a.e in s as is shown in Appendix 4. Thus, $\int_0^s\int_0^1 1_{\dot{b}>0}\dot{\Gamma}$ is equal to the expected expression.

The inequality holds a.e in s. Since $\int_0^1 \dot{b}^+(s)$ is continuous to the right, it holds everywhere. The introduction of ω does not change the basic estimates. The result follows. The computations for $\frac{\partial}{\partial s}\int(|b|-\nu)^+$ follow the same type of arguments. q.e.d.

Proof of Lemma L. We consider two distinct solutions $b, \tilde{b}$ of the evolution equation. They are both bounded in $H^1(S^1)$ for every s and they both satisfy $\int_0^s\int_0^1 |\ddot{b}|^2 \le C$. Since the equation holds in L^2, a.e. in s, we compute:

$$\int_0^1 \frac{\partial}{\partial s}(b-\tilde{b})(b-\tilde{b}) = \frac{1}{2}\frac{\partial}{\partial s}\int_0^1 (b-\tilde{b})^2 \tag{84}$$

$$= \int_0^1 \ddot{\left(\frac{1}{a}N_0(b) - \frac{1}{\tilde{a}}N_0(\tilde{b})\right)}(b-\tilde{b}) + \int_0^1 \overline{\left(\frac{\lambda_0 b}{a} - \frac{\tilde{\lambda}_0\tilde{b}}{\tilde{a}}\right)}^{\,\cdot}(b-\tilde{b}) +$$

$$+\int_0^1 \left(aN_0(b)\tau - bN_0(b)\bar{\mu}_\xi(x) - \tilde{a}N_0(\tilde{b})\tilde{\tau} + \tilde{b}N_0(\tilde{b})\bar{\mu}_\xi(\tilde{x})(b-\tilde{b})\right).$$

Observe that a satisfies

$$\frac{\partial a}{\partial s} = -\int_0^1 bN_0(b).$$

Thus,

$$|\frac{\partial}{\partial s}(a-\tilde{a})| \leq \int_0^1 |bN_0(b)-\tilde{b}N_0(\tilde{b})|.$$

We can assume that b and $\tilde{b}$ are both in V.
Then,

$$|bN_0(b)-\tilde{b}N_0(\tilde{b})| \leq C|b-\tilde{b}|.$$

Thus,

$$|a-\tilde{a}|(s) \leq |a-\tilde{a}|(0) + C\int_0^s \left(\int_0^1 |b-\tilde{b}|^2\right)^{1/2}.$$

Since $b, \tilde{b}, N_0(b), N_0(\tilde{b})$ are bounded H^1, the terms in $(a-\tilde{a})(b-\tilde{b})$ are good terms in (84). Also $(\tau-\tilde{\tau})(b-\tilde{b})$ and $(\bar{\mu}_\xi(x)-\bar{\mu}_\xi(\tilde{x}))(b-\tilde{b})$ depend on $d(x,\tilde{x})$, hence are also good terms and so are the terms involving $\int (N_0(b)-N_0(\tilde{b}))(b-\tilde{b})$.

Observe that every term of the type $0\left(\int |b-\tilde{b}| 0(\psi_{\theta,\varepsilon}(b)-\psi_{\tilde{\theta},\tilde{\varepsilon}}(\tilde{b}))\right)$ is $0\left(\int_0^1 |b-\tilde{b}|\,|b-\tilde{b}|_{H^1}\right) \leq \frac{c}{1000}|b-\tilde{b}|^2_{H^1} + \frac{C}{c}|b-\tilde{b}|^2_{L^2}$, because $|\psi_{\theta,\varepsilon}(\tilde{b})-\psi_{\tilde{\theta},\tilde{\varepsilon}}(\tilde{b})| \leq C(b)(|\theta-\tilde{\theta}|+|\varepsilon-\tilde{\varepsilon}) \leq C'|b-\tilde{b}|_{H^1}$. All these terms are therefore good terms. As pointed out above $|b-\tilde{b}|_{H^1}$ could be replaced by $|\dot{b}-\dot{\tilde{b}}|_{L^0}$ here.

Observe that: (see (59) - (61))

$$\tilde{\eta}_0 - \eta_0 = \eta_0\left(e^{\int_{t_1^-}^t -\bar{\mu}\tilde{b} - \int_{t_1^-}^t \bar{\mu}b} - 1\right), \tag{85}$$

$$\dot{\overline{\tilde{\eta}_0 - \eta_0}} = \dot{\eta}_0\left(e^{\int_{t_1^-}^t \bar{\mu}\tilde{b} - \int_{t_1^-}^t \bar{\mu}b} - 1\right) \tag{86}$$

$$+\eta_0\left(\bar{\mu}\tilde{b} - \bar{\mu}b\right)e^{\int_{t_1^-}^t \bar{\mu}\tilde{b} - \int_{t_1^-}^t \bar{\mu}b},$$

$$\ddot{\overline{\tilde{\eta}_0 - \eta_0}} = \ddot{\eta}_0\left(e^{\int_{t_1^-}^t \bar{\mu}\tilde{b} - \int_{t_1^-}^t \bar{\mu}b} - 1\right) + \tag{87}$$

$$+2\dot{\eta}_0(\bar{\mu}\tilde{b} - \bar{\mu}b)e^{\int_{t_1^-}^t \bar{\mu}\tilde{b} - \int_{t_1^-}^t \bar{\mu}b} +$$

$$+\eta_0\left((\bar{\mu}\tilde{b} - \bar{\mu}b)^2 e^{\int_{t_1^-}^t \bar{\mu}\tilde{b} - \bar{\mu}b} + \dot{\overline{\bar{\mu}\tilde{b}}} - \dot{\overline{\bar{\mu}b}}\right).$$

All these expressions are multiplied, in the terms we have to estimate, by $(b-\tilde{b})\zeta$ where ζ is a suitable function and integrated over $[0,1]$. The presence of $\ddot{\eta}_0$ in (87), which presents Diracs at t_1^-, t_2^- is not a problem. $\ddot{\eta}_0$ is multiplied by $\psi_{\theta,\varepsilon}(b)\zeta_1$ or $\psi_{\tilde{\theta},\tilde{\varepsilon}}(\tilde{b})\tilde{\zeta}_1$. We have seen that this yields a contribution only on $[t_1^-, t_2^-]$, where $\ddot{\eta}_0$ is bounded.

All terms of the type $\int(|\ddot{b}|+|\ddot{\tilde{b}}|)|\tilde{\eta}_0-\eta_0||b-\tilde{b}|$ are

$$0((\int(|\ddot{b}|+|\ddot{\tilde{b}}|)|b-\tilde{b}|)\int_0^1|\bar{\mu}\tilde{b}-\bar{\mu}b|)=0\left((1+\int\ddot{b}^2+\int\ddot{\tilde{b}}^2)^{1/2}\times\right.$$

$$\left.\times\left(\int_0^1|b-\tilde{b}|^2\right)^{1/2}\int_)^1|\bar{\mu}\tilde{b}-\bar{\mu}b|\right).$$

It is the only term where $\tilde{\eta}_0-\eta_0$ appears (or $\dot{\overline{\tilde{\eta}_0-\eta_0}}$ (or $\ddot{\overline{\tilde{\eta}_0-\eta_0}}$) with second derivatives on b or $\tilde{b}$ multiplying it.

We encounter expressions of the type below and show how to estimate them appropriately, quite easily. The only new fact here, which we encounter in other expressions, is due to the factor $\bar{\mu}$. We were careful not to factorize $\bar{\mu}$ because $\bar{\mu}b$ and $\bar{\mu}\tilde{b}$ are the $\bar{\mu}$'s related to two different curves, i.e., the first one is $\bar{\mu}(x(s,t))$. The second one is $\bar{\mu}(\tilde{x}(s,t))$. The same phenomenon appears for $\dot{\bar{\mu}}$, also in all expressions involving $\bar{\mu}_\xi$ or τ. We compute this contribution as follows:

For any function $\omega=\pi^3\to\mathbb{R}$, we have:

$$|\omega(x(s,t))-\omega(\tilde{x}(s,t))|\leq C_\omega|x(s,t)-\tilde{x}(s,t)|\leq \tag{88}$$

$$\leq C_\omega(\int_0^1|x(s,z)-\tilde{x}(s,z)|dz+\int_0^1|a-\tilde{a}|+|b-\tilde{b}|)\text{ for any }t.$$

On the other hand, since $\frac{\partial x}{\partial s}=\lambda\xi+\frac{\dot{\eta}+\lambda b}{a}v+\eta w$, with $\eta=\sum\alpha_i\eta_{12,i}+cb$, using our bounds on $b,\tilde{b},\dot{b},\dot{\tilde{b}}$. (For example, $\int_0^1|\frac{\dot{\eta}+\lambda b}{a}||v(x)-v(\tilde{x})|\leq C(1+|\ddot{b}|_{L^2})(\int_0^1|x-\tilde{x}|)$.)

$$\frac{\partial}{\partial s}\int_0^1|x(s,z)-\tilde{x}(s,z)|dz\leq C\left(\int_0^1|b-\tilde{b}|+|a-\tilde{a}|+\right. \tag{89}$$

$$\left.+(1+|\ddot{b}|_{L^2})\int_0^1|x(s,z)-\tilde{x}(s,z)|dz+\int_0^1|\dot{b}-\dot{\tilde{b}}|\right).$$

In order to obtain (89), there is a small difference with the arguments below in the way we handle $\theta,\varepsilon,\tilde{\theta},\tilde{\varepsilon}$. Namely the difference $\dot{\overline{\psi_{\theta,\varepsilon}(b)-\psi_{\tilde{\theta},\tilde{\varepsilon}}}}=\dot{\tilde{\theta}}-\dot{\theta}+\frac{\partial\psi_{\theta,\varepsilon}}{\partial b}\dot{b}-\frac{\partial\psi_{\tilde{\theta},\tilde{\varepsilon}}}{\partial\tilde{b}}\dot{\tilde{b}}=\dot{\tilde{\theta}}-\dot{\theta}-+\left(\frac{\partial\psi_{\theta,\varepsilon}}{\partial b}-\frac{\partial\psi_{\tilde{\theta},\tilde{\varepsilon}}}{\partial\tilde{b}}\right)\dot{b}+0(\dot{b}-\dot{\tilde{b}})$ is studied below.

A slight difference which we use here is the following: we have as below:

$$\left|\frac{\partial\psi_{\theta,\varepsilon}}{\partial b}-\frac{\partial\psi_{\tilde{\theta},\tilde{\varepsilon}}}{\partial\tilde{b}}\right|\leq C|b-\tilde{b}|+C(|\theta-\tilde{\theta}|+|\varepsilon-\tilde{\varepsilon}|),$$

where C is related to the partition of unity where $\psi_{\theta,\varepsilon}$ is built.

Below, $|\theta-\tilde\theta|+|\varepsilon-\tilde\varepsilon|$ is upperbounded by $|b-\tilde b|_{H^1}$.

Here we use $C\int_0^1|\dot b-\dot{\tilde b}|$, i.e. a $W^{1,1}$ proximity which is enough for the construction of $\tilde\theta,\tilde\varepsilon$, since we only need for these that b and $\tilde b$ are close L^∞.

Coming back to (89), we derive that (we use the fact that $\int_0^1(1+|\ddot b|_{L^2})<+\infty$:

$$\int_0^1|x(s,z)-\tilde x(s,z)|dz\le C\left(\int_0^s\int_0^1|b-\tilde b|+\int_0^s\int_0^1(a-\tilde a)+\right. \tag{90}$$

$$\left.+\int_0^s\int_0^1|\dot b-\dot{\tilde b}|\right)+\int_0^1|x(0,z)-\tilde x(0,z)|dz.$$

Thus, combining with (88):

$$|\omega(x(s,t))-\omega(\tilde x(s,t))|_\infty\le \tag{91}$$

$$\le C''_\omega\left(\int_0^s\int_0^1|b-\tilde b|+\int_0^s\int_0^1|a-\tilde a|+\int_0^s\int_0^1|\dot b-\dot{\tilde b}|+\right.$$

$$\left.+\int_0^1|a-\tilde a|+\int_0^1|b-\tilde b|+\int_0^1|x(0,z)-\tilde x(0,z)|\right).$$

Thus, for example, $\int|\ddot b|+|\ddot{\tilde b}|\,|\tilde\eta_0-\eta_0|\,|b-\tilde b|$ which we estimated by $0\left(\left(1+\int\ddot b^2+\int\ddot{\tilde b}\right)^{1/2}\left(\int|b-\tilde b|^2\right)^{1/2}\int_0^1|\bar\mu\tilde b-\bar\mu b|\right)$, with different $\bar\mu$'s, becomes:

$$0\left(\left(1+\int\ddot b^2+\int\ddot{\tilde b}^2\right)^{1/2}\left(\int|b-\tilde b|^2\right)^{1/2}\left(\left(\int(b-\tilde b)^2\right)^{1/2}+\left(\int_0^1|a-\tilde a|\right)\right)+\right.$$

$$\left.+\left(\int_0^s\int_0^1|b-\tilde b|^2\right)^{1/2}+\left(\int_0^s\int_0^1|\dot b-\dot{\tilde b}|^2\right)^{1/2}+\int_0^1|x(0,z)-\tilde x(0,z)|\right)$$

$$=0\left(\left(1+\int\ddot b^2+\int\ddot{\tilde b}^2\right)\int|b-\tilde b|^2+\right.$$

$$+\int|b-\tilde b|^2+\left(\int_0^1|a-\tilde a|\right)^2+\int_0^s\int_0^1|b-\tilde b|^2+$$

$$\left.+\int_0^s\int_0^1|\dot b-\dot{\tilde b}|^2+\left(\int_0^1(x(0,z)-\tilde x(0,z)|\right)^2\right).$$

To the usual expressions which we obtain, below, we add $\int_0^s \int_0^1 |\dot{b} - \dot{\tilde{b}}|^2$. This is the worst case involving such differences.

We are thus left with $\ddot{\overline{N_0(b) - N_0(\tilde{b})}} + \int_0^1 \dot{\overline{(\lambda_0 b - \tilde{\lambda}_0 \tilde{b})}}$ where we do not count anymore the terms involving $\eta_0 - \tilde{\eta}_0, \dot{\eta}_0 - \dot{\tilde{\eta}}_0, \ddot{\overline{\eta_0 - \tilde{\eta}_0}}$.

In $\int_0^1 \ddot{\overline{N_0(b) - N_0(\tilde{b})}}(b - \tilde{b})$, there is a $-c\int_0^1 |\dot{\overline{b - \tilde{b}}}|^2$. This allows to upperbound $-\int_0^1 (\lambda_0 b - \tilde{\lambda}_0 \tilde{b})\dot{\overline{(b - \tilde{b})}}$ by $\frac{c}{10}\int_0^1 \left(\dot{\overline{b - \tilde{b}}}\right)^2 + \frac{10}{c}\int_0^1 \left(\lambda_0 b - \tilde{\lambda}_0 \tilde{b}\right)^2$.

The term $\frac{c}{10}\int_0^1 \left(\dot{\overline{b - \tilde{b}}}\right)^2$ is absorbed in $-c\int_0^1 |\dot{\overline{b - \tilde{b}}}|^2$. The term $\int_0^1 (\lambda_0 b - \tilde{\lambda}\tilde{b})^2$ is a good term.

The partition of unity introduces coefficients α_i in N_0.

They contribute in the above terms with terms of the type:

$$R_i = \left|\alpha_i(b) - \alpha_i(\tilde{b})\right| \int_0^1 \left(\left|\ddot{\overline{\psi_{\theta_i,\varepsilon_i}(b)\eta_0^i}}\right| + \left|\ddot{\overline{\psi_{\tilde{\theta}_i,\tilde{\varepsilon}_i}(\tilde{b})\eta_0^i}}\right| \right) |b - \tilde{b}| \tag{92}$$

and other terms of the same type.

The i's to consider are in finite number, since we can assume we are in V. Then (we discussed in the proof of Lemmas 3 (iii) why the Diracs of $\ddot{\eta}_0^i$ do not matter)

$$\begin{aligned} &\left|\ddot{\overline{\psi_{\theta_i,\varepsilon_i}(b)\eta_o^i}}\right| + \left|\ddot{\overline{\psi_{\tilde{\theta}_i,\tilde{\varepsilon}_i}(\tilde{b})\eta_0^i}}\right| \leq \\ &\leq C_i \left(|\ddot{b}| + |\ddot{\tilde{b}}| + |\dot{b}|^2 + |\dot{\tilde{b}}|^2 + 1 \right). \end{aligned}$$

Thus, using the bounds on $b, \tilde{b}$ in H^1 and the facts that there are at most N_i coefficients α_i involved:

$$\begin{aligned} |R_i| &\leq \tilde{C}_i |b - \tilde{b}|_{H^1} \int_0^1 |b - \tilde{b}|(|\ddot{b}| + |\ddot{\tilde{b}}| + 1 + |\dot{b}|^2 + |\dot{\tilde{b}}|^2) \leq \\ &\leq \tilde{C}_i |b - \tilde{b}|_{H^1} |b - \tilde{b}|_{L^2} \left(1 + \left(\int_0^1 |\ddot{b}|^2 \right)^{1/2} + \left(\int_0^1 |\ddot{\tilde{b}}|^2 \right)^{1/2} \right) \leq \\ &\leq \frac{c}{1000 N_i} |b - \tilde{b}|_{H^1}^2 + \tilde{\tilde{C}}_i \frac{1000 N_i}{c} |b - \tilde{b}|_{L^2}^2 \left(1 + \int_0^1 |\ddot{b}|^2 + \int_0^1 |\ddot{\tilde{b}}|^2 \right). \end{aligned} \tag{93}$$

We can then absorb $\frac{c}{1000N_i}|b - \tilde{b}|_{H^1}^2$ in $-c\int_0^1 |\dot{\overline{|b - \tilde{b}}}|^2$. The term which is left is a good term. The coefficients of the partition of unity can be tackled. We are left with the contribution of $\psi_{\theta_i,\varepsilon_i}(b) sgnb\,\eta_0$ for those i's for which $\alpha_i(\tilde b)$ and $\alpha_i(b)$ are non zero. We consider these terms below.

They are of the type, (with $\alpha_i(\tilde b)$ and $\alpha_i(b)$ non zero)

$$\int_0^1 \ddot{\overline{\psi_{\theta,\varepsilon}(b)\eta_0\, sgnb - \psi_{\tilde\theta,\tilde\varepsilon}(\tilde b)\eta_0\, sgn\tilde b}}(b-\tilde b).$$

$sgnb$ and $sgn\tilde b$ are the same along the oscillation (i.e on the support of $\psi_{\theta,\varepsilon}$). We can assume they are both positive. Then, $\int_0^1(\psi_{\theta,\varepsilon}(b)-\psi_{\tilde\theta,\tilde\varepsilon}(\tilde b))\ddot\eta_0(b-\tilde b)$ is a good term in particular because $\tilde\theta,\tilde\varepsilon$ can be choosen in C^2 to depend in a Lipshitz way on $b\in L^\infty$ i.e. $|\theta-\tilde\theta|_{C^2}+|\varepsilon-\tilde\varepsilon|_{C^2}\le C|b-\tilde b|_\infty$ where $C=C_i$ depends on α_i.

Thus, $\int_0^1\left(\psi_{\theta,\varepsilon}(b)-\psi_{\tilde\theta,\tilde\varepsilon}(\tilde b)\right)\ddot\eta_0(b-\tilde b) = 0(|b-\tilde b|_\infty|b-\tilde b|_2) =$
$0\left(\frac{c}{1000}|b-\tilde b|_{H^1}^2+\frac{C}{c}\int(b-\tilde b)^2\right)$.

The term $\int_0^1 \dot{\overline{\psi_{\theta,\varepsilon}(b)-\psi_{\tilde\theta,\tilde\varepsilon}(\tilde b)}}\dot\eta_0(b-\tilde b)$ is equal to

$$\int_0^1\left(\frac{\partial\psi_{\theta,\varepsilon}}{\partial x}(b)\dot b-\frac{\partial\psi_{\tilde\theta,\tilde\varepsilon}}{\partial x}(\tilde b)\dot{\tilde b}+\dot{\tilde\theta}-\dot\theta\right)\dot\eta_0(b-\tilde b).\ \text{Since}$$

$$\frac{\partial\psi}{\partial\theta}\dot\theta+\frac{\partial\psi}{\partial\varepsilon}\dot\eta=-\dot\theta;\ \frac{\partial\psi_{\tilde\theta,\tilde\varepsilon}}{\partial\theta}\dot{\tilde\theta}+\frac{\partial\psi_{\tilde\theta,\tilde\varepsilon}}{\partial\tilde\varepsilon}\dot{\tilde\varepsilon}=-\dot{\tilde\theta}\ \text{we split it into}$$

$$\int_0^1\left(\frac{\partial\psi_{\theta,\varepsilon}}{\partial x}(b)-\frac{\partial\psi_{\theta,\varepsilon}}{\partial x}(\tilde b)\right)\dot{\tilde b}\dot\eta_0(b-\tilde b)+\int_0^1\left(\frac{\partial\psi_{\theta,\varepsilon}}{\partial x}(b)\right)\dot{\overline{b-\tilde b}}\dot\eta_0(b-\tilde b)+$$

$$+\int_0^1\left(\dot{\tilde\theta}-\dot\theta\right)\dot\eta_0(b-\tilde b)+\int_0^1\left(\frac{\partial\psi_{\theta,\varepsilon}}{\partial x}(\tilde b)-\frac{\partial\psi_{\tilde\theta,\tilde\varepsilon}}{\partial x}(\tilde b)\right)\dot{\tilde b}\dot\eta_0(b-\tilde b).$$

The third term can be upperbounded by $\frac{c}{100}\int_0^1(\dot{\overline{b-\tilde b}})^2+\frac{C}{c}\int_0^1(b-\tilde b)^2$.

The second term can be handled as $\int_0^1(\lambda_0 b-\tilde\lambda_0\tilde b)\dot{\overline{(b-\tilde b)}}$.

The first term is bounded by

$$C\int_0^1|\dot{\tilde b}|\,|b-\tilde b|^2\le C|\ddot{\tilde b}|_{L^2}\int_0^1|b-\tilde b|^2\le$$

$$\le C\left(1+|\ddot b|_{L^2}^2\right)\int_0^1|b-\tilde b|^2.$$

The last term is, using the estimates on $\int_0^1 \dot{\dot{b}}^2$ and on $|\theta - \tilde{\theta} + |\varepsilon - \tilde{\varepsilon}|, 0\left(\left(\int_0^1 (b-\tilde{b})^2\right)^{1/2} |b - \tilde{b}|_\infty\right) \leq \frac{c}{1000}\int_0^1 |\overline{\dot{b - \tilde{b}}}|^2 + \frac{C}{c}\int_0^1 (b-\tilde{b})^2$.

Finally, we are left with:

$$\int_0^1 \overline{\psi_{\theta,\varepsilon}(b) - \psi_{\tilde{\theta},\tilde{\varepsilon}}(\tilde{b})}^{\cdot\cdot}\eta_0(b-\tilde{b}).$$

Observe that $-\dot{\theta} = \frac{\partial\psi}{\partial\theta}\dot{\theta} + \frac{\partial\psi}{\partial\varepsilon}\dot{\varepsilon}$, by ($\gamma$). Thus, $\overline{\psi_{\theta,\varepsilon}(b)}^{\cdot} = \frac{\partial\psi_{\theta,\varepsilon}}{\partial x}\dot{b} - \dot{\theta}, \overline{\psi_{\theta,\varepsilon}(b)}^{\cdot\cdot} = -\ddot{\theta} + \frac{\partial^2\psi_{\theta,\varepsilon}}{\partial x^2}\dot{b}^2 + \frac{\partial\psi_{\theta,\varepsilon}}{\partial x}\ddot{b} + \frac{\partial}{\partial x}(-\dot{\theta})\dot{b} = -\ddot{\theta} + \frac{\partial^2\psi_{\theta,\varepsilon}}{\partial x^2}\dot{b}^2 + \frac{\partial\psi_{\theta,\varepsilon}}{\partial x}\ddot{b}$.

Hence,

$$\int_0^1 \overline{\psi_{\theta,\varepsilon}(b) - \psi_{\tilde{\theta},\tilde{\varepsilon}}(\tilde{b})}^{\cdot\cdot}\eta_0(b-\tilde{b}) = \int_0^1 \left(\frac{\partial^2\psi_{\theta,\varepsilon}}{\partial x^2}(b)\dot{b}^2 - \frac{\partial^2\psi_{\tilde{\theta},\tilde{\varepsilon}}}{\partial x^2}(\tilde{b})\dot{\tilde{b}}^2 + \ddot{\tilde{\theta}} - \ddot{\theta} + \right. \tag{94}$$

$$\left. + \frac{\partial\psi_{\theta,\varepsilon}}{\partial x}(b)\ddot{b} - \frac{\partial\psi_{\tilde{\theta},\tilde{\varepsilon}}}{\partial x}(\tilde{b})\ddot{\tilde{b}}\right)\eta_0(b-\tilde{b}).$$

The term $\int_0^1 (\ddot{\tilde{\theta}} - \ddot{\theta})\eta_0(b-\tilde{b})$ is easily handled.

Replacing $\frac{\partial^2\psi_{\tilde{\theta},\tilde{\varepsilon}}}{\partial x^2}(\tilde{b})$ and $\frac{\partial\psi_{\tilde{\theta},\tilde{\varepsilon}}}{\partial x}(\tilde{b})$ by $\frac{\partial^2\psi_{\theta,\varepsilon}}{\partial x^2}(\tilde{b})$ and $\frac{\partial\psi_{\theta,\varepsilon}}{\partial x}(\tilde{b})$ yield a term of the type:

$$\int_0^1 \left(\frac{\partial^2\psi_{\tilde{\theta},\tilde{\varepsilon}}}{\partial x^2}(\tilde{b}) - \frac{\partial^2\psi_{\theta,\varepsilon}}{\partial x^2}(\tilde{b})\right)\dot{\tilde{b}}^2\eta_0(b-\tilde{b}) +$$

$$+ \int_0^1 \left(\frac{\partial\psi_{\tilde{\theta},\tilde{\varepsilon}}}{\partial x}(\tilde{b}) - \frac{\partial\psi_{\theta,\varepsilon}}{\partial x}(\tilde{b})\right)\ddot{\tilde{b}}\eta_0(b-\tilde{b}).$$

Integration by parts yields:

$$-\int_0^1 \left(\frac{\partial\psi_{\tilde{\theta},\tilde{\varepsilon}}}{\partial x}(\tilde{b}) - \frac{\partial\psi_{\theta,\varepsilon}}{\partial x}(\tilde{b})\right)\dot{\tilde{b}}\overline{\eta_0(b-\tilde{b})}^{\cdot} -$$

$$-\int_0^1 \dot{\tilde{b}}\eta_0(b-\tilde{b})\left(\frac{\partial^2\psi_{\tilde{\theta},\tilde{\varepsilon}}}{\partial x\partial\tilde{\theta}}\dot{\tilde{\theta}} + \frac{\partial^2\psi_{\tilde{\theta},\tilde{\varepsilon}}}{\partial x\partial\tilde{\varepsilon}}\dot{\tilde{\varepsilon}} - \frac{\partial^2\psi_{\theta,\varepsilon}}{\partial x\partial\theta}\dot{\theta} - \frac{\partial^2\psi_{\theta,\varepsilon}}{\partial x\partial\varepsilon}\dot{\varepsilon}\right).$$

Since $|\tilde{\theta} - \theta|_{C^2} + |\tilde{\varepsilon} - \varepsilon|_{C^2}$ is upperbounded by $C|b - \tilde{b}|_{H^1}$, the second term can be tackled as other terms above.

(Also, we use $\left(\left|\frac{\partial^2\psi_{\tilde\theta,\tilde\varepsilon}}{\partial x\partial\tilde\theta} - \frac{\partial^2\psi_{\theta,\varepsilon}}{\partial x\partial\theta}\right| + \left|\frac{\partial^2\psi_{\tilde\theta,\tilde\varepsilon}}{\partial x\partial\tilde\varepsilon} - \frac{\partial^2\psi_{\theta,\varepsilon}}{\partial x\partial\varepsilon}\right|\right)(\tilde b) \le C_i(|\tilde\theta-\theta| + |\tilde\varepsilon-\varepsilon|))$

In the first term, $\int_0^1 \left(\frac{\partial\psi_{\tilde\theta,\tilde\varepsilon}}{\partial x}(\tilde b) - \frac{\partial\psi_{\theta,\varepsilon}}{\partial x}(\tilde b)\right)\dot{\tilde b}\eta_0(b-\tilde b)$ has been upperbounded above. We are left with

$$\int_0^1 \left(\frac{\partial\psi_{\tilde\theta,\tilde\varepsilon}}{\partial x}(\tilde b) - \frac{\partial\psi_{\theta,\varepsilon}}{\partial x}(\tilde b)\right)\dot{\tilde b}\eta_0\overline{(b-\tilde b)}^{\cdot}.$$

Using the estimates (ψ), we have:

$$\left|\frac{\partial\psi\tilde\theta,\tilde\varepsilon}{\partial x}(\tilde b) - \frac{\partial\psi_{\theta,\varepsilon}}{\partial x}(\tilde b)\right| \le \mathrm{Sup}_{\substack{y\in[\tilde\theta,\theta]\\ z\in[\tilde\varepsilon,\varepsilon]}}\left[\left|\frac{\partial^2\psi_{y,z}}{\partial x\partial y}\right||\tilde\theta-\theta| + \left|\frac{\partial^2\psi_{y,z}}{\partial x\partial z}\right||\tilde\varepsilon-\varepsilon|\right]$$

$\tilde b$ and b both belong to the neighborhood U_i when α_i is non zero.

We can assume that U_i has been built so small so that ε and $\tilde\varepsilon$ are positive, of the same order. Thus, we can assure that

$$\frac{\varepsilon}{2} < \tilde\varepsilon < 2\varepsilon. \tag{95}$$

Using (ψ), we have:

$$\left|\frac{\partial\psi_{\tilde\theta,\tilde\varepsilon}}{\partial x}(\tilde b) - \frac{\partial\psi_{\theta,\varepsilon}}{\partial x}(\tilde b)\right| \le \frac{C}{\varepsilon\theta}|\tilde\theta-\theta| + \frac{C}{\varepsilon}|\tilde\varepsilon-\varepsilon|. \tag{96}$$

We know that

$$\left|\frac{\eta_0}{\varepsilon\theta}\right| \le \frac{C}{\gamma} \text{ on } [t_1^-,t_1^+]\cup[t_2^+,t_2^-]. \tag{97}$$

where C is a fixed constant on V (C depends on the initial partition of unity only).

$$\left|\int_{[t_1^-,t_1^+]\cup[t_2^-,t_2^+]} \left(\frac{\partial\psi_{\tilde\theta,\tilde\varepsilon}}{\partial x}(\tilde b) - \frac{\partial\psi_{\theta,\varepsilon}}{\partial x}(\tilde b)\right)\dot{\tilde b}\eta_0\overline{(b-\tilde b)}^{\cdot}\right| \le \tag{98}$$

$$\le \frac{C}{\gamma}\int_{[t_1^-,t_1^+]\cup[t_2^+,t_2^-]} |\dot{\tilde b}|\,|\overline{b-\tilde b}^{\cdot}|(|\theta-\tilde\theta| + |\tilde\varepsilon-\varepsilon|) \le \frac{C'}{\gamma}|b-\tilde b|_\infty|b-\tilde b|_{H^1}\times$$

$$\times\left(\int_{[t_1^-,t_1^+]\cup[t_2^+,t_2^-]} |\dot{\tilde b}|^2\right)^{1/2} < \frac{c}{1000}|b-\tilde b|^2_{H^1} + \frac{C'''}{\gamma c}|b-\tilde b^2_{L^2}(1+\int_0^1 \ddot{\tilde b}^2)$$

where C'' depends on the bound we have on $\int_0^1 \dot b^2 + \int_0^1 \dot{\tilde b}^2$. It is this bound which allows to replace $|b-\tilde b|^2_\infty$ by $C''\int_0^1(b-\tilde b)^2$.

Given ε, θ, we can ask, on the other hand, that our partition of unity is so refined that

$$|\left(\frac{\partial\psi_{\tilde{\theta},\tilde{\varepsilon}}}{\partial x}(\tilde{b}) - \frac{\partial\psi_{\theta,\varepsilon}}{\partial x}(\tilde{b})\right) 1_{[t_1^+, t_2^+]}| \leq o(1)(|\tilde{\theta} - \theta| + |\tilde{\varepsilon} - \varepsilon|). \tag{99}$$

where $o(1)$ is a function small L^∞. Indeed, on $[t_1^+, t_2^+]$, $|b(t)|$ is larger than μ_2, thus $|\tilde{b}|(t)$ larger than $\mu_2 - \delta$ and $\tilde{\mu}_2 - \delta$, where δ is as small as we wish, even with respect to $\tilde{\varepsilon}_1 = \frac{\tilde{\mu}_1 - \tilde{\mu}_1^-}{\tilde{\mu}_1}$ and $\tilde{\varepsilon}_2 = \frac{\tilde{\mu}_2 - \tilde{\mu}_2^-}{\tilde{\mu}_2^-}$ (which are close to ε_1 and ε_2).

Thus,

$$|\frac{\partial^2\psi_{\theta,\varepsilon}}{\partial x \partial \theta}(\tilde{b})| + |\frac{\partial^2\psi_{\theta,\varepsilon}}{\partial x \partial \varepsilon}(\tilde{b})| = o(1) \text{ on } [t_1^+, t_2^+] \tag{100}$$

and $o(1)$ is as small as we wish in this interval, and this holds for θ, ε replaced by $\theta', \varepsilon' \in [\theta, \tilde{\theta}], [\varepsilon, \tilde{\varepsilon}]$, since

$$\psi_{\theta',\varepsilon'}(\tilde{b}) \text{ is nearly } |\tilde{b}| - \theta' \text{ on } [t_1^+, t_2^+]. \tag{101}$$

($\psi_{\theta,\varepsilon}(b)$ is equal to $|b| - \theta$ on $[t_1^+, t_2^+]$, see (γ), (θ', ε') are very close to (θ, ε) and $\tilde{b}$ is very close to b).

Thus, the second derivatives are very small, by a simple continuity argument. (101) yields easily the needed estimate for these terms, on $[t_1^+, t_2^+]$, hence on $[t_1^-, t_2^-]$ and [0,1].

Coming back to (84), we are left with terms which can be brought back to expressions of the type:

$$\int_0^1 0(1)(\dot{b}^2 - \dot{\tilde{b}}^2)(b - \tilde{b}), \int_0^1 0(1)(\ddot{b} - \ddot{\tilde{b}})(b - \tilde{b}), \tag{102}$$

$$\int_0^1 \eta_0(\omega(b) - \omega(\tilde{b}))\ddot{\tilde{b}}(b - \tilde{b}), \int_0^1 \eta_0(\omega(b) - \omega(\tilde{b}))\dot{\tilde{b}}^2(b - \tilde{b})$$

with various functions $0(1)$ and ω.

$|\int_0^1 \eta_0(\omega(b) - \omega(\tilde{b}))\dot{\tilde{b}}^2(b - \tilde{b})|$ can be upperbounded by $C|\ddot{b}|^2_{L^2} \int_0^1 |b - \tilde{b}|^2$.

$|\int_0^1 0(1)(\dot{b}^2 - \dot{\tilde{b}}^2)(b - \tilde{b})|$ can be upperbounded by $\delta \int_0^1 |\dot{b} - \dot{\tilde{b}}|^2 + C(\delta) \int_0^1 |b - \tilde{b}|^2(\dot{b}^2 + \dot{\tilde{b}}^2) \leq \delta \int_0^1 (\dot{b} - \dot{\tilde{b}}|^2 + C(\delta)\left(|\ddot{b}|^2_{L^2} + |\ddot{\tilde{b}}|^2_{L^2}\right) \int_0^1 |b - \tilde{b}|^2 . \delta \int_0^1 |\dot{b} - \dot{\tilde{b}}|^2$ can be absorbed in $-c\int_0^1 \dot{b} - \dot{\tilde{b}}|^2$.

$\int_0^1 \eta_0(\omega(b) - \omega(\tilde{b}))\ddot{\tilde{b}}(b - \tilde{b})$ is equal, after integration by parts to $\int_0^1 0(1)(\omega(b) - \omega(\tilde{b}))(b - \tilde{b})\dot{\tilde{b}} + \int_0^1 \eta_0(\omega'(b)\dot{b} - \omega'(b)\dot{\tilde{b}})\dot{\tilde{b}}(b - \tilde{b}) + \int_0^1 \eta_0(\omega(b) - \omega(\tilde{b}))\dot{\tilde{b}}(\overline{\dot{b} - \dot{\tilde{b}}})$.

The first term is handled in the same way than above. The second term behaves like $\int_0^1 \eta_0(\omega(b)-\omega(\tilde{b}))\dot{\tilde{b}}^2(b-\tilde{b})$ which we have already upperbounded in a suitable way, plus another term of the type $\int_0^1 0(1)\dot{\tilde{b}}(\dot{\overline{b-\tilde{b}}})$, which we upperbounded with $\delta\int_0^1(\dot{\overline{b-\tilde{b}}})^2 + C(\delta)\int_0^1 \dot{\tilde{b}}^2(b-\tilde{b})^2 \leq \delta\int_0^1(\dot{\overline{b-\tilde{b}}})^2 + C(\delta)|\ddot{\tilde{b}}|^2_{L^2}\int_0^1(b-\tilde{b})^2$, a suitable bound.

The term $\left|\int_0^1 \eta_0(\omega(b)-\omega(\tilde{b}))\dot{\tilde{b}}(\dot{\overline{b-\tilde{b}}})\right|$ is upperbounded by

$$C\int_0^1 |b-\tilde{b}|\,|\dot{\tilde{b}}|\,|\dot{\overline{b-\tilde{b}}}| \leq C(\delta)\int_0^1 |\dot{\tilde{b}}|^2|b-\tilde{b}|^2 + \delta\int_0^1(\dot{\overline{b-\tilde{b}}})^2 \leq$$

$$\leq C(\delta)|\ddot{\tilde{b}}|^2_{L^2}\int_0^1 |b-\tilde{b}|^2 + \delta\int_0^1(\dot{\overline{b-\tilde{b}}})^2.$$

We are left with

$$\int_0^1 0(1)(\ddot{b}-\ddot{\tilde{b}})(b-\tilde{b}) =$$

$$= \int_0^1 0(1)(\dot{b}-\dot{\tilde{b}})(b-\tilde{b}) + \int_0^1 0(1)(\dot{b}-\dot{\tilde{b}})(b-\tilde{b}) +$$

$$+ \int_0^1 0(1)\dot{\tilde{b}}(\dot{b}-\dot{\tilde{b}}(b-\tilde{b}) + \int_0^1 0(1)(\dot{\overline{b-\tilde{b}}})^2.$$

The three first terms have already been studied. We are left with $\int_0^1 0(1)(\dot{\overline{b-\tilde{b}}})^2 = -\int_0^1 \frac{\partial\psi_{\theta,\epsilon}}{\partial x}(\tilde{b})(\dot{\overline{b-\tilde{b}}})^2 < 0$ since $\int_0^1 0(1)(\ddot{b}-\ddot{\tilde{b}})(b-\tilde{b}) = \int_0^1 \frac{\partial\psi_{\theta,\epsilon}}{\partial x}(\tilde{b})(\ddot{b}-\ddot{\tilde{b}})(b-\tilde{b})$.

Thus, we need not be worried by this term.

Summing up, we have:

$$\frac{c}{2}\int_0^1(\dot{\overline{b-\tilde{b}}})^2 + \frac{\partial}{\partial s}\int_0^1(b-\tilde{b})^2 \leq C\left(|a-\tilde{a}|^2(0) + \int_0^s\left(\int_0^1 |b-\tilde{b}|^2\right)\right) +$$

$$+C\int_0^s\int_0^1(\dot{b}-\dot{\tilde{b}})^2 + C\int_0^1 |x(0,z)-\tilde{x}(0,z)|^2 + C\left(1+|\ddot{b}|^2_{L^2}+|\ddot{\tilde{b}}|^2_{L^2}\right)\int_0^1 |b-\tilde{b}|^2.$$

Thus, upperbounding the right hand side:

$$\frac{c}{2}\frac{\partial}{\partial s}\int_0^s\int_0^1(\dot{\overline{b-\tilde{b}}})^2 + \frac{\partial}{\partial s}\int_0^1(b-\tilde{b})^2 \tag{103}$$

$$\leq C\left(|a-\tilde{a}|^2+\int_0^1 |x(0,z)-\tilde{x}(0,z)|^2\right)+C'\left(1+|\ddot{b}|^2_{L^2}+|\ddot{\tilde{b}}|^2_{L^2}\right)\times$$

$$\times\left(\frac{c}{2}\int_0^s\int_0^1 \overline{(b-\tilde{b})}^{\cdot 2}+\int_0^1 |b-\tilde{b}|^2\right)+C''\int_0^s\int_0^1 |b-\tilde{b}|^2.$$

Setting $w=\frac{c}{2}\int_0^s\int_0^1 \overline{(b-\tilde{b})}^{\cdot 2}+\int_0^1 (b-\tilde{b})^2$ and using the fact that $1+|b\ddot{b}|^2_{L^2}+|\ddot{\tilde{b}}|^2_{L^2}$ is bounded, we derive (c is constant in (103), we have lowerbounded the initial c locally):

$$w(s)\leq C'''\left(w(0)+|a-\tilde{a}|^2+\int_0^1 |x(0,z)-\tilde{x}(0,z)|^2\right)+C^{(iv)}\int_0^s\int_0^\tau\int_0^1 |b-\tilde{b}|^2. \tag{104}$$

Taking $s\in[0,s_1]$, we then derive:

$$w(s)\leq C^{(iv)}s_1^2 \operatorname*{Sup}_{z\in[0,s_1]} w(z)+C'''\left(w(0)+|a-\tilde{a}|^2+\int_0^1 |x(0,z)-\tilde{x}(0,z)|^2\right)\forall s\in[0,s_1]. \tag{105}$$

Thus

$$\left(1-C^{(iv)}s_1^2\right)\operatorname*{Sup}_{z\in[0,s]} w(z)\leq C'''\left(w(0)+|a-\tilde{a}|^2+\int_0^1 |x(0,z)-\tilde{x}(0,z)|^2\right). \tag{106}$$

If $C^{(iv)}s_1^2$ is less than $1/2$, we derive that:

$$\int_0^1 |b-\tilde{b}|^2+\int_0^s\int_0^1 |\overline{b-\tilde{b}}^{\cdot}|^2 \text{ tends to zero when } |a(0)-\tilde{a}(0)|+$$

$$+|b(0,t)-\tilde{b}(0,t)|^2_{L^2}+d(x(0),\tilde{x}(0)) \text{ tend to zero.}$$

It is easy to derive, then, that $\operatorname*{Sup}_{\tau\in[0,1]}|b(\tau,t)-\tilde{b}(\tau,t)|^2_\infty$ tends to zero, since, for any $\alpha>0$:

$$b(\tau,t)-\tilde{b}(\tau,t)=\frac{1}{2\alpha}\int_{t-\alpha}^{t+\alpha}(b(\tau,x)-\tilde{b}(\tau,x))dx+0\left(\sqrt{\alpha}\left(\int \dot{b}^2+\int \dot{\tilde{b}}^2\right)^{1/2}\right)$$

$$=\frac{1}{2\sqrt{\alpha}}0\left(\left(\int_0^1 |b-\tilde{b}|^2(\tau,t)\right)^{1/2}\right)+0(\sqrt{\alpha}).$$

Since $\int_0^1 |b-\tilde{b}|^2$ tends to zero,

$$\overline{\lim_{\tilde{b}\to b}}\operatorname*{Sup}_{\tau\in[0,1]}|b(\tau,t)-\tilde{b}(\tau,t)|^2_{L^\infty(S^1)}\leq C\sqrt{\alpha}$$

for any $\alpha > 0$.

$i)$ follows. The continuity from $H^1(S^1)$ into $H^1(S^1)$ almost holds - at this stage, since $\int_0^s \int_0^1 \ddot{b}^2 + \dddot{b}^2 < +\infty$, thus $\int_0^1 \ddot{b}^2(s) + \int_0^1 \dddot{b}^2(s)$ is finite a.e in s. Since $\tilde{b}$ converges to b in $L^2(S^1)$, such a bound, which holds only a.e. in s at this point, would imply the H^1- continuity if it held everywhere.

$ii)$ is an immediate consequence of $i)$ q.e.d.

Coming back to $\phi_\varepsilon(b)$ and (78), we can compute, for $m \in \mathbb{N}$, $\frac{\partial}{\partial s} \int \left(b_\varepsilon^{(m)} \phi_\varepsilon^{(m)}(b)\right)^2$ and go to the limit as ε goes to zero. We have established an H^1- bound on b, for a small, uniform, s-time. Therefore, the non zero coefficients of the partition of unity stabilize, the $\psi_{\theta,\varepsilon}$ are in finite number, they are bounded in any C^k. It is quite easy to see then, from the formula for $\frac{\partial}{\partial s} \int (b^{(m)})^2$ that b is H^m for any m, for $s > o$, if b is C^m. Combining this with Lemmas K and L, we derive:

Lemma M. *$b(s,t)$ is C^k in t and s a.e in s and the map $b_0 \longmapsto b(s,t,b_0)$ is continuous from every C^k to C^k a.e in s.*

I.2. Limit - behavior analysis

Let now 0 be an oscillation of type 1), with $\bar{\bar{\mu}}$ its maximum.

Let

$$\begin{cases} \tilde{\nu} \text{ be a given number} \\ \theta_1 \text{ be a small positive number.} \end{cases} \tag{107}$$

Assume that we have:

$$\bar{\bar{\mu}} \geq \tilde{\nu} + 2\theta_1, \underset{\sim}{\mu} \leq \tilde{\nu} + \frac{\theta_1}{2}. \tag{108}$$

Observe that, in the choice of μ_1, μ_2 related to Lemma 1, Definition 2 and Lemma 3, we can choose, on all the η_{12} 's we will build on 0:

$$\mu_1 = \tilde{\nu} + \frac{5\theta_1}{6} + o(1),\ \mu_2 \text{ will be close as we wish} \tag{109}$$

to μ_1, so that μ_2 is almost $\tilde{\nu} + \frac{5\theta_1}{6}$ as well. o(1) is as small as we wish.

Let I be the interval of the oscillation 0, $|I|$ be its measure.

We then have:

Lemma 4. *We use here only the η_{12} provided by $i)$ of Definition 2, on intervals of type 1).*

For any $\varepsilon_1 > 0$ there exists $\tilde{c}(\varepsilon_1) > 0$ such that, glueing up a finite number of pieces of this type on various disjoint subintervals of I, we can build an η_{12} on I such that

$$c(\bar{\tilde{\mu}}) \times c_0 \int_I b\eta_{12} \leq \bar{\tilde{\mu}} \int_I (|b| - \tilde{\nu})^+ \leq \frac{1}{c_0}\left(2\varepsilon_1 \bar{\tilde{\mu}}|I| + \frac{4\bar{\tilde{\mu}}}{\theta_1^2 \tilde{\nu}\tilde{c}(\varepsilon_1)} \int_I b\eta_{12} + \frac{3\theta_1}{2}|I|\right) \bar{\tilde{\mu}}$$

where c_0 is a universal constant depending only on α and $c(\bar{\tilde{\mu}})$ is another continuous positive function.

Proof of Lemma 4.

Remark 1. The formula of Lemma 4 can be added up for several intervals I satisfying (59). It keeps the same form. We will use this below.

Remark 2. Taking $c_0 \leq \frac{1}{4}$, we can drop in (59) the requirement $\bar{\tilde{\mu}} \geq \tilde{\nu} + 2\theta_1$. It suffices to set $\eta_{12} = 0$ on an interval where $\bar{\tilde{\mu}} < \tilde{\nu} + 2\theta_1$. Lemma 4's statement will still hold.

Proof of 1. Let us first assume that

$$\left|\left\{t \in I / |b(t)| > \tilde{\nu} + \frac{3\theta_1}{2}\right\}\right| \leq \varepsilon_1 |I|.$$

We then have, for any η_{12} built through Definition 2 on I:

$$\int_I (|b| - \tilde{\nu})^+ \leq \bar{\tilde{\mu}}\varepsilon_1|I| + \int_{\{t \in I/|b(t)| \leq \tilde{\nu} + \frac{3\theta_1}{2}\}} (|b| - \tilde{\nu})^+ \leq \left(\bar{\tilde{\mu}}\varepsilon_1 + \frac{3\theta_1}{2}\right)|I|.$$

On the other hand, using the construction of the η_{12} of Definition 2, even on several subintervals, with μ_1 and μ_2 as in (109) and $o(1)$ very small, we have:

$$c(\bar{\tilde{\mu}}) \int_I b\eta_{12} \leq \frac{1}{c_0}\bar{\tilde{\mu}} \int_I (|b| - \tilde{\nu})^+$$

Since the interval satisfies 1), $\int_{t_1^-}^t \bar{\mu} b$ is uniformly bounded. η_0 satisfies (10), hence is bounded by $\frac{1}{c(\bar{\tilde{\mu}})}$, where $c(\bar{\tilde{\mu}})$ is an appropriate function. The lemma follows under this condition. Our argument establishes the left hand inequality without using this condition. Observe that we could have replaced $\tilde{\nu}$ by $\tilde{\nu} + \theta_{1/2}$ with above argument, using (109). The same conclusion holds.

Let us now assume that

$$\left|\{t \in I / |b(t)| > \tilde{\nu} + \frac{3\theta_1}{2}\}\right| > \varepsilon_1 |I|.$$

The set $\{t \in I / |b(t)| > \frac{4.5}{6}\theta_1 + \tilde{\nu}\}$ is a countable reunion of open intervals $\cup_{j=1}^{+\infty} F_j$. All F_j's are of type 1. We can keep a finite number of them and absorb the integral of $\tilde{\bar{\mu}}(|b| - \tilde{\nu})^+$ on the other ones in $\frac{2\varepsilon_1\tilde{\bar{\mu}}^2|I|}{c_0}$. We can also assume that their total measure is less than $\varepsilon_{1/10}|I|$. The integral of $\bar{\mu}(|b| - \tilde{\nu})^+$ on

$$\{t \in I s.t\, t \notin \bigcup_{j=1}^{+\infty} F_j\} \subset \{t \in I s.t (|b| - \tilde{\nu})^+ \leq \frac{4.5}{6}\theta_1\}$$

can be absorbed in $\frac{1}{c_0}\frac{3\theta_1}{2}|I|\tilde{\bar{\mu}}$.

We are left with a finite number of intervals (disjoint) $F_1, --, F_{\ell_0}$. Those such that

$$\{t \in F_j/|b(t)| > \tilde{\nu} + \frac{3\theta_1}{2}\} \leq \frac{\varepsilon_1}{10}|F_j|$$

can be handled as in the beginning of our proof and their total contribution is therefore absorbed in

$$\frac{1}{c_0}\left(2\varepsilon_1\tilde{\bar{\mu}}|I| + \frac{3\theta_1}{2}|I|\right)\tilde{\bar{\mu}}.$$

We are left with $F_1, --, F_{\ell_1}$ (disjoint) and we can assume that

$$\left|\{t \in F_j, j = 1, --, \ell_1 s.t|b|(t)| > \tilde{\nu} + \frac{3\theta_1}{2}\}\right| > \frac{\varepsilon_1}{2}|I|. \tag{110}$$

Indeed, since:

$$|\{t \in I/|b(t)| > \tilde{\nu} + \frac{3\theta_1}{2}\}| > \varepsilon_1|I|$$

and

$$\sum_{\ell_0+1}^{+\infty} |F_j| < \frac{\varepsilon_1}{10}|I|$$

we have:

$$\left|\{t \in F_1, --, F_{\ell_0} s.t|b(t)| > \tilde{\nu} + \frac{3\theta_1}{2}\}\right| > \frac{9\varepsilon_1}{10}|I|.$$

Now

$$\sum \left|\{t \in F_{\ell_1+1}, --, F_{\ell_0} s.t|b(t)| > \tilde{\nu} + \frac{3\theta_1}{2}\}\right| \leq \frac{\varepsilon_1}{10}\sum_{\ell_1+1}^{\ell_0} |F_j| \leq \frac{\varepsilon_1}{10}|I|.$$

Thus

$$\left|\left\{t \in F_1, --, F_{\ell_1} s.t|b(t)| > \tilde{\nu} + \frac{3\theta_1}{2}\right\}\right| > \frac{8\varepsilon_1}{10}|I| > \frac{\varepsilon_1}{2}|I|.$$

Furthermore, on each $F_j, j = 1, --, \ell_1$, we have:

$$\left|\{t \in F_j/|b(t)| > \tilde{\nu} + 3\theta_{1/2}\}\right| > \frac{\varepsilon_1}{10}|F_j|. \tag{111}$$

On each F_j, we introduce

$$E_j = \{t \in F_j s.t|b(t)| > \tilde{\nu} + \frac{3\theta_1}{2}\}. \tag{112}$$

which we can decompose in a countable union of disjoint intervals $\bigcup_{k=1}^{+\infty} E_j^k$. We can throw out but a finite number of them and still keep (112) with $\varepsilon_{1/20}$ instead of $\varepsilon_{1/10}$. Starting from each of these intervals which are left, and where $|b(t)| > \tilde{\nu} + \frac{3\theta_1}{2}$, we can build an oscillation where an η_{12} can be set. Indeed, any of these intervals is contained in F_j. $|b|$ at its boundaries takes the value $\frac{4.5}{6}\theta_1 + \tilde{\nu}$, by construction.

Hence, we can define for each of these intervals a μ_1, μ_2 satisfying (109) and a related interval $[t_1^-, t_2^-]$, with an η_{12}. (Some of the E_j^k's might yield the same interval). We thus have now a finite number of disjoint intervals $\mathcal{L}_j^k = [t_{1,j,k}^-, t_{2,j,k}^-]$, where we have an $\eta_{12,j,k}$. The intervals can be made disjoint using the $o(1)$ in (109). Furthermore

$$\left|\left\{t \in U\mathcal{L}_{j,k} \text{ s.t. } |b(t)| > \tilde{\nu} + \frac{3\theta_1}{2}\right\}\right| > \varepsilon_1/4|I| \tag{113}$$

and, since we can get rid of the other ones as we did previously, we can assume:

$$|\{t \in \mathcal{L}_j^k / |b(t)| > \tilde{\nu} + 3\theta_1/2\}| > \varepsilon_1/20|\mathcal{L}_{j,k}|. \tag{114}$$

We then have, denoting $F_{j,k} = \{t \in \mathcal{L}_{j,k} \text{ s.t. } |b(t)| > \tilde{\nu} + 3\theta_1/2\}$. Observe that, by (10), $\eta_{0,j,k}$ is larger than $\frac{(\tilde{c}_0 t - t_{1,j,k}^-)(t_{2,j,k}^- - t)}{(t_{1,j,k}^- - t_{2,j,k}^-)^2}$. Thus:

$$\int_{\mathcal{L}_{j,k}} \eta_{0,j,k} dt \geq \tilde{c}_0 \int_{t_{1,j,k}^-}^{t_{2,j,k}^-} \frac{(t - t_{1,j,k}^-)(t_{2,j,k}^- - t)}{(t_{1,j,k}^- - t_{2,j,k}^-)^2} dt \tag{115}$$

$$= \tilde{c}_0 \int_0^1 s(1-s) ds |\mathcal{L}_{j,k}| = \tilde{c}_1 |\mathcal{L}_{j,k}|$$

and

$$\int_{F_{j,k}} \eta_{0,j,k} dt \geq \tilde{c}_0 \int_{F_{j,k}} \frac{(t - t_{1,j,k}^-)(t_{2,k,k}^- - t)}{(t_{1,j,k}^- - t_{2,j,k}^-)^2} dt \tag{116}$$

$$= \tilde{c}_0 \int_{\widetilde{F}_{j,k}} s(1-s) ds |\mathcal{L}_{j,k}|$$

where

$$|\widetilde{F}_{j,k}| > \frac{\varepsilon_1}{20(t_{2,j,k}^- - t_{1,j,k}^-)} |\mathcal{L}_j^k| = \varepsilon_1/20. \tag{117}$$

Thus

$$\int\limits_{F_{j,k}} \eta_{0,j,k} dt \geq c(\varepsilon_1)|\mathcal{L}_{j,k}| \tag{118}$$

where $c(\varepsilon_1)$ is a precise function of ε_1 which we can take to be ε_1^3 for ε_1 small.

Thus, by (113):

$$\frac{\varepsilon_1}{4}|I| < \sum_{j,k} |\mathcal{L}_{j,k}| \leq \sum_{j,k} \frac{\tilde{c}_2}{c(\varepsilon_1)} \int\limits_{F_{j,k}} \eta_{0,j,k} dt. \tag{119}$$

Hence, using (119) and the fact that $|b(t)| > \tilde{\nu} + 3\theta_1/2$, hence $|b(t)| - \mu_1^-$ and $\psi_{\theta_1}(|b(t)|) \geq \frac{\theta_1}{6}$ or

$$\int\limits_I (|b| - \tilde{\nu})^+ \leq \tilde{\bar{\mu}}|I| < \frac{4\tilde{\bar{\mu}}\tilde{c}_2}{\varepsilon_1 c(\varepsilon_1)} \sum_{j,k} \int\limits_{F_{j,k}} \eta_{0,j,k} dt \tag{120}$$

$$\leq \frac{4\tilde{\bar{\mu}}}{\varepsilon_1 c(\varepsilon_1)} \frac{36\tilde{c}_2}{\theta_1^2} \sum_{j,k} \int\limits_{F_{j,k}} |b|\eta_{12,j,k} \leq \frac{C\tilde{\bar{\mu}}}{\theta_1^2 \tilde{c}(\varepsilon_1)} \int\limits_I b\eta_{12}$$

1) follows. Again, we could have worked with $\tilde{\nu} + \theta_1/2$ instead of $\tilde{\nu}$.

If we come back to the proof of Lemma 4, we can see that we also have:

Lemma 5. *For any $\varepsilon_1 > 0$, there exists $\tilde{c}(\varepsilon_1) > 0$ such that, gluing up a finite number of pieces of type η_{12} on various subintervals of I, we can build η_{12} on I such that:*

$$\tilde{\bar{\mu}} \int\limits_I \left(|b| - \left(\tilde{\nu} + \frac{3\theta_1}{2} \right) \right)^+ \leq \frac{1}{c_0} \left(2\varepsilon_1 \tilde{\bar{\mu}}|I| + \frac{4\tilde{\bar{\mu}} \int_I b\eta_{12}}{\theta_1^2 \tilde{\nu}\tilde{c}(\varepsilon_1)} \right).$$

The proof of Lemma 5 follows readily from the proof of Lemma 4. We, furthermore, have:

Lemma 6. *Assume $\theta_1 < \frac{1}{2}$ and $|x_1^- - x_2^-| \leq \bar{c}/|b|_\infty$, where $\bar{c}$ is an appropriate constant. Then, η_{12} can be built so that*

$$\frac{\partial}{\partial s} \int\limits_I (|b| - \tilde{\tilde{\nu}})^+ < 0 \quad \textit{for any} \quad \tilde{\tilde{\nu}} \geq \tilde{\nu} + 1$$

for the flow generated by $\eta_{12} + cb$.

Proof of Lemma 6. We write I as UI_ℓ, where the I_ℓ's are the disjoint intervals containing the various pieces of η_{12}. The maximum of b (we assume b to be positive on I) is larger than $\tilde{\nu} + 2\theta_1$, the minimum, at the boundaries of I_ℓ, is $\tilde{\nu} + \frac{5}{6}\theta_1 + o(1)$.

The gluing function θ, satisfying (γ), can be assumed to be constant on $\{t \in I_\ell$ such that $b(t) > \tilde{\bar{\nu}}\}$.

We then have:

$$\frac{\partial}{\partial s}\int_I \left(|b| - \left(\tilde{\nu} + \frac{5}{3}\theta_1\right)\right)^+ = \sum \left(\frac{\overline{\psi_{\theta,\varepsilon}\eta_0 + cb + \lambda b}^{\,\cdot}}{a}\Bigg|_{\tilde{t}_j^-}^{\tilde{t}_j^+} \right. \tag{121}$$

$$\left. + \int_{\tilde{t}_j^-}^{\tilde{t}_j^+} \tau a(\psi_{\theta,\varepsilon}\eta_0 + cb) - \int_{\tilde{t}_j^-}^{\tilde{t}_j^+} b(\psi_{\theta,\varepsilon}\eta_0 + cb)\bar{\mu}_\xi \right) = (1)$$

where $(\tilde{t}_j^-, \tilde{t}_j^+)$ are the various intervals where $|b|$ is larger than $\tilde{\bar{\nu}}$, in I.

Again, $\overline{\psi_{\theta,\varepsilon}}^{\,\cdot} = -\dot{\theta} + \frac{\partial \psi_{\theta,\varepsilon}}{\partial b}\dot{b} = \frac{\partial \psi_{\theta,\varepsilon}}{\partial b}\dot{b}$ on these intervals, and is negative at $\tilde{t}_j^+$, positive at $\tilde{t}_j^-$; $\dot{b}$ also. Thus, since $\psi_{\theta,\varepsilon}(b(\tilde{t}_j^-)) = \psi_{\theta,\varepsilon}(b(\tilde{t}_j^+)) = \tilde{\nu} + \frac{5}{3}\theta_1 - \theta$

$$(1) \leq \sum \frac{(\tilde{\bar{\nu}} - \theta)}{a}\int_{\tilde{t}_j^-}^{\tilde{t}_j^+} \overline{\dot{\eta}_0 - \bar{\mu}\eta_0 b}^{\,\cdot} + \tag{122}$$

$$+ \sum \frac{1}{a}\tilde{\bar{\nu}} \int_{\tilde{t}_j^-}^{\tilde{t}_j^+} \left[b(\psi_{\theta,\varepsilon}\eta_0 + cb) - \int_0^1 b(\psi_{\theta,\varepsilon}\eta_0 + cb)\right] +$$

$$+ \sum \int_{\tilde{t}_j^-}^{\tilde{t}_j^+} (\psi_{\theta,\varepsilon}\eta_0 + cb)(a\tau - b\bar{\mu}_\xi) \tag{123}$$

$\overline{\dot{\eta}_0 - \bar{\mu}\eta_0 b}^{\,\cdot}$ is equal to

$$\frac{d}{dt}\left(e^{\int_{t_1^-}^t \bar{\mu} b}\frac{d}{dt}\left(e^{-\int_{t_1^-}^t \bar{\mu} b}\eta_0\right)\right) = e^{-\int_{t_1^-}^t \bar{\mu} b}\frac{d^2}{dx^2}\left(e^{-\int_{t_1^-}^t \bar{\mu} b}\eta_0\right)$$

$$= -C(1+b^2)\eta_0 e^{-2\int_{t_1^-}^t \bar{\mu} b} - \frac{2}{|x_1^- - x_2^-|^2}e^{-\int_{t_1^-}^t \bar{\mu} b}.$$

Thus, using our bound on $\int_0^1 |b|$, upper bounding $\psi_{\theta,\varepsilon}$ by b, c by 1, and lower bounding $\tilde{\bar{\nu}} - \theta$ by $\frac{1}{2}$, taking C large enough

$$(1) \leq \sum \int_{t_j^-}^{\tilde{t}_j^+} \left[-Ce^{-2\int_{t_1^-}^t \bar{\mu} b}(1+b^2) + \right. \tag{124}$$

$$\left. + \overline{C}\left(1 + \frac{1}{a} + a\right)(|b|^2 + 1)\right]\eta_0 dt + \sum \left(-\frac{\tilde{c}}{|x_1^- - x_2^-|^2} + \tilde{\bar{c}}|b|_\infty^2\right)(\tilde{t}_j^+ - \tilde{t}_j^-).$$

There are no extra-terms involving $\tilde{t}_j^+ - \tilde{t}_j^-$, due to cb, where there is no companion term $-\frac{\tilde{c}}{|x_1^- - x_2^-|^2}$ because $[\tilde{t}_j^-, \tilde{t}_j^+]$ is contained in I. The result also extends to $\widetilde{\psi}_{\tilde{\theta},\tilde{\varepsilon}}\tilde{\eta}_0$. Our conclusion follows.

For $b \in H^1$, we introduce:

$$\sum\nolimits_{\tilde{\varepsilon}_0}(b) = \{\psi \text{ nonnegative step-functions on } S^1 \text{ such that} \tag{125}$$

$(|b| - \psi)^+$ is continuous and such that $C\left(|I| + \int_I b^2\right)|I| < \tilde{\varepsilon}_0$ for any interval I where $|b| > \psi\}$ $\tilde{\varepsilon}_0$ is dominated, in the sequel, by the constant ε_0 introduced in Case 1 and Case 2. Unless b is constant, $\sum_{\varepsilon_0}(b)$ is not empty. Let then:

$$\mathcal{O}_{\tilde{\varepsilon}_0}(b) = \operatorname*{Sup}_{\psi \in \sum_{\tilde{\varepsilon}_0}(b)} \int (|b| - \psi)^+. \tag{126}$$

We then have:

Lemma 7. *We take* $\tilde{\varepsilon}_0 < \frac{1}{4}$. *Then:* $\mathcal{O}_{\tilde{\varepsilon}_0}(b') - \mathcal{O}_{2\tilde{\varepsilon}_0}(b) \leq 2|b - b'|_{L^\infty}$, *if* $2C(1 + |b|_\infty)|b - b'|_{L^\infty} < \sqrt{\frac{\tilde{\varepsilon}_0}{2}}$.

Proof. Let $\psi' \in \sum_{2\tilde{\varepsilon}_0}(b')$. Let

$$\psi = \psi' + |b - b'|_{L^\infty}. \tag{127}$$

Then, since

$$|b| - |b'| \leq |b - b'|_{L^\infty}. \tag{128}$$

We have:

$$|b| - \psi' - |b - b'|_{L^\infty} > 0 \Rightarrow |b'| - \psi' > 0. \tag{129}$$

Thus, any interval I of positivity of $|b| - \psi$ is contained in an interval of positivity I' of $|b'| - \psi'$. Since $(|b'| - \psi')^+$ is continuous, ψ' is constant on I, hence ψ is constant on I and $(|b| - \psi)^+$ is constant. Furthermore, since $I \subset I'$:

$$\begin{aligned} C\left(|I| + \int_I b^2\right)|I| &\leq C\left(|I'| + \int_I b'^2 + (1 + |b|_\infty)|b - b'|_{L^\infty}\right)|I'| \\ &\leq \tilde{\varepsilon}_0 + 2C|b - b'|_{L^\infty}|I'|(1 + |b|_\infty). \end{aligned} \tag{130}$$

Thus, if $2C|b - b'|_{L^\infty} \leq \frac{\tilde{\varepsilon}_0}{1 + |b|_\infty}$ or, more generally – and we will use this fact later – if $2(1 + |b|_\infty)|b - b'|_{L^\infty}|I'|C < \tilde{\varepsilon}_0$, we have:

$$C\left(|I| + \int_I |b|^2\right)|I| \leq 2\tilde{\varepsilon}_0. \tag{131}$$

Thus, ψ belongs to $\sum_{2\tilde{\varepsilon}_0}(b)$ and

$$\int (|b'| - \psi')^+ \leq \int (|b| - \psi)^+ + 2|b - b'|_{L^\infty} \leq \mathcal{O}_{2\tilde{\varepsilon}_0}(b) + 2|b - b'|_{L^\infty}. \tag{132}$$

Taking the Sup on ψ' in $\sum_{\tilde{\varepsilon}_0}(b)$, we conclude.

We assume in the sequel that:

$$\tilde{\varepsilon}_0 = \varepsilon_0/10. \tag{133}$$

Parameters and restrictions in the construction of the cancellation flows When constructing and patching several η_{12} corresponding to the various parts of Definition 2, also along the lines described in Lemma 4, we have to proceed step by step in order to obtain a clear result. We need four sets of parameters (plus the parameters ε_{15} and ε_{16} we started with)

a) ε_{10} is a positive parameter, tending to zero. We ask that the support of the η_{12}'s is contained $\{t/|b|(t) \geq \varepsilon_{10}\}$

b) ε_{11} is another positive parameter, tending to zero. We ask that any oscillation which yields an η_{12} has $\bar{\tilde{\mu}} - \mu_2 \geq \varepsilon_{11}$ (i.e. Γ of Lemma 3 and its extensions is larger than ε_{11}). ε_{11} will be take less than $\frac{\beta_{k,q}}{40}$ defined below

c) k is an integer-parameter tending to $+\infty$. We ask that

$$\nu \neq \frac{j}{k} \qquad \forall j \in \mathbb{N}.$$

Let $\alpha_k = \inf\limits_{j \in \mathbb{N}} \left|\nu - \frac{j}{k}\right| > 0$. We ask that the step functions involved in $\sum_{\tilde{\varepsilon}_0}(b)$ be further restricted by the requirement that they assume values in $\left\{\frac{j}{k}, j \in \mathbb{N}\right\}$. The related set of functions ψ is denoted $\sum^k_{\tilde{\varepsilon}_0}(b)$.

We clearly have:

Lemma 8.

$$|\mathcal{O}_{\tilde{\varepsilon}_0}(b) - \mathcal{O}^k_{\tilde{\varepsilon}_0}(b)| < \frac{1}{k}$$

where $\mathcal{O}^k_{\tilde{\varepsilon}_0}(b) = \underset{\psi \in \sum^k_{\tilde{\varepsilon}_0}(b)}{Sup} \int (|b| - \psi)^+$.

Proof. Any $\psi \in \sum^k_{\tilde{\varepsilon}_0}(b)$ is contained in $\sum_{\tilde{\varepsilon}_0}(b)$. Thus

$$\mathcal{O}^k_{\tilde{\varepsilon}_0}(b) < \mathcal{O}_{\tilde{\varepsilon}_0}(b).$$

On the other hand, for any $\psi \in \sum_{\tilde{\varepsilon}_0}(b)$, there exists ψ',

$$\psi \leq \psi' \leq \psi + \frac{1}{k}$$

such that ψ' takes values in $\left\{\frac{j}{k}, j \in \mathbb{N}\right\}$. Furthermore any interval of positivity J of $|b| - \psi'$ is contained in an interval I of positivity of $|b| - \psi$. Therefore

$$\left(|I| + \int_I |b|^2\right) |J| < \left(|I| + \int_I |b|^2\right) |I| \leq \tilde{\varepsilon}_0.$$

Therefore, $\psi' \in \sum^k_{\tilde{\varepsilon}_0}(b)$ and

$$\int (|b| - \psi')^+ \geq \int (|b| - \psi)^+ - \frac{1}{k}.$$

Thus,

$$\mathcal{O}^k_{\bar{\varepsilon}_0}(b) \geq \mathcal{O}_{\bar{\varepsilon}_0}(b) - \frac{1}{k}.$$

Our claim follows.

We proceed now with our last requirement.

d) Let q be another integer parameter, tending also to $+\infty$. We assume that k and q are prime one to the other. Let C_{41} be a bound on $|b|_\infty$ all along our deformations to come. By Lemma 3 and its extensions, we know that such a bound exists at each stage where we use the cancellation flow. C_{41} depends on the stage of the process. But, at each stage, along the cancellation process, C_{41} exists, C_{41} might change when we apply the second stage of the process, to come, i.e. the creation/cancellation of oscillations.

Let then

$$\bar{\mathcal{L}}_{k,q}(C_{41}) = \Big\{(j,m) \in \mathbb{N} \times \mathbb{N} \text{ such that} \tag{134}$$

$$\frac{j}{k} \leq C_{41}+1, \frac{m}{q} \leq C_{41}+1, \left|\frac{j}{k} - \frac{m}{q}\right| > 0\Big\}$$

and

$$\delta_{k,q} = \inf_{\mathcal{L}_{k,q}} \left|\frac{j}{k} - \frac{m}{q}\right| > 0. \tag{135}$$

Observe, since k and q are prime to each other, that

$$\frac{j}{k} = \frac{m}{q} \quad \text{in} \quad \bar{\mathcal{L}}_{k,q}(C_{41}) \tag{136}$$

implies that $\frac{m}{q}$ and $\frac{j}{k}$ are an integer, less than or equal $C_{41}+1$. Let

$$\beta_{k,q} = \inf\left(\frac{\alpha_k}{2}, \frac{\delta_{k,q}}{10}\right). \tag{137}$$

We require that the construction of $\mu_1, \mu_1^-, \mu_2, \mu_2^-$ for each oscillatory piece η_{12} takes place in $\{\frac{j}{k} + \beta_{k,q},\ j \in \mathbb{N}\}$. This does not yield restrictions on μ_1^- or μ_2^- of the oscillation since μ_1^- and μ_2^- are as close as we wish respectively to μ_1 and μ_2; however, μ_1 and μ_2 have to be very close, both of them to j/k. Using Lemma 4, we derive from Lemma 8 that this is only possible if the oscillation has an amplitude larger than $1/k$. Hence, our flow will screen the oscillations of b of size larger than ε_{11} and observe also that, since $\beta_{k,q}$ is less than $\alpha_{k/2}$ and since all the curves which we will deform have $|b|_\infty$ less than C_{41}, the intervals $[\mu_1^-, \mu_1]$, $[\mu_2^-, \mu_2]$ where we start the construction of the η_{12}'s will be far away from ν on our deformation classes. Therefore, (γ) will be satisfied and all our previous claims will hold for this flow on our deformation classes.

Let $\varepsilon_{12} > 0$ be given

$$0 < \varepsilon_{12} < \beta_{k,q}/40.$$

Under those restrictions, our flows will be denoted

$$Z_0^\nu = Z^\nu_{\varepsilon_{10},\varepsilon_{11},k,q,\varepsilon_{12}}. \tag{138}$$

We are going to obtain a good picture of the curves which impede the deformation downwards of a for these ν-flows of iterating Z_0^ν, with parameters which will tend to zero and $+\infty$ along the iteration.

We first have:

Lemma 9. *If we choose our parameters relative orders appropriately, then: i) At each stage of the iteration process; $b(s,t)$ converges weakly in L^2 to $\varphi.$ when s tends to the explosion time or to $+\infty$, where $\varphi \in L^\infty$.*

ii) Any $J = UJ_\ell, J_\ell$ disjoint intervals where $\varphi.$ has oscillations above any values $\mu_{1,\ell}$ larger than 4 $Sup(\varepsilon_{11}, 1/k)$ has one of its intervals J_ℓ such that $C(|J_\ell| + \int_{J_\ell} \psi^2_\cdot)|J_\ell| \geq \varepsilon_0/20$ or $\sum \int_{J_\ell} (|\varphi.| - \mu_{1,\ell})^+ < \varepsilon_{12} + \frac{12}{k} + 8\varepsilon_{10}$.

iii) $\overline{lim}_{\substack{s\to+\infty \\ \text{or} \\ s\to \text{explosion time}}} \int_0^1 |b(s,t) - \varphi.| \leq C2000\ Sup\left(\frac{1}{q}, q\varepsilon_{11}, q\varepsilon_{10}, \frac{1}{k}\right).$

Remark. $\varphi.$ is not necessarily continuous. However, $\varphi.$ is a uniform lowest lower-bound, almost everywhere, of continuous functions. The word "oscillation above μ_1 of $\varphi.$" has to be understood with respect to this sequence of continuous functions converging to $\varphi.$.

Proof of Lemma 9. Proof of i): Let f be any C^∞-function on S^1. We compute:

$$\frac{\partial}{\partial s}\int_0^1 bf = -\int_0^1 \dot{\eta} + \frac{\lambda b}{a}\dot{f} + \int_0^1 (a\eta\tau - b\eta\bar{\mu}_\xi)f \tag{138'}$$
$$= O_f\left(-\frac{\partial a}{\partial s}\right) + \int_0^1 \frac{\eta_{12} f}{a} + O_f\left(\int_0^1 |\eta_{12}|\right)\left(1 + \int_0^1 |b|\right).$$

We know that $\int_0^1 |b|$ is bounded along the deformation.

We also know that

$$\int_0^1 |\eta_{12}| \leq \frac{1}{\varepsilon_{10}}\int_0^1 b\eta_{12}. \tag{139}$$

Thus

$$\frac{\partial}{\partial s}\int_0^1 bf = O_f\left(-\frac{\partial a}{\partial s}\right). \tag{140}$$

Since a converges, $\int_0^1 bf$ converges. b is bounded L^2 since b is bounded L^∞ i) follows.

Proof of ii) and iii): Lemma 3 and Lemma 3′ imply that, if s tends $+\infty$ or s tends to the explosion time, since Γ is lower bounded, c (the viscosity coefficient introduced in (15) in the construction of Z.) has to tend to zero at least along a subsequence. We will analyze when c can tend to zero, which is equivalent to the fact that $\int_0^1 b\eta_{12}$ tends to zero and see that ii) and iii) have then to occur along subsequences. In order to make ii) and iii) global, we need a further observation, which is the following:

As pointed out earlier and can be checked easily, the cancellation flow built following i) of Definition 2 has the property of controlling $\frac{\partial}{\partial s}\int(|b|-\nu)^+$ by $-C\frac{\partial a}{\partial s}$. This property, due to our construction and the careful choice of our parameters (see the definition of $\beta_{k,q}$) extends to all $\frac{m}{q}$, $m \in \mathbb{N}$, $\frac{m}{q} < C_{41}+1$ which are not integers. Thus,

$$\frac{\partial}{\partial s}\int\left(|b|-\frac{m}{q}\right)^+ \leq -C\frac{\partial a}{\partial s} \qquad \forall m \in \mathbb{N} \quad \text{such that} \tag{141}$$

$\frac{m}{q} < C_{41}+1$, $\frac{m}{q}$ not an integer.

Assume now that we have proved, on a subsequence s_r, that:

$$\int_0^1 |b(s_k,t)-\varphi.| \leq C100 \text{ Sup}\left(\varepsilon_{11}, \frac{1}{k}, \varepsilon_{10}\right). \tag{142}$$

We wish then to prove iii).

Using (141) and (142), we have; for any $\frac{m}{q}$ not an integer a $s \geq s_r$:

$$\begin{aligned}\int\left(|b|-\frac{m}{q}\right)^+(s) &\leq \int\left(|b|-\frac{m}{q}\right)^+(s_r) - C(a(s)-a(s_r)) \\ &\leq \int_0^1\left(|\varphi.|-\frac{m}{q}\right)\chi_{|\varphi.|>\frac{m}{q}} + 200 \text{ Sup}(\varepsilon_{11},\varepsilon_{10},1/k).\end{aligned} \tag{143}$$

The above inequality holds for k large enough $(a(s)-a(s_r))$ tends to zero).

Thus, since $x^+ = x + x^-$:

$$\begin{aligned}&\int\left(|b|-\frac{m}{q}\right)^-\chi_{|\varphi.|>\frac{m}{q}}(s) + \int\left(|b|-\frac{m}{q}\right)\chi_{|\varphi.|>\frac{m}{q}}(s)+ \\ &+\int\left(|b|-\frac{m}{q}\right)^+\chi_{|\varphi.|\leq\frac{m}{q}}(s) \leq \int_0^1\left(|\varphi.|-\frac{m}{q}\right)\chi_{|\varphi.|>\frac{m}{q}}+ \\ &+300 \text{ Supp}(\varepsilon_{11},\varepsilon_{10},1/k).\end{aligned} \tag{144}$$

Since b has an upper bounded number of zeros and b and $\varphi.$ are bounded L^∞, $|b|$ converges to $|\varphi.|$ weakly and

$$\int\left(|b|-\frac{m}{q}\right)\chi_{|\varphi.|>\frac{m}{q}}(s) - \int\left(|\varphi.|-\frac{m}{q}\right)\chi_{|\varphi.|>\frac{m}{q}} \tag{145}$$

tends to zero.

Thus, for s large enough:

$$\int \left(|b| - \frac{m}{q}\right)^{-} \chi_{|\varphi.| > \frac{m}{q}}(s) + \int \left(|b| - \frac{m}{q}\right)^{+} \chi_{|\varphi.| \le \frac{m}{q}}(s) \tag{146}$$
$$\le C\ 1000\ \mathrm{Sup}(\varepsilon_{10}, \varepsilon_{11}, 1/k).$$

We now consider $\int |\ |b| - |\varphi.|\ |$. If this is suitably upper bounded then so will $\int |b - \varphi.|$ be since b has an upper bounded number of zeros.

We write

$$\int |\ |b| - |\varphi.|\ | \le \sum_{\substack{m, \frac{m}{q} < C_{41}+1 \\ \frac{m}{q} \text{ not an integer} \\ \frac{m+1}{q} \text{ not an integer}}} \int_{\frac{m}{q} < |\varphi| \le \frac{m+1}{q}} |\ |b| - |\varphi.|\ | + \tag{147}$$
$$+ \sum_{\substack{\ell, \ell < C_{41}+1 \\ \ell \text{ integer}}} \int_{\ell - \frac{1}{q} < |\varphi| < \ell + \frac{1}{q}} |\ |b| - |\varphi.|\ |.$$

We have, using (146):

$$\int_{\frac{m}{q} < |\varphi.| \le \frac{m+1}{q}} |\ |b| - \varphi.|\ | \le \int_{\substack{\frac{m}{q} < |\varphi.| \le \frac{m+1}{q} \\ \frac{m-1}{q} \le |b| \le \frac{m+2}{q}}} |\ |b| - |\varphi.|\ | + \tag{148}$$
$$+ 2\left(\int (|b| - \frac{m+1}{q}\right)^{+} \chi_{|\varphi. \le \frac{m+1}{q}} +$$
$$+ \int \left(|b| - \frac{m}{q}\right)^{-} \chi_{|\varphi.| > \frac{m}{q}}$$
$$\le \frac{2}{q} \left| \frac{m}{q} < |\varphi.| \le \frac{m+1}{q} \right| + C2000\ \mathrm{Sup}(\varepsilon_{10}, \varepsilon_{11}, 1/k).$$

Thus

$$\int |\ |b| - |\varphi.|\ |(s) \le \frac{2}{q} + 2000qC\ \mathrm{Sup}(\varepsilon_{11}, \varepsilon_{10}, 1/k) \tag{149}$$

for $s \ge s_r$.

iii) follows under (142).

We now prove (142):

$\varphi.$ is an integrable function. Therefore, taking a partition of [0,1] in tiny intervals and replacing $\varphi.$ by its average $\frac{1}{|I|} \int_I |\varphi.|$ on each of these intervals, we obtain step-functions which converge to $\varphi.$ in L^1. These functions are all bounded by $C_{41} + 1$.

We can replace each $\frac{1}{|I|}\int\limits_I |\varphi.|$ by an $\frac{j}{k}, j$ by an integer. We avoid the integers values of j/k, replacing them by $\frac{j+1}{k}$ or $\frac{j-1}{k}$. We then obtain a step-function on S^1, γ_k such that if we have refined enough on subintervals:

$$\int | \, |\varphi.| - \gamma_k| < \frac{3}{k}. \tag{150}$$

We can assume that the intervals I are so tiny that we have

$$(|I| + |I|(C_{41}+1)^2)|I| \le \tilde{\varepsilon}_0 = \frac{\varepsilon_0}{20}. \tag{151}$$

Thus, $\gamma_k \in \sum_{\tilde{\varepsilon}_0}^k(b)$, for any $s \ge 0$ such that $(|b| - \gamma_k)^+$ is continuous. Let us study the continuity of $(|b| - \gamma_k)^+$: this functions' only discontinuity points are contained in the discontinuity points of γ_k, which are in finite number. Let z be such a point and let

$$I(z) \text{ be the maximal interval containing } z \text{ such that } |b| \text{ is larger than } \gamma_k. \tag{152}$$

$I(z)$ depends on s. There are finitely many such intervals. On each $I(z)$, we have:

$$\sum \int\limits_{I(z)} (|b| - \gamma_k)^+ = \sum \int\limits_{I(z)} (|b| - \gamma_k) = \sum \int\limits_{I(z)} (|b| - |\varphi.|) + \tag{153}$$

$$+ \sum \int\limits_{I(z)} (|\varphi.| - \gamma_k) \le \sum \int\limits_{I(z)} (|b| - |\varphi.|) + \frac{3}{k}.$$

As we pointed out earlier, $|b|$ tends weakly to $|\varphi.|$. The intervals $I(z)$ are in finite number. We can assume that they "converge". Thus,

$$\sum_j \left| \int\limits_{I(z)} (|b| - |\varphi.|) \right| \tag{154}$$

is as small as we wish and thus, removing these intervals by replacing γ_k there by $\frac{\ell}{k}$, $\frac{\ell}{k} > C_{41} + 1$, we obtain $\tilde{\gamma}_k$ which satisfies

$$\int (|b| - \tilde{\gamma}_k)^+ \ge \int\limits_{(UI(z))^c} (|b| - |\varphi.|)^+ - \int\limits_{(UI(z))^c} | \, |\varphi.| - \tilde{\gamma}_k| \tag{155}$$

$$\ge \int (|b| - |\varphi.|)^+ - \int\limits_{(UI(z))^c} | \, |\varphi.| - \gamma_k| - \int\limits_{UI(z)} (|b| - |\varphi.|)^+$$

$$\ge \int (|b| - |\varphi.|)^+ - \int | \, |\varphi.| - \gamma_k| - \sum \int\limits_{I(z)} (|b| - |\varphi.|) - \frac{3}{k}.$$

Thus, for s large enough:

$$\int (|b| - \tilde{\gamma}_k)^+ \geq \int (|b| - |\varphi.|)^+ - \frac{6}{k} \tag{156}$$

$(|b| - \tilde{\gamma}_k)^+$ is now continuous, so that $\tilde{\gamma}_k \in \sum_{\tilde{\varepsilon}_0}^k(b)$. Eliminating the oscillations above $\tilde{\gamma}_k$ of size less than $\varepsilon_{11} + \frac{1}{k}$, also those where $\tilde{\gamma}_k$ takes values less than ε_{10} – on there intervals UL_j, which are in finite number, $\sum \int_{L_j} |b|$ converges to $\sum \int_{L_j} |\varphi.|$ which is less than $\varepsilon_{10} + \frac{3}{k}$ we obtain, after denoting J the collection of intervals which remain:

$$\int_J (|b| - \tilde{\gamma}_k)^+ \geq \int (|b| - |\varphi.|)^+ - \frac{10}{k} - \varepsilon_{11} - 2\varepsilon_{10} \tag{157}$$

J is made of intervals I, which all satisfy (151). Hence, the oscillations are of type i), Lemma 4 applies. Coming back to Lemma 4 and (109), we can take above the values $\tilde{\nu} = \frac{j}{k}$ assumed by $\tilde{\gamma}_k$, $\theta_1 = \frac{\beta_{k,q}}{20}$ so that the construction of η_{12} according to Lemma 4 does not interfere with $\{t/|b|(t) = \frac{m}{q}, \frac{m}{q} \leq C_{41} + 1\}$. $\varepsilon_1 > 0$ is chosen in the sequel. Observe that

$$\beta_{k,q} < \frac{1}{k}. \tag{158}$$

Observe also that we can take the neighborhood in our partition of unity so small so that if b' belongs to the same basic neighborhood, some common $U_i$then (we might have to ask that $\tilde{\gamma}_k$ avoids integers values $\pm\frac{2}{k}$ which is easy to fulfill)

$$\int_J \left(|b'| - \left(\tilde{\gamma}_k + \frac{1}{k}\right)\right)^+ \geq \int (|b| - \varphi.)^+ - \frac{14}{k} - \varepsilon_{11} - 2\varepsilon_{10}. \tag{159}$$

This is a weak-H^1 requirement.

Using Lemma 7 and 8, we clearly have:

$$\begin{aligned} O_{2\tilde{\varepsilon}_0}(b') &\geq O_{\tilde{\varepsilon}_0}(b) - \frac{1}{k} \geq O^k_{\tilde{\varepsilon}_0}(b) - \frac{1}{k} \\ &\geq \int (|b| - \varphi.)^+ - \frac{16}{k} - \varepsilon_{11} - 2\varepsilon_{10}. \end{aligned} \tag{160}$$

If, instead of b, $b^1 = b_i$ is the center point of U_i, then the construction of η_{12} for b_i is again completed using Lemma 4 with $\tilde{\nu} = \frac{j}{k}$, $\theta_1 = \frac{\beta_{k,q}}{40}$.

We then have, by Lemma 4, for any $\varepsilon_1 > 0$, the existence of $\eta_{12,0}$ such that

$$\int_J b_i \eta_{12,i} \geq \frac{\beta^2_{k,q}}{10{,}000} \frac{\varepsilon_{10}\tilde{c}(\varepsilon_1)}{1 + C_{41}} \left(\frac{c_0}{2} \int \left(|b_i| - \tilde{\gamma}_k - \frac{1}{k}\right)^+ - 3\beta_{k,q} - 2\varepsilon_1(1 + C_{41})\right) \tag{161}$$

where the $\eta_{12,i}$ are built on oscillations of amplitude larger than $\frac{3\theta_1}{2}$, hence than ε_{11}. With respect to 1) of Lemma 4, we upper bounded $\tilde{\bar{\mu}}$ by $1+C_{41}$ and lower bounded $\tilde{\nu}$ by ε_{10}. Thus:

$$\int b_i\eta_{12,i} \geq \frac{\beta_{k,q}^2}{10,000}\frac{\varepsilon_{10}\bar{c}(\varepsilon_1)}{1+C_{41}}\left(\frac{c_0}{2}\int(|b|-|\varphi.|)^+ - \right. \tag{162}$$
$$\left. -\frac{c_0}{2}\frac{14}{k}-\frac{c_0}{2}\varepsilon_{11}-c_0\varepsilon_{10}-3\beta_{k,q}-2\varepsilon_1(1+C_{41})\right).$$

We can assume (162) to hold on the global η_{12} which we build. Indeed, we construct η_{12} by taking, at each function b a function $\psi\in\sum_{2\tilde{\varepsilon}_0}^k(b)$ which achieves nearly $O_{2\tilde{\varepsilon}_0}^k(b)$. Using Lemma 4, we build a related η_{12}. This η_{12} can be extended to a tiny weak neighborhood of b contained in a ball $B(o,n)$ so as to satisfy an inequality such as (161) above, with $\int\left(|b_i|-\tilde{\gamma}_k-\frac{1}{k}\right)^+$ replaced by $\frac{1}{2}O_{\tilde{\varepsilon}_0}^k(b)$, at any $\tilde{b}$ in this neighborhood. Taking then a related partition of unity and extracting a locally finite one we derive that our final η_{12} must satisfy (161) above with $O_{\tilde{\varepsilon}_0}^k(b)$ replacing $\int\left(|b|-\tilde{\gamma}_k-\frac{1}{k}\right)^+$. Using (159) and (160) we derive (162) for η_{12}. This is entirely compatible with the H^1-weak construction which we are following.

Then, would $\int(|b|-|\varphi.|)^+$ remain always larger than $\frac{28}{k}+\varepsilon_{11}+2\varepsilon_{10}+\frac{c_0}{kc_0}$, then choosing $\varepsilon_1=\frac{\frac{28}{k}+\varepsilon_{11}+2\varepsilon_{10}}{100(1+C_{41})}$, quoting that $\beta_{k,q}<\frac{1}{k}$, we derive that:

$$\int b_i\eta_{12,i}\geq d(\varepsilon_{10},\varepsilon_{11},C_{41},k)>0. \tag{163}$$

Taking U_i very small (depending only on the parameters of the problem), we can ask that, at b:

$$\int b\bar{\eta}_{12}\geq\frac{d(\varepsilon_0,\varepsilon_{11},C_{41},k)}{2} \tag{164}$$

for any s, which contradicts the fact that c has to tend to zero along a subsequence. Therefore, we must have, on a subsequence s_r:

$$\int(|b|-|\varphi.|)^+(s_r)<\bar{c}_0\frac{28}{k}+\varepsilon_{11}+2\varepsilon_{10} \tag{165}$$

as claimed. iii) follows.

We now prove ii): Since $\varphi.$ has an oscillation larger than 4 $\text{Sup}(\varepsilon_{10},\varepsilon_{11},1/k)$ on $\bigcup_{\ell=1}^{r}J_\ell$, b, which converges weakly to $\varphi.$ must have an oscillation larger than 2 $\text{Sup}(\varepsilon_{10},\varepsilon_{11},1/k)$ on J_ℓ, $1\leq\ell\leq r$ of "strength" at least equal to half the "strength" of the oscillation of $\varphi.$, where the strength of the oscillation of $\varphi.$ designates $\int_{J_\ell}(\varphi.-\mu_{1,\ell})^+$ on the interval J_ℓ where this oscillation of amplitude larger than

4 Sup$(\varepsilon_{10}, \varepsilon_{11}, 1/k)$ is taking place. $\mu_{1,\ell}$ is the essential value (one of the essential values of $\varphi.$) at the boundaries of J. Assume that

$$C\left(|J_\ell| + \int_{J_\ell} |\varphi.|^2\right)|J_\ell| < \varepsilon_0/20 \qquad \ell = 1, \dots, r \tag{166}$$

b has a related oscillation, of "strength" $\sum_i \int_{\widetilde{J}_{\ell,i}} (b - \mu_{1,\ell} + o(1))^+ \geq \frac{1}{2} \int_{J_\ell} (\varphi. - \mu_{1,\ell})^+$ on a finite collection of intervals $J_{\ell,j}(s)$ such that $U\widetilde{J}_{\ell,j}$ is very close to J_ℓ. The reason for the collection rather than a single interval lies in the fact that b might change sign very rapidly on $J_{\ell,j}$. Thus, using the bound of $|b|_\infty$ and iii):

$$\sum |\widetilde{J}_{\ell,i}(s)| = |J_\ell| + o(1) \tag{167}$$

$$\sum \int_{\widetilde{J}_{\ell,i}(s)} b^2 = \int_{J_\ell} b^2 + o(1)$$

$$\leq \int_{J_\ell} \varphi.^2 + \overline{C}2000(C_{41}+1) \text{ Sup}\left(\frac{1}{q}, q\varepsilon_{11}, q\varepsilon_{10}\right) \tag{168}$$

and

$$C\left(|J_{\widetilde{\ell},i}(s)| + \int_{\widetilde{J}_{\ell,i}(s)} b^2\right)|\widetilde{J}_{\ell,i}(s)| < \tag{169}$$

$$M < \frac{\varepsilon_0}{20}(1 + o(1)) + 4000(1 + C_{41})\overline{C} \text{ Sup}\left(\frac{1}{q}, q\varepsilon_{11}, q\varepsilon_{10}, \frac{q}{k}\right).$$

We can choose q very large, $q\varepsilon_{11}$, $q\varepsilon_{10}$ and q/k very small is that:

$$4000(1 + C_{41})\overline{C} \text{ Sup}\left(\frac{1}{q}, q\varepsilon_{11}, q\varepsilon_{10}, \frac{q}{k}\right) < \frac{\varepsilon_0}{10}. \tag{170}$$

Thus:

$$C\left(|\widetilde{J}_{\ell,j}(s)| + \int_{\widetilde{J}_{\ell,j}(s)} b^2\right)|\widetilde{J}_{\ell,j}(s)| < \varepsilon_0/5 \tag{171}$$

with, for r large enough:

$$\sum_{\ell=1}^{r}\sum_i \int_{\widetilde{J}_{\ell,j}(s)} (b - \mu_{i,\ell} - o(1))^+ \geq \frac{1}{2}\sum_{\ell=1}^{r} \int_{J_\ell} (\varphi. - \mu_{1,\ell})^+ \geq \frac{1}{4}\varepsilon_{12} + \frac{5}{k} \tag{172}$$

$$b = \mu_{1,\ell} + o(1) \text{ on } \partial \widetilde{J}_{\ell,j}(s) \text{ and } b \text{ has an} \tag{173}$$

oscillation larger than 2 $\mathrm{Sup}(\varepsilon_{11}, 1/k, \varepsilon_{10})$ or $\widetilde{J}_{\ell,j}$, $\ell = 1, \dots, r$.

We can then repeat the previous argument with $\frac{\beta_{k,q}}{40(1+2c_0)} = \theta_1$: adding to $\mu_{1,\ell}1/k$ will reduce, but not kill the oscillation of $\varphi.$, since its amplitude is at least 4 $\sup(\varepsilon_{11}, 1/k, \varepsilon_{10})$. We can also assume that it is above ε_{10}.

J_ℓ becomes J_ℓ^1, $\widetilde{J}_{\ell,j} = \widetilde{J}_{\ell,j}(s)$ becomes $\widetilde{J}_{\ell,1}^1(s)$, which bears an oscillation of type 1), since (171) holds. We can apply Lemma 4 to b or rather to b_i the center of U_i, a ball of very small radius related to ε_{12} (so small that b_i has an oscillation of size at least $\varepsilon_{11} + \frac{2}{k} + \varepsilon_{10}$ above a value larger than ε_{10} and with a strength at least $\varepsilon_{12/4} + \frac{5}{k} + \varepsilon_{10}$). We again derive that c cannot tend to zero, a contradiction ii) follows.

One can check easily that ii) and iii) carry to sequences b_r bounded by $1 + C_{41}$ in the L^∞-norm, continuous i.e. we have:

Lemma 10. *Let (b_r) be a sequence of continuous functions bounded by $C_{41} + 1$. Then, i) up to extraction of a subsequence, (b_r) converges weakly to a function $\varphi.$.*

Furthermore, assuming that $O_{\bar{\varepsilon}_0}(b_r)$ tends to zero, then:

ii) Let $J = UJ_\ell, J_\ell$ disjoint intervals, be a collection of intervals where $\varphi.$ has oscillations above values $\mu_{1\ell}$ of size $\hat{\delta} > 0$, $\hat{\delta} > 0$ a fixed given number. Then, either there exists an interval J_ℓ such that $C(|J_\ell| + \int_{J_\ell} \varphi_\cdot^2)|J_\ell| \geq \varepsilon_0/20$. Or

$$\sum \int_{J_\ell} (|\varphi.| - \mu_{1,\ell})^+ < 12\hat{\delta}$$

iii) Under the same assumptions than ii), $\varlimsup_{r \to +\infty} \int_0^1 |b_r - \varphi.| < 100\hat{\delta}$.

The proof of Lemma 10 is embedded in the proof of ii) and iii) of Lemma 9.

As pointed out earlier, the flow built following i) of Definition 2 has the property of controlling $\frac{\partial}{\partial s} \int (|b| - \nu)^+$ by $-C\frac{\partial a}{\partial s}$. This property, due to our construction and the careful choice of our parameters (see the definition of $\beta_{k,q}$) extends to all $\frac{m}{q}$, $m \in \mathbb{N}$, $\frac{m}{q} < C_{41} + 1$ which are not integers. Thus, if we only used the cancellation flow, we would have:

$$\frac{\partial}{\partial s} \int \left(|b| - \frac{m}{q} \right)^+ \leq -C \frac{\partial a}{\partial s} \qquad \forall m \in \mathbb{N} \tag{174}$$

such that $\frac{m}{q} < C_{41} + 1$, $\frac{m}{q}$ not an integer. However, other parts of our flow are built following ii) or iii) of Definition 2. We claim that for these, hence for the convex combination of those and our cancellation flow, we have:

$$\begin{aligned} \frac{\partial}{\partial s} \int \left(|b| - \frac{m}{q} \right)^+ &\leq -C\left(\varepsilon_{10}, \varepsilon_{15}, \varepsilon_{16}, m, |b|_\infty q\right) \frac{\partial a}{\partial s} \\ &= -C(\varepsilon_{10}, \varepsilon_{15}, \varepsilon_{16}, m, C_{41} + 1, q) \frac{\partial a}{\partial s} \end{aligned} \tag{175}$$

which will satisfy our needs at this point.

We, indeed, observe that with any η of ii) or iii) of Definition 2, we have:

$$\int_0^1 b\eta \geq \varepsilon_{10} c(\varepsilon_{15}, \varepsilon_{16}, |b|_\infty) = d(\varepsilon_{10}, \varepsilon_{15}, \varepsilon_{16}, |b|_\infty) > 0 \tag{176}$$

since the functions η or ω (see ii) of Definition 2 and iii), also (18)) are built on intervals of length at least ε_{16} or $c(|b|_\infty) > 0$ and on intervals where $|b|$ is larger than ε_{10} (requirement a) of our present construction). We then have with such a flow:

$$\frac{\partial}{\partial s} \int_0^1 \left(|b| - \frac{m}{q}\right)^+ = \sum \frac{\dot{\eta} + \lambda b}{a} \operatorname{sgn} b \Big|_{z_j^-}^{z_j^+} + \int_{z_j^-}^{z_j^+} (a\eta\tau - b\eta\bar{\mu}_\xi) \operatorname{sgn} b \tag{177}$$

where z_j^-, z_j^+ are the various values such that $|b|$ assume the value $\frac{m}{q}$ · sgn b does not change on $[z_j^-, z_j^+]$. Clearly, using the lower bound on the length of our intervals which yields upper bounds on $|\eta|_{C^2}$:

$$\begin{aligned} \sum_j |\dot{\eta}(z_j^+) - \dot{\eta}(z_j^-)| &\leq \sum_j \left| \int_{z_j^-}^{z_j^+} \ddot{\eta} \right| \leq \int_0^1 |\ddot{\eta}| \leq C(\varepsilon_{16}, \varepsilon_{15}, |b|_\infty) \\ &\leq C_1(\varepsilon_{10}, \varepsilon_{15}, \varepsilon_{16}, |b|_\infty) \int_0^1 b\eta = -C_1 \frac{\partial a}{\partial s}. \end{aligned} \tag{178}$$

Also

$$\sum \left| \frac{\lambda + \bar{\mu} b}{a} \operatorname{sgn} b \Big|_{z_j^-}^{z_j^+} \right| \leq -C \frac{\partial a}{\partial s} \tag{179}$$

and, since $b\eta \geq 0$:

$$\sum \left| \int_{z_j^-}^{z_j^+} (a\eta\tau - b\eta\bar{\mu}_\xi) \operatorname{sgn} b \right| \leq \frac{C}{\varepsilon_{10}} \int_0^1 b\eta \leq -C_2 \frac{\partial a}{\partial s}. \tag{180}$$

We are thus left with

$$\begin{aligned} \sum \left| \bar{\mu} \frac{m}{q} \Big|_{z_j^-}^{z_j^+} \right| &= \frac{m}{q} \sum \int_{z_j^-}^{z_j^+} |\dot{\bar{\mu}}| \leq \frac{m}{q}(1 + |b|_\infty) \leq \frac{m}{q} \frac{(1 + |b|_\infty)}{d} \int_0^1 b\eta \\ &\leq -C_3 \frac{\partial a}{\partial s} \end{aligned} \tag{181}$$

Thus, using this flow, we have:

$$\frac{\partial}{\partial s} \int_0^1 \left(|b| - \frac{m}{q}\right)^+ \leq -C_4 \frac{\partial a}{\partial s} \tag{182}$$

and, if we combine with the cancellation flow, we obtain a similar estimate with a multiplicative coefficient depending on $(\varepsilon_{16}, \varepsilon_{15}, C_{41}, m, q, \varepsilon_{10})$. C_{41} has replaced $|b|_\infty$, since it upper bounds it.

As Lemma 9 shows, we have, at the end of a cancellation process, an interesting phenomenon:

Given an oscillation of $\varphi.$, a weak limit, of amplitude larger than $4\sup(\varepsilon_{11}, 1/k, \varepsilon_{10})$, we can lower μ_1, i.e. the constant above which $\varphi.$ (we assume for example that μ_1 is positive) has developed an oscillation

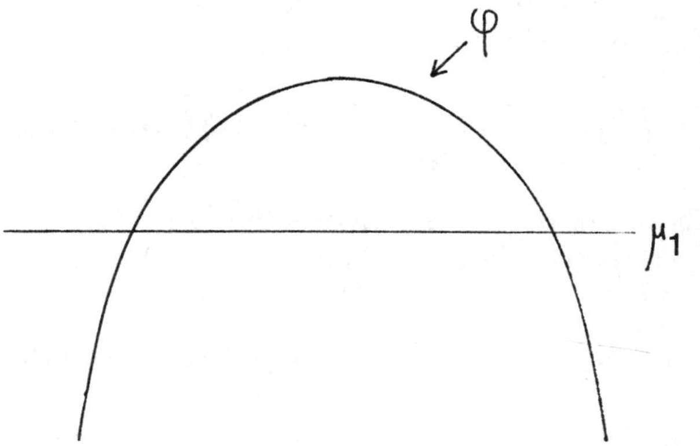

FIGURE F

Lowering μ_1, either we reach a value such that

$$\sum \int_{J_\ell} (|\varphi.| - \mu_1)^+ > 2\varepsilon_{12} + 4\ \mathrm{Sup}(\varepsilon_{11}, 1/k, \varepsilon_{10}) + \frac{24}{k} + 10\varepsilon_{10}. \tag{183}$$

(The reason for the discontinuity is that new J_ℓ's might appear when we lower μ_1.) Again all these oscillations, on each J_ℓ, will have amplitude at least 4 $\mathrm{Sup}(\varepsilon_{11}, 1/k, \varepsilon_{10})$, since we started with an oscillation of this type and only lowered μ_1, throwing away all the new oscillations which might develop and whose size is less than 4 $\mathrm{Sup}(\varepsilon_{11}, 1/k, \varepsilon_{10})$. We can then move μ_1 up, in $\mu_{1,\ell}$ – i.e. the movement depends on the interval J_ℓ – until we reach values $\mu_{1,\ell}$ such that (this time, the movement is continuous. No new interval J_ℓ develops suddenly yielding jumps)

$$\sum \int_{J_\ell} (|\varphi.| - \mu_{1,\ell})^+ = 2\varepsilon_{12} + 30\ \mathrm{Sup}(\varepsilon_{11}, 1/k, \varepsilon_{10}). \tag{184}$$

Throwing out the intervals where the oscillation is less than 4 $\mathrm{Sup}(\varepsilon_{11}, 1/k, \varepsilon_{10})$, we have:

$$2\varepsilon_{12} + 30\ \mathrm{Sup}(\varepsilon_{11}, 1/k, \varepsilon_{10}) \geq \sum \int_{J_\ell} (|\varphi.| - \mu_{1\ell})^+ \geq 2\varepsilon_{12} + 26\ \mathrm{Sup}\left(\varepsilon_{11}, \varepsilon_{10}, \frac{1}{k}\right). \tag{185}$$

This, in particular, proves, that some J_ℓ's are left; there exists then an interval J_ℓ such that:

$$C\left(|J_\ell| + \int_{J_\ell} \varphi_{\cdot}^2\right)|J_\ell| \geq \varepsilon_0/20 \tag{186}$$

by (ii) of Lemma 9. We will study this later.

Or, (183) is never satisfied, until $\mu_1 = \varepsilon_{10}$. Then, for all oscillations above μ_1 of size 4 $\mathrm{Sup}(\varepsilon_{11}, 1/k, \varepsilon_{10})$, we have

$$\sum \int_{J_\ell} (|\varphi.| - \varepsilon_{10})^+ \leq 2\varepsilon_{12} + 30\ \mathrm{Sup}(\varepsilon_{11}, 1/k, \varepsilon_{10}). \tag{187}$$

The oscillations above ε_{10} of size less than 4 $\mathrm{Sup}(\varepsilon_{11}, 1/k, \varepsilon_{10})$ will contribute to $\int_J (|\varphi.| - \varepsilon_{10})^+$ by 4 $\mathrm{Sup}(\varepsilon_{11}, 1/k, \varepsilon_{10})$. Thus:

$$\int_J (|\varphi.| - \varepsilon_{10})^+ < 2\varepsilon_{12} + 34\ \mathrm{Sup}(\varepsilon_{11}, 1/k, \varepsilon_{10}). \tag{188}$$

Using then iii), we derive:

$$\int_J (|b| - \varepsilon_{10})^+ \leq 2\varepsilon_{12} + 34\ \mathrm{Sup}(\varepsilon_{11}, 1/k, \varepsilon_{10}) + \\ + 2000\ C\ \mathrm{Sup}(1/q, q\varepsilon_{11}, q\varepsilon_{10}, q/k) \tag{189}$$

where J designates the maximal collection of intervals containing the initial oscillations such that $|\varphi.|$ remains above ε_{10}. We can even introduce $\widetilde{J}$, the maximal related collection of intervals such that $|b|$ remains above ε_{10} and derive:

$$\int_{\widetilde{J}} (|b| - \varepsilon_{10})^+ \leq 2\varepsilon_{12} + 34\ \mathrm{Sup}(\varepsilon_{11}, 1/k, \varepsilon_{10}) + \\ + 2000\ C\ \mathrm{Sup}(1/q, q\varepsilon_{11}, q\varepsilon_{10}, q/k). \tag{190}$$

Therefore, denoting $\widetilde{\widetilde{J}}$ the related connected component of positivity of $|\varphi.|$, we have:

$$\int_{\widetilde{\widetilde{J}}} |b| \leq 2\varepsilon_{12} + 50\ \mathrm{Sup}(\varepsilon_{11}, 1/k, \varepsilon_{10}) + 2000\ C\ \mathrm{Sup}(1/q, q\varepsilon_{11}, q\varepsilon_{10}, q/k). \tag{191}$$

Thus, we have:

Lemma 11. *We assume that* $|b|_\infty$ *is very large with respect to* $\int_0^1 |b|$, *so that* b *and* $\varphi.$ *are not close to a constant in the* L^1*-sense. Given a connected component* $\tilde{\tilde{J}}$ *of positivity of* $|\varphi.|$, *either*

(i) $$\overline{\lim} \int_{\tilde{\tilde{J}}} |b| \leq 2\varepsilon_{12} + 50\ Sup(\varepsilon_{11}, 1/k, \varepsilon_{10}) + 2000\ C\ Sup(1/q, q\varepsilon_{11}, q\varepsilon_{10}, q/k).$$

or

(ii) $|\varphi.|$ *has an oscillation of size at least* $4\ Sup(\varepsilon_{11}, 1/k, \varepsilon_{10})$ *above an appropriate value* $\mu_{1,\ell}$ *on an interval* J_ℓ *satisfying:*

a) $C\left(|J_\ell| + \int_{J_\ell} \varphi_.^2\right)|J_\ell| \geq \varepsilon_0/20$

b) $\int_{J_\ell}(|\varphi.| - \mu_{1,\ell})^+ \leq 2\varepsilon_{12} + 50\ Sup(\varepsilon_{11}, 1/k, \varepsilon_{10})$

c) $J_\ell \subset \tilde{\tilde{J}}$.

iii), i) and ii) extend to sequences (b_r) *under the same assumptions than ii) and iii) of Lemma 9.*

Proof. It is almost contained in the previous arguments. Starting with an oscillation of amplitude larger than 4 $\text{Sup}(\varepsilon_{11}, 1/k, \varepsilon_{10})$, we can first move μ_1 up so that the amplitude of the oscillation of $|\varphi.|$ above μ_1 is at most 5 $\text{Sup}(\varepsilon_{11}, 1/k, \varepsilon_{10})$ and at least 4 $\text{Sup}(\varepsilon_{11}, 1/k, \varepsilon_0)$. Of course, $\varphi.$ might have discontinuities. But any continuous function close enough to $\varphi.$ in a weak sense will have to fill the gaps, almost. This was the argument in order to prove ii) of Lemma 9. Thus, by oscillation on $\varphi.$ above a value μ_1, we include all values μ_1 which are intermediate values between essential values of $\varphi.$ at given points. Then, our claim makes sense: There exists a value of μ_1 so that the amplitude of the oscillation of $|\varphi.|$ above μ_1 is at most 5 $\text{Sup}(\varepsilon_{11}, 1/k, \varepsilon_{10})$ and at least 4 $\text{Sup}(\varepsilon_{11}, 1/k, \varepsilon_{10})$. (The values considered to compute the amplitude are the convex hull of the essential values of $\varphi.$.)

For this value of μ_1, we apply ii) of Lemma 9. Either, we have $C\left(|J_\ell| + \int_{J_\ell} |\varphi.|^2\right)|J_\ell| \geq \varepsilon_0/20$ with also:

$$\int_{J_\ell} (|\varphi.| - \mu_1)^+ \leq 5\ \text{Sup}(\varepsilon_{11}, 1/k, \varepsilon_{10}) \tag{192}$$

by construction. Or

$$\int_{J_\ell} (|\varphi.| - \mu_1)^+ < \varepsilon_{12} + \frac{12}{k} + 8\varepsilon_{10}. \tag{193}$$

We can then start to decrease μ_1 until we have

$$\sum \int_{J_\ell} (|\varphi.| - \mu_1)^+ > 2\varepsilon_{12} + 4\ \text{Sup}(\varepsilon_{11}, 1/k, \varepsilon_{10}) + \frac{12}{k} + 8\varepsilon_{10}, \tag{194}$$

where the sum is taken over all the intervals in $\tilde{\tilde{J}}$, containing the initial oscillation and where $|\varphi.| > \mu_1$. We can argue as above; (see (183)–(191)). Applying ii) of Lemmas 9 or 10, we then also have a), c) is immediate. Lastly, it is possible that, decreasing μ_1 to ε_{10}, we never have (194). (i) then holds, as seen above. qed

In case (i), there is no "energy" on $\tilde{\tilde{J}}$, for $\varphi.$ as well as for b. In case (ii), either we have:

$$|J_\ell| \geq \varepsilon_0/40C \tag{195}$$

ε_0 is a fixed constant, as well as C. Using b) of (ii), we then see – we let $\varepsilon_{12}, \varepsilon_{11}, \varepsilon_{10}$ go to zero, k go to $+\infty$ – that, on J_ℓ, $\varphi.$ is almost constant equal to μ_1.

Or, we have:

$$C \int_{J_\ell} \varphi.^2 |J_\ell| \geq \varepsilon_0/40. \tag{196}$$

Thus, using b) and the fact that $|\varphi.|$ is less than $C_{41}+1$:

$$C \int_{J_\ell} \mu_1^2 |J_\ell| \geq \varepsilon_0/40 - C(C_{41}+1)\left(2\varepsilon_{12} + 50 \text{ Sup}\left(\varepsilon_{11}, \varepsilon_{10}, \frac{1}{k}\right)\right) \tag{197}$$

hence, taking

$$C(C_{41}+1)\left(2\varepsilon_{12} + 50 \text{ Sup}\left(\varepsilon_{11}, \varepsilon_{10}, \frac{1}{k}\right)\right) < \frac{\varepsilon_0}{80} \tag{198}$$

$$\int_{J_\ell} |\varphi.| \geq \sqrt{\varepsilon_0/80C}. \tag{199}$$

Since we have a global bound, which we will preserve on $\int |b|$, hence on $\int |\varphi.|$, we have only a finite number of such intervals J_ℓ. This holds also in case of (195).

We thus have established that there are a finite number of such intervals J_ℓ corresponding to (ii) of Lemmas 9 or 10. The bound on this number depends only on our initial map which we are trying to deform and on nothing else.

We can try to repeat the argument of (ii) of Lemma 11 for several intervals J_ℓ, i.e. after finding the first one, or several of the same type, we could still have oscillations, on this connected component from which we remove the intervals J_ℓ already singled out, of amplitude larger than 4 $\text{Sup}(\varepsilon_{10}, \varepsilon_{11}, 1/k)$. We can then try to repeat the same argument. Either we find new intervals satisfying a), b) and c). Their total number remains bounded by the same bound. Or, and this is what will happen in the end, we find that the total strength $\sum \int_{J'_p} (|\varphi.| - \mu'_{1,p})^+$ on the intervals J'_p where there is an oscillation of amplitude at least 4 $\text{Sup}(\varepsilon_{10}, \varepsilon_{11}, 1/k)$, outside the ones singled out, is at most $\varepsilon_{12} + \frac{12}{k} + 4\varepsilon_{10}$. Outside the J_ℓ and J'_p, there is no oscillation of amplitude at least 4 $\text{Sup}(\varepsilon_{10}, \varepsilon_{11}, 1/k)$.

Thus we have:

Lemma 12. *i)* $\varphi.$ *has a finite number* ℓ_1*, where* ℓ_1 *depends on the initial map we are deforming and on nothing else, of oscillations of amplitude at least* 4 *Sup*$(\varepsilon_{11}, 1/k, \varepsilon_{10})$ *on intervals* $J_1, \dots, J_{\ell_1}$ *above values* $\mu_{1,1}, \dots, \mu_{1,\ell_1}$ *satisfying a) and b) of Lemma 11.*
ii) On [0,1] – UJ_ℓ*, there exists a function* $\psi.$*, having no oscillation of amplitude* 4 *Sup*$(\varepsilon_{11}, 1/k, \varepsilon_{10})$ *or more on each connected component of* $[0,1] - UJ_\ell$*, such that*

α) $\varphi.$ *and* $\psi.$ *have the same sign*

β) $|\varphi.| \geq |\psi.|$

γ) $$\int_{[0,1]-UJ_\ell} (|\varphi.| - |\psi.|) = \int_{[0,1]-UJ_\ell} |\varphi. - \psi.| < 2\varepsilon_{12} + 50 \ \mathit{Sup}\left(\varepsilon_{11}, \frac{1}{k}, \varepsilon_{10}\right).$$

Proof. We will argue as if $\varphi.$ was continuous. Otherwise, using arguments similar to the ones used in the proof of Lemma 4, or approximating $\varphi.$ with a continuous function bounded by $C_{41} + 1$, very close to $\varphi.$ in L^1 and L^2, we derive a similar result, which passes to the limit.

Let us consider the intervals I_ℓ, I_ℓ being an interval of [0,1] – UJ_ℓ. If on I_ℓ, $\varphi.$ has no oscillation of amplitude 4 Sup$(\varepsilon_{11}, \varepsilon_{10}, 1/k)$, we are done. We set $\psi. = \varphi.$ on I_ℓ. Otherwise, $\varphi.$ has an oscillation on J'_p, of amplitude 4 Sup$(\varepsilon_{11}, \varepsilon_{10}, 1/k)$ above $\mu'_{1,p}$. Replacing $\varphi.$ by φ'_0, where

$$\begin{aligned} \varphi'_0 &= \varphi. \quad \text{on} \quad I_\ell - J'_p \\ \varphi'_0 &= \mu'_{1,p} \quad \text{on} \quad J'_p \end{aligned} \tag{200}$$

we can restart our process and iterate.

We then have a decreasing sequence of functions, which has to converge to a function $\psi.$, having no oscillation of amplitude 4 Sup$(\varepsilon_{11}, 1/k, \varepsilon_{10})$ or more. α) and β) are obviously satisfied. If γ) holds (considering all intervals I_ℓ together), we are done. Let us assume that γ) does not hold.

Observe that, in all this process, we can assume that the oscillations we are removing are such that $\psi'. = \varphi.$ on the boundaries of the related intervals, where $\psi'.$ stands for the function involved in that step, from which we remove the oscillation. This is established step by step, by induction: as we remove an oscillation, if the boundaries of an interval where it takes place are not such that $\psi'. = \varphi.$ at these boundaries, then the interval is not maximal, because $\psi'.$ is not equal to $\varphi.$, at each stage, only on intervals J where $\psi'.$ is constant, equal to d_J. If the oscillation we are removing has a boundary inside J, then the related constant μ_1 is d_J and we can extend the interval of the oscillation and include J in it. The process could be infinite but modifying a tiny bit $\varphi.$ into an analytic function, we can assume that it is a finite process. Passing to the limit will yield the result.

We can rebuild $\varphi.$ from the final function $\psi.$ by adding oscillations of amplitude 4 Sup$(\varepsilon_{11}, 1/k, \varepsilon_{10})$ or more. Each of these oscillations has a "strength" less that

$\int_J(|\varphi.| - \mu_1)^+$: if $\psi'_.$ oscillates above μ_1 on $J, \varphi.$, which dominates $\psi'_.$, oscillates above μ_1 and if J is maximal, $\varphi.$ is equal to $\psi'_.$ on ∂J, thus to μ_1.

This strength $\int_J(|\varphi.|-\mu_1)^+$, hence $\int_J(|\psi'_.|-\mu_1)^+$ has to be less than $\varepsilon_{12}+\frac{12}{k}+8\varepsilon_{10}$, otherwise by ii) of Lemmas 9 or 10, there would be an additional interval J_ℓ. If we assumed that our initial family were maximal, this would be impossible. Working in this way, we move back from $\psi.$ to $\varphi.$ by tiny bits less than $\varepsilon_{12}+\frac{12}{k}+8\varepsilon_{10}$. At each step, we have a new function, satisfying α), β), having now, possibly, oscillations of amplitude 4 $\mathrm{Sup}(\varepsilon_{11}, 1/k, \varepsilon_{10})$ or more and satisfying:

$$\int_{[0,1]-UJ_\ell} (|\varphi.| - |\psi'_.|) = \int_{[0,1]-UJ_\ell} |\varphi. - \psi'_.| \geq 2\varepsilon_{12} + 50 \ \mathrm{Sup}\left(\varepsilon_{11}, \frac{1}{k}, \varepsilon_{10}\right) \tag{201}$$

(201) cannot hold forever, since when we have added up all the removed oscillations, $\psi'_.$ becomes $\varphi.$ contradicting (201). Thus, there is a step where (201) does not hold anymore. Taking the first such step, (201) holds a step earlier. With the addition of the new oscillation, we have a drop of at most $\varepsilon_{12}+\frac{12}{k}+8\varepsilon_{10}$. Thus

$$\begin{aligned} \varepsilon_{12} + 30 \ \mathrm{Sup}\left(\varepsilon_{11}, \frac{1}{k}, \varepsilon_{10}\right) < \int_{[0,1]-UJ_\ell} (|\varphi.| - |\psi'_.|) &= \int_{[0,1]-UJ_\ell} |\varphi. - \psi'_.| \\ &\leq 2\varepsilon_{12} + 50 \ \mathrm{Sup}\left(\varepsilon_{11}, \frac{1}{k}, \varepsilon_{10}\right). \end{aligned} \tag{202}$$

By construction – we have removed only inside oscillations – $\varphi.$ is equal to $\psi'_.$ on the boundaries of J_ℓ. $\varphi.$ is obtained from $\psi'_.$ by adding oscillations of amplitude 4 $\mathrm{Sup}(\varepsilon_{10}, \varepsilon_{11}, 1/k)$ or more.

Clearly,

$$\int_{[0,1]-UJ_\ell} (|\varphi.| - |\psi'_.|) = \sum \int_{J'_p} (|\varphi.| - \mu'_{1,p})^+ \tag{203}$$

where $|\varphi.|$ takes the value $\mu'_{1,p}$ on $\partial J'_p$. Using (202) and (203), we can repeat the argument of (186)–(191), apply ii) of Lemmas 9 or 10 (or 11). We derive the existence of another interval of the type of J_ℓ. This cannot hold more than ℓ_1 times. After, ii) of Lemma 12 must hold. q.e.d.

Remark. When $\varphi.$ is not analytic, we can approximate it with analytic functions which will have property ii) of Lemma 12, with slightly modified constants. This is mainly due to the finiteness of the intervals J_ℓ for $\varphi.$, which holds by (ii) of Lemma 12. Applying Lemma 12 to these analytic functions, we derive the result for $\varphi.$ through a limit process. $\psi.$ will satisfy α), β), γ) of Lemma outside a set N of zero measure.

$\varphi.$ has now a clear behavior: a few intervals J_ℓ where it is almost constant. The J_ℓ's are either large enough (of length larger than δ_3, a fixed constant) or $\int|\varphi.|$ is large enough (larger than ε_0^3, ε_0 a fixed constant). Outside of these intervals, $\varphi.$ oscillates very little in average over a function $\psi.$ which has no oscillation of size $4\ \mathrm{Sup}(\varepsilon_{10}, \varepsilon_{11}, \frac{1}{k})$.

b is very close to $\varphi.$, hence follows the same pattern. For such functions and curves, we are very much tempted to use the η_{12} of ii) and iii) of Definition 2 on the intervals J_ℓ such that $\mu_{1,\ell}$ is away from $|b|_\infty$ and ν (as required by ii) and iii) of Definition 2). We have to keep away from these two values, otherwise we might loose control of $|b|_\infty$ or of $\int(|b| - \nu)^+$. Observe that we can do this on J_ℓ because, either $|J_\ell| \geq \delta_3$ or $\int_{J_\ell} |\varphi.|$ or $\int_{J_\ell} |b|$ is larger than ε_0^3, thus $|J_\ell| \geq \varepsilon_0^3/1 + C_{41}$. Choosing ε_{16} in Definition 2 less than $\frac{\varepsilon_0^3}{2(1+C_{41})}$, we can fit in our framework. C_{41} is related to $|b|_\infty$. If $\mu_{1,\ell}$ is away from $|b|_\infty$, such a flow does not change $|b|_\infty$. We thus can use $C_{41} + 1$ as a bound on $|b|_\infty$.

Between two consecutive intervals J_ℓ and $J_{\ell+1}$, $\varphi.$ oscillates less, in average, than $4\ \mathrm{Sup}(\varepsilon_{10}, \varepsilon_{11}, \frac{1}{k})$. Therefore, always in average i.e. for $\psi.$ of Lemma 12, if $|\varphi.|$ falls below $|\varphi.|_\infty - 50\ \mathrm{Sup}(\varepsilon_{10}, \varepsilon_{11}, \frac{1}{k})$, it cannot come back above $|\varphi.|_\infty - 45\ \mathrm{Sup}(\varepsilon_{10}, \varepsilon_{11}, \frac{1}{k})$, unless it does not fall back later below $|\varphi.|_\infty - 50\ \mathrm{Sup}(\varepsilon_{10}, \varepsilon_{11}, \frac{1}{k})$. The same result holds for $\nu \pm 50\ \mathrm{Sup}(\varepsilon_{10}, \varepsilon_{11}, \frac{1}{k})$ instead of $|\varphi.|_\infty$. And a similar result holds for b in average. That is, starting from $\mu_{1,\ell}$ and going to $\mu_{1,\ell+1}$, and working in a limit process as if ε_{10}, ε_{11} were zero and k were infinite, $\psi.$ would either be a decreasing or an increasing function from $\mu_{1,\ell}$ to $\mu_{1,\ell+1}$ or would fall (in absolute value) to a minimum and increase then (in absolute value) to $\mu_{1,\ell+1}$.

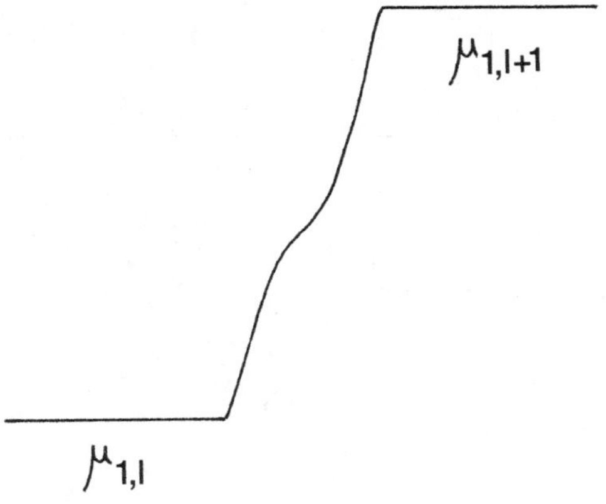

FIGURE G

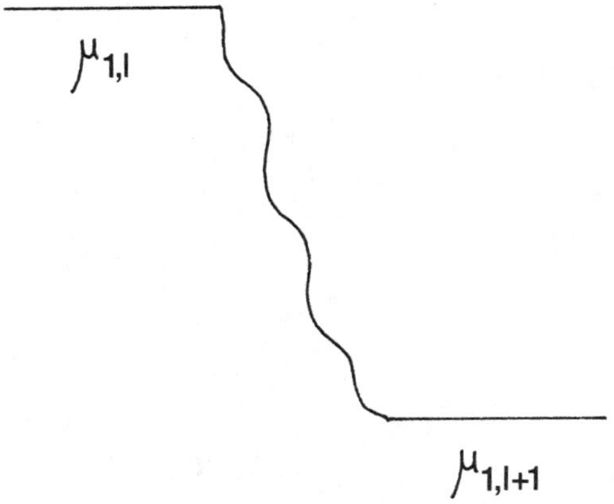

FIGURE H

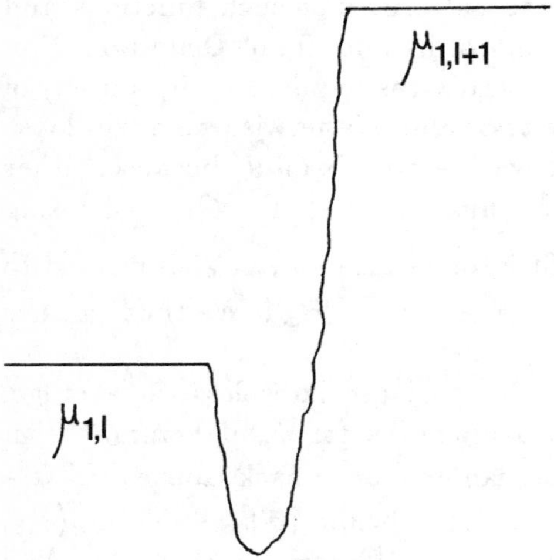

FIGURE I

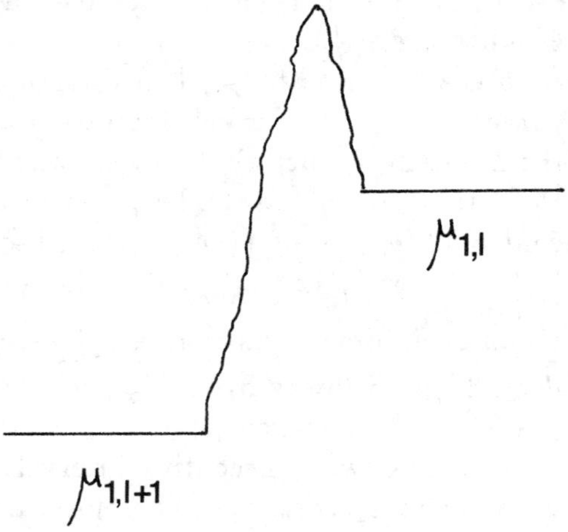

FIGURE J

In the third case, μ_ℓ and $\mu_{1,\ell}$ must have the same sign, because, if there is a change of sign, there is an oscillation of $|\psi.|$ above $|\mu_\ell|$. Such oscillations, in the limit, must be zero, thus Figure I must be:

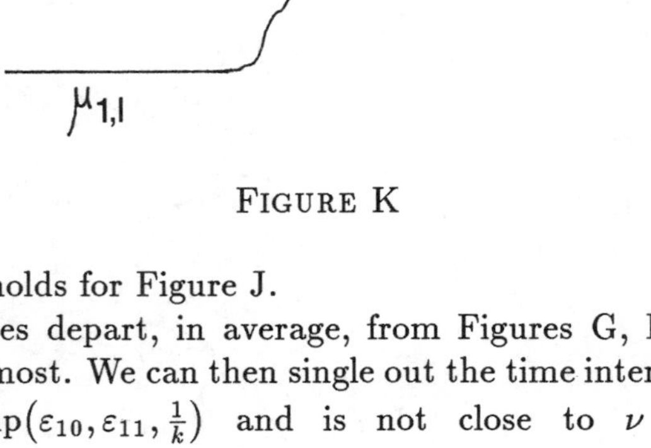

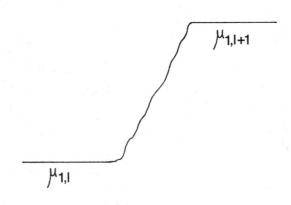

FIGURE K

The same argument holds for Figure J.

Our actual pictures depart, in average, from Figures G, H, I and J only by $4\ \mathrm{Sup}(\varepsilon_{10}, \varepsilon_{11}, \frac{1}{k})$ at most. We can then single out the time intervals where $|\psi.|$ is not above $|\psi.|_\infty - 4\ \mathrm{Sup}(\varepsilon_{10}, \varepsilon_{11}, \frac{1}{k})$ and is not close to ν or to ε_{10} up to

4 $\text{Sup}(\varepsilon_{10}, \varepsilon_{11}, 1/k)$. In the limit picture, there are at most four of these intervals. In the actual picture, a similar process happens: there cannot be more than two back and forth movements between $|\psi.|_\infty - 4\ \text{Sup}(\varepsilon_{10}, \varepsilon_{11}, \frac{1}{k})$ and $|\psi.|_\infty - 8\ \text{Sup}(\varepsilon_{10}, \varepsilon_{11}, \frac{1}{k})$ and the same result holds for $\nu \pm 8\ \text{Sup}(\varepsilon_{10}, \varepsilon_{11}, \frac{1}{k})$ and $\nu \pm 4\ \text{Sup}(\varepsilon_{10}, \varepsilon_{11}, \frac{1}{k})$. In fact, there cannot be more than two movements back and forth between $|\psi.|_\infty - 8\ \text{Sup}(\varepsilon_{10}, \varepsilon_{11}, \frac{1}{k})$ and $\nu + 8\ \text{Sup}(\varepsilon_{10}, \varepsilon_{11}, \frac{1}{k})$ (we are assuming $|\psi.|_\infty$ is large enough) or between $\nu - 8\ \text{Sup}(\varepsilon_{10}, \varepsilon_{11}, \frac{1}{k})$ and $\varepsilon_{10} + 8\ \text{Sup}(\varepsilon_{10}, \varepsilon_{11}, 1/k)$. Therefore, we have finitely many of such intervals. Outside of them, the function $\psi.$ is close to $\pm|\psi.|_\infty$ or to $\pm\nu$ or to zero. $\pm|\psi|_\infty$ occurs only near $\mu_{1,\ell}$ or $\mu_{1,\ell+1}$, including the case where it occurs all over $\psi.^{-1}[\mu_{1,\ell}, \mu_{1,\ell+1}]$. The two intervals J_ℓ and $J_{\ell+1}$ can then be somewhat compounded.

$\pm\nu$ and 0 occur in the middle, at most twice for ν and once for zero; otherwise, again, sizable oscillations develop.

If the intermediate intervals, which are thus in finite number, four or six at most, do not collapse below ε_{12}, we are again tempted to use the η_{12} of ii) or iii) of Definition 2. The same problem develops: $\varphi.$ and b might oscillate and reach $|b|_\infty$ or ν above these intermediate intervals, ruining thereby our control on $|b|_\infty$ and $\int(|b| - \nu)^+$. We need some intermediate step to take care of this problem.

The idea for this step is the following: when we started our cancellation process we have decreased a at the beginning a tiny bit $\delta a > 0$. This small amount can be taken to be uniform on all the compact sets which we are trying to deform. We choose our parameters $q, \varepsilon_{11}, \varepsilon_{10}, \varepsilon_{12}, k$ so that

$$(1 + (C_{41} + 2)^3)\ \text{Sup}\left(\frac{1}{q}, q\varepsilon_{11}, q\varepsilon_{10}, \frac{q}{k}, \varepsilon_{12}\right) = o(\delta a). \tag{204}$$

How small o should be, we will see later.

We then introduce the

Definition 3. *Regularizing flow to be the flow corresponding to* $\eta = \frac{b}{1+|b|^3_\infty}$ *or* $\eta = b$.

Clearly, the use of this flow makes us loose control on $|b|_\infty$. However, designating $\frac{\partial}{\partial s}$ this flow, we have:

$$\frac{\partial}{\partial s}|b|_\infty \leq C\left(\text{for } \eta = \frac{b}{1+|b|^3_\infty}\right); \quad \frac{\partial}{\partial s}|b|_\infty \leq C(1+|b|^3_\infty)(\text{for } \eta = b) \tag{205}$$

where C does not depend on anything, but the geometry in the problem. Therefore, if we use $\eta = \frac{b}{1+|b|^3_\infty}$ during a time

$$s \leq \delta a < \frac{1}{C} \tag{206}$$

we still keep control on $|b|_\infty$ ($|b|_\infty \le C_{41}+2$). If we study what happens with respect to $\eta = b$ rather than $\eta = \frac{b}{1+|b|_\infty^3}$, the time allowed will be:

$$s \le \frac{\delta a}{1+(C_{41}+2)^3} = \tilde{s}_0 \tag{207}$$

as any easy argument will show (take $\delta a < \frac{1}{C+1}$. In the sequel δa will be the absolute value of a previous decrease in a, using, for example, the cancellation flow.) C_{41} depends only on our initial compact and on the geometric data of the problem, even after iterating this process several times, since i) of Lemma 3 holds at each stage for the cancellation flow and since the other steps which follow – those already defined and those to come-keep this control. This time s_0 which we are given, we use it, under (207), in order to average out b, hence φ. and bring it to be close to its average values, hence to $\mu_{1,\ell}$ on J_ℓ and to ψ. on the intermediate intervals. We can then apply the flow η_{12} of 1) and ii) of Definition 2 on the intervals where b, averaged, is between $\pm\nu + o(1)$ and $\pm|b|_\infty - o(1)$ or $\nu - o(1)$ and $o(1)$. This yields a three stage process, which can be iterated, providing control on $|b|_\infty$ and on $\int(|b|-\nu)^+$. The iteration yields the fact that the intervals between $\pm|b|_\infty$ and $\pm\nu$, $\pm\nu$ and zero, have to be tiny, less than ε_{16}. b then becomes made up of pieces, J_ℓ where $b = \pm|b|_\infty \pm o(1)$ with $\int_{J_\ell}|b| \ge \delta_0$, pieces $J_{\ell'}$ where $b = \pm\nu \pm o(1)$, two at most between J_ℓ and $J_{\ell+1}$, and pieces $K_{\ell'}$ when $b = o(1)$, one at most between J_ℓ and $J_{\ell+1}$.

The crucial step in this iterative process is the regularizing process. The other steps follow naturally. We start, thus, by describing the effect of this regularizing process. We then show how to take all these processes together, including the two next steps which we have not yet described. A crucial fact, here, relies in the fact that, as long as we are "away" from the curves built out of sizable $\pm v$-pieces followed by ξ-pieces, a will always decrease in this iterative process of at least a fixed amount of δa. This will allow to keep the parameters $\varepsilon_{10}, \varepsilon_{11}, \varepsilon_{12}, \varepsilon_{16}$ away from zero as long as we are not in the "critical zone" for the curves behaving as those of Lemmas 9, 10 etc. there will be one additional problem, which we will take care of in time, when we will globalize the whole process. We have:

Lemma 13. *Let $\varepsilon_{17} > 0$ be given small enough in function of C_{41}. Let Z be the vector-field generated by $\eta = b$*
i) there exists $s_0(C_{41}) > 0$, $s_1(K)$ – where K is the initial compact we are trying to deform – such that if we use Z, during the time $\inf(s_0(C_{41}), \varepsilon_{17}^3, s_1(K))$ on curves $x_0(t)$ which component $b_0(t)$ on v behaves as the functions $b(s,t)$ of Lemma 9 when s tends to the explosion time or to $+\infty$ – i.e. $b_0(t)$ satisfies iii) of Lemma 9 with respect to a function φ., bounded by C_{41}, satisfying ii) α), β), γ) of Lemma 12 and if $\varepsilon_{11}, \varepsilon_{10}, \frac{1}{k}, \frac{1}{q}$ are small enough in function of ε_{17} and of $s_1(K)$, the resulting curve $x(s,t)$ has a component $b(s,t)$ which behaves like ψ. i.e. $b(s,t)$ is nearly constant on a finite number of intervals J_ℓ. The bound on the number depends only on the initial compact set of curves we started with. Between these intervals J_ℓ, $b(s,t)$ behaves

like the function ψ. *of Figures G, H, I, J if the length of this interval is larger than* $2\sqrt{\varepsilon_{17}}$. $b(s,t)$ *is bounded by* $C_{41}+1$.
ii) After applying Z *during the time* $\inf(s_0(C_{41}), s_1(K), \varepsilon_{17}^3)$ *we can use the vector-field* $Z_{\varepsilon_{17}}$ *of ii) or iii) of Definition 2 (with* $\varepsilon_{16} = 2\sqrt{\varepsilon_{17}}$ *for example) on* $x(s,t)$. *If the intermediate intervals, between the* J_ℓ, *where* $|b(s,t)|$ *belongs to* $[\nu + o(1), |b|_\infty - o(1)]$ *or* $[o(1), \nu - o(1)]$, *where* $o(1)$ *is small given and* $\varepsilon_{11}, \varepsilon_{12}, \varepsilon_{10}, \frac{1}{k}, \frac{1}{q}$ *are chosen small enough with respect to this* $o(1)$, *are of size larger than* $2\sqrt{\varepsilon_{17}}$, *the energy level* a *of the curve (the process is not globalized; we are speaking of a given curve, which started with a special* $b_0(t)$, *behaving as* $b(s,t)$ *of Lemma 9) will decrease in this process by a fixed amount* $\beta(o(1), \varepsilon_{17}) > 0$.
iii) Finally, throughout this problem (with Z *and* $Z_{\varepsilon_{17}}$ *combined),* $|b|_\infty \le |b|_\infty(o) - C(a(\tau) - a(o))$, *where* C *is a universal constant.*

Remark. $s_1(K)$ is provided by (207): δa is related to K. This will be studied in the sequel.

Proof of Lemma 13. We compare the solution of the evolution equation:

$$\begin{cases} \dfrac{\partial \bar{b}}{\partial s} = \ddot{\bar{b}} \\ \bar{b}(0,t) = b_0 \end{cases} \tag{208}$$

with the one we obtain in our deformation process when $\eta = b$

$$\begin{cases} \dfrac{\partial b}{\partial s} = \ddot{b} + \dot{\overline{\lambda b}} + a(ab\tau - b^2_{\bar{\mu}_\xi}) \\ b(0,t) = b_0. \end{cases} \tag{209}$$

The first one has "mixing" properties, which will bring b_0 to look like its "weak limit", we will see this later. If the second one remains close to the first one, it will have then the same "mixing" properties.

Let

$$w = b - \bar{b}. \tag{210}$$

We assume that b_0 satisfies:

$$|b_0|_\infty \le C_{41} + 1. \tag{211}$$

We then know that:

$$|\dot{\bar{b}}|_\infty \le \frac{C(C_{41}+1)}{\sqrt{s}} \tag{212}$$

where C is a universal constant. (212) comes from the explicit representation formula for the solution of (208).

We assume that s is small enough so that the solutions of (208) and (209) satisfy:

$$\sup_{\tau \in [0,s]} |b(\tau,t)| \le C_{41} + 2, \quad \sup_{\tau \in [0,s]} |\bar{b}(s,t)| \le C_{41} + 2. \tag{213}$$

This provides an upper bound on s in function of C_{41} and of nothing else, (see, for example, for (209) the proof of Lemma 16, below; for (208),it is obvious) thus

$$0 < s \leq s_0(C_{41}). \tag{214}$$

We then have, with constants which depends on C_{41} and a_0, a_0 being the maximum of a on our initial compact

$$\frac{1}{2}\frac{\partial}{\partial s}\int_0^1 w^2 \leq -\int_0^1 \dot{w}^2 + A(C_{41}, a_0)\left(1 + \left(\int_0^1 \dot{w}^2\right)^{1/2}\right). \tag{215}$$

Thus

$$\frac{1}{2}\frac{\partial}{\partial s}\int_0^1 w^2 \leq -\frac{1}{2}\int_0^1 \dot{w}^2 + C(C_{41}, a_0) \tag{216}$$

which yields:

$$\frac{1}{2}\int_0^s\int_0^1 \dot{w}^2 + \frac{1}{2}\int_0^1 w^2 \leq C(C_{41}, a_0)s. \tag{217}$$

On the other hand, using (209):

$$\begin{aligned}\frac{1}{2}\frac{\partial}{\partial s}\int_0^1 \dot{w}^2 &\leq -\int_0^1 \ddot{w}^2 + A(C_{41}, a_0)\left(\int \ddot{w}^2\right)^{1/2}\left(1 + \left(\int_0^1 \dot{b}^2\right)^{1/2}\right) \\ &\leq -\frac{1}{2}\int_0^1 \ddot{w}^2 + C(C_{41}, a_0)\left(1 + \int_0^1 \dot{b}^2 + \int_0^1 \dot{w}^2\right) \\ &\leq -\frac{1}{2}\int_0^1 \ddot{w}^2 + C'(C_{41}, a_0)\left(1 + \frac{1}{s} + \int_0^1 \dot{w}^2\right).\end{aligned} \tag{218}$$

Thus, using (217):

$$\frac{1}{2}\int_0^1 \dot{w}^2(s) \leq \left(\frac{1}{2}\int_0^1 \dot{w}^2(\tau) + C'\left(s - \tau + \mathrm{Log}\,\frac{s}{\tau}\right)\right)e^{C''(s-\tau)}. \tag{219}$$

Using (217), we can find $\tau \in [s/2, s]$ such that

$$\int_0^1 \dot{w}^2(\tau) \leq 4C(C_{41}, a_0). \tag{220}$$

Plugging this τ in (219) and using the fact that $\frac{s}{\tau} \in \left[\frac{1}{2}, 1\right]$ yields:

$$\frac{1}{2}\int_0^1 \dot{w}^2(s) \leq C'''(C_{41}, a_0). \tag{221}$$

Combining (221) and (217) yields:

$$|w(s,t)|_\infty(s) \le \inf_{\alpha,\beta[0,1]} \left(C^{(iv)}(C_{41},a_0)\sqrt{|\beta-\alpha|} + \frac{\sqrt{2C(C_{41},a_0)s}}{\sqrt{|\beta-\alpha|}} \right)$$
$$\le C^{(5)}(C_{41},a_0)s^{1/4}. \tag{222}$$

In order to prove (212) (this is classical), we have:

$$\bar{b} = \sum_{-\infty}^{+\infty} \frac{1}{\sqrt{t}} \int_0^1 e^{-\frac{(x-y-k)^2}{4t}} b_0(x)dx. \tag{223}$$

Thus:

$$\dot{\bar{b}}(y) = -\sum_{-\infty}^{+\infty} \frac{1}{\sqrt{t}} \int_0^1 \frac{(x-y-k)}{2t} e^{-\frac{(x-y-k)^2}{4t}} b_0(x)dx \tag{224}$$
$$= -\sum_{-\infty}^{+\infty} \frac{1}{\sqrt{t}} \int_0^1 \frac{x-y-k}{2\sqrt{t}} b_0(x) e^{-\frac{(x-y-k)^2}{4t}} d\left(\frac{x-y-k}{2\sqrt{t}}\right).$$

Thus

$$|\dot{\bar{b}}(y)| \le \frac{2}{\sqrt{t}} |b_0|_\infty \sum_{-\infty}^{+\infty} \int_{\frac{y+k}{2\sqrt{t}}}^{\frac{y+k+1}{2\sqrt{t}}} |z|e^{-z^2}dz \le \frac{2|b_0|_\infty}{\sqrt{t}} \int_{-\infty}^{+\infty} |z|e^{-z^2}dz \tag{225}$$

(212) follows.

We now quantify the mixing properties of the heat flow. In our framework, b_0 is a $b(s,t)$, very close L^1 to $\varphi.$ and $\varphi.$ is very close L^1 to $\psi.$, where $\psi.$ is constant, equal to $\mu_{1,\ell}$ on the intervals J_ℓ and $\psi.$, up to 4 $\text{Sup}(\varepsilon_{10},\varepsilon_{11},1/k)$, obeys the pattern described in Figures G, H, I and J above.

We thus have:

$$|b_0 - \psi.|_{L^1} \le 2000 \text{ Sup}\left(\frac{1}{q}, q\varepsilon_{11}, q\varepsilon_{10}, \frac{q}{k}\right) + 4\varepsilon_{12} + \tag{226}$$
$$+ 20 \text{ Sup}\left(\varepsilon_{10},\varepsilon_{11},\frac{1}{k}\right) = \delta(k,q,\varepsilon_{11},\varepsilon_{10},\varepsilon_{12}) = \delta.$$

$$|b_0|_\infty \le C_{41}+1 \qquad |\psi.|_\infty \le C_{41}+1. \tag{227}$$

Let

$$\tilde{\psi}.(s,t) \text{ be the solution of } \begin{cases} \dfrac{\partial\tilde{\psi}.}{\partial s} = \ddot{\tilde{\psi}}. \\ \tilde{\psi}(0,t) = \psi.(t). \end{cases} \tag{228}$$

We thus have, using (222) and (223):

$$|\tilde{\psi}.(s,t) - b(s,t)|_\infty(s) \le C(C_{41}, a_0)s^{1/4} + \frac{C}{\sqrt{s}}\delta. \tag{229}$$

If we require:

$$\delta \le s \tag{230}$$

i.e. if we can wait long enough, we will have

$$|\tilde{\psi}.(s,t) - b(s,t)|_\infty(s) \le C\sqrt{\delta} + C(C_{41}, a_0)s^{1/4}. \tag{231}$$

This requirement (230), together with (214) and (207) yields:

$$\delta \le \inf\left(s_0(C_{41}), \frac{\delta a}{1 + (C_{41}+2)^2}\right). \tag{232}$$

Since C_{41} is given once and for all and δ tends to zero with ε_{10}, ε_{11}, ε_{12}, $\frac{1}{q}$ and $\frac{1}{k}$, (232) can be fulfilled. $b(s,t)$ becomes very close to $\tilde{\psi}.(s,t)$.

We can work some more and replace $\tilde{\psi}.(s,t)$ by $\tilde{\tilde{\psi}}_0(s,t)$, where $\tilde{\tilde{\psi}}_0(0,t)$ is close to $\psi.(t)$ and has the precise behavior of Figures G, H, I and J. The price to pay is some change of δ, an increase in it, of an amount tending to zero when ε_{10}, ε_{11}, $\frac{1}{k}$ tend to zero. We will work with $\tilde{\tilde{\psi}}_0(s,t)$.

Suppose that we want to analyze the behavior of $\tilde{\tilde{\psi}}_0(s,t)$ at t_0. Since $\tilde{\tilde{\psi}}_0(0,t)$ is bounded by C_{41}, corresponding to a parameter $\varepsilon_{17} > 0$, the contribution of the complement of $(t_0 - \varepsilon_{17}, t_0 + \varepsilon_{17})$ to $\tilde{\tilde{\psi}}_0(s,t_0)$ is bounded by

$$CC_{41}\frac{e^{-\varepsilon_{17}^2/4s}}{\sqrt{s}}. \tag{233}$$

Thus, if

$$0 < s \le \varepsilon_{17}^3 \quad \text{and} \quad \varepsilon_{17} \text{ is small enough} \tag{234}$$

this contribution is tiny, called $\gamma_{C_{41}}(\varepsilon_{17})$, $\tilde{\tilde{\psi}}_0(s,t_0)$ is then essentially equal to $\frac{1}{\sqrt{s}}\int_{t_0-\varepsilon_{17}}^{t_0+\varepsilon_{17}} e^{-(\frac{t-t_0}{4s})^2}\tilde{\tilde{\psi}}_0(0,t)dt = \frac{1}{\sqrt{s}}\int_{-\varepsilon_{17}}^{\varepsilon_{17}} e^{-z^2/4s}\tilde{\tilde{\psi}}_0(0,t_0+z)dz = \Gamma(s,t_0)$.

On the intervals where $\tilde{\tilde{\psi}}_0(0,t)$ is decreasing, Γ will be decreasing on the same intervals contracted of ε_{17} from both sides. The same thing happens on the intervals where $\tilde{\tilde{\psi}}_0$ is increasing. On the intervals where $\tilde{\tilde{\psi}}_0$ is constant, on the intervals J_ℓ in particular, if they are of size larger than $2\varepsilon_{17}$, $\Gamma(s,t)$ is nearly constant, since $\varepsilon_{17}/\sqrt{s} \le \varepsilon_{17}/\varepsilon_{17}^{3/2}$ tends to $+\infty$ when ε_{17} tends to zero. Observe that:

$$(1 + (C_{41}+1)^2|J_\ell|)|J_\ell| \ge \varepsilon_0/40. \tag{235}$$

Thus

$$|J_\ell| \geq \gamma(\varepsilon_0, C_{41}) > 0. \tag{236}$$

Taking ε_{17} small enough, $\Gamma(s,t)$ will be essentially constant on J_ℓ.

All the intervals, intervals J_ℓ, intervals where $|\tilde{\tilde{\psi}}_0(0,t)|$ lies between $[\nu + o(1), |\tilde{\tilde{\psi}}_0(0,t)|_\infty - o(1)]$ or $[o(1), \nu - o(1)]$, are in finite number $3(\ell_1 + 1)$. $\Gamma(s,t)$ has the same behavior than $\tilde{\tilde{\psi}}_0(0,t)$ on these intervals, if their size is larger than $\sqrt{\varepsilon_{17}}$. The total loss in $\int |b|$ due to the fact that we neglect those of size less than $\sqrt{\varepsilon_{17}}$ is $3(\ell_1 + 1)\sqrt{\varepsilon_{17}} \times C_{41}$, an amount which is tiny if ε_{17} is small enough.

On the remaining intervals, if $o(1)$ is small but still large enough with respect to $\gamma_{C_{41}}(\varepsilon_{17})$, $\tilde{\tilde{\psi}}_0(s,t)$ will belong to intervals of the same type than $\tilde{\tilde{\psi}}_0(0,t)$. For example, if $|\tilde{\tilde{\psi}}_0(0,t)|$ belongs to $[\nu + o(1), |\tilde{\tilde{\psi}}_0(0,t)|_\infty - o(1)]$, $|\tilde{\tilde{\psi}}_0(s,t)|$ will belong to $\left[\nu + \frac{o(1)}{2}, |\tilde{\tilde{\psi}}_0(0,t)|_\infty - \frac{o(1)}{2}\right]$. Taking $\frac{o(1)}{4}$ larger than $C\sqrt{\delta} + C(C_{41}, a_0)s^{1/4} + O\big(\mathrm{Sup}(\varepsilon_{10}, \varepsilon_{11}, \frac{1}{k})\big)$, $b(s,t)$, by (231), will belong to the same range of values, on the same intervals contracted of ε_{17} on both sides. Hence, if $b_0(t) = b(0,t)$ were, in average (i.e. for its weak limit φ., when we think of $b(0,t)$ as $b(s,t)$ of Lemma 9, with s close to $+\infty$ or to the blow-up time so that Lemma 9 i), ii), iii) hold), between $\nu + 2o(1)$ and $|b|_\infty(0,t) - 2o(1)$ on an interval of size larger than $\sqrt{\varepsilon_{17}}$ with $\varepsilon_{17} > \varepsilon_{16}^2$ for example, we can apply the η of ii) and iii) of Definition 2 to $b(s,t)$, for s satisfying (234), provided our parameters $\varepsilon_{11}, \varepsilon_{10}, \varepsilon_{12}$, $\frac{1}{k}, \frac{1}{q}$ are small enough i.e. satisfy (230). The decrease of a in this process depends then only on $o(1)$ and ε_{17}, provided $\varepsilon_{17}, \varepsilon_{10}, \varepsilon_{12}$, $\frac{1}{k}, \frac{1}{q}$ are small enough.

The requirements are:

a) on ε_{17}, ε_{17} small enough, given C_{41}

b) on s, (207), (214) and (234)

c) on $\varepsilon_{11}, \varepsilon_{10}, \frac{1}{k}, \frac{1}{q}$, (232)

d) on o(1): o(1) should be large with respect to
$$C\sqrt{\delta} + C(C_{41}, a_0)s^{1/4} + O\big(\mathrm{Sup}(\varepsilon_{10}, \varepsilon_{11}, \tfrac{1}{k})\big)$$

as stated. The decrease in a is a function $\beta(o(1), \varepsilon_{17}) > 0$, which tends to zero only if o(1) ad ε_{17} tends to zero. iii) is immediate.

Corollary 14. *If we iterate the process on a given curve having a b behaving as $b(s,t)$ of Lemma when s tends to $+\infty$ or to the explosion time, we do not need to change $\varepsilon_{10}, \varepsilon_{11}, \varepsilon_{12}$, $\frac{1}{k}, \frac{1}{q}$ at each step, if o(1) and ε_{17} are given.*

Proof. It is the above requirement c) which might force us to do so, when combined with (207). Indeed, the time allowed for the regularizing flow cannot be too large in function of the previous decrease δa – it has to satisfy (207) – if we want to keep control on $|b|_\infty$. However, this imposes a restriction of $\varepsilon_{10}, \varepsilon_{11}, \varepsilon_{12}$, $\frac{1}{k}, \frac{1}{q}$ if we want it to have an effect on b_0. Of course, we cannot use the same δa repeatedly i.e. if we wish to repeat our procedure – cancellation, regularization, flow of ii) or iii) of Definition 2 – we need to use the decrease δa of the previous step and only it, once. The control on $|b|_\infty$ will then be kept. However, here, the decrease in a is $\beta(o(1), \varepsilon_{17}) > 0$, a fixed amount, depending only on $o(1)$ and ε_{17} and on nothing else, once $\varepsilon_{11}, \varepsilon_{10}, \frac{1}{k}, \frac{1}{q}$ satisfy (232). Therefore, the next time, we only need to have also (207) with $\delta a - \beta(o(1), \varepsilon_{17})$ which, when combined with (232), will provide a new constraint on $\varepsilon_{11}, \varepsilon_{10}, \frac{1}{k}, \frac{1}{q}$ and which will remain the same in the next steps. Namely, this constraint is:

$$\delta(k, q, \varepsilon_{11}, \varepsilon_{10}, \varepsilon_{12}) < \inf\left(\frac{\beta(o(1), \varepsilon_{17})}{1 + (C_{41} + 2)^3}, s_0(C_{41})\right). \tag{237}$$

As long as we keep the same $o(1)$ and ε_{17}, (237) will be enough. Thus, there is no need to change $k, q, \varepsilon_{10}, \varepsilon_{11}, \varepsilon_{12}$ at each step, as claimed. q.e.d.

Another way to proceed, which we will rather use in some way is, given a compact K_j, such that we decreased of at least $\delta a_j > 0$ in the previous step, to choose $\varepsilon_{10}, \varepsilon_{11}, \varepsilon_{12}, \frac{1}{k}, \frac{1}{q}$ so small so that, to any curve which will satisfy Lemma 11 ii), a), b) and c), if we apply the regularizing flow during the time δa_j (or $\inf(\delta a_j, s_0(C_{41}))$, we can apply thereafter our flow of Definition 2, ii) or iii). If we proceed in this way, we will have:

Corollary 15. *Any curve which will not have, at a certain time of the deformation, a component $b(t)$ such as described in ii) of Lemma 9, has decreased in the whole process of $\beta(o(1), \varepsilon_{17})$.*

Proof. It follows immediately from the previous arguments. Again, this process can be iterated thereafter.

A problem in the deformation process

These two corollaries have to be read well, otherwise they are misleading:

Clearly, either a curve is such that, at $\bar{s}$, $b(\bar{s}, t)$ behaves as in Lemma 9, then Corollary 14 applies: the curve can be regularized and $\varepsilon_{10}, \varepsilon_{11}, \varepsilon_{12}, \frac{1}{k}, \frac{1}{q}$ need not to be changed. Or it is not the case, i.e. $b(\bar{s}, t)$ does not behave as in Lemma 9. If, in addition, it does not behave in this way in the future ($s \geq \bar{s}$), then Corollary 15 applies and a will eventually decrease substantially.

However, there is some time between the two processes and, in a deformation process, we encounter intermediate situations where, for example, in the past, b could have been regularized or has been regularized, in the future a will eventually decrease substantially; meanwhile, we do not know exactly where we stand; that the situations which impede the deformation process can be assumed to be intermediate between the two ones described above follows from an easy form of the deformation lemma, which we will establish in time.

We will need some specific work to handle these intermediate situations. This will be completed later, once we have described the last stages of the deformation process.

I.3. Globalization of the deformation

Definition 4. A stretched ν-curve. *We start now the final stages of our deformation process. Using the global vector-field, made of cancellations and stretching processes introduced in Definition 2, we can stretch the curves so that the top ones – i.e. the ones impeding the deformation to proceed downwards, the ones whose unstable manifolds are the key pieces of the deformed object – are made of almost flat pieces when $|b|$ is almost equal to $|b|_\infty$, of L^1-length at least $\delta_0 > 0$, a universal constant, depending only on α, followed by pieces where $|b|$ decreases extremely fast, as fast as we may prescribe from $|b|_\infty$ to ν, then possibly, after some flat ν-piece to zero, then moves upwards from zero to ν and from ν to $|b|_\infty$, as fast as it decreased. The new – almost flat $|b|_\infty$-piece is also of L^1-size at least δ_0. There might be some crossings of zero (the number of which is controlled by the maximal number of zeros of b in our initial topological class which we are deforming) which occur in the flat zero pieces. These stretched curves can be assumed to be C^∞ through a tiny regularization using the flow having $\eta = cb$, during a tiny time, so small that any increment in $|b|_\infty$ is controlled by previous decreases in a. This is very easy to construct: first use the cancellation and stretching flow, hence a decrease in a and a controlled increase in $|b|_\infty$ (by i) of Lemma 3), then once a has decreased even a tiny bit, use $\eta = cb$ with c so small and the time so small that all increases in $|b|_\infty$ are controlled by the previous decrease in a. The curves we obtain are then stretched and C^∞. The norm $|b|_{C^4}$ is controlled by a function $|b|_\infty$, which tends to $+\infty$ when the regularization time tends to zero.*

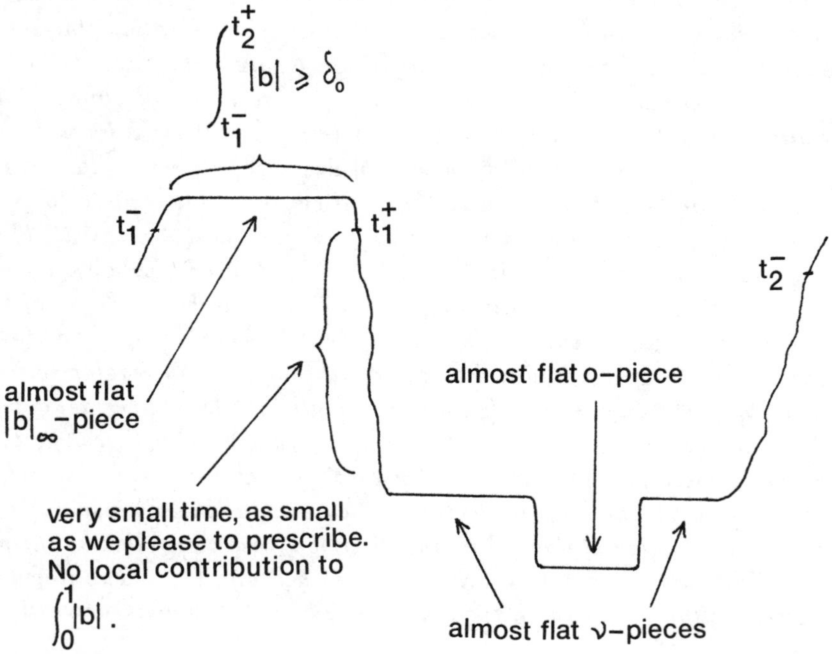

FIGURE L

We thus have:

$$\int_{t_1^+}^{t_2^-} (|b| - \nu)^+ \textit{ very small, as small as we wish}$$

where t_1^+, t_2^- *are points on the curve where* $|b|$ *is larger that* $3|b|_\infty/4$ *and less than* $\frac{4}{5}|b|_\infty$*, for example, before and after one flat* $\nu - 0 - \nu$ *piece.*

We wish now to decrease a even starting from these stretched ν-curves. However, any such decrease involves a loss of control on $\int(|b|-\nu)^+$, hence a loss of compactness. We want to control this problem, and we want to build a deformation process, step by step, which will lead us to the only serious compactness problems, which cannot be avoided in this variational problem (see [1]), that is to the curves where the flat ν-pieces are not there anymore. For this purpose, we take a flat ν-piece which is sizable (i.e. $\int |b|$ on this ν-piece $> \varepsilon$, $\varepsilon > 0$ given). If there is no such ν-piece, we are done.

Otherwise, we introduce on this flat ν-piece an oscillation by applying an $\eta = \varphi(t)$, which is C^∞, with support inside this flat ν-piece. Such an η provides us a local vector-field $\overline{Z}$ which decreases a if φ has the sign of b on this flat ν-piece.

$\overline{Z}$ yields a deformation, this is not difficult to see, in H^1: φ is C^∞; in the a evolution equation, everything is smooth. In the b-equation, we have:

$$\frac{\partial b}{\partial s} = \frac{\ddot{\eta} + \overline{\lambda \dot{b}}}{a} + a\eta\tau - b\eta\bar{\mu}_\xi.$$

Since η is equal to $\sum w_i(x)\varphi_i(t)$ where w_i is a C^∞-positive function having support near the stretched curves ($\sum \omega_i = 1$) and φ_i is C^∞, the only possibly bothering term comes from $\lambda\dot{b}$. However, easy computations show that it is neither a problem for the existence proof, nor for the uniqueness. We skip this detail, here. The deformation extends to curves such that $|b|$ has a flat piece which is $0(\delta_{10})$ close to ν, where δ_{10} is a small (with respect to C_{41}) fixed parameter, instead of having a completely flat piece. We designate by flat piece in the sequel this $0(\delta_{10})$ proximity to a completely flat piece.

Definition 5. *We deform, using $\overline{Z}$, as long as $|b|$ remains larger than $\nu/2$ on the flat-ν-pieces and less than 2ν. In fact, it will be enough to deform a very tiny bit, however of a constant size if the flat ν-piece remains of size $> \varepsilon$. In doing this, we have decreased a of an amount equal to $\delta(\varepsilon)$, we have kept $|b|$ above the level $\nu/2$ on the support of η, but we have lost control of $\int(|b|-\nu)^+$ in function of the decrease in a. If we complete this procedure repeatedly, $\int|b|$ might increase beyond control, which we want to avoid. The details about the assertions of this description/definition are easy and left to the reader (decrease of $\delta(\varepsilon)$, while $|b|$ stays above the level $\nu/2$).*

We wish now to build a second deformation process which will allow us to retake control on $\int(|b| - \nu)^+$, the control which we started to loose through the $\overline{Z}$-deformation. This deformation has introduced two oscillations on a flat ν-piece, above and below ν, which we wish to cancel now, in order to regain control on $\int(|b| - \nu)^+$. Hence, we want to come back to the stretched curves of Definition 4; however, since we want to regain control on $\int(|b| - \nu)^+$, what we really want is to stretch back the curve between t_1^+ and t_2^- (see Fig. L), keeping control on $\int(|b|-\nu)^+$ outside the interval $[t_1^+, t_2^-]$, where t_1^+, t_2^- are defined, in this first stage of the deformation process as points on the curves on each side of the sink made of flat ν or 0-pieces satisfying:

$$\begin{cases} \dfrac{\partial\tau(s)}{\partial s} = -\dfrac{\lambda + \bar{\mu}\eta}{a}(\tau(s)) \\ \tau(0) = \begin{matrix} t_1^+(0) \\ t_2^-(0) \end{matrix} \end{cases} \qquad \tau(s) = t_1^+(s), t_2^-(s)$$

where $t_1^+(0), t_2^-(0)$ are points on each side of the sink where $\frac{4|b|_\infty}{5} \geq |b| \geq 3|b|_\infty/4$. The nearly non-oscillatory behavior of $\widetilde{\psi}$. or b allows to choose $t_1^+(0)$, $t_2^-(0)$ depending differentiably on the curves which are nearly stretched.

We then have the following lemmas which shows that we have kept, through this first deformation process, a good control on $\int_{t\notin[t_1^+,t_2^-]} (|b| - \nu)^+$.

Lemma 16. *Assume that* $a(s) \geq 3\frac{a(0)}{4}$ *and* $\nu < \frac{3}{16}|b|_\infty$. *Then:*
i) $\frac{4|b|_\infty}{5a(0)} \geq \frac{|b|(s,\tau(s))|}{a(s)} \geq \frac{3|b|_\infty}{8a(0)}$, *for* $\tau(s) = t_1^+, t_2^-$
ii) $\int_{t \notin [t_1^+, t_2^-]} (|b| - \nu)^+(s) \leq \int_{t \notin [t_1^+, t_2^-]} (|b| - \nu)^+(0) - C(a(s) - a(0))$
where C *is a universal constant depending only on* α *and* ν.
iii) Assume that $|b|$ *remains larger than* $\nu/2$ *on the support of* η. *Then* $\int_{t_1^+}^{t_2^-} |b|(s) \leq 2\nu - C_3(\nu)(a(s) - a(0))$.

Proof. We have:

$$\frac{\partial}{\partial s} b(s, \tau(s)) = \frac{\overline{\dot{\eta} + \lambda b}}{a} + a\eta\tau - b\eta\bar{\mu}_\xi + \tau'(s)\dot{b}. \tag{238}$$

Since ν is very small with respect to $|b|_\infty$ – we assume that $\frac{3}{8}|b|_\infty > 2\nu - \eta$ is zero at $\tau(s) = t_1^+, t_2^-$, and near t_1^+, t_2^- at least for small s

$$\left(|b(0, t_1^+(0))| \geq \frac{3|b|_\infty}{4}, \qquad |b(0, t_2^-(0))| \geq \frac{3|b|_\infty}{4}\right).$$

Thus, using (238):

$$\begin{aligned} \frac{\partial}{\partial s} b(s, \tau(s)) &= \left(\frac{\lambda}{a} + \tau'(s)\right)\dot{b} + \frac{\dot{\lambda} b}{a} = \frac{\overline{(\lambda + \bar{\mu}\eta)b}}{a} = \frac{b(b\eta - \int_0^1 b\eta)}{a} \\ &= -b\frac{\int_0^1 b\eta}{a}. \end{aligned} \tag{239}$$

Thus:

$$b(s, \tau(s))/a \quad \text{is constant.} \tag{239'}$$

Since $a(s)$ is assumed to be larger than $a(0)/2$, we have:

$$|b(s, \tau(s))| > \frac{1}{2}|b(0, \tau(0))| = \frac{3}{8}|b|_\infty. \tag{240}$$

Thus η remains zero at $\tau(s)$ beyond s as long as $a(s)$ remains above $3a(0)/4$ and i) holds.

For ii), we compute as in ii) of Lemma 3 (up to the new boundaries t_2^-, t_1^+ to take into account):

$$\begin{aligned} \frac{\partial}{\partial s} \int_{t \notin [t_1^+, t_2^-]} (|b| - \nu)^+ \leq \operatorname{sgn} b \frac{\dot{\eta} + \lambda b}{a}\Big|_{t_2^-}^{t_1^+} - \int_{t_2^-}^{t_1^+} (a\eta\tau - b\eta\bar{\mu}_\xi) \operatorname{sgn} b \chi_{|b| \geq \nu} - \\ - C(\nu)\frac{\partial a}{\partial s} + (|b| - \nu)\tau'\Big|_{t_2^-}^{t_1^+} \end{aligned} \tag{241}$$

$-C(\nu)\frac{\partial a}{\partial s}$ is due to the contribution of the points where $|b| = \nu$ in $[t_2^-, t_1^+]$). Thus, using (241) and i) ($\eta, \dot{\eta} = 0$ at t_2^-, t_1^+)

$$\frac{\partial}{\partial s} \int_{t \notin [t_1^+, t_2^-]} (|b| - \nu^+) \le -\nu\tau' \Big|_{t_2^-}^{t_1^+} - C(\nu)\frac{\partial a}{\partial s} \tag{242}$$

$$= \nu \frac{\lambda + \bar{\mu}\eta}{a}\Big|_{t_2^-}^{t_1^+} - C(\nu)\frac{\partial a}{\partial s} \le -C'(\nu)\frac{\partial a}{\partial s}$$

ii) follows.

Proof of iii): The contribution of the boundary terms in $\frac{\partial}{\partial s}\int_{t_1^+}^{t_2^-} |b|$ is dominated by $-C(\nu)\frac{\partial a}{\partial s}$ as we just saw in (242) for another expression, η has support in the almost flat ν-pieces, where $|b|$ remains above $\nu/2$. Thus:

$$\frac{\partial}{\partial s}\int_{t_1^+}^{t_2^-} |b| \le -C(\nu)\frac{\partial a}{\partial s} + \int_{t_1^+}^{t_2^-} |a\eta\tau - b\eta\bar{\mu}_\xi| \tag{243}$$

$$\le -C_3(\nu)\frac{\partial a}{\partial s}.$$

Thus

$$\int_{t_1^+}^{t_2^-} |b|(s) \le \int_{t_1^+(0)}^{t_2^-(0)} |b|(0) - C_3(\nu)(a(s) - a(0)) \tag{244}$$

$$\le 2\nu - C_3(\nu)(a(s) - a(0))$$

if the curve is stretched enough.

Hence, after this first stage of the deformation – $a(s) \ge 3a(0)/4$ is not a problem; if we decreased of $a(0)/4$, we decreased substantially; the next deformation section shows how to globalize all those local deformations – we have kept control on $\int_{t \notin [t_1^+, t_2^-]}(|b| - \nu)^+$. If we are able to stretch back the curve between t_1^+, t_2^- as it was at time zero or better, decreasing by the same token a, our deformation could proceed with a good control on $\int(|b| - \nu)^+$.

Definition 6. *Let 0 be the final time of the previous deformation. We scale, therefore, back the time. $t_1^+(0), t_2^-(0)$ will be now the values of t_1^+, t_2^- at the end of the previous deformation.*

We wish to stretch back the curves between $t_1^+(0)$ and $t_2^-(0)$. i.e. given two times t_1 and t_2, we want to stretch the piece of curve between t_1 and t_2. For this purpose, we complete the following construction:

We observe that the cancellation/stretching flow built out pieces of η_{12} to which some cb is added, as in Definition 2, and later in section I.2, Lemmas 5–13, through *the regularization and use of the flow of ii) and iii) of Definition 2, all this construction can be completed with the further restriction that η_{12} has support in $[t_1+\delta, t_2-\delta]$, where $\delta > 0$ is some tiny number which we will be using for some estimates and has no importance whatsoever here. Patching together through partitions of unity, we obtain a vector-field:*

$$Z_\delta(t_1, t_2, x)$$

(that is a three stage process: cancellation, regularization, flow of ii)–iii) of Definition 2. It can be made in a vector-field of the above form, see Remark in the global globalization section) Before $t_1+\delta$ and after $t_2-\delta$, η of this flow is equal to cb. Thus, in $[t_1+\delta, t_2-\delta]$, this vector-field will enjoy all the stretching and cancelling properties of the general flow that we built earlier. In particular, Lemma 9 (the stretching lemma) will hold for this flow between $t_1+\delta$ and $t_2-\delta$. Also, all the lemmas of section I.1, about existence and uniqueness, as well as all the estimates that we established for the previous flow will hold here.

We couple this flow with differential equations on t_1, t_2:

$$\begin{cases} \dfrac{\partial x}{\partial s} = Z_\delta(t_1, t_2, x) & x(0) = x_0 \\ \dfrac{\partial t_1}{\partial s} = -\dfrac{\lambda+\bar{\mu}\eta}{a}(t_1) & t_1(0) = t_1^+(0) \\ \dfrac{\partial t_2}{\partial s} = -\dfrac{\lambda+\bar{\mu}\eta}{a}(t_2) & t_2(0) = t_2^-(0). \end{cases} \tag{245}$$

The first equation of (245) has been thoroughly studied previously, when we built the cancellation flow on S^1. It requires only minor adaptations at this stage. The second and third equation define obviously Lipschitz vector-fields in (t_1, t_2, x).

Therefore (245) defines a semi-flow which is continuous from $H^2 \times \mathbb{R} \times \mathbb{R}$ into $W^{1,\infty} \times \mathbb{R} \times \mathbb{R}$, a decreases along this flow.

We want to prove that, with a proper choice of δ, Z_δ will stretch back our curves between $t_2^-(s)$ and $t_1^+(s)$ and restore our control on $\int(|b|-\nu)^+$. However, we will run into a problem: Z_δ will indeed not create any new oscillation between $t_2^-(s)$ and $t_1^+(s)$ but could add to the strength of each oscillation on the edges (near t_2^- and t_1^+). This leads us to new points $\tilde{t}_2^-$ and $\tilde{t}_1^+$ analogous t_2^- and t_1^+, with a loss of control, of a specific type, on $\int(|b|-\nu)^+$, see Lemma 17 and 18 below. Later, we prove that we can control $\int(|b|-\nu)^+$. At this point, we study thoroughly the effect of Z_δ. We first need the following Lemma :

Lemma 17. *We assume throughout this lemma that the origin of time in the definition of $\lambda+\bar{\mu}\eta$ is taken between t_1^+ and t_2^-, i.e. in the sink. Given δ, we assume c of (15) to be small enough; how small will be made clear from the proof.*

Let, then, $w(s,t)$, in $[t_2^-(0)-\frac{3\delta}{4}, t_1^+(0)+\frac{3\delta}{4}]$, be equal to $t_2^-(s)+\frac{a_0}{a}(t-t_2^-(0))$. Assuming a_0/a is close to 1:

i) w can be extended into a diffeomorphism of S^1 which has all t-derivatives and inverse derivatives up to the third order bounded independently of s and t by a fixed constant M independent of c, bounded b C/δ^{10}, C a universal constant *ii)* $|\varphi(s,t)-w(s,t)| = \int_0^s O(c|b|_\infty^2) = o(1)$ *uniformly in s and t for* $t \in [t_2^-(0)-\frac{3\delta}{4}, t_1^+(0)+\frac{3\delta}{4}]$

iii) *Let* $B(s,t) = b(s,w(s,t)), N_{12}(s,t) = \eta_{12}(s,w(s,t))$. *Then,*

$$a\frac{\partial}{\partial s}\left(\frac{B(s,t)}{a}\right) = \frac{c}{aw_t^2}\frac{\partial^2}{\partial t^2}B + ca\left(B\tau - \frac{B^2}{a}\bar{\mu}_\xi - \frac{\frac{\partial}{\partial t}(\bar{\mu}B^2)}{aw_t} + \frac{B^3}{a^2} - \right.$$

$$\left. -\frac{1}{a}\frac{w_{tt}}{w_t^3}B_t\right) + DB_t + \frac{1}{a}w_t\frac{\partial}{\partial t}\left(\frac{1}{w_t}\frac{\partial}{\partial t}N_{12} - \bar{\mu}N_{12}B\right) + \frac{B^2}{a}N_{12}$$

where $D = \frac{1}{w_t}\int_{t_2^-(s)}^{w} b\eta$ *if* $t \in [t_2^-(0)-\frac{3\delta}{4}, t_1^+(0)+\frac{3\delta}{4}]$, D, D_t *are bounded uniformly in function of* C_{41} *and* δ *otherwise*

iv) $B(s,t) = B(0,t)+o(1)$ *uniformly in s for* $t \in [t_2^-(0)-\frac{\delta}{4}, t_1^+(0)+\frac{\delta}{4}]$. *Finally, the argument extends to a combination of several of these flows, provided the origin of time in the definition of* $\lambda + \bar{\mu}\eta$ *is taken in one of the sinks.*

We now have the following:

Lemma 18. *Assume that* $\frac{3(|b|_\infty(0)-C_1a_0)}{16} > \nu$. *If c in (15) is chosen small enough also in function of* δ, *then:*

i)
$$\left|\frac{|b|_\infty(s)}{a(s)} - \frac{|b|_\infty(0)}{a_0}\right| \le C_1a_0 \qquad \forall s$$

where C_1 *is a universal constant depending only on* α.

ii)
$$\int_{t_2^-}^{t_1^+}(|b|-\nu)^+(s) \le \int_{t_2^-}^{t_1^+}(|b|-\nu)^+(0) - C(\nu)(a(s)-a(0)) + C\delta(|b|_\infty(0)+C_1a_0)$$

where $C(\nu)$ *depends only on* ν *and* α, *as well as C.*

Furthermore, no new oscillation develops between t_1^+ *and* t_2^- *or outside this interval (i.e. no new bump of size* δ_0 *to the least) unless a has decreased of a fixed amount* $\gamma(\nu) > 0$ *in the full first and second steps. Also, although it is not necessary,* $|b|(t_1^+)$ *and* $|b|(t_2^-)$ *remain larger than* $3|b|_\infty/16$. *The only possible phenomenon is the creation of two new* $|b|_\infty$*-nearly flat pieces after* t_1^+ *and before* t_2^-, *of size at most* ν, *i.e. the oscillations before* t_1^+ *and after* t_2^- *pick up some strength. Defining two new points* $\tilde{t}_1^+$ *and* $\tilde{t}_2^-$, *we can quietly stretch back the curve using the global flow of Definition 2 (see the next section for the globalization of this process), and we have*

iii)
$$\overline{\lim_{\substack{s\to\text{explosion time}\\ \text{or}\\ s\to+\infty}}}\int_{\tilde{t}_1^+}^{\tilde{t}_2^-}(|b|-\nu)^+(s) \le C\delta(|b|_\infty(0)+C_1a_0).$$

Thus, if δ has been chosen less than $\left(\frac{|b|_\infty(0)+C_1a_0}{2}\right)^{-1} \times \int_{\tilde{t}_1^+}^{\tilde{t}_2^-}(|b|-\nu)^+(-1)$ *and less than* $(C|b|_\infty(0)+C_1a_0))^{-1}|\Delta a(-2)|$, *where* $\int_{\tilde{t}_1^+}^{\tilde{t}_2^-}(|b|-\nu)^+(-1)$ *is the value of* $\int_{\tilde{t}_1^+}^{\tilde{t}_2^-}(|b|-\nu)^+$ *before even the first step of this deformation process (the one involving* $\overline{Z}$ *and* $\Delta a(-2)$ *is the associated decrease in* a*), we have restored the control on* $\int(|b|-\nu)^+$ *between* t_2^- *and* t_1^+, *as well as between* $\tilde{t}_1^+$ *and* $\tilde{t}_2^-$. *Between* t_2^- *and* $\tilde{t}_2^-$ *and between* $\tilde{t}_1$ *and* t_1^+, *the curve might have developed, up to* $0(\delta)$, *two nearly flat* $|b|_\infty$*-pieces.*

Proof of Lemma 18. When we compute $\frac{\partial}{\partial s}\int_{t_2^-}^{t_1^+}(|b|-\nu)^+$ or $\frac{\partial}{\partial s}\int_{t_1^+}^{t_2^-}(|b|-\nu)^+$ we find terms such as in the proof of ii) of Lemma 3 but for the contribution of the boundary term where $\eta = cb$ here, by construction of the deformation.

Hence, we have to take care of ($\tau(s) = t_1^+(s)$ or $t_2^-(s)$)

$$\operatorname{sgn} b\frac{\dot{\overline{cb}}+\lambda b}{a} + (|b|-\nu)^+\tau'(s) = \Delta(\tau(s)) \tag{246}$$

where $\tau' = -\frac{\lambda+\bar{\mu}\eta}{a} = -\frac{\lambda+\bar{\mu}cb}{a}$. Thus, since $|b|(\tau(s))$ will be larger than $3|b|_\infty/16 > \nu$ (otherwise, $|b|_\infty(0)$ is bounded by $C_1a_0 + \frac{16\nu}{3}$. We can continue to deform through our global flow), $(|b|-\nu)^+(\tau) = |b|-\nu$ and

$$\Delta(\tau(s)) = c\left[(\operatorname{sgn} b)\frac{\dot{b}-\bar{\mu}b}{a} - \nu\tau'\right]. \tag{247}$$

Thus, computing for example for $\int_{t_1^+}^{t_2^-}(|b|-\nu)^+$ and assuming for simplicity that b is positive, we have:

$$\begin{aligned}\frac{\partial}{\partial s}\int_{t_1^+}^{t_2^-}(|b|-\nu)^+ &\le C\frac{c}{a}|b|_\infty + \frac{c}{a}(\dot{b}(t_2^-(s)) - \dot{b}(t_1^+(s)) + \nu|t^{+\prime}_1 - t^{-\prime}_2| - \\ -\tilde{C}(\nu)\frac{\partial a}{\partial s} &\le C\frac{c}{a}|b|_\infty + \frac{c}{a}(\dot{b}(t_2^-(s)) - \dot{b}(t_1^+(s)) - C(\nu)\frac{\partial a}{\partial s}.\end{aligned} \tag{248}$$

We can absorb $\frac{Cc}{a}|b|_\infty$ into $-C(\nu)\frac{\partial a}{\partial s}$, changing the value of $C(\nu)$. However, we are bothered by $\frac{c}{a}(\dot{b}(t_2^-(s)) - \dot{b}(t_1^+(s)))$ and we want to get rid of it. This will be completed by averaging over $t_1^+(0)$ and $t_2^-(0)$ running in small intervals of size $C\delta, C$ a small fixed constant.

We introduce the change of variables:

$$\begin{cases}\frac{\partial\varphi}{\partial s}(s,t) = -\frac{\lambda+\bar{\mu}\eta}{a}(\varphi(s,t)) \\ \varphi(0,t) = t.\end{cases} \tag{249}$$

Clearly:

$$\begin{cases} \dfrac{\partial}{\partial s}\varphi_t(s,t) = -\dfrac{(b\eta - \int_0^1 b\eta)}{a}\varphi_t \\ \varphi_t(0,t) = 1. \end{cases} \tag{250}$$

If $\varphi(s,t)$ belongs to $[t_1^+(s), t_1^+(s)+\delta]$ or $[t_s^-(s)-\delta, t_2^-(s)]$, the $\eta = cb$, thus:

$$\begin{cases} \dfrac{\partial}{\partial s}(a\varphi_t(s,t)) = O(c|b|_\infty^2)a\varphi_t(s,t) \\ \varphi_t(0,t) = 1 \end{cases} \tag{251}$$

where O, in (251), does not depend on c, under (15).

Observe that if the decrease in a is not large, which we can assume, we have

$$\varphi_t(s,t) \in \left[\frac{1}{2}, \frac{3}{2}\right] \text{ if } t \in \left[t_1^+(0), t_1^+(0) + \frac{\delta}{16}\right] \text{ or} \tag{252}$$

$$t \in \left[t_2^-(0) - \frac{\delta}{16}, t_2^-(0)\right] \quad \text{and}$$

$$\varphi(s,t) \in [t_1^+(s), t_1^+(s)+\delta] \text{ if } t \in \left[t_1^-(0), t_1^+(0) + \frac{\delta}{16}\right], \tag{253}$$

$$\varphi(s,t) \in [t_2^-(s)-\delta, t_2^-(s)] \text{ if } t \in \left[t_2^-(0) - \frac{\delta}{16}, t_2^-(0)\right].$$

Computing as in (248), we have:

$$\frac{\partial}{\partial s}\int_{\varphi(s,t^+)}^{\varphi(s,t^-)}(|b|-\nu)^+ \leq \frac{c}{a}(\dot{b}(\varphi(s,t^-)) - \dot{b}(\varphi(s,t^+)) - C(\nu)\frac{\partial a}{\partial s}. \tag{254}$$

Introducing:

$$\underline{w} = \frac{1}{\varphi(s,t_2^-(0)) - \varphi\left(s, t_2^+(0) - \frac{\delta}{16}\right)} \times \frac{1}{\varphi\left(s, t_1^+(0) + \frac{\delta}{16}\right) - \varphi(s, t_1^+(0))} \times$$
$$\times \int_{t_2^-(0)-\delta/16}^{t_2^-(0)} \varphi_t(s,t^-)dt^- \int_{t_1^+(0)}^{t_1^+(0)+\frac{\delta}{16}} \varphi_t(s,t^+)dt^+ \int_{\varphi(s,t^+)}^{\varphi(s,t^-)}(|b|-\nu)^+ \tag{255}$$

we have, using (251), (253) and the positivity of φ_t:

$$\frac{\partial \underline{w}}{\partial s} \leq -C(\nu)\frac{\partial a}{\partial s} + \frac{c}{a}\frac{b(t_2^-(s)) - b\left(\varphi\left(s, t_2^-(0) - \frac{\delta}{16}\right)\right)}{t_2^-(s) - \varphi\left(s, t_2^-(0) - \frac{\delta}{16}\right)} -$$

$$- \frac{c}{a}\frac{b\left(\varphi\left(s, t_1^+(0) + \frac{\delta}{16}\right)\right) - b(t_1^+(s))}{\varphi\left(s, t_1^+(0) + \frac{\delta}{16}\right) - t_1^+(s)} + Cc|b|_\infty^3 -$$

$$- \left(\frac{\frac{\partial}{\partial s}\left(a\left(t_2^-(s) - \varphi\left(s, t_2^-(0) - \frac{\delta}{16}\right)\right)\right)}{a\left(t_2^-(s) - \varphi\left(s, t_2^-(0) - \frac{\delta}{16}\right)\right)} \right.$$

$$\left. + \frac{\frac{\partial}{\partial s}\left(a\left(\varphi\left(s, t_1^+(0) + \frac{\delta}{16}\right) - t_1^+(s)\right)\right)}{a\left(\varphi\left(s, t_1^+(0) + \frac{\delta}{16}\right) - t_1^+(s)\right)} \right) \underline{w} \tag{256}$$

Observe that, using (251):

$$\frac{\partial}{\partial s}\left(a\left(t_2^-(s) - \varphi\left(s, t_2^-(0) - \frac{\delta}{16}\right)\right)\right) = \frac{\partial}{\partial s}\int_{t_2^-(0)-\frac{\delta}{16}}^{t_2^-(0)} a\varphi_t(s,t)dt$$

$$= \int_{t_2^-(0)-\frac{\delta}{16}}^{t_2^-(0)} O(c|b|_\infty^2)a\varphi_t dt \tag{257}$$

$$= O(c|b|_\infty^2)\left(a\left(t_2^-(s) - \varphi\left(s, t_2^-(0) - \frac{\delta}{16}\right)\right)\right).$$

A similar relation holds for $\varphi(s, t_1^+(0) + \frac{\delta}{16}) - t_1^+(s)$. Thus, combining (256) and (257), with an appropriate constant C:

$$\frac{\partial \underline{w}}{\partial s} \leq -C(\nu)\frac{\partial a}{\partial s} + Cc|b|_\infty^3 + 2\frac{c}{a}(|b|_\infty(0) + C_1 a_0)\left(\frac{1}{t_2^-(s) - \varphi\left(s, t_2^-(0) - \frac{\delta}{16}\right)}\right.$$

$$\left. + \frac{1}{\varphi\left(s, t_1^+(0) + \frac{\delta}{16}\right) - t_1^+(s)}\right) \tag{258}$$

Using then (252),

$$t_2^-(s) - \varphi\left(s, t_2^-(0) - \frac{\delta}{16}\right) \geq \frac{\delta}{B^2}, \quad \varphi\left(s, t_1^+(0) + \frac{\delta}{16}\right) - t_1^+(s) \geq \frac{\delta}{32}. \tag{259}$$

Thus:

$$\frac{\partial \underline{w}}{\partial s} \leq -C(\nu)\frac{\partial a}{\partial s} - Cc|b|_\infty^3 + \frac{128c}{a}\frac{(|b|_\infty(0) + C_1 a_0)}{\delta}. \tag{260}$$

Choosing c in (15) so that

$$Cc|b|_\infty^3 + \frac{128c}{a}\frac{(|b|_\infty(0) + C_1 a_0)}{\delta} < -\frac{\delta a}{\partial s}|[t_1^+, t_2^-]| \tag{261}$$

we derive:

$$\frac{\partial \underline{w}}{\partial s} \leq -C'(\nu)\frac{\partial a}{\partial s}. \tag{262}$$

Thus

$$\underline{w}(s) \leq \underline{w}(0) - C'(\nu)(a(s) - a(0)). \tag{263}$$

Hence, since by (253):

$$\left| \underline{w}(s) - \int_{t_1^+(s)}^{t_2^-(s)} (|b| - \nu)^+ \right| \leq \delta |b|_\infty(s) \leq \delta(|b|_\infty(0) + C_1 a_0) \tag{264}$$

we have:

$$\int_{t_1^+(s)}^{t_2^-(s)} (|b| - \nu)^+(s) \leq \int_{t_1^+(0)}^{t_2^-(0)} (|b| - \nu)^+(0) + 2\delta(|b|_\infty(0) + C_1 a_0) - C'(\nu)(a(s) - a(0)). \tag{265}$$

The above arguments can be used, in a very parallel way, in order to prove ii). If we combine ii) and (265), it is clear that, unless a has decreased of $\gamma(\nu)$, no new oscillation will develop, in particular no new oscillation between t_1^+ and t_2^-. Indeed, for such a new oscillation to develop, we would need that, between t_1^+ and t_2^-, we create a mass of at least $\sqrt{\frac{\varepsilon_0}{4c_0}} = \delta_0$, where C_0 is a universal constant used to characterize cases 1) and 2). Outside of $[t_1^+, t_2^-]$, if we are careful enough in the computations and if c is small, things change very little as we will see. No new oscillation develops. Inside, the creation of an oscillation requires that $\int_{t_1^+}^{t_2^-} (|b| - \nu)^+$ becomes larger than $\delta_0/2$. By iii) of Lemma 16, it started below $2\nu - C_3(\nu)\Delta a(-1)$, where $\Delta a(-1)$ is the total decrease in the first step of the process. During the second step, by (265), it will vary of at most $-C_4(\nu)\Delta a(1)$, where $\Delta a(1)$ is the new decrease. (iv) of Lemma 17 implies that $|b(s, t_2^-(s))|$ and $|b(s, t_1^+(s))|$ remain below $\frac{6}{7}|b|_\infty$. The sink between t_1^+ and t_2^- therefore remains.

Combining the two processes, we obtain our claim that there is no new oscillation. The only phenomenon which might occur is that some flat $|b|_\infty$-pieces are added to each of the two oscillations, the one to the right and the one to the left. In I.4, we show how to take care of this phenomenon. Thus, the curve has to stretch back between t_1^+ and t_2^-, (iii) follows easily. We thus see that all rests on proving that b changes very little between t_2^- and t_1^+, i.e. on proving Lemma 17.

Proof of Lemma 17. We first observe that $\eta = cb$ between t_2^- and t_1^+, by construction.

We show below, in the proof of (iv) that as long as $\frac{a_0}{a}$ is close to 1, i) of Lemma 17 holds. For ii), we observe that (extending w into $\tilde{w}$ by the same formula):

$$\varphi(s, t_2^-(0)) = t_2^-(s) = \tilde{w}(s, t_2^-(0)) \tag{266}$$

$$\begin{cases} \dfrac{\partial}{\partial s}(a(\varphi_t - \tilde{w}_t)) = O(c|b|_\infty^2) \text{ if } \varphi(s,t), w(s,t) \in [t_2^-(s) - \delta, t_1^+(s) + \delta] \\ a(\varphi_t - \tilde{w}_t)(0, t) = 0. \end{cases} \tag{267}$$

Assuming that $O(c|b|^2_\infty)$ is very small and has an s-integral very small with respect to δ and using our previous arguments on φ, φ_t, ii) follows easily ($a(\varphi - \tilde{w}) = o(1)$ as long as $\varphi(s,t), \tilde{w}(s,t) \in [t_2^-(s) - \delta, t_1^+(s) + \delta]$, we then use (251) and some iterative argument. This proof is independent of i)). For iii), we compute

$$\begin{aligned} a\frac{\partial}{\partial s}\left(\frac{B(s,t)}{a}\right) &= \frac{\partial b}{\partial s}(s,w(s,t)) + \frac{\partial w}{\partial s}\frac{\partial b}{\partial t}(s,w) + \frac{\int_0^1 b\eta}{a} b(s,w) \\ &= \frac{c}{a}\frac{\partial^2 b}{\partial t^2}(s,w) + \left(\frac{1}{a}(\lambda + \bar{\mu}\eta) + \frac{\partial w}{\partial s}\right)\frac{\partial b}{\partial t}(s,w) + \\ &+ ca\left(b(s,w)\tau - \frac{b^2}{a}(s,w)\bar{\mu}_\xi - \frac{\dot{\overline{\bar{\mu}b^2}}}{a^2}(s,w) + \frac{b^3(s,w)}{a^2}\right) + \\ &+ \frac{\dot{\overline{\eta_{12} - \bar{\mu}\eta_{12}b}}}{a}(s,w) + (a\eta_{12}\tau - b\eta_{12}\bar{\mu}_\xi)(s,w) + \frac{b^2(s,w)}{a}\eta_{12}. \end{aligned} \tag{268}$$

Observe that

$$\left.\begin{aligned} \frac{\partial b}{\partial t}(s,w) &= \frac{1}{w_t}\frac{\partial B}{\partial t} \\ \frac{\partial^2 b}{\partial t^2}(s,w) &= \frac{1}{w_t^2}\frac{\partial^2 B}{\partial t^2} - \frac{w_{tt}}{w_t^3}\frac{\partial B}{\partial t} \end{aligned}\right\} \tag{269}$$

$$\left.\begin{aligned} \frac{\dot{\overline{\bar{\mu}b^2}}}{a^2}(s,w) &= \frac{1}{w_t a^2}\frac{\partial}{\partial t}(\bar{\mu}B^2) \\ \frac{\dot{\overline{\eta_{12} - \bar{\mu}\eta_{12}b}}}{a}(s,w) &= \frac{1}{w_1}\frac{\partial}{\partial t}\frac{\left(\frac{1}{w_t}\frac{\partial}{\partial t}N_{12} - \bar{\mu}N_{12}B\right)}{a} \end{aligned}\right\} \tag{270}$$

$$\begin{aligned} w_t D &= \frac{\lambda + \bar{\mu}\eta}{a}(w) + \frac{\partial w}{\partial s} \\ &= \frac{(\lambda + \bar{\mu}\eta)(w)}{a} - \frac{(\lambda + \bar{\mu}\eta)(t_2^-(s))}{a} + \frac{1}{a}\int_0^1 b\eta\frac{a_0}{a}(t - t_2^-(0)) \\ &= \frac{\int_{t_2^-(s)}^w b\eta - (w - t_2^-(s))\int_0^1 b\eta}{a} + \frac{1}{a}\int_0^1 b\eta\frac{a_0}{a}(t - t_2^-(0)) \\ &= \frac{1}{a}\int_{t_2^-(s)}^w b\eta \end{aligned} \tag{271}$$

provided $t \in \left[t_2^-(0) - \frac{3\delta}{4}, t_1^+(0) + \frac{3\delta}{4}\right]$.

If $t \notin \left[t_2^-(0) - \frac{3\delta}{4}, t_1^+(0) + \frac{3\delta}{4}\right]$, D and D_t are obviously bounded in function of C_{41} independently of s and t. $\left(D = \frac{1}{w_t}\left(\frac{(\lambda+\bar{\mu}\eta)}{a}(w) + \frac{\partial w}{\partial s}\right)\right)$. (iii) follows immediately from (268)–(271).

We now prove (iv).

We introduce the fundamental solution G_c for the Cauchy parabolic problems:

$$\frac{\partial u}{\partial s} = \frac{ca}{a_0^2}\frac{\partial^2 u}{\partial t^2} \tag{272}$$

for periodic boundary conditions.

Denoting

$$\tilde{s} = \int_0^s \frac{ca}{a_0^2}(z)dz \tag{273}$$

we have, up to an irrelevant multiplicative constant:

$$G_c(x,y,s,s') = \sum_{-\infty}^{+\infty} \frac{e^{-\frac{(x-y-k)^2}{4(\tilde{s}-\tilde{s}')}}}{2\sqrt{\tilde{s}-\tilde{s}'}}, \quad \text{for } s \geq s'. \tag{274}$$

We rewrite iii) under the form:

$$\begin{aligned}\frac{\partial}{\partial s}\left(\frac{B(s,t)}{a}\right) &= \frac{ca}{a_0^2}\frac{\partial^2}{\partial t^2}\left(\frac{B}{a}\right) + ca\left(B\tau - \frac{B^2}{a}\bar{\mu}_\xi - \frac{\frac{\partial}{\partial t}(\bar{\mu}B^2)}{a^2 w_t} + \frac{B^3}{a^2} - \frac{1}{a}\frac{w_{tt}B_t}{w_t^3}\right) + \\ &+ DB_t + \frac{1}{a}w_t\frac{\partial}{-1\partial t}\left(\frac{1}{w_t}\frac{\partial}{\partial t}N_{12} - \bar{\mu}N_{12}B\right) + \\ &+ c\left(\frac{1}{aw_t^2} - \frac{a}{a_0^2}\right)\frac{\partial^2}{\partial t^2}\left(\frac{B}{a}\right) + \frac{B^2 N_{12}}{a} + a''\tau'' N_{12}.\end{aligned} \tag{275}$$

Observe that $\frac{1}{aw_t^2} - \frac{a}{a_0^2}$ is zero in $\left[t_2^-(0) - \frac{3\delta}{4}, t_1^+(0) + \frac{3\delta}{4}\right]$. $B(s,t)$ then reads, up to irrelevant multiplicative constants:

$$\begin{aligned}B(s,t) &= \frac{1}{2\sqrt{\tilde{s}}}\int_{-\infty}^{+\infty} e^{-\frac{(x-t)^2}{4\tilde{s}}} B(0,x)dx + \\ &+ \int_0^s d\tau_1 \int_{-\infty}^{+\infty} \frac{1}{2\sqrt{\tilde{s}-\tilde{\tau}_1}} e^{-\frac{(x-t)^2}{4(\tilde{s}-\tilde{\tau}_1)}} \left[ca\left(B\tau - \frac{B^2}{a}\bar{\mu}_\xi - \right.\right. \\ &\left. - \frac{\frac{\partial}{\partial x}(\bar{\mu}B^2)}{a^2 w_x} + \frac{B^3}{a^2} - \frac{1}{a}\frac{w_{xx}B_x}{w_x^3}\right) + B^2 N_{12} \\ &+ DB_x + \frac{1}{a}w_x\frac{-1\partial}{\partial x}\left(\frac{1}{w_x}\frac{\partial}{\partial x}N_{12} - \bar{\mu}N_{12}B\right) + a''\tau''N_{12} \\ &\left. + c\left(\frac{1}{aw_x^2} - \frac{a}{a_0^2}\right)\frac{\partial^2}{\partial x^2}\left(\frac{B}{a}\right)\right](\tau_1, x)dx.\end{aligned} \tag{276}$$

The first term in (276) tends to $B(0,t)$ uniformly, since $|B(0,t)|_{L^\infty} \leq C_{41} + 1$, for $\tilde{s}$ small, i.e. under suitable constraints on c, to be incorporated in (15). In the second term, all the expressions with no x-derivative are easily seen to be $O(\tilde{s})$ besides those involving N_{12} which are studied below, hence they can be made to be as small as we wish. When we consider expressions bearing an x-derivative, but multiplied by c, such as typically $\int_0^s d\tau_1 \int_{-\infty}^{+\infty} \frac{c}{\sqrt{\tilde{s}-\tilde{\tau}_1}} e^{-\frac{(x-t)^2}{4(\tilde{s}-\tilde{\tau}_1)}} \frac{w_{xx}}{w_x^3} B_x dx = (1)$, we have, after integration by parts ($\frac{1}{\delta^{100}}$ is added to take care of all derivatives of w):

$$
\begin{aligned}
(1) &= \frac{1}{\delta^{100}} O\left(\int_0^s d\tau_1 \int_{-\infty}^{+\infty} \frac{c}{\sqrt{\tilde{s}-\tilde{\tau}_1}} e^{-\frac{(x-t)^2}{4(\tilde{s}-\tilde{\tau}_1)}} \left(O(1) + \frac{|x-t|}{\tilde{s}-\tilde{\tau}_1}\right) dx\right) \\
&= \frac{1}{\delta^{100}} O\left(\int_0^s c\left(1 + \frac{1}{\sqrt{\tilde{s}-\tilde{\tau}_1}}\right) d\tau_1\right) = \frac{O(\tilde{s}+\sqrt{\tilde{s}})}{\delta^{100}} = o(1), (\tilde{s} \text{ will be } O(\delta^{300})).
\end{aligned}
\tag{277}
$$

We thus see that these terms fit our estimates. We are left with the contribution of DB_x, of $\frac{1}{a} w_x \frac{\partial}{\partial x}\left(\frac{1}{w_x}\frac{\partial}{\partial x} N_{12} - \bar{\mu} N_{12} B\right)$ and of $c\left(\frac{1}{aw_x^2} - \frac{a}{a_0^2}\right)\frac{\partial^2}{\partial x^2}\left(\frac{B}{a}\right)$ and of $B^2 N_{12} + a''\tau'' N_{12}$.

We integrate by parts, once or twice, in all the related expressions, in order to get rid of the x-derivatives on B or on N_{12}. We observe that, by construction:

$$
\left.
\begin{aligned}
&|N_{12}|_\infty \le C(|b|_\infty) \le C(1 + C_{41}) \\
&|D|_\infty + |D_x|_\infty \le C(|b|_\infty, \delta) \le C(\delta)(1 + C_{41})
\end{aligned}
\right\}.
\tag{278}
$$

Therefore, the three related expressions can be written as:

$$
\begin{aligned}
&\frac{1}{\delta^{100}} O\left(\left(\int_0^s d\tau_1 \int_{-\infty}^{+\infty} \frac{e^{-\frac{(x-t)^2}{8(\tilde{s}-\tilde{\tau}_1)}}}{\sqrt{\tilde{s}-\tilde{\tau}_1}} \left(\frac{|D| + |D_x| + |N_{12}|}{\sqrt{\tilde{s}-\tilde{\tau}_1}} + \frac{c}{\sqrt{\tilde{s}-\tilde{\tau}_1}} + \right.\right.\right. \\
&\left.\left.\left. + \frac{1}{\tilde{s}-\tilde{\tau}_1}\left(|N_{12}| + c\left|\frac{1}{aw_x^2} - \frac{a}{a_0^2}\right|\right) dx\right)\right)\right).
\end{aligned}
\tag{279}
$$

If $t \in \left[t_2^-(0) - \frac{\delta}{2}, t_1^+(0) + \frac{\delta}{2}\right]$, then, using (ii) and the estimates on φ and φ_t, we will have:

$$
w(s,t) \in \left[t_2^-(s) - \frac{3\delta}{4}, t_1^+(s) + \frac{3\delta}{4}\right] \qquad \forall s.
\tag{280}
$$

Thus, if $t \in \left[t_2^-(0) - \frac{\delta}{4}, t_2^+(0) + \frac{\delta}{4}\right]$, either $|x-t|$ is less than $\frac{\delta}{2}$, then $w(\tau_1, x) \in \left[t_2^-(s) - \frac{3\delta}{4}, t_1^+(s) + \frac{3\delta}{4}\right]$, thus, using also the fact that the origin of time for $\lambda + \bar{\mu}\eta$ is between t_1^+ and t_2^-:

$$
N_{12}(x) = 0, \quad \frac{1}{aw_x^2} - \frac{a}{a_0^2} = 0, \quad |D| + |D_x| = O(c).
\tag{281}
$$

The contribution of such values of x in (280) is then $\frac{1}{\delta^{100}} O(\tilde{s} + \sqrt{\tilde{s}})$. Or, $|x-t|$ is larger than $\delta/4$. $\tilde{s} - \tilde{\tau}_1$ can be assumed to be very small with respect to δ^{10}. We need, before proceeding with the proof of ii) an estimate on $|D| + |D_x|$. We observe that $D = \frac{1}{w_x}\frac{\lambda+\bar\mu\eta}{a}(w) + \frac{\partial w}{\partial s}$, $D_x = -\frac{w_{xx}}{w_x^2}\frac{\lambda+\bar\mu\eta}{a} + \frac{B\eta - \int_0^1 b\eta}{a} + \frac{\partial}{\partial s}w_x$ are both equal to $\frac{1}{\delta^{10}}O\big(-\frac{\partial a}{\partial s}\big) + b\eta_{12}$ where O does not depend on δ. This is obvious for $\frac{\lambda+\bar\mu\eta}{aw_x}$ and its derivative. w is easily constructed on $\big[t_1^+(0) + \frac{3\delta}{4}, t_2^-(0) - \frac{3\delta}{4}\big]$, using only $t_2^-(s)$ and functions of a: we need to go from $-1 + \big(t_2^-(s) + \frac{a_0}{a}(t_1^+(0) + \frac{3\delta}{4} - t_2^-(0))\big) = -1 + \big(t_1^+(s) + \frac{3\delta}{4}\frac{a_0}{a} + o(1)\big)$ (by ii)) to $t_2^-(s) - \frac{3\delta}{4}\frac{a_0}{a}$ on the interval $\big[-1 + t_1^+(0) + \frac{3\delta}{4}, t_2^-(0) - \frac{3\delta}{4}\big]$, which is tantamount to going from $-1 + t_1^+(s)$ to $t_2^-(s)$, since δ and $o(1)$ are very small. This involves a linear function of t and some regularization, to make $w \in C^2$ in t at the boundaries. Observe that w_x and w_x^{-1} are bounded away from zero because, on $\big[t_2^-(0) - \frac{3\delta}{4}, t_1^+(0) + \frac{3\delta}{4}\big]$, the formula is explicit and, on $\big[-1 + t_1^+(0) + \frac{3\delta}{4}, t_2^-(0) - \frac{3\delta}{4}\big]$, we can argue as follows: Taking the origin of time in the formula for $\lambda + \bar\mu\eta$ to be between $t_1^+(s)$ and $t_2^-(s)$, we have:

$$t_1^{+\prime}(s) - t_2^{-\prime}(s) = -\frac{1}{a}\left(\int_{t_2^-}^{t_1^+} b\eta - (t_1^+ - t_2^-)\int_0^1 b\eta\right)$$

$$= -\frac{1}{a}\left(\int_{t_2^-}^{t_1^+} cb^2 - (t_1^+ - t_2^-)\int_0^1 b\eta\right) = (t_1^+ - t_2^-)O\left(-\frac{\partial a}{\partial s}\right).$$

Thus, $t_1^+(s) - t_2^-(s) = (t_1^+(0) - t_2^-(0)(1 + O(\Delta a))$. Unless Δa is of the same order of magnitude than $t_2^-(0) - t_1^+(0) + 1$ (the complement of $t_1^+(0) - t_2^-(0)$ to 1), $t_2^-(s) - t_1^+(s) + 1$ will be of the same order of magnitude than $t_2^-(0) - t_1^+(0) + 1$, hence w_x and w_x^{-1} will be bounded. Would Δa be of the same order of magnitude than $t_2^-(0) - t_1^+(0) + 1$, then, since the loss of control on $\int(|b| - \nu)^+$ was at most $\nu(t_2^-(0) - t_1^-(0) + 1)$ in the creation of oscillation process, this control has been restored. Otherwise, $|1 + t_2^-(s) - t_1^+(s)|$ will be the same order than $|1 + t_2^-(0) - t_1^+(0)|$ and our claim holds. This works if our initial interval was large enough: $-\Delta a$ is then sizable. However, if $-\Delta a$ is not sizable, we might face through iteration of the use of the Z_δ's a loss of control on the curve. This does not happen for the following reason: If the length of $[t_1^+(s), t_2^-(s)]$ collapses below $2\delta, \eta_{12}$, which has support in $[t_1^+(s) + \delta, t_2^-(s) - \delta]$ becomes zero. This forces Z_δ to vanish. The curve does not move anymore. All estimates hold. Thus, $[t_1^+(0), t_2^-(0)]$ as well as $[t_1^+(s), t_2^-(s)]$ are at least of length 2δ. Thus, assuming that $\frac{a_0}{a} \le \frac{8}{7}$ and $o(1)$ is small enough, $-1 + \big(t_1^+(s) + \frac{3\delta}{4}\frac{a_0}{a} + o(1)\big)$ is away from $t_2^-(s) - \frac{3\delta}{4}\frac{a_0}{a}$ by at least $\frac{\delta}{100}$. $-1 + t_1^+(o)$ is away from $t_2^-(0)$ by at least 2δ. w is then made of two linear pieces, which we regularize into a C^3 function. Since $|1 - t_2^-(0) + t_1^+(0)|$ is larger than 2δ, we can regularize in $\big[-1 + t_1^+(0) + \frac{3\delta}{4}, t_2^-(0) - \frac{3\delta}{4}\big]$. The two slopes of the linear functions are $\frac{a_0}{a}$ on one interval, $\frac{\frac{a_0}{a}(t_1^+(0) - t_2^-(0)) - 1}{t_2^-(0) - t_1^+(0) + 1}$ on the other one. $\frac{a_0}{a}(t_1^+(0) - t_2^-(0)) - 1$ is equal to $-1 - t_2^-(s) + t_1^-(s) + o(1)$ by ii) (whose proof is independent of i)) where $o(1)$ can be made small with respect to $1 - (t_1^+(0) - t_2^-(0))$ and $1 - (t_1^-(s) - t_2^+(s))$ as well. Therefore, the two slopes are bounded and bounded away from zero. We can regularize w easily. The x-derivatives of w up to the order 3 will be $O(1/\delta^{10})$ at most. (The interval of regularization is of size larger than $\delta/2$). i) follows.

The s-derivative of these slopes is $O(-\frac{\partial a}{\partial s})$ for the first one, $\frac{O(-\frac{\partial a}{\partial s})}{t_2^-(0)-t_1^+(0)+1} = \frac{1}{\delta}O(-\frac{\partial a}{\partial s})$ for the second one.

Regularizing, we have $\frac{\partial}{\partial s}w_x = \frac{1}{\delta^{10}}O(-\frac{\partial a}{\partial s})$. Thus

$$|D|+|D_x| = \frac{1}{\delta^{10}}O\left(-\frac{\partial a}{\partial s}+b\eta_{12}\right). \tag{282}$$

We then come back to the proof of iv). $|x-t|$ is larger than $\delta/4$. We then have

$$\begin{aligned}(2) &= \frac{1}{\delta^{100}}O\left(\int_0^s d\tau_1 \int_{|x-t|\geq\frac{\delta}{4}} \frac{e^{-\frac{(x-t)^2}{4(\tilde{s}-\tilde{\tau}_1)}}}{\sqrt{\tilde{s}-\tilde{\tau}_1}}\left(\frac{|D|+|D_x|+|N_{12}|}{\sqrt{\tilde{s}-\tilde{\tau}_1}}+\right.\right.\\ &\quad \left.\left. +\frac{c}{\sqrt{\tilde{s}-\tilde{\tau}_1}}+\frac{1}{\tilde{s}-\tilde{\tau}_1}(|N_{12}|+O(c)\right)\right)dx\\ &= \frac{C(|b|_\infty)}{\varepsilon_{10}\delta^{200}}O\left(\int_0^s \frac{-\frac{\partial a}{\partial s}(\tau_1)}{\tilde{s}-\tilde{\tau}_1}e^{-\frac{\delta^2}{100(\tilde{s}-\tilde{\tau}_1)}}d\tau_1\right),\end{aligned}$$

where ε_{10} is a lower bound on b on the oscillation where η_{12} is built.

(283)

Indeed, c is bounded by $-\frac{\partial a}{\partial s}$, $\tilde{s}-\tilde{\tau}_1$ is small, $|D|+|D_x|$ is bounded by $\frac{1}{\delta^{10}}O(-\frac{\partial a}{\partial s})+|b|_\infty N_{12}$ and $\int_{|x-t|\geq\frac{\delta}{4}} e^{-\frac{(x-t)^2}{8(\tilde{s}-\tilde{\tau}_1)}}|N_{12}|dx$ is upper bounded by $\frac{C}{\varepsilon_{10}}\int_{|x-t|\geq\frac{\delta}{4}} e^{-\frac{(x-t)^2}{8(\tilde{s}-\tilde{\tau}_1)}}BN_{12}(w)dx$, where ε_{10} lower bounds globally B on the oscillations where N_{12} is nonzero. This integral is easily seen to be $O(\frac{C}{\varepsilon_{10}}\times e^{-\frac{\delta^2}{100(\tilde{s}-\tilde{\tau}_1)}}\int_0^1 BN_{12}(w$

Using our lower bound on w_x, we derive (283), since $\int_0^1 b\eta_{12}$ is less than $-\frac{\partial a}{\partial s}$.

c, in (15), can be taken of the type $-\varepsilon_{100}\delta^{300}\int_0^1 b\eta_{12}$ where ε_{100} is a small fixed constant, depending on C_{41}; δ^{300} takes care of all our other requirements. Thus c is larger than $\frac{1}{2}\varepsilon_{100}\delta^{300}\times(-\frac{\delta a}{\partial s})$.

Then, (2) in (283) can be rewritten:

$$\begin{aligned}(2) &= O\left(\frac{C_{41}}{\varepsilon_{100}\varepsilon_{10}\delta^{600}}\right)\int_0^s \frac{c}{\tilde{s}-\tilde{\tau}_1}e^{-\frac{\delta^2}{100(\tilde{s}-\tilde{\tau})}}d\tau_1\\ &= O\left(\frac{C_{41}}{\varepsilon_{100}\varepsilon_{10}\delta^{600}}\right)\int_0^{\tilde{s}} \frac{e^{-\delta^2/100z}}{z}dz\end{aligned}$$

$\tilde{s}$ is, by (273), of the order of $\varepsilon_{100}\delta^{300}(a(0) - a(s))$, which we can assume to be less δ^{100}. Then $\frac{e^{-\delta^2/100z}}{z}$ is upper bounded by $e^{-\delta^2/200z}$, for δ small enough and $z \in [0, \tilde{s}]$. Thus (2) $= O(\frac{C_{41}}{\varepsilon_{100}\varepsilon_{10}\delta^{600}})\tilde{s}e^{-\delta^2/200\tilde{s}}$. Since $\tilde{s}$ is of the order of $\varepsilon_{100}\delta^{300}(a(0) - a(s))$, (2) is $o(1)$ as claimed, for δ small. The proof of Lemma 17 is thereby complete.

The above construction and Lemmas 16–18 can be repeated with other intervals $[t_i^-, t_{i+1}^+]$. We can patch the various $\overline{Z}$ obtained in this way. The resulting $\overline{Z}$ behaves as the above $\overline{Z}$ does. We will use this global $\overline{Z}$ in an global deformation.

I.4. The convergence theorems

We need now to take care of the new flat $|b|_\infty$ pieces near t_2^- and t_1^+ which might have been introduced by the repetitive use of $\bar{Z}$. In addition to this, we will describe another interesting phenomenon, which yields a similar problem: Namely, up to this point, we have not used any hypothesis on v, and we have brought our curves to be ν-stretched with a good control, depending on ν, on $\int_0^1 |b|$. (We could have used, as we will now, $\tilde{\nu} = \frac{\nu}{\int_0^1 |b|+1}$, instead of ν, we would have kept this bound). We introduce here one hypothesis, namely hypothesis (H) below. Under this hypothesis we will transform these ν-(or $\tilde{\nu}$) stretched curves into critical points at infinity. We will indicate however how to get rid of (H) later at the expense of deforming only on $\underset{k}{U} W_u(\Gamma_{2k})$, where Γ_{2k} is the set of curves alternatively tangent to ξ and $\pm v, k$ times, rather than on $\bigcup_{\bar{x}_\infty} W_u(\bar{x}_\infty)$, where $\bar{x}_\infty$ are the critical points at infinity.

Later also, through a milder hypothesis than (H) (hypothesis (A1), α turns well along v) we will derive the statement on $\bigcup_{\bar{x}_\infty} W_u(\bar{x}_\infty)$. (H) is not needed in the convergence process, but has the advantage to immediately point out the critical points at infinity.

Due to the fact that the arguments below were written at the very end, we will use some Definitions of sections II and III which we briefly recall. These definitions were introduced in [1]. We apologize to the reader for this inconvenience and the absence of numbering of the formulae.

We consider in the sequel a $\tilde{\nu}$-stretched curve, i.e. we free ourselves here of the precise value of ν. We will use $\tilde{\nu}$ in lieu of ν. $\tilde{\nu}$ will assume various values later.

We start with:

Definition of the conjugate points flow

Let us consider an oscillation of b preceded and followed by two pieces where $|b|$ falls from $|b|_\infty$ to $\tilde{\nu}$ or zero, in a very short time, i.e. a subpiece of a stretched $\tilde{\nu}$-curve, of the type:

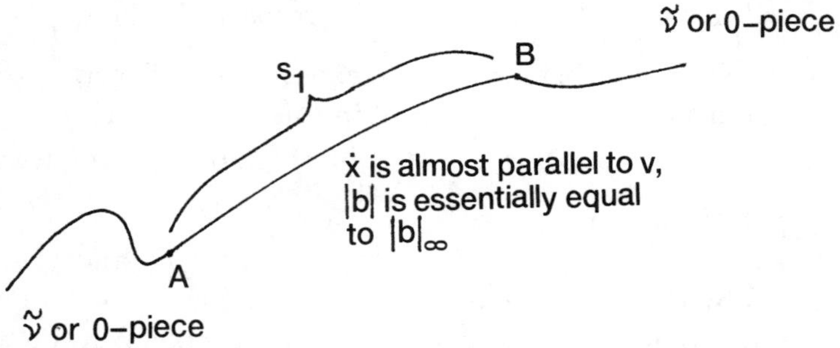

FIGURE M

Our arguments above show that we can assume that the curve is indeed $\tilde{\nu}$-stretched. Using the regularizing flow with $\eta = b$ during a small time, we can assume that $|b|_\infty$ and C_{41} are so large with respect to $\int_0^1 |b|$ (which is controlled up to now and will also be later) that the system of differential equations:

$$(\omega)\begin{cases} \dot{\overline{\lambda + \bar{\mu}}} = b\eta \\ \dot{\eta} = -\lambda b + \mu a \\ \dot{\mu} = -b\eta\bar{\mu}_\xi + a\eta\tau \end{cases} \qquad z = \lambda\xi + \mu v + \eta w \qquad (\omega)$$

behaves just as if a were zero and $A - B$ were ab s_1-piece of v-orbit out of A. The above system of differential equations is therefore very close to the one defined by $D\phi_s$, reparametrized by $s = \tilde{s}|b|_\infty(b = |b|_\infty$ between A and $B)$, where ϕ_s is the one-paramete group generated by v. $s = 0$ is $A, s = s_1$ is very close to B. A and B are chosen so that

$$\int_0^1 |b| + 1 \leq |b(A)|, |b(B)| \leq 2(\int_0^1 |b| + 1 \qquad (\tilde{\omega})$$

(Both satisfy both inequalities).

Since our curves will all be nearly $\tilde{\nu}$-stretched, the definition of A and B will be ambiguous only up to $0(\delta)$ in $\int |B|$. Convex-Combination allows to define A and B varying C^∞ with the curve.

Let us assume:

$$|D\phi_s|_{C^1} \leq K, \text{ where } K \text{ is independent of } s. \qquad (H)$$

(H) implies that $\det D\phi_s$ is bounded above and below, independently of s; thus, using Proposition A3 and it proof, see Appendix 2, $\ker\alpha$ turns in the transport along v, A has, under (H), infinitely many coincidence points i.e. infinitely many points x_{s_i} on the v-orbit through A such that $\ker\alpha$ has complete k_i revolutions between A and x_{s_i}. Coincidence points have been introduced in [1]; their definition is recalled in Definition 9, below. Among these coincidence points, there might be some conjugate points i.e. coincidence points such that the form α, not only the plane $\ker\alpha$, has been transported by v between A and x_{s_i}. Conjugate points have also been introduced in [1]. Their definition is recalled in Definition 9', below;

Let us assume that:

$$B \text{ is at least } \delta s_1\text{-away from any conjugate point of } A. \qquad (\theta)$$

Conjugate points are precisely those such that α is v-transported from A to any of them. Therefore, $\lambda + \bar{\mu}(A)$ is equal to $\lambda + \bar{\mu}\eta(C)$, where C is conjugate point of A, and (λ, μ, η) are the components of a vector z transported by $D\phi_s$. Since (ω) is very close to $D\phi_s$ as long as the length is of the order of s_1 and since (H) holds and B is at least $0(\delta s_1)$-away from any conjugate point of A, there exists an initial condition $z(A)$ for (ω), with $z(B)$ the value at B through $(\omega), z(A)$ non zero, such that:

$$|(\lambda + \bar{\mu}\eta)(B) - \lambda + \bar{\mu}\eta)(A)| \geq \theta\delta s_1(|z(A)| + |z(B)|) \qquad (\rho)$$

where $\theta = \theta(K) > 0$ is a uniform positive constant, which depends only on K in (H). This claim, in view of (H), is easy to prove.

z has now been defined on $[A, B]$. By the first and third equations of (ω), b and a are unchanged on this segment. We extend z to the two nearly $\bar{\nu}$ or 0-pieces, the one before A and the one after B as follows, let us consider the piece after B, for example:

Assume that such a piece is occuring on a length of time $\Delta t(B)$. Then, we can extend η into a C^∞-function, which dies in the middle of this piece. Since $\dot{\eta} = \mu a - \lambda b = 0((1 + \int_0^1 |b|)|z(B)|)$ at B, we can require that

$$\dot{\eta} = \left((1 + \int_0^1 |b|)|z(B)| + \frac{|\eta(B)|}{|\Delta t(B)|}\right), \eta = 0(|\eta(B)|)$$

on this piece. We can also assume that $\ddot{\eta}$ changes at most three times its sign on this piece. λ is also extended on the same piece and dies also in the middle, with $\lambda = 0(\lambda(B)), \dot{\lambda} = 0\left((1 + \int_0^1 |b|)|\eta(B)|\right)$.

We thus have defined a variation which does not keep a to be a constant. In order to obtain this property, we add to z $\lambda_2\dot{x}$, where

$$\lambda_2 = -(\lambda + \bar{\mu}\eta) + \int_{t_A}^t b\eta - (t - t_A)\int_0^1 b\eta + (\lambda + \bar{\mu}\eta)(A).$$

Let z_1 be the vector-field corresponding to $\eta_1 = cb, c$ such as in (15). Let $\tilde{z} = z + \frac{\lambda_2}{a}\dot{x} + z_1$. The variation $\tilde{z}$ keeps a to be a constant and can be extended to a suitable small $W^{1,\infty}$-neighborhood of these curves. $\tilde{z}$ has $\tilde{\eta}$ equal to $\eta + cb$, $\tilde{\lambda}$ equal to $\lambda + \lambda_2 - \bar{\mu}cb + c\left(\int_0^t b^2 - t\int_0^1 b^2\right)$, $\tilde{\mu}$ equal to $\mu + (\lambda_2 + \lambda_1)b$ (λ_1 is the component of z_1 on ξ).

In view of (ω), we have at every t such that $|b(t)| = |b|_\infty$ ($\tilde{z}$ reduces to cb outside of the b-oscillation followed by the $\tilde{\nu}$-piece which we are considering), for example if $b(t) = |b|_\infty$ so that $\ddot{b}$ is negative:

$$\dot{\tilde{\mu}} + a\tilde{\eta}\tau - b\tilde{\eta}\bar{\mu}_\xi \leq \dot{\mu} + a\eta\tau - b\eta\bar{\mu}_\xi + \frac{\dot{\lambda}_2 b}{a} + 0(c|b|^3_\infty) =$$

$$= -\frac{b}{a}(\overline{\lambda + \bar{\mu}\eta}^{\cdot} - b\eta + \int_0^1 b\eta) + 0(c|b|^3_\infty) = -\frac{b}{a}\int_0^1 b\eta + 0(c|b|^3_\infty).$$

Thus

$$\frac{\partial}{\partial s}|b|_\infty \leq -\frac{|b|_\infty}{a}\int_0^1 b\eta + 0(\int_0^1 b\eta).$$

The flow of $\tilde{z}$ controls $|b|_\infty$. In fact, on the almost v-pieces, $b/_a$ is basically constant, the variation depends only on c, see the proof of Lemma 17 iv). The introduction of the small viscosity cb allows to derive easily local existence, uniqueness etc., as long as c remains away from zero. c tends to zero only if $-\int_0^1 b\eta$ tends to zero for z. We will, using $\delta s_1 > 0$, keep away from these regions. Clearly, using $\tilde{z}$, the $|b|_\infty$- pieces remain basically $|b|_\infty$-pieces (up to $0(c)$). Starting from $\tilde{\nu}$-stretched curves, with small δ-pieces, we will have existence, for this flow, for a uniform time $\delta t > 0$, as long as δs_1 remains away from zero.

We are going to use this time δt.

We continue exploring the properties of the flow of $\tilde{z}$:

This flow might destroy our control on the zeros of b; but, using the cancellation flow as soon as a small oscillation of b above or below zero develops, we can restore this control on the number of zeros: the only oscillations above zero which we create and which escape such a control must be of type 2). We can assume that $|b|$, on each of them; remains small, less than a fixed constant ε_{10}. (We are using here the property of the cancellation flow to decrease the maxima. It can be combined with the use of ii) or iii) of Definition 2). Then, such oscillations have to take place on a large interval. They are, therefore, in finite number. Thus, if we do increase the number of zeros of b, it is of a fixed amount. However, the cancellation flow perturbs all the curves beyond control, at this point. We thus have to use the local form of it, such as Z_δ, i.e cancel and stretch in the sink with the $\tilde{\nu}$-flow. The drawback is that we, now, can create small $\pm|b|_\infty$ -pieces. We will take care of this later and estimate their contribution.

Each single oscillation which is created is small, unless the decrease in a is substantial. Therefore, at the end of the whole process, (we prove that algebraically, the edges move very little; hence the oscillations above zero have little strength), when the curves are made only of ξ and v-pieces, we can use the cancellation flow on the whole curve and cancel the oscillations, or, since they are extremely stretched, use the regularization flow, taking them by packs near an edge, or the local cancellation flow around one pack, and cancel them. The bounds will persist. The top curves will be made of ξ or $\pm v$- jumps, the $\pm v$-jumps being nearly conjugate.

The average length of a time-interval on a $\tilde{\nu}$-piece is. $\frac{\delta_0}{1+\int_0^1(|b|-\tilde{\nu})^+}$, since each jump in v contributes at least δ_0 to $\int_0^1 |b|$. Let $\varepsilon_0 > 0$ be a small constant. Using (H), we can neglect all the intervals of time-length less than $\frac{\varepsilon_0\delta_0}{1+\int_0^1(|b|-\nu)^+}$. Indeed, their combined contribution even after mapping through ϕ_{s_1} etc., i.e. adjusting them to become precisely consecutive v-jumps as follows:

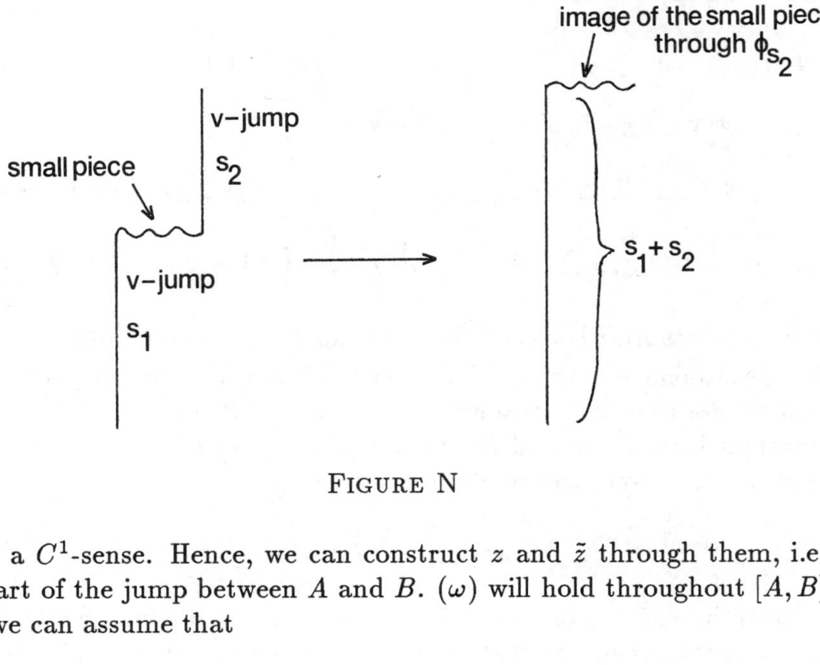

FIGURE N

is $0(\varepsilon_0)$ in a C^1-sense. Hence, we can construct z and $\tilde{z}$ through them, i.e. count them as part of the jump between A and B. (ω) will hold throughout $[A, B]$.

Then we can assume that

$$\Delta t(A) \text{ and } \Delta t(B) \text{ are larger than } \frac{\varepsilon_0\delta_0}{1+\int_0^1(|b|-\tilde{\nu})^+}.$$

Control on $\int(|b|-\tilde{\nu})^+$ along flow-lines and on $\int(|b|-\tilde{\nu})^+$ in the sinks

We compute then $\frac{\partial}{\partial s}\int(|b|-\tilde{\nu})^+.\dot{\bar{\mu}} + a\eta\tau - b\eta\bar{\mu}_\xi$ is equal to $\frac{\dot{\overline{\lambda_2 b}}}{a}$ on $[A, B]$. Thus, since we know how to take care of the contribution of cb:

$$\frac{\partial}{\partial s}\int(|b|-\tilde{\nu})^+ \le \left.\frac{\lambda_2|b|}{a}\right|_A^B + \int_{t\notin[A,B]} sgnb\left(\frac{\dot{\overline{\eta-\bar{\mu}\eta b}}}{a} + a\eta\tau - b\eta\bar{\mu}_\xi\right)\chi_{|b|>\tilde{\nu}}+$$

$$+\int_{t\notin[A,B]} sgnb\overline{\frac{(\tilde{\lambda}+\bar{\mu}\eta)b}{a}}^{\displaystyle\cdot}\chi_{|b|>\tilde{\nu}}+0\left(\frac{-\partial a}{\partial s}\right)$$

$\overline{\dot{\eta}-\bar{\mu}\eta b}^{\cdot}$ might yield a non standard contribution when intergrated between A and B and the first time where $|b|$ takes the value $\tilde{\nu}$ before or after. However $(\dot{\eta}-\bar{\mu}\eta b)(A)$ and $(\dot{\eta}-\bar{\mu}\eta b)(B)$ are $0\left((1+\int_0^1|b|)(|z(A)|+|z(B)|)\right)$ and $(\dot{\eta}-\bar{\mu}\eta b)(\bar{t})$, if $|b(\bar{t})|=\tilde{\nu}$ is $0\left((1+\int_0^1|b|)(|z(A)|+|z(B)|+\frac{|\eta(B)|}{\Delta t(B)}+\frac{|\eta(A)|}{\Delta t(A)}\right)$. Because of the sign requirements on $\ddot{\eta}$, $\int|\ddot{\eta}|$ is at most equal to $6\underset{t\notin[A,B]}{\mathrm{Sup}}|\dot{\eta}|$, η is globally $0(|\eta(A)+\eta(B))$, $\eta,\dot{\eta}.\ddot{\eta}$ are zero outside intervals of length $\Delta t(A),\Delta t(B)$ in $[B,A]$. $\int\overline{\bar{\mu}\eta b}^{\cdot}sgnb$ between two consecutive times where $|b|=\tilde{\nu}$ is $\tilde{\nu}\int\overline{\bar{\mu}\eta}^{\cdot}$ between the same times. Thus, with $\tilde{\nu}$ less than 1 and using (ρ):

$$(\tilde{\psi})\begin{cases}\frac{\partial}{\partial s}\int(|b|-\tilde{\nu})^+\leq 0\left(\frac{|\eta(A)|+|\eta(B)|}{\varepsilon_0}+|z(A)|+|z(B)|\right)(1+\int_0^1(|b|-\tilde{\nu})^+)+\\ +0\left(-\frac{\partial a}{\partial s}\right)+\frac{\lambda+\bar{\mu}\eta}{a}|b|\Big|_A^B+\frac{\tilde{\nu}}{a}\int_B^a\frac{\overline{|\lambda+\bar{\mu}\eta|}^{\cdot}}{a}\leq\\ \leq 0\left(\frac{|\eta(A)|+|\eta(B)|}{\varepsilon_0}+|z(A)|+|z(B)|\right)(1+\int_0^1(|b|-\tilde{\nu})^+)+0\left(-\frac{\partial a}{\partial s}\right)\leq\\ \leq 0\left(-\frac{\partial a}{\partial s}\right)+\frac{1}{\delta s_1}\left(\frac{1}{\varepsilon_0}+1\right)|(\lambda+\bar{\mu}\eta)(B)-(\lambda+\bar{\mu}\eta)(A)|(1+\int_0^1(|b|-\tilde{\nu})^+)\end{cases}$$

Picking up points A_i, B_i on each jump, we can do also better and control $\int_{B_i}^{A_{i+1}}(|b|-\tilde{\nu})^+$. We indeed point out the similarity with Lemma 17: Between A_i and B_i, our flow, up to cb, does not change much $b/_a$. In fact, if it were not for cb, $b/_a$ would remain constant between A_i and B_i: We transform $\tilde{z}$ into a time dependent flow by following A_i and B_i along an evolution equation such as

$$t'(s)=-(\widetilde{\lambda+\bar{\mu}\eta})(t(s))$$

ask then (ω) to be satisfied on $[A-\delta,B+\delta]$, where δ is a tiny number to be changed at each stage of the process so that the comulative contribution of $(C_{41}+1)\sum\delta_k$ is $o(1)$. Arguing as in the proof of Lemma 17 iv), $\frac{b}{a}(\varphi(s,t))$ is changed very little on $[A_i-\frac{3}{4}\delta,B_i+\frac{3}{4}\delta]$, A_i and B_i can be used forever i.e. $(\tilde{\omega})$ is satisfied.

We can then define a quantity similar to $\underline{w}$ of (255) in order to estimate the evolution of $\sum_i\int_{B_i}^{A_{i+1}}(|b|-\tilde{\nu})^+dt$. Computing as in (256) - (265), we obtain $(\lambda_2(A_i)=0,\lambda_2(B_i)=0(\int_0^1 b\eta),\lambda(C)=0(\int_0^1 b\eta)$ for $C\in[A_i-\delta,B_i+\delta]$, since (ω) holds on $[A_i-\delta,B_i+\delta])$

$$\frac{\partial\underline{w}}{\partial s}=\frac{\partial}{\partial s}\left(\int_{B_i}^{A_{i+1}}(|b|-\tilde{\nu})^+dt+0(\delta)\right)\leq 0\left(-\frac{\partial a}{\partial s}\right)+\frac{1}{\delta s_1}0\left(\frac{1}{\varepsilon_0}+1\right)\times$$

$$\times\left(\sum(\lambda+\bar{\mu}\eta)(B_j)-(\lambda+\bar{\mu}\eta)(A_j)(1+\int_0^1(|b|-\tilde{\nu})^+\right).$$

Lastly, we have:

$$\begin{aligned}\int_0^1 b\eta &= \sum(\lambda+\bar{\mu})(B_j)-(\lambda+\bar{\mu}\eta)(A_j)\\ &\quad+\sum_j 0(|\eta(A_j)|+|\eta(B_j)|\times\left(\tilde{\nu}+\int_{B_j}^{A_{j+1}}(|b|-\tilde{\nu})^+\right)\\ &=\sum((\lambda+\bar{\mu}\eta)(B_j)-(\lambda+\bar{\mu}\eta)(A_j))\left(1+\frac{1}{\delta s_1}0\left(\tilde{\nu}+\int_{B_j}^{A_{j+1}}(|b|-\tilde{\nu})^+\right)\right).\end{aligned}$$

We will take δs_1 to be a small fixed number such that $\frac{1}{\delta s_1}0\left(\tilde{\nu}+\int_{B_j}^{A_{j+1}}(|b|-\tilde{\nu})^+\right)$ is $o(1)$. We will check the coherence of there choices later. Then, $\sum(\lambda+\bar{\mu}\eta)(B_j)-(\lambda+\bar{\mu}\eta)(A_j)$ is of the same order than $-\frac{\partial a}{\partial s}$.

Thus, using our differential equations above:

$$(\tilde{\varphi})\begin{cases}\int(|b|-\tilde{\nu})^+\leq -2\Delta a+e^{-\frac{C(\varepsilon_0)}{\delta s_1}\Delta a}\int(|b|-\tilde{\nu})^+(0)\\ \int_{B_i}^{A_{i+1}}(|b|-\tilde{\nu})^+(s)\leq -C\Delta a-Ce^{-\frac{C(\varepsilon_0)}{\delta s_1}\Delta a}\Delta a\int(|b|-\tilde{\nu})^+(0)\\ \qquad+\int_{B_i}^{A_{i+1}}(|b|-\tilde{\nu})(0)+0(\delta).\end{cases}$$

Thus, if $\Delta a=o(\delta s_1),\Delta a\int(|b|-\tilde{\nu})^+(0)=o(\delta s_1),\delta=0(\delta s_1)$ and $\int_{B_i}^{A_{i+1}}(|b|-\tilde{\nu})(0)=o(\delta s_1)$ (which is the case for a $\tilde{\nu}$-stretched curve), all our above arguments apply. (φ) holds.

We, of course, choose $(\lambda+\bar{\mu}\eta)(B_j)-(\lambda+\bar{\mu}\eta)(A_j)\geq 0$, $\sum_j((\lambda+\bar{\mu}\eta)(B_j)-(\lambda+\bar{\mu}\eta)(A_j))>0$, as soon as one B_j is not δs_1-close to a conjugate point of A_j. Our vector-field is now a decreasing pseudo-gradient for a.

The transversal displacement of the nearly v-jumps is $0((-\Delta a)_i)$

Another remarkable property of this flow is its effect on the nearly v-jumps. We have seen already that, if it were not for cb, b and a would be constant, b very large on the nearly v-jumps. The influence of cb is very small, if c is chosen appropriately, this has already been shown in the proof of Lemma 17, if we are careful and pick a tiny $\delta>0$ before B and after A where $|b(o,t)|$ is less than $3\left(+\int_0^1|b|\right)$. Then, between $A+\delta$ and $B-\delta$, things vary very little as far as $b/_a$ is concerned. The time-span of such time-intervals is changed very little. There is at most a nearly uniform time dilation $\frac{a_0}{a}$.

Picking up a time t_0 in this interval and assuming there is no time shift, i.e. assuming for example that A occurs always at time zero, we have:

$$\frac{\partial}{\partial s}x\left(s, \frac{a_0}{a}t_0\right) = \tilde{\lambda}\xi + \tilde{\mu}v + \tilde{\eta}w + 0(c)|b|^2_\infty + c\dot{b}v + \frac{a_0}{a^2}t_0\int_0^1 b\eta(a\xi + bv).$$

Using the fact that the time-span is lower bounded, we can average on a fixed size interval around t_0. The size of the interval is tiny equal to $\Delta(0)$. c will be taken small even with respect to this interval. We then see that ($\lambda_2, \tilde{\lambda}$ and $\tilde{\eta}$ are $0\left(\int_0^1 b\eta\right)$ see above)

$$\frac{1}{\Delta(0)}\frac{\partial}{\partial s}\int_{t_0-\frac{\Delta(0)}{2}}^{t_0+\frac{\Delta(0)}{2}} x(s, \frac{a_0}{a}t)dt = \frac{1}{\Delta(0)}|b|^2_\infty 0(c) + 0\left(\int_0^1 b\eta\right)$$
$$+\frac{0}{\Delta(0)}\left(\int_{t_0-\frac{\Delta(0)}{2}}^{t_0+\frac{\Delta(0)}{2}}(\tilde{\mu} + \frac{a_0}{a^2}t_0\int_0^1 b\eta b)v\right).$$

Observe now that:

$$\tilde{\mu} - \frac{a_0}{a^2}t_0\int_0^1 b\eta b = \frac{\dot{\eta} + \tilde{\lambda}b}{a} + t_0\frac{a_0}{a^2}b\int_0^1 b\eta = \mu + \frac{(\tilde{\lambda} - \lambda)b}{a} + t_0\frac{a_0}{a^2}b\int_0^1 b\eta$$
$$= \mu + \frac{(\lambda_2 + 0(c)|b|^2_\infty)b}{a} + \frac{a_0}{a^2}t_0\int_0^1 b\eta$$

$\lambda_2(A)$ is zero, $\dot{\lambda}_2$ is equal to $-\int_0^1 b\eta$ from A to B. Thus, $\lambda_2(s, \frac{a_0}{a}t_0) = -\frac{a_0}{a}t_0\int_0^1 b\eta$. and

$$\frac{1}{\Delta(0)}\frac{\partial}{\partial s}\int_{t_o-\frac{\Delta(0)}{2}}^{t_0+\frac{\Delta(0)}{2}} x(s, \frac{a_0 t}{a})dt = 0\left(\int_0^1 b\eta\right) = 0\left(-\frac{\partial a}{\partial s}\right).$$

Thus,

$$d\left(\frac{1}{\Delta(0)}\int_{t_0-\frac{\Delta(0)}{2}}^{t_0+\frac{\Delta(0)}{2}} x(s, \frac{a_0 t}{a})dt, \frac{1}{\Delta(0)}\int_{t_0-\frac{\Delta(0)}{2}}^{t_0+\frac{\Delta(0)}{2}} x(s', \frac{a_0 t}{a})dt\right) = 0(-\Delta a).$$

Since $\int_{t_0-\frac{\Delta(0)}{2}}^{t_0+\frac{\Delta(0)}{2}}|\dot{x}| = 0(|b|_\infty\Delta(0))$, if we choose $|b|_\infty\Delta(0) = o(1)$, we obtain that the two nearly v-branches are distant of $0(-\Delta a)+o(1)$, where $o(1)$ is a fixed quantity which can be choosen to be small from the beginning, at any rate, we can require that $o(1) = 0\left(\frac{1}{L^2}\right)$.

We single out here that the same result holds for the flow $\tilde{Z}$, since this flow reduces to $\eta = cb$ on the almost v-jumps,

Thus, if we combine these two flows, the v-branches remain close up to $0(-\Delta a)+o(1)$ globally. Furthermore, this $-\Delta a$ can be transformed in $(-\Delta a)_i + o\left(\frac{-\Delta a}{L}\right)$ because, as the arguments above show, the $-\Delta a$ involved in the estimate of the transversal displacement is related to the conjugate points flow on the jump order i. Otherwise, the displacement is simple due to c, hence can be included in $o\left(\frac{-\Delta a}{L}\right)$.

The deformation process combining three flows

As long as $\tilde{\nu}$ is less than 1, $(\tilde{\varphi})$ holds and the conjugate points flow controls $\int(|b| - \tilde{\nu})^+(s)$ up to an exponential factor.

Another flow has the same property.

Namely, let L be a global bound on $\int_0^1 |b|$ in our initial compact which we are deforming. After all the ν-flows, this bound might have been transformed onto $L+1$, if $-\Delta a$ is less than a fixed constant which depends only on ν.

We take then

$$\tilde{\nu} = \frac{\nu}{L+10}.$$

Using the $\tilde{\nu}$- cancellation and stretching flows, we have: (see the computation in (29) - (33)):

$$\begin{aligned}\frac{\partial}{\partial s}\int(|b| - \tilde{\nu})^+ &\le -C\frac{\partial a}{\partial s} + \int |a\eta\tau| \le -C\frac{\partial a}{\partial s} - C\tilde{\nu}\frac{\partial a}{\partial s} \le \\ &\le -C\left(1 + \frac{L+10}{\nu}\right)\frac{\partial a}{\partial s}.\end{aligned}$$

We will assume that

$$-\Delta a(L+10) < \varepsilon_{30}$$

where ε_{30} is fixed small constant.

Then, coming back to the conjugate points flow, we have, by $(\tilde{\varphi})$:

$$\int(|b| - \tilde{\nu})^+(s) \le 0(\varepsilon_{30}) + \int(|b| - \tilde{\nu})^+(0) \le L + 2$$

if $-\Delta a(L+10) < \varepsilon_{30}$. Thus, using this bound in $(\tilde{\psi})$:

$$\frac{\partial}{\partial s}\int(|b| - \tilde{\nu})^+(s) \le -C\left(1 + \frac{1}{\delta s_1}\right)(L+2)\frac{\partial a}{\partial s}.$$

Thus, if $-\Delta a(L+10) < \varepsilon_{30}$, both flows satisfy the same differential equation on $\int(|b| - \tilde{\nu})^+(s)$. They can be convex-combined or used repeatedly one after the one; a good control on $\int(|b| - \tilde{\nu})^+(s)$ will be kept, as well as on $\int |b|$. ε_{30} has to depend on δs_1, but δs_1 is a fixed positive small number.

Using repeatedly this combination flow on our compact $\bar{K}$, we derive that

$$\bar{K} \cup J_c \text{ deforms onto a subset of } \left(\bigcup_{\bar{x}} W_u(\bar{x}) \cup J_{c - \frac{\varepsilon_{30}}{L+10}}\right)$$

where $\bar{x}$ is a $\tilde{\nu}$-stretched curve having its nearly v-jumps conjugate up to δs_1 i.e. B_i is at most $0(\delta s_1)$ away from a conjugate point of A_i. Furthermore, denoting S_1 the point of the deformed set above the level $c - \frac{\varepsilon_{30}}{L+10}$, the curves of S_1 satisfy:

$$\int |b| \le L + 3.$$

We then use the $2\tilde{\nu} = \frac{2\nu}{L+10}$ cancellation/ stretching flow on S_1 in the sink, i.e. between B_i and A_{i+1} in order to lower it a little bit (lower the top curves of it, $\bar{x}$) without touching at $S_1 \cap J_{c-\frac{\varepsilon_{30}}{L+10}}$.

Of course, this is tantamount to the flow of creation of oscillations / cancellations: we loose progressively control on $\int_0^1 |b|$, because by Lemmas 17 and 18, we do not create oscillations in the sinks, there is not enough mass there; but we might add flat $|b|_\infty$-pieces on each side, of strength at most $2\tilde{\nu} = \frac{2\nu}{L+10}$, i.e. of a small amount, adding length to the almost v-jumps.

We can then restart from scratch, use the two $\tilde{\nu}$-flows, the one of conjugate points and the one of cancellations/stretching etc.

At each step k, a small constant δ_k are chosen, which plays the role of δ of Lemmas 17 and 18. We can assume that their cumulated effect is extremely small. The time span at each step on each nearly ξ-piece (up to $2\tilde{\nu}$) which we are perturbing cannot change very much because the v-jumps are dilated by a factor at most $1 + 0(\Delta a) = 1 + 0\left(\frac{\varepsilon_0}{L}\right)$ hence the reduction of the other pieces, the nearly ξ-pieces, is at most $0\left(\frac{\varepsilon_0}{L}\right)$. As we iterate, the reduction might pick up. It does not. We show below that the cumulated time taken by the new $\pm|b|_\infty$- flat pieces is $o\left(\frac{1}{L}\right)$ and that these $\pm|b|_\infty$- flat pieces when added globally at one edge contibute algebraically only $0(1)$. We derive in fact much stronger estimates, see below. We should be careful to keep a continuous evolution for t_2^- and t_1^+, throughout this repeated process. This allows, by Lemmas 17 and 18 (which extend easily to the framework of this repeated process), to keep nearly $\pm|b|_\infty - v$-pieces all along the process before t_1^+ and after t_2^- and also to end up with curves made of $\pm v$-jumps alternated with ξ-pieces as top curves at each step of this process. This can be done easily.

The problem which we face is the loss of control on $\int_0^1 |b|$, which is equivalent to a loss of control of $\int_0^1 |b|$ on the iterated curves $\bar{x}_k$ in this repeated process. Would these ones be controlled, then since the iterated S_1^k (analogous to S_1 but at the kth-step) are only flow-lines out of $\bar{x}_k$ of the $\tilde{\nu}$-flows, all the curves of S_1^k will have their L^1-length in b controlled.

The striking fact is that such a bound holds. In fact:

Lemma N. *If ε_{30} is small enough, $\int_0^1 |b|$ on $\bar{x}_k$ is bounded by $L+10$. $\bar{x}_k$ converges when k tends to $+\infty$ to a curve $\bar{x}_\infty$ made of almost v-jumps of size at least δ_0 alternated with ξ-jumps of size at least $\frac{\varepsilon_0}{L+10}$. The above statement still holds if we remove (H) and do not use the conjugate points flow.*

Observe that $\bar{x}_\infty$ depends slightly on ε_0 due to these small time intervals, of size at most $\frac{\varepsilon_0}{L+10}$ where $|b|$ can be assumed, if we start by a procedure of $\tilde{\nu}$-cancellation stretching, before the whole process, (we still will have a good control on $\int_0^1 |b|$) to be less than $\frac{\nu}{L} = \tilde{\nu}$.

The cumulated effect of all these pieces is thus

$$L \times 0\left(\frac{\varepsilon_0}{L+10}\right) \times \frac{\nu}{L} = 0\left(\frac{\varepsilon_o}{L^2+1}\right) \times L = 0\left(\frac{\varepsilon_0}{L}\right)$$

if we replace them by ξ-pieces. It is small, but it is there. In order to get rid of them, we have to use the process, described below, which does not make use of (H). If, at any time, the big jumps, i.e. those we reduced previously move away more than δs_1 from being conjugate, we can come back to our former process. In this way, we have a very strong control on our curves, in particular on the first stage, i.e. when we set the smaller pieces to be tangent to ξ, we keep the larger ones also to be almost conjugate. There is a perturbation of $0\left(\frac{1}{L}\right)$ of the larger one due to the fact that we are working on two pieces which were part of it. This will yield a possible increase of $0\left(\frac{1}{L}\right)$ in length, L-times at most i.e. $0(1)$ for the total reduction. S_1^∞ deforms then into S_2^∞, with the same bound on $\int_0^1 |b|$. S_2^∞ might have some flat pieces of flow, which we arrange to be decreasing. Iteratively, we reach $\bar{S}_1^\infty$ and the curves $\bar{x}_1^\infty$ on top of $\bar{S}_1^\infty$ have only ξ-pieces alternated with at least δ_0-almost v-jumps.

Deformation Theorem 1. *Assume (H). Then*

1) $\bar{K} \cap J_c$ depends onto $\bar{S}_1^\infty \cup D$, where $D \subset J_{c-\frac{\varepsilon_{30}}{L+10}}$. All the curves at the end of the deformation have $\int_0^1 |b|$ less than $L+10$. $\bar{S}_1^\infty$ is $\bigcup_{\bar{x}_1^\infty} W_u(\bar{x}_1^\infty)$, where the $\bar{x}_1^\infty$'s are made of almost-v-jumps of size at least δ_0 between δs_1 almost conjugate points alternated with ξ-jumps.

2) Iterating, $\bar{K} \cap J_c$ deforms onto a subset of $\underset{\bar{x}_1^\infty}{U} W_u(\bar{x}_1^\infty)$.

3) If we remove (H), we have to remove from the above statement the fact that the almost v-jumps are between δs_1-almost conjugate points.

Proof. The Deformation Theorem 1 follows almost immediately from Lemma N

For 2), we just observe that, in order to decrease by $\frac{\varepsilon_{30}}{L+10}$ (outside of $W_u(\bar{x}_1^\infty)$), we increased L at most by 10. Iterating,

$$\sum_k \frac{\varepsilon_{30}}{L+10k} = +\infty. \text{ hence 2).}$$

Some additional explanations and ingredients are needed for the proof of the deformation theorem 1: the global cancellation / stretching flow, which we built in the previous sections, is the one which allows to bring us near the $\tilde{\nu}$-stretched curves. The flows $Z_\delta(t_1, t_2, x)$ which we are using now, with the additional provision that t_1 and t_2 should be chosen to vary continuously through our iterative process, is well adapted to the sinks. But, the cancellation / stretching flow has also an action on the v-pieces. If we are to replace it by our iterative process in some regions, we have to check that the cancellation / stretching flow had no action on these v-pieces in these regions. Under (H), we are completing such a change near curves which nearly v-branches are $O(\delta s_1)$-conjugate, hence have a minimal length along v, $\theta > 0$. If we take C large enough in the definition of case 1) and 2), then the cancellation / stretching flow which we build has no action on the nearly float v-pieces of size larger than $\theta/_2$, hence will have no action on a nearly flat v-piece of a curve close enough to our critical curves. In other words, one can show that the unstable manifold of such critical curves, with such a choice of C, does not touch the nearly v-pieces, but for the use of the conjugate-points flow, which we are incorporating, under (H), in our iterative process.

Thus, since, along the unstable manifolds of such critical curves, the nearly v-pieces are changed only by the action of the conjugate-points flow, we can use the flows $Z_\delta(t_1, t_2, x)$ in lieu of our cancellation / stretching flow. The final result is the same.

If we do not use (H), then, the same argument works for "large" nearly v-branches. For others, which are "small" i.e of a size which is upperbounded by a universal constant, and lowerbounded also (because our flows cancel small oscillations), we need some more work. We observe that the argument below, in the proof of Lemma N, provides us with an algorithm which allows us to evolve from the $\tilde{\nu}$-stretched curves and the parts of their unstable manifolds along which the nearly v-pieces are untouched to curves made to $\pm v$ and $\pm\xi$ pieces and the parts of their unstable manifolds along which the nearly v-pieces are again untouched. This process moves the nearly v-branches only very slightly, although it might change their size. The change in the size can be checked to be $O\left(\frac{\nu}{L}\right)$ for each nearly v-branch. Hence, the fact that the cancellation / stretching flow could have an action on these v-branches is perturbed very little. This algorithm can be easily changed into a retraction by deformation. When the v-branches are large enough, we thus truly have a local retraction by deformation on $\bigcup_k W_u(\Gamma_{2k})$, since, then, the cancellation / stretching flow has no action on these v-branches. When they become small, more work is needed. They are at any rate sizable, in function of C. The estimates established in the proof of Lemma N, below, show that the $\tilde{\nu}$-stretched curves and the corresponding limit curves made of ξ and $\pm v$-pieces are very close C^0. Every variation which does not touch the v-branches retracts by deformation on a variation on the corresponding limit curve. Section II shows that a Morse Lemma is available near these limit-curves. This Morse Lemma implies, when the ξ-pieces before and after a v-branch are not small and are not characteristic (see section II), that the action of the cancellation / stretching flow can be split into independent actions on a neighborhood of this v-branch and on the corresponding ξ-pieces.

Thus, we again have a local retraction by deformation on $\bigcup W_u(\Gamma_{2k})$. When the ξ-pieces are small or characteristic, some more technical work is needed, which does not contain any new fact. Simply, when the ξ-pieces are small, the cancellation / stretching flow reorders them, until it cannot have any action, because the v-pieces have become large. This is easy to see. When the ξ-pieces are characteristic, there is a natural folding of Γ_{2k} which is the reason for this technical problem. This again can be handled after some technical work, which will be written elsewhere, when a more systematic study of a Morse Lemma at infinity will be completed. Such a study is very natural in this type of problems. Our present work sets the perspective and the parameters. There is no hidden surprise to come, only some technical work to display. q.e.d.

Proof of Lemma N. Observe that the $\bar{x}_k$'s are derived step by step by small changes in the length of the almost v-jumps, which have points on them which are distant at most of $0(-\Delta a) + o(1)$. We can assume that $o(1) = 0\left(\frac{\varepsilon}{L^2}\right)$.

Our argument has two steps combined in one: first, by a very crude argument, on each v-jump, since the end-points are conjugate up to δs_1, the number of revolutions of $ker\alpha$ (see Proposition A3 of Appendix 2) is the same from $\bar{x}_k$ to $\bar{x}_{k+1}$, since the lengths of these v-jumps are only slightly changed. Therefore, it is also the same for $\bar{x}_k$ and $\bar{x}_q, k \neq q$ arbitrary.

In fact, for the jump of order i, we can find points A_i^k and B_i^k for $\bar{x}_k$, near both ends of the jump (up to $o(1)$ by averaging), and close points of the v-jump of $\bar{x}_q, \tilde{A}_i^q$ and $\tilde{B}_i^k$, close up to $0((-\Delta a)_i) + o(1)$ to the one of $\bar{x}_k$. Both are nearly conjugate if we use the conjugate points flow, under (H). Therefore, the length of the v-jump for $\bar{x}_q$ is close, up to $0((-\Delta a)_i) + o(1)$, to the one for $\bar{x}_k$, if $q \geq k$. The two v-jumps are close.

We will prove that both jumps have about the same size. Indeed, we can pick up, in the middle of these jumps two points

$$C_{k,i} \text{ and } C_{q,i} \text{ close up to } 0(-\Delta a)_i + o(1).$$

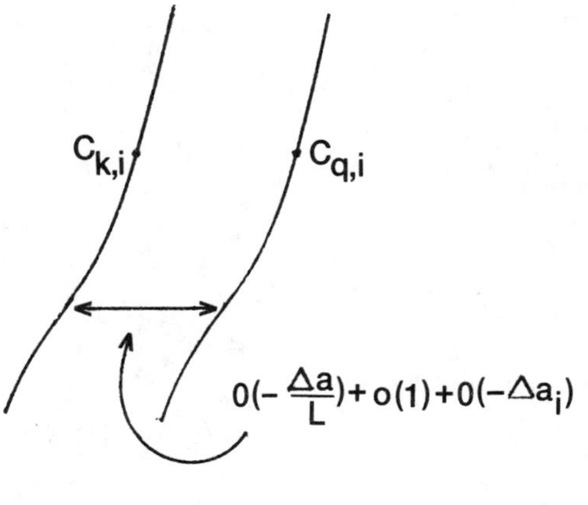

FIGURE O

and we can do the same thing for the next jump with two points $C_{k,i+1}, C_{q,i+1}$. The figure then looks like:

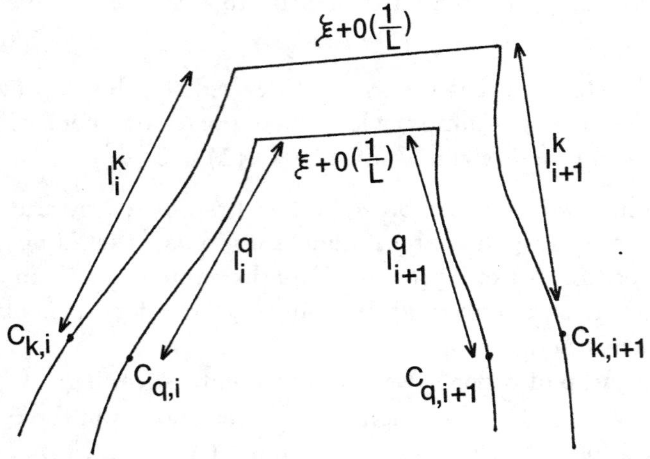

FIGURE P

ℓ_i^k, ℓ_i^q etc. are the length along v from $C_{k,i}$ or $C_{q,i}$ to the edges of the v-jumps. $\dot{x}_k$ and $\dot{x}_q$ are parallel to v up to $0\left(\frac{1}{|b|_\infty}\right)$. There are some pieces where $\dot{x}_k$ or $\dot{x}_q$ is rather parallel to ξ, but there are of length at most $0\left(\frac{1}{L}\right)$. $|b|_\infty$ is large with respect to L. Hence, we can bring everything back to one v-orbit through $C_{q,i}$ and another one through $C_{q,i+1}$, using (H). If $C_{q,i}, C_{q,i+1}$ are near edges, which we can assume, (H) is not needed. We then have:

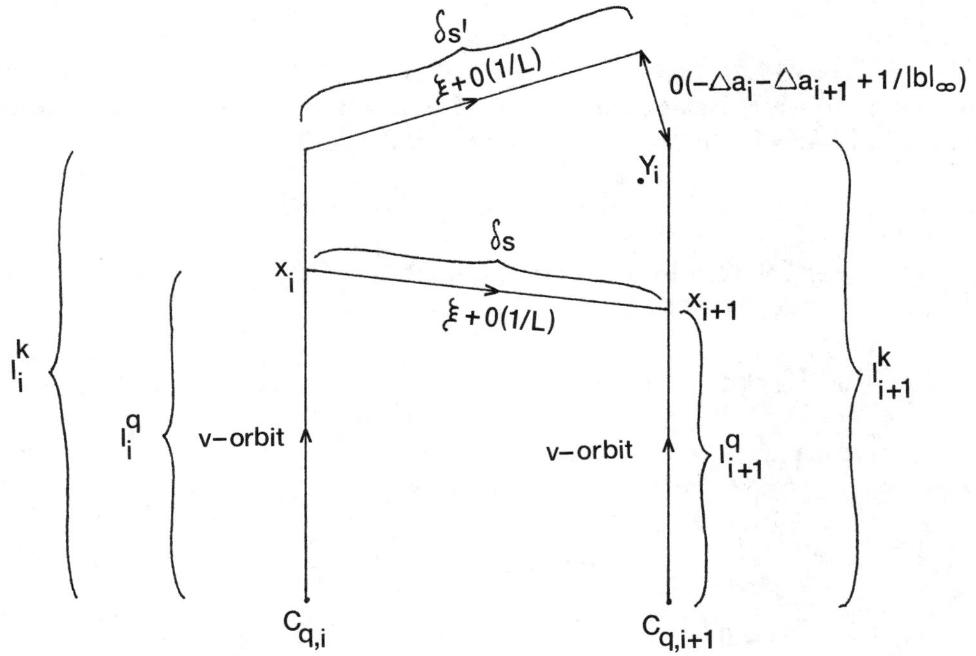

FIGURE Q

If $q < k$ and for example $k = q + 1, l_i^k - \ell_i^q$ as well as $\ell_{i+1}^k - \ell_{i+1}^q$ are small and this will continue for a number of steps, because each change is small, at most $C\frac{\nu}{L}$ in $\int |b|$ at the edges, also in $\delta s'$ with respect to δs for each step. Assuming the changes are small, we write the equation expressed in Figure Q and prove that they remain small at every iteration. We derive then precise estimates on the evolution. ℓ_i^k and ℓ_i^q are algebraic lengths. As we pointed out above, we might now have movements back and forth, due to the fact that the conjugate points flow might create zeros, which will be transformed by the $\tilde{\nu}$-flow into flat $\pm|b|_\infty$ pieces. Assuming the changes are small, denoting ψ_s the one-parameter group generated by ξ and taking a connection where v is constant, we have: $Y_i = \psi_{\delta s}(x_i + (\ell_i^k - \ell_i^q)v) = x_{i+1} + (\ell_{i+1}^k - \ell_{i+1}^q)v - (\delta s' - \delta s)\xi(Y_i) + 0((\delta s' - \delta s)^2) + 0\left(\frac{1}{|b|_\infty} + \frac{1}{L^2} - \Delta a_i - \Delta a_{i+1} + \frac{\delta s}{L} + \frac{\delta s'}{L}\right) \qquad (\mu)$

This expresses the fact that
$\psi_{\delta s'}(x_i+(\ell_i^k-\ell_i^q)v) = x_{i+1}+(\ell_{i+1}^k-\ell_{i+1}^q)v+0\left(\frac{\delta s}{L} + \frac{\delta s'}{L} + \frac{1}{|b|_\infty} + \frac{1}{L^2} - \Delta a_i - \Delta a_{i+1}\right).$

We then observe, since $\psi_0 = Id, D\psi_0 = Id, D^2\psi_0 = 0$, that we have:

$$\psi_{\delta s}(x_i + (\ell_i^k - \ell_i^q)v) = \psi_{\delta s}(x_i) + D\psi_{\delta s}((\ell_i^k - \ell_i^q)v) + 0(|\delta s||\ell_i^k - \ell_i^q|^2) =$$

$$= x_{i+1} + 0\left(\tfrac{\delta s}{L}\right) + D\psi_{\delta s}(v)(\ell_i^k - \ell_i^q) + 0(|\delta s||\ell_i^k - \ell_i^q|^2).$$

Thus,

$$(\ell_i^k - \ell_i^q)D\psi_{\delta s}(v) + 0(|\delta s||\ell_i^k - \ell_i^q|^2) + (\delta s' - \delta s)\xi(\psi_{\delta s}(x_i)) + 0((\delta s' - \delta s)(\ell_i^k - \ell_i^q))$$

$$+0((\delta s' - \delta s)^2) + 0\left(\tfrac{1}{|b|_\infty} + \tfrac{1}{L^2} - \Delta a_i - \Delta a_{i+1} + \tfrac{\delta s}{L} + \tfrac{\delta s'}{L}\right) = (\ell_{i+1}^k - \ell_{i+1}^q)v(\psi_{\delta s}(x_i)).$$

Now, v represents β along the ξ-orbit. β is a contact form. ξ is in its kernel. Therefore, v rotates in the transport along ξ, see Proposition A3 of Appendix 2

$$D\psi_{\delta s}(v) = (1 + 0(\delta s)) - \bar{c}\delta s[\xi, v]$$

where $|\bar{c}|$ is lower bounded by a positive constant $\bar{c}_0$.

Thus:

$$(\ell_i^k - \ell_i^q)(1 + 0(\delta s) + 0(\delta s')) + 0\left((\delta s' - \delta s)^2 + \frac{1}{|b|_\infty} + \frac{1}{L^2} - \Delta a_i - \Delta a_{i+1} + \frac{\delta s}{L} + \frac{\delta s'}{L}\right)$$

$$= \ell_{i+1}^k - \ell_{i+1}^q$$

$$\bar{c}\delta s(\ell_i^k - \ell_i^q) = 0(\delta s|\ell_i^k - \ell_i^q|^2) + |\delta s' - \delta s||\ell_i^k - \ell_i^q| +$$

$$+ 0\left((\delta s' - \delta s)^2 + \frac{1}{|b|_\infty} + \frac{1}{L^2} - \Delta a_i - \Delta a_{i+1} + \frac{\delta s}{L} + \frac{\delta s'}{L}\right)$$

$$(\delta s' - \delta s)(1 + o(1)) = 0\left(\delta s(\ell_i^k - \ell_i^q)^2 + \frac{1}{|b|_\infty} + \frac{1}{L^2} - \Delta a_i - \Delta a_{i+1} + \frac{\delta s}{L}\right).$$

Using the fact that $\ell_i^k - \ell_i^q$ is small, we derive, since δs is larger than $\frac{\varepsilon_0}{L}$.

$$\delta s' = \delta s(1 + o(1) + 0)\frac{1}{L}) + 0(\frac{1}{|b|_\infty} + \frac{1}{L^2} - \Delta a_i - \Delta a_{i+1}) =$$

$$= \delta s(1 + o(1) + o(\frac{1}{L})) + o(\frac{1}{L}) = \delta s(1 + o(1)).$$

Thus, as long as the changes are small, $\delta s'$ is larger than $\frac{\varepsilon_0}{2L}$.

We also have:

$$\delta s(\ell_i^k - \ell_i^q) = 0\left(\frac{\delta s}{L} - \Delta a_i - \Delta a_{i+1} + \frac{1}{|b|_\infty} + \frac{1}{L^2}\right).$$

Thus,

$$\ell_i^k - \ell_i^q = 0\left(\delta s - L\Delta a_i - L\Delta a_{i+1} + \frac{L}{|b|_\infty} + \frac{1}{L}\right) = o(1)$$

and so is $\ell_{i+1}^k - \ell_{i+1}^q$. This also shows that, after the continuous choice of t_2^- and t_1^+ in the iterated process, the b-length between t_1^+ and t_2^- remains small, hence does not allow the creation of any new oscillation. The arguments of Lemmas 17 and 18, which easily extend, show that, through this repeated process, we only might add $\pm|b|_\infty$-pieces on each side.

These three equations show that, if we start with a small δs, i.e. $|\delta s| < c_0$ a fixed constant, we will have for all iterates, because these iterates only come by small changes, a δs of the same order (which was a claim we made earlier), with variations in the v-jumps which are $o(1)$ and more precisely:

$$\ell_i^k - \ell_i^q = 0(\delta s_i - L\Delta a_i - L\Delta a_{i+1} + \frac{1}{L}).$$

Since $\sum \delta s_i = 0(1)$ and $-L(\sum \Delta a_i) = 0(1)$, the total variation in the length due to the small pieces is $0(1)$.

We are left with $|\delta s| \geq c_0$, where c_0 is a fixed positive constant. $\psi_{\delta s}$ preserves the curves whose tangent vector is in $\ker\alpha$. x_{i+1} is $\psi_{\delta s}(x_i)+0(\frac{\delta s}{L}) = \psi_{\delta s}(x_i)+0(\frac{1}{L})$. Therefore, (μ) implies that

$$\delta s' - \delta s = 0(\frac{1}{L}) = o(1).$$

Would $\ell_i^k - \ell_i^q$ become larger, it would have to cross values where it is sizable, but small. For these values, $\psi_{\delta s}$, up to $0(\frac{1}{L})$, would map a small orbit of v onto a small orbit of v. This small orbit is small, but sizable.

Using a genericity argument, this can be ruled out for $L = +\infty$, because such a piece is small, but sizable. At one point, we can map through $\psi_s, s > 0, v$ onto λv with a given length i.e. at $x, s = s(x)$. However, this length varies when the point varies. Using its Taylor expansion, it cannot take the same value locally, on a small piece of v-orbit for more than five points, generically.

However, in our case, if $\frac{\bar{\nu}}{L}$ is small and $\ell_i^k - \ell_i^q$ become large, this is what should happen up to $0(\frac{1}{L})$ between the time where $\ell_i^k - \ell_i^q$ is tiny and the time where it becomes large; a contradiction.

Thus, $\ell_i^k - \ell_i^q$ remains equal to $o(1)$. Since $|\delta s| \geq c_0$, there are finitely many of such pieces. Hence, L has at most changed by $0(1)$ as claimed.

The above argument extends when, in the second stage, we touch at intervals with time-span $0(\frac{1}{L^2})$. We have to increase $|b|_\infty$ to be $0(\frac{1}{L^2})$ on the top of S_1^∞ which we are changing, using the regularizing flow. Also, $o(1)$ should now become $0(1/_{L^3})$.

Coming back to our proof above, we modify it slightly: $0\left(\frac{\delta s}{L}\right)$ is in fact $\frac{C\nu\delta s}{L}$. It contributes $\frac{C\nu}{L}$ to the addition in the total length. They are at most, assuming a bound equal to $L+10$, $\frac{L+10}{\delta_o}$ such $\delta s's$. Thus L is increased in the process of $\frac{C\nu}{L} \times \frac{L+10}{\delta_o} < 10$ if ν is small enough.

The contribution of this term is globally controlled. The other terms behave as $0\left(\frac{1}{L} + \varepsilon_{30}^2\right)$, where 0 is bounded independently of the stage.

Inductively, we have

$$0(\sum \varepsilon_{30}^k) + 0(\sum \delta s_i) + 0(1) + 0\left(\frac{1}{L}\right) + - - - + 0\left(\frac{1}{L^n}\right) + - - - = 0(1).$$

Our proof is now complete. (A regularization or cancellation procedure gets rid of the oscillation above zero which we created). In the beginning of Appendix 4, we derive a much better estimate on this increase, which is clearly independent of ν, C, ε_0 etc.

(H) is not needed for the argument after Figure Q

An inspection of the proof above shows that the argument following Figure Q can be applied right after Lemmas 17 and 18, without making use of the conjugate points flow which required (H). When (H) is removed and we then do not use the conjugate points flow, then the transversal displacement is tiny and is due to c. The argument eases then because $(-\Delta a)_i$ is very easy to control, with an a priori control of c by $\frac{1}{L^{10}}$ for example. Then, without (H), the proof is rather straightforward, based on the argument of Figure Q. Thus, (H) is not needed. The flat $|b|_\infty$-pieces of Lemma 18 are controlled at every order of iteration of the $\tilde{\nu}/2\tilde{\nu}$ flow. However, with (H), since the points have to remain nearly conjugate, the proof seems somewhat to be more natural and leads us to the critical points at infinity directly. Without (H), we are left with $\bigcup_k W_u(\Gamma_{2k})$, i.e. the unstable manifolds of the curves having $k \pm v$-jumps alternated with $k\xi$-jumps.

We have had oscillations, during the very end of the writing of this paper, between the need for (H) and some additional hypothesis involving the conjugate hypothesis and a result which would hold in full generality.

We end up now after the argument of Figure Q with the full generality.

If we drop this argument, the result still holds under (H) plus some other mild hypothesis on the behavior of ξ with respect to the hypersurfaces of conjugacy.

I.5. Global globalization of the deformation

We have now five steps in our deformation process. The steps 2) and 3) are used in a tied way i.e. given $\delta a > 0$ from a previous decrease of a, which we will use only once, we choose ε_{10}, ε_{11}, ε_{12}, $\frac{1}{k}, \frac{1}{q}$ very small, apply 2) and then regularize deriving the time $O\left(\frac{\delta a}{(C_{41}+2)^2+1}\right)$.

1) $|b|_\infty$ is globally controlled, throughout all deformations by $C_{41}+1$, where C_{41} depends only on the initial compact K $\left(\text{at each stage, } \frac{\partial}{\partial s}|b|_\infty \leq -C\frac{\partial a}{\partial s}\right)$.

2) We have a cancellation process, which we can push up to the point when b enters a neighborhood, in L^1, controlled by parameters ε_{10}, ε_{11}, ε_{12}, $\frac{1}{k}, \frac{1}{q}$ of a function ψ. having a finite number of flat pieces J_ℓ, the number depending only on K and behaving between the J_ℓ's as the curves of Figures G, H, I and J.

3) We have a regularizing process which can be applied during a time $s = O\left(\frac{\delta a}{1+(C_{41}+2)^3}\right)$, where δa is a previous decrease, during a sequence of previous steps, which we will use only once. As explained in Corollary 15, given $\delta a > 0$ from a previous decrease, we can modulate ε_{10}, ε_{11}, ε_{12}, $\frac{1}{k}, \frac{1}{q}$ i.e. make them so small

that the regularization timing $O\left(\frac{\delta a}{2(1+(C_{41}+2)^2)}\right)$ will bring the curves to satisfy i) of Lemma 13. However, adjusting ε_{10}, ε_{11}, $\frac{1}{k}$, $\frac{1}{q}$ to the decrease δa allowed impedes a full deformation lemma, because the rate of decrease depends on the δa allowed. We will show (Lemma 24, below) how a crucial property allows to overcome this problem.
4) We then have the flow of ii) and iii) of Definition 2 if the component b of x has too much mass between $\nu + o(1)$ and $|b|_\infty - o(1)$ or $o(1)$ and $\nu - o(1)$ in absolute value (the mass is measured by the length of these intervals, which are in finite number depending only on K). This flow allows a decrease of $\beta(o(1), \varepsilon_{17}) > 0$. The curves which will not decrease of this amount are the curves such that, taking $o(1)$ even very small, have these intervals of size less than $2\sqrt{\varepsilon_{17}}$. Thus, b has no "mass" between 0 and ν or ν and $|b|_\infty$.
5) The last step is made of two sub-steps: if a curve is such that b has no mass between o and ν and ν and $|b|_\infty$ (if it is a stretched new curve) and if the ν-pieces are not $o_1(1)$, we decrease a of a fixed δa depending only on $o_1(1)$ by creating an oscillation. We then restore, using a local form of our cancellation/regularization/flow of ii) and iii) of Definition 2 flow, our control on $\int(|b| - \nu)^+$. We can iterate the whole process.

The curves which will not decrease of a fixed amount during this last stage are the curves such that

1) they are ν-stretched
2) they are such that the ν-pieces – which are in finite number, a number depending only on K – are occurring on $o_1(1)$ intervals.

In fact, all the above process is applied with ν replaced by $\tilde{\nu} = \frac{\nu}{L}$, if $\int_0^1 |b| \leq L$, for the curves whose energy level is between c and $c - \frac{\varepsilon_{30}}{L}$ (see I.4). Lemma N and the deformation theorem which follows show that iterating these flows, we will still control $\int_0^1 |b|$ globally.
K being our initial compact, there is a related maximal number ℓ_1 of intervals J_ℓ which we will encounter in our process. ℓ_1 depends only on K, we will denote it k_0 in the sequel, and on the choice of C in (10). C will eventually be a constant, chosen once and for all in our deformation process. We will refine our process by changing the value of C. However, we will see that ℓ_1 stabilizes then. Thus, ℓ_1 will be also given, once and for all depending on K.

$$\ell_1 = k_0 \tag{284}$$

a_0 is a bound on a on the curves of K. We denote (see also the next section)

$$\Gamma_{2k_0} \text{ the curves made of } k_0 \pm v\text{-pieces of} \tag{285}$$

length at least $\varepsilon_0/100$, alternated with $k_0\xi$-pieces.

$$\Gamma_{2k_0}(a_0), \text{ those such that the total length along } \xi \tag{286}$$

is less than a_0.

Let C_β^+ be the curves of $\mathcal{L}_\beta$ having

$$\alpha_x(\dot{x}) \geq 0. \tag{287}$$

Let

$$W \text{ be a neighborhood of } \Gamma_{2k_0} \text{ in } C_\beta^+, \text{ equipped with the} \tag{288}$$

graph topology (i.e. the L^∞-topology when parameterized by arc-length).
Let

$$W(a_0) \text{ be a neighborhood of the same type of } \Gamma_{2k_0}(a_0) \tag{289}$$

Let

$$\widetilde{W}(a_0) \text{ be the curves of } C_\beta \text{ whose graph lies in } W(a_0). \tag{290}$$

Let

$$C_{41} \text{ be a bound on } |b|_\infty, \text{ for } x \in K. \tag{291}$$

Observe that, if we are careful in the regularizing process which we use only during a small time depending on a previous decrease in a, δa, $C_{41}+1$ can be kept as a bound in all our deformation process, even if we iterate it, at least as long as a has no decreased on a fixed amount Δa, independent of C_{41}. Indeed, the various parts of the deformation satisfy i) of Lemma 3, but for the regularizing process. For this process, with $\eta = \frac{b}{1+|b|_\infty^3}$, we have:

$$\frac{\partial}{\partial s}|b|_\infty \leq 1.$$

If we allow a time δs equal to $-\delta a$, a previous decrease, we have

$$|b|_\infty(s) \leq |b|_\infty(s') - \delta a \qquad \forall s \geq s'.$$

Combining with the other processes, we will have:

$$|b|_\infty(s) \leq |b|_\infty(0) + \widetilde{C}(a(0) - a(s)) \tag{292}$$

where $\widetilde{C}$ is a universal constant.

The claim follows.

We wish to prove:

Lemma 19. *There exists a $\bar{\delta} = \bar{\delta}(W(a_0), C_{41})$ such that, for any $c_1 > 0$, for any compact K having $|b|_\infty \leq C_{41}+1$, $J_{c_1} \cap K$ deforms continuously onto a subset of $(J_{c_1} \cap \widetilde{W}(a_0)) \cup J_{c_1 - \bar{\delta}}$.*

Let us assume that we start with K satisfying:

$$|b|_\infty(0) \leq C_{41}. \tag{293}$$

Using (292), the subset of $\widetilde{W}(a_0) \cup J_{c_1-\bar{\delta}}$ on which $J_{c_1} \cap K$ deforms, which is a new compact $K_1 \cap J_{c_1}$, has

$$|b|_\infty(0) \le C_{41} + \widetilde{C}\bar{\delta}. \tag{294}$$

If

$$\widetilde{C}\bar{\delta} < 1 \tag{295}$$

we can iterate the argument and deform K_1 etc. Using the general idea of the deformation theory of Proposition 4.21 of [9], we explain below how Conley's theory leads to:

Lemma 20. *$J_{c_1} \cap K$ deforms continuously onto a subset of $W_u(J_{c_1} \cap \widetilde{W}(a_0)) \cup J_{c_1-1/\widetilde{C}}$ where $W_u(\widetilde{W}(a_0))$ is the unstable manifold of $\widetilde{W}(a_0) \cap J_{c_1}$ for this sequence of processes. (We give a precise meaning to this object below, it will eventually fall in the classical meaning of unstable manifolds for flows.)*

Lemma 20 follows naturally from Lemma 19 through an iteration of the process: we deform and redeform etc. The tip of the deformed part, the one lying above $J_{c_1-\bar{\delta}}$, will then be attached in the next step to $J_{c_1-2\bar{\delta}}$ through the sequence of flow-lines (we have various flows) which bring $J_{c_1-\bar{\delta}}$ onto $W_u(J_{c_1-\bar{\delta}} \cap \widetilde{W}(a_0)) \cup J_{c_1-2\bar{\delta}}$ etc.

This sequence of flow-lines build $W_u(J_{c_1} \cap \widetilde{W}(a_0))$, which is made of pieces of the flow-lines for the various parts of the deformation. These pieces can be regularized to become C^∞. In order to obtain a flow, we can argue as in [7], (A7)–(A41): if we only consider compact subsets of $J_{c_1} \cap \widetilde{W}(a_0)$, we can assume, using genericity, that these pieces, after regularization, do not intersect i.e. each deformation line is isomorphic to $\mathbb{R}^+$ and two distinct deformation lines do not intersect. In [7], we have shown how this yields a flow. The same argument, with some care, extends here.

We thus want to prove Lemma 19. The proof of Lemma 19 is long because, as we have explained after Corollary 14 and Corollary 15, we need to do something about the regularizing process in order to avoid having to change $\varepsilon_{11}, \varepsilon_{10}$ etc. repeatedly or, more precisely, in order to avoid that $\varepsilon_{11}, \varepsilon_{10}$ etc. depend on δa, thus on K. We allow a dependence on C_{41}, but C_{41} is very well controlled, while δa is not. Using C_{41}, we can go quite low, up to $J_{c_1-1}/\widetilde{C}$. δa depends on K and does not allow, at least in an obvious way, such a decrease.

A first step of the proof of Lemma 19 is to prove that a similar version of this lemma holds for each part of the deformation process, but for the regularizing piece i.e. for example for the cancellation flow that we have:

Lemma 21. *Let $W_{\varepsilon_{18}}(a_0, C_{41}) = \{x$ s.t. $|b|_\infty \le 1 + C_{41}$, $a \le a_0$, and $|b - \psi.|_{L^1} < \varepsilon_{18}\}$ for a curve $\psi.$ behaving as the ones of Figures G, H, I and J. There exists $\bar{\bar{\delta}}(\varepsilon_{18}, C_{41}) > 0$ such that, for any compact K such that $|b|_\infty \le C_{41}$ for any $x \in K$, for any $c_1 > 0$, $J_{c_1} \cap K$ retracts by deformation onto a subset of $J_{c_1-\bar{\bar{\delta}}} \cup W_{\varepsilon_{18}}(a_0, C_{41})$. $\bar{\bar{\delta}}(\varepsilon_{18}, C_{41})$ is independent of c of (15), of $\mu_1 - \mu_1^-$, $\mu_2 - \mu_2^-$, $\mu_1 - \mu_2$ very small and also of ε_{11}, $\frac{1}{q}$, $\frac{1}{k}$, ε_{12}, provided these are small enough; it depends only on ε_{18} and*

C_{41}. *A similar lemma can be stated for the flow of ii) and iii) of Definition 2, in which case the proof is very easy. The part which emerges from $J_{c_1-\delta}$ in all cases is the part $W_\varepsilon(a_0, C_{41})$ where the flow is small. We therefore will provide one proof for all these cases. $\widetilde{W}_\varepsilon(a_0, C_{41})$ will stand for this tip, where the scalar-product of the flow with J' is small*

$$\widetilde{W}_\varepsilon(a_0, C_{41}). \tag{296}$$

Proof of Lemma 21 and the related statements. Classically, the argument to prove such claims runs as follows: one chooses a smaller neighborhood $\widetilde{W}_{\tilde{\varepsilon}}(a_0, C_{41}) \subset \widetilde{W}_\varepsilon(a_0, C_{41})$ such that

$$\mathrm{dist}(\partial\widetilde{W}_{\tilde{\varepsilon}}(a_0, C_{41}), \partial\widetilde{W}_\varepsilon(a_0, C_{41})) > 0. \tag{297}$$

If, during the deformation, a flow-line does not enter $\widetilde{W}_{\tilde{\varepsilon}}(a_0, C_{41})$, it does not penetrate a neighborhood of the critical set, hence it has to decrease of at least a fixed amount δ_1, δ_1 depending only on $\widetilde{W}_{\tilde{\varepsilon}}(a_0, C_{41})$ hence or $\widetilde{W}_\varepsilon(a_0, C_{41})$.

This step extends readily here, but for the fact that $\widetilde{W}_\varepsilon$ and $\widetilde{W}_{\tilde{\varepsilon}}$ are not given for the whole sequence of vector-fields, but have to change at each step. Since, in this lemma, we are focusing on a single part of the deformation, this step applies readily. The second step runs as follows: if a flow-line enters $\widetilde{W}_{\tilde{\varepsilon}}$, then it cannot exit $\widetilde{W}_\varepsilon$ without inducing a substantial decrease δ_2 in the functional, because – usually – the vector-field is uniformly bounded below a_0 and takes a certain time, therefore, to go from $\partial\widetilde{W}_{\tilde{\varepsilon}}$ to $\partial\widetilde{W}_\varepsilon$. During this time, since we are outside of $\widetilde{W}_{\tilde{\varepsilon}}$, we are away from the critical set, the rate of decrease is lower bounded by $\varepsilon_1 > 0$. Hence, the claim. Combining the two steps, one then sees easily that either the flow-line has decreased of $\inf(\delta_1, \delta_2) = \delta$, or it is in $\widetilde{W}_\varepsilon$. Hence, the claim of Lemma 21, would this second step extend.

The second step extends easily for the flow of ii) or iii) of Definition 2. But parts of our process involve a diffusion flow, which is not bounded, hence the second step becomes less obvious. We nevertheless claim that it holds. Namely, we are faced with the following situation: between $\widetilde{W}_{\tilde{\varepsilon}}$ and $\widetilde{W}_\varepsilon$, or rather the ones relevant to these parts of the flow, $\int_0^1 b\tilde{\eta}_{12} + c\int_0^1 b^2$ which is lower bounded by $\int_0^1 b\tilde{\eta}_{12}$, the oscillatory part, for the flow built with i) of Definition 2 – the cancellation flow – and $\int_0^1 b^2 dt$ for the "regularizing" flow ($\eta = b$), both quantities are lower bounded by $\theta > 0$, which depends only on $\widetilde{W}_{\tilde{\varepsilon}}$ i.e. on $\widetilde{W}_\varepsilon$, i.e. on how far we are from the "critical curves" for this flow (the curves having b equal to $\psi_\cdot, \psi_\cdot$ having flat pieces of intervals J_ℓ alternating with the behavior of Figures G, H, I and J, this provided $\varepsilon_{11}, \frac{1}{k}, \frac{1}{q}$ are small enough. Therefore, the total decrease in a when travelling from $\partial\widetilde{W}_{\tilde{\varepsilon}}$ to $\partial\widetilde{W}_\varepsilon$ is lower bounded by θs_0 (at $s = 0$, the curve is on $\partial\widetilde{W}_1(a_0)$, at $s = s_0$, it is on $\partial\widetilde{W}(a_0)$), where $\int_0^1 b^2$ is lower bounded by $\tilde{\theta} > 0$, because of the contribution of at least one interval J_ℓ. Also, if $\int_0^1 b^2$ could be as small as we want, we would be near a periodic orbit, a direct, different analysis, could be introduced. We eliminate from $\widetilde{W}_\varepsilon$ such neighborhoods.) Therefore, this total decrease tends to zero only if the time s_0 needed to travel tends to zero.

Observe that all our curves have $|b|_\infty$ less than $C_{41}+1$. Let us consider a sequence (b_0^ℓ, s_0^ℓ) of this type, with $b_0^\ell \in \partial\widetilde{W}_{\bar\varepsilon}$ and s_0^ℓ tending to zero. Using the fact that $|b|_\infty$ is bounded by $C_{41}+1$, it is easy to see that, for any C^∞ function φ:

$$\frac{\partial}{\partial s}\int_0^1 b^\ell \varphi = O_\varphi(1).$$

Thus, $b^\ell(s_k^0)(t) - b_0^\ell(t)$ tends weakly to zero when ℓ tends to $+\infty$. (We use the fact that the cancellation flow is built of convex combination of sums of $\psi_{\theta_i,\varepsilon_i}\eta_{0,i}$, where the $\eta_{0,i}$ have, for given sum, disjoint supports.)

Reducing $\widetilde{W}_{\bar\varepsilon}$ very much, b_0^ℓ, which lies on $\partial\widetilde{W}_{\bar\varepsilon}$, will be as close as we wish from a function $\psi_\cdot^\ell$ having flat pieces J_ℓ and intermediate pieces such as in Figures G, H, I and J; these pieces are in finite number, $\psi_\cdot^\ell$ is nonoscillatory (it is decreasing, increasing or flat, a finite number of times, which depends only on the initial compact K). They are also bounded by $C_{41}+1$. Therefore, combining with the weak convergence to zero of $b^\ell(s_0^\ell) - b_0^\ell$ and their boundedness by $2(C_{41}+1)$:

$$\int_0^1 (b^\ell(s_0^\ell)(t) - b_0^\ell(t))\psi_\cdot^\ell \to 0. \tag{298}$$

Thus, as $\widetilde{W}_1(a_0)$ becomes smaller and smaller:

$$\int_0^1 (b^\ell(s_0^\ell)(t) - b_0^\ell(t))b_0^\ell(t) \to 0. \tag{299}$$

Thus

$$\int_0^1 (b^\ell(s_0^\ell)(t) - b_0^\ell(t))^2 = \int_0^1 b^{\ell^2} - \int_0^1 b_0^{\ell^2} + o(1). \tag{300}$$

We now claim that

$$\int_0^1 b^{\ell^2}(s_0^\ell) - \int_0^1 b_0^{\ell^2} \le o(1). \tag{301}$$

Thus, $\int_0^1 (b^\ell(s_0^\ell)(t) - b_0^\ell(t))^2$ tends to zero, yielding a contradiction, since $\partial\widetilde{W}_\varepsilon$ has a positive distance to $\partial\widetilde{W}_{\bar\varepsilon}$ (i.e. a positive L^2-distance to the functions $\psi_\cdot$). We are thus left with proving (298) and (301). For (298), we need only to prove that an increasing sequence ω^ℓ, which is bounded L^∞ on [0,1], converges strongly in L^2 (up to extraction of a subsequence). We have, denoting ω its weak limit:

$$\int_0^1 (\omega^\ell - \omega)\omega^\ell = \left(\int_0^1 \omega^\ell - \omega\right)\omega^\ell(1) - \int_0^1 \left(\int_0^\tau \omega^\ell - \omega\right)\dot{\omega}^\ell \tag{302}$$

$$= o(1) + \operatorname*{Sup}_{\tau\in[0,1]} \left|\int_0^\tau \omega^\ell - \omega\right| \int_0^1 |\dot{\omega}^\ell|$$

$$= o(1) + O\left(\operatorname*{Sup}_{\tau\in[0,1]} \left|\int_0^\tau \omega^\ell - \omega\right|\right)$$

$\int_0^\tau \omega^\ell$ is bounded in H^1, hence converges in L^∞, our claim follows. The proof of (301) is trickier. It is obvious for the "regularizing" flow, where $\eta = b$, less for the cancellation flow. Indeed, for the "regularizing" flow, we have:

$$\frac{\partial}{\partial s}\int_0^1 b^2 = 2\int_0^1 b\left(\frac{\dot{\overline{\dot{b} + \lambda b}}}{a} + ab\tau - b\eta\bar{\mu}_\xi\right) \tag{303}$$

$$= O(1) - \frac{2}{a}\int_0^1 \dot{b}^2 - 2\int_0^1 \dot{b}b\frac{(\lambda + \bar{\mu}b)}{a} + 2\int_0^1 \frac{\dot{b}b^2\bar{\mu}}{a}$$

$$= O(1) - \frac{2}{a}\int_0^1 \dot{b}^2 + \frac{1}{a}\int_0^1 b^2\left(b^2 - \int_0^1 b^2\right) - \frac{1}{a}\int_0^1 b^3\dot{\bar{\mu}} \le O(1).$$

Thus

$$\int_0^1 b^2(s) \le \int_0^1 b^2(0) + O(1)s. \tag{304}$$

From (304) with $b(s) = b^\ell(s_0^\ell)$ and $b(0) = b_0^\ell$, we derive (301) for the "regularizing" flow.

The proof is more involved for the cancellation flow. We will complete it for $\eta = \psi_{\theta,\varepsilon}\eta_0 + cb$. Using the fact that the sums $\sum \psi_{\theta_i,\varepsilon_i}\eta_{0,i}$ have disjoint support and that we later only convex combine such sums, the claim follows for a general η built after i) of Definition 2. For $\eta = \psi_{\theta,\varepsilon}\eta_0 + cb$, we have:

$$\frac{\partial}{\partial s}\int_0^1 b^2 \le O(1) - \frac{2c}{a}\int_0^1 \dot{b}^2 + 2\int_0^1 b\frac{\dot{\overline{\dot{\overline{\psi_{\theta\varepsilon}\eta_0}} + \lambda b}}}{a} \tag{305}$$

$$= O(1) - \frac{2c}{a}\int_0^1 \dot{b}^2 - 2\int_0^1 \dot{b}\frac{\dot{\overline{\psi_{\theta,\varepsilon}\eta_0}} + \lambda b}{a}$$

$$= O(1) - 2\int_0^1 \frac{\dot{b}b(\lambda + \bar{\mu}\eta)}{a} + 2\int_0^1 \frac{\dot{b}b\bar{\mu}\eta}{a} +$$

$$+ \frac{2}{a}\int_0^1 b\ddot{\overline{\psi_{\theta,\varepsilon}\eta_0}} - \frac{2c}{a}\int_0^1 \dot{b}^2$$

$\int_0^1 \frac{\dot{b}b}{a}(\lambda+\bar{\mu}\eta)$ is equal to $-\int_0^1 \frac{b^2}{2a}(b\eta-\int_0^1 b\eta)=O(1)$. On the other hand, $2\int_0^1 \frac{\dot{b}b\bar{\mu}\eta}{a}=O(a)-\frac{1}{a}\int_0^1 b^2\bar{\mu}\overline{\psi_{\theta,\varepsilon}\eta_0}^{\cdot}$.

As before: (we will work as if b were positive. It does not reduce the generality of the argument)

$$\overline{\psi_{\theta,\varepsilon}}^{\cdot} = -\dot{\theta} + \frac{\partial\psi_{\theta,\varepsilon}}{\partial b}\dot{b} \tag{306}$$

$$\overline{\psi_{\theta,\varepsilon}}^{\cdot\cdot} = -\ddot{\theta} + \frac{\partial^2\psi_{\theta,\varepsilon}}{\partial b^2}\dot{b}^2 + \frac{\partial\psi_{\theta,\varepsilon}}{\partial b}\ddot{b} \tag{307}$$

$$\left|\int_0^1 b\frac{\partial^2\psi_{\theta,\varepsilon}}{\partial b^2}\dot{b}^2\eta_0\right| + \left|\int_0^1 b\ddot{\theta}\eta_0\right| + \left|\int_0^1 b\dot{\theta}\dot{\eta}_0\right| + \int_0^1 b^2|\bar{\mu}|\,|\dot{\theta}|\eta_0 \tag{308}$$

can be made as small as we wish by choosing $\mu_1^- - \mu_1$ and $\mu_2^- - \mu_2$ very small (the argument is similar to the one used in (47)–(50) and related estimates). We can assume that our flow is built so that they are absorbed in $-\frac{c}{100a}\int_0^1 \dot{b}^2$.

We are thus left with:

$$\frac{4}{a}\int_0^1 b\frac{\partial\psi_{\theta,\varepsilon}}{\partial b}\dot{b}\dot{\eta}_0 + \frac{2}{a}\int_0^1 b\frac{\partial\psi_{\theta,\varepsilon}}{\partial b}\ddot{b}\eta_0 + \tag{309}$$
$$+\frac{2}{a}\int_0^1 b\psi_{\theta,\varepsilon}\ddot{\eta}_0 - \frac{1}{a}\int_0^1 b^2\bar{\mu}\frac{\partial\psi_{\theta,\varepsilon}}{\partial b}\dot{b}\eta_0 - \frac{1}{a}\int_0^1 b^2\bar{\mu}\psi_{\theta,\varepsilon}\dot{\eta}_0 - \frac{c}{a}\int_0^1 \dot{b}^2.$$

We have:

$$\frac{2}{a}\int_0^1 b\frac{\partial\psi_{\theta,\varepsilon}}{\partial b}\ddot{b}\eta_0 = -2\int_0^1 b\frac{\partial\psi_{\theta,\varepsilon}}{\partial b}\dot{b}\dot{\eta}_0 - \frac{2}{a}\int_0^1 \dot{b}^2\frac{\partial\psi_{\theta,\varepsilon}}{\partial b}\eta_0 - \tag{310}$$
$$-\frac{2}{a}\int_0^1 b\overline{\frac{\partial\psi_{\theta,\varepsilon}}{\partial b}}^{\cdot}\eta_0$$
$$= -\frac{2}{a}\int_0^1 b\frac{\partial\psi_{\theta,\varepsilon}}{\partial b}\dot{b}\dot{\eta}_0 - \frac{2}{a}\int_0^1 \dot{b}^2\frac{\partial\psi_{\theta,\varepsilon}}{\partial b}\eta_0 -$$
$$-\frac{2}{a}\int_0^1 b\frac{\partial^2\psi_{\theta,\varepsilon}}{\partial b^2}\dot{b}^2\eta_0.$$

The last term is as small as we wish again. We are thus left with:

$$\frac{2}{a}\int_0^1 b\frac{\partial\psi_{\theta,\varepsilon}}{\partial b}\dot{b}\dot{\eta}_0 - \frac{2}{a}\int_0^1 \dot{b}^2\frac{\partial\psi_{\theta,\varepsilon}}{\partial b}\eta_0 - \frac{1}{a}\int_0^1 b^2\bar{\mu}\frac{\partial\psi_{\theta,\varepsilon}}{\partial b}\dot{b}\eta_0 - \tag{311}$$
$$-\frac{1}{a}\int_0^1 b^2\bar{\mu}\psi_{\theta,\varepsilon}\dot{\eta}_0 + \frac{2}{a}\int_0^1 b\psi_{\theta,\varepsilon}\ddot{\eta}_0 - \frac{c}{a}\int_0^1 \dot{b}^2 = \Delta$$

We will go to the limit when $\mu_1^- \to \mu_1$, $\mu_1 \to \mu_2$, $\mu_2^- \to \mu_2$. Their common value is denoted $\underline{\mu}$. $\frac{\partial \psi_{\theta,\varepsilon}}{\partial b}$ tends to $1_{[t_1^+,t_2^+]}$, $\psi_{\theta,\varepsilon}$ to $(b-\underline{\mu})^+ 1_{[t_1^+,t_2^+]}$, $b(t_1^+) = b(t_2^+) = \underline{\mu}$ in the limit

$$\Delta_{limit} = \frac{2}{a}\int_{t_1^+}^{t_2^+} b\dot{b}\dot{\eta}_0 - \frac{2}{a}\int_{t_1^+}^{t_2^+} \dot{b}^2\eta_0 - \frac{1}{a}\int_{t_1^+}^{t_2^+} \bar{\mu}b^2\dot{b}\eta_0 - \tag{312}$$

$$-\frac{1}{a}\int_{t_1^+}^{t_2^+} b^2\bar{\mu}(b-\underline{\mu})\dot{\eta}_0 + \frac{2}{a}\int_{t_1^+}^{t_2^+} b(b-\underline{\mu})\ddot{\eta}_0 - \frac{c}{a}\int_0^1 \dot{b}^2$$

$(b > \underline{\mu}$ on $[t_1^+, t_2^+])$. $\frac{2}{a}\int_{t_1^+}^{t_2^+} b\dot{b}\dot{\eta}_0$ is equal to

$$\frac{1}{a}\underline{\mu}^2(\dot{\eta}_0(t_2^+) - \dot{\eta}_0(t_1^+)) - \frac{1}{a}\int_{t_1^+}^{t_2^+} b^2\ddot{\eta}_0$$

$$= -\frac{1}{a}\int_{t_1^+}^{t_2^+} (b^2 - \underline{\mu}^2)\ddot{\eta}_0$$

$$\left|\int_{t_1^+}^{t_2^+} \bar{\mu}b^2\dot{b}\eta_0\right| \le \frac{1}{2}\int_0^1 \dot{b}^2\eta_0 + O(1)(t_2^+ - t_1^+)$$

$$\left|\int_{t_1^+}^{t_2^+} b^2\bar{\mu}(b-\underline{\mu})\dot{\eta}_0\right| = \left|\int_{t_1^+}^{t_2^+} \overline{b^2\bar{\mu}(b-\underline{\mu})\eta_0}^{\cdot}\right| \le \frac{1}{2}\int_0^1 \dot{b}^2\eta_0 + O(1)(t_2^+ - t_1^+).$$

Thus

$$\Delta_{lim} \le -\frac{1}{a}\int_{t_1^+}^{t_2^+} \dot{b}^2\eta_0 + O(1)(t_2^+ - t_1^+) + \tag{313}$$

$$+\frac{2}{a}\int_{t_1^+}^{t_2^+}\left(b(b-\underline{\mu}) + \frac{\underline{\mu}^2}{2} - \frac{1}{2}b^2\right)\ddot{\eta}_0$$

$$= -\frac{1}{a}\int_{t_1^+}^{t_2^+} \dot{b}^2\eta_0 + O(t_2^+ - t_1^+) + \frac{2}{a}\int_{t_1^+}^{t_2^+}\left(\frac{b^2}{2} - \underline{\mu}b + \frac{\underline{\mu}^2}{2}\right)\ddot{\eta}_0.$$

As noted previously (see (10) and (59) for example)

$$\ddot{\eta}_0 = -\frac{2}{(x_1^- - x_2^-)^2}e^{-\int_{t_1^-}^{t}\bar{\mu}bd\tau} - C(1+b^2)\eta_0 e^{-2\int_{t_1^-}^{t}\bar{\mu}bd\tau} + \overline{\bar{\mu}\eta_0 b}^{\cdot} \tag{314}$$

$$\le \overline{\bar{\mu}\eta_0 b}^{\cdot}$$

$\int_{t_1^+}^{t_2^+}(b^2 - 2\underline{\mu}b + \underline{\mu}^2)\overline{\bar{\mu}\eta_0 b}^{\cdot} = -2\int_{t_1^+}^{t_2^+}(b-\underline{\mu})\dot{b}\bar{\mu}\eta_0 b$ which can be tackled as similar terms above. Thus, since $(b-\underline{\mu})^2 = b^2 - 2\underline{\mu}b + \underline{\mu}^2$ is positive:

$$\Delta_{lim} \le O(t_2^+ - t_1^+). \tag{315}$$

Hence, for μ_1 and μ_2 close enough, μ_1^- close enough to μ_1 and μ_2^- close enough to μ_2:

$$\Delta \leq O(t_2^+ - t_1^+). \tag{316}$$

Hence:

$$\frac{\partial}{\partial s}\int_0^1 b^2 \leq O(1). \tag{317}$$

This result extends to sums with disjoint supports and convex combinations of such sums. We thus have also for this flow

$$\int_0^1 b^{\ell^2}(s_0^\ell) - \int_0^1 b_0^{\ell^2} \leq O(1)s_0^\ell \tag{318}$$

hence the claim. The argument above does not depend on c, under (15), hence our claim holds for the whole family of cancellation flows uniformly on c satisfying (15). It does not also depend on the proximity of μ_1, μ_2, $\mu_1\mu_1^-$ and μ_2, μ_2^-. This argument depends only on a lower bound on $\int_0^1 b\eta_{12}$ between $\widetilde{W}_{\varepsilon/2}(a_0, C_{41})$ and $\widetilde{W}_\varepsilon(a_0, C_{41})$, thus only on having ε_{11} $\frac{1}{k}, \frac{1}{q}$ small enough. The proof of Lemma 21 is thereby complete.

Let $\psi.$ be a function with bounded variation behaving as the functions of Figures G, H, I and J:

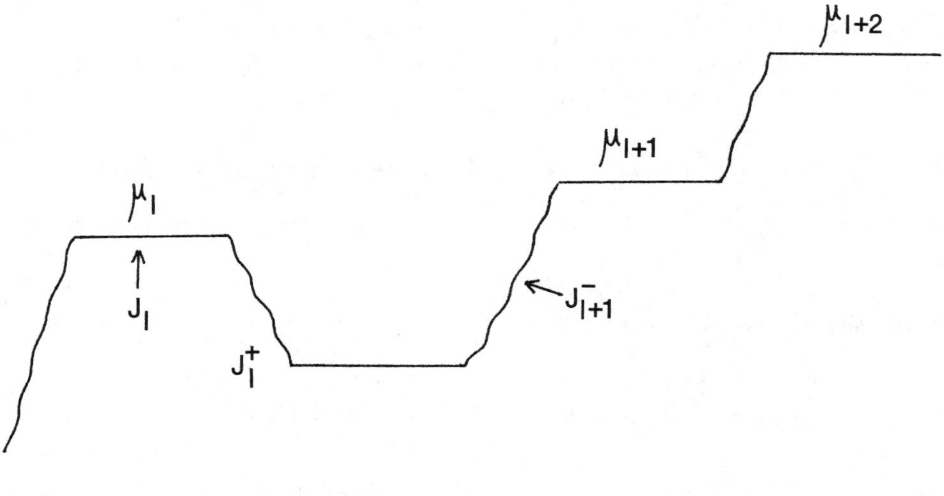

FIGURE R

i.e. $\psi.$ has flat pieces where it takes the value μ_ℓ, on intervals J_ℓ with

$$C(|J_\ell| + |\mu_\ell|^2|J_\ell|) \times |J_\ell| \geq \frac{\varepsilon_0}{20} \tag{319}$$

and intermediate pieces where it behaves like the functions of Figures G, H, I and J.

In the intermediate pieces we distinguish a decreasing or increasing piece:

$$J_\ell^+ \tag{320}$$

and an increasing or decreasing piece:

$$J_{\ell+1}^- \tag{321}$$

J_ℓ^+ and $J_{\ell+1}^-$ are time intervals, which have a common intersection. J_ℓ^+ or $J_{\ell+1}^-$ might not exist. They might also, both of them, collapse if $\psi.$ has a jump from μ_ℓ to $\mu_{\ell+1}$, both satisfying (319).

We introduce:

Definition 7. *Let $\psi.$ be a function of the above type. Let $\varepsilon_{13} > 0$ be a small parameter. The characteristic values of $\psi.$, up to ε_{13}, are the μ_ℓ's and all the other values $\mu_\ell + k\varepsilon_{13}$, $k \in \mathbb{Z}$ or $\mu_{\ell+1} + k\varepsilon_{13}$, $k \in \mathbb{Z}$ such that: denoting*

$$\begin{aligned} z_\ell &= |\{t \in J_\ell^+ \ s.t. \ \psi.(t) \in [\mu_\ell + k\varepsilon_{13}, \mu_\ell + (k-1)\varepsilon_{13})\}| \\ C(1 + (\mu_\ell &+ k\varepsilon_{13})^2 z_\ell) z_\ell \geq \frac{\varepsilon_0}{140} \end{aligned} \tag{322}$$

(resp. $z_{\ell+1} = |\{t \in J_{\ell+1}^- \ s.t. \ \psi.(t) \in [(\mu_{\ell+1} + (k-1)\varepsilon_{13}, \mu_{\ell+1} + k\varepsilon_{13})\}|(1 + (\mu_{\ell+1} + k\varepsilon_{13})^2 z_{\ell+1}) z_{\ell+1} \geq \varepsilon_0/140$). We will often consider in the sequel characteristic values as above satisfying furthermore (319). These are not the same (but almost the same) than the values of $\psi.$ on its flat pieces.

Since J_ℓ^+ and $J_{\ell+1}^-$ have at most one flat piece in common, we have:

Lemma 22. *The number of characteristic values of $\psi.$ is upper bounded by $16 \times 10^{20}(\int |\psi.| + 1)\sqrt{\frac{C}{\varepsilon_0}}$.*

In the sequel, we will have

$$\int |\psi.| \leq C_{100} - 1, \quad |\psi.| \leq 1 + C_{41} \tag{323}$$

where C_{100} is a given constant, which will depend ultimately only on an initial compact K.

Let m be an integer larger than

$$\sqrt{C}\,\frac{16 \times 10^{20}}{\sqrt{\varepsilon_0}} C_{100} < m < \frac{\sqrt{C}}{\sqrt{\varepsilon_0}} 16 \times 10^{20} C_{100} + 1. \tag{324}$$

All the characteristic values of $\psi.$ have their absolute value $|\mu'_1(\psi.)|$ bounded by $1 + C_{41}$. There are m of them at most. Let therefore γ be a value in

$$\left(\frac{C_{41}}{2}, \frac{3C_{41}}{4}\right) \tag{325}$$

such that

$$\inf_i |\gamma - |\mu'_i(\psi.)|| > \frac{C_{41}}{8m} \geq \frac{\gamma}{40m} \tag{326}$$

γ, at later stages, will lie in intervals

$$\left((1-\delta_{10})C_{41}, C_{41}\left(1 - \frac{\delta_{10}}{2}\right)\right) \text{ or } (\tilde{\alpha} - \delta_{10}\tilde{\alpha}, \tilde{\alpha} + \delta_{10}\tilde{\alpha}), \tilde{\alpha} - \delta_{10}\tilde{\alpha} > \nu \tag{327}$$

where δ_{10} is small and given. This will be when we eliminate the oscillations of size $\delta_{10}C_{41}$ over $(1-\delta_{10})C_{41}$ or of size $2\delta_{10}\tilde{\alpha}_0$ over $\tilde{\alpha} - \delta_{10}\tilde{\alpha}$.

$$(326) \text{ will be replaced by } \inf_i |\gamma - \mu'_i(\psi)|| \geq \frac{\delta_{10}\gamma}{80m} \text{ or more generally} \geq d\gamma. \tag{328}$$

For the moment, we are trying to eliminate oscillations of size $a\gamma/400m$ around $a\gamma$. That is, we are studying all oscillations of b having:

$$\bar{\tilde{\mu}} \geq a(\gamma + d\gamma - \delta a) \tag{329}$$

$$\underset{\sim}{\bar{\mu}} \leq a(\gamma - d\gamma + \delta a) \tag{330}$$

where (331) $d\gamma = \gamma/400m$ ($d\gamma = \delta_0\gamma/800m$ in general) δ is a very small positive parameter which we will choose later:

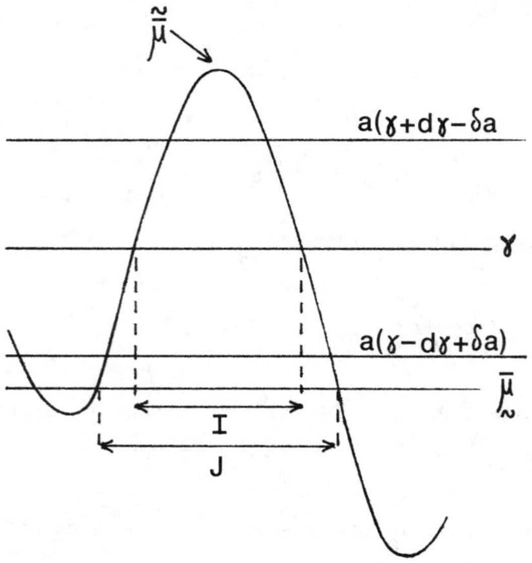

FIGURE S

Let b be a v-component of a curve $x(t)$ nearly critical for the cancellation flow i.e. $\mathcal{O}_{\bar{\varepsilon}_0}(b)$ is very small. b is assumed to be bounded by $C_{41}+1$. By Lemmas 9, 10 and 11, there exists a function $\widetilde{\psi}.$ similar to $\psi.$ such that

$$\frac{b}{a} = \widetilde{\psi}. + o(1) \tag{332}$$

$o(1)$ is small in the L^1-sense, i.e. we can include in $o(1)$ all the $\varepsilon_{11}+\frac{1}{k}+\varepsilon_{10}$-oscillations of b or $\hat{\delta}$-oscillations of b. The arguments used in the proofs of Lemmas 9, 10 and 11 prove that if b is not as in (332), with a $\psi.$ as indicated above, then

$$\int b\eta_{12} \text{ is not } o(1) \text{ i.e. } \mathcal{O}_{\bar{\varepsilon}_0}(b) \text{ is not } o(1) \tag{333}$$

(The previous $o(1)$ is function of the present one.) We will use this fact later.

(332) implies that b has – if we extend appropriately the definition of characteristic values – characteristic values close to the $\mu_i'(\widetilde{\psi}.) = \tilde{\mu}_i'$.

We then have:

Definition 8. *An oscillation over an interval J having $\underset{\sim}{\bar{\mu}} \leq (\gamma - d\gamma + \delta a)a$ and having $\tilde{\bar{\mu}} \geq (\gamma + d\gamma - \delta a)a$ is said to be of the family θ_γ of oscillations if*

$$C\left(|I| + \int_I b^2\right)|I| < \varepsilon_0/800 \quad \textit{where}$$

$I = \{t \in J \ s.t. \ |b|(t)| > a\gamma\}$.

We then have the following:

Lemma 23. *If an oscillation over an interval J having $\underset{\sim}{\bar{\mu}} \le (\gamma - d\gamma + \delta a)a$ and having $\tilde{\bar{\mu}} \ge (\gamma + d\gamma - \delta a)a$ is such that $C(|I| + \int_I b^2)|I| = \varepsilon_0/800$ where I can be equal to $\{t \in J \ s.t. \ |b(t)| > a\gamma\}$ or $I = \{t \in J \ s.t. \ |b(t)| \ge a\gamma\}$, then $|\gamma - \tilde{\mu}'_{i_0}| = o(1)$, for an appropriate $\tilde{\mu}'_{i_0}$, a characteristic value of $\widetilde{\psi}.$ satisfying furthermore (319), where $o(1)$ tends to zero when $O_{\bar{\varepsilon}_0}(b)$ tends to zero.*

Remark. $\tilde{\mu}'_{i_0}$ is not the value of $\psi.$ on a flat piece as in Figure R. It is a characteristic value such as in Definition 7, which satisfies furthermore (319).

Proof. Taking $2\tilde{\theta}_1 < d\gamma - \delta a$, we have, since $(|I| + \int_I b^2)|I| = \varepsilon_0/800$:

$$\int_J (|b| - a\gamma + \tilde{\theta}_1 a)^+ \ge \frac{\tilde{\theta}_1}{2}|I|a \ge \frac{\tilde{\theta}_1}{2} a \inf\left(\sqrt{\varepsilon_0}/800, \frac{1}{C_{41}+1}\sqrt{\varepsilon_0}/800\right). \tag{334}$$

The lower-bound tends to zero only with $\tilde{\theta}_1$. Thus, for a given $\tilde{\theta}_1 > 0$, if $\{t \in J$ s.t. $|b|(t) > a(\gamma - \tilde{\theta}_1)\}$, is only made of intervals of type 1), we can use Lemma 4, with an appropriate $\theta_1 < \frac{\tilde{\theta}_1}{100\bar{c}_0}$, $\bar{\nu}$ of Lemma 4 equal to $\gamma - \tilde{\theta}_1$ and $\varepsilon_1 > 0$, we can build an η_{12}. This yields a lower bound on $O_{\bar{\varepsilon}_0}(b)$, which is only function of $\tilde{\theta}_1$, hence a contradiction.

Observe that, since the oscillation has $\underset{\sim}{\bar{\mu}}$ less than $a(\gamma - d\gamma + \delta a)$, $\tilde{\theta}_1$ can be taken to be any value less than $d\gamma/2$.

We thus must have an interval of type 2) among the intervals of $\{t \in J/|b|(t) > a(\gamma - \tilde{\theta}_1)\}$. Thus, using iii) of Lemma 10 combined with iii) of Lemma 11, there exists an interval J_ℓ in J where b is very close to a constant μ''_ℓ, J_ℓ is again an interval of type 2) for b. Since $\frac{b}{a} = \widetilde{\psi}. + o(1)$, where $o(1)$ is small L^1 and b is bounded L^∞ by $1 + C_{41}$ as well as $a\widetilde{\psi}.$, J_ℓ is also an interval of type 2) for $\widetilde{\psi}.$, and $\widetilde{\psi}.$ is very close to the constant $a\mu'_\ell$ on the interval, in the L^1-sense. Using then the special form of $\widetilde{\psi}.$, we derive that μ''_ℓ is very close to a characteristic value $a\mu'_\ell(\widetilde{\psi}.)$ satisfying furthermore (319). In all the above arguments, we use the fact that $|J_\ell|$ is lower bounded independently of b when $O_{\bar{\varepsilon}_0}(b)$ tends to zero. The characteristic values of $\widetilde{\psi}.$ depend on a small parameter $\varepsilon_{13} > 0$, which we can make as small as we want. (We can choose it small, then have $O_{\bar{\varepsilon}_0}(b)$ tend to zero.) We thus have:

$$\mu'_\ell(\widetilde{\psi}.) > (\gamma - \tilde{\theta}_1) + o(1) \tag{335}$$

where $o(1)$ tends to zero when ε_{13} and $O_{\bar{\varepsilon}_0}(b)$ tend to zero. On the other hand, taking $\hat{\delta}_1 > 0$, a small parameter, we claim that we must have, for $O_{\bar{\varepsilon}_0}(b)$ and ε_{13} small enough:

$$\mu'_\ell(\widetilde{\psi}.) < (\gamma + \hat{\delta}_1). \tag{336}$$

Indeed, otherwise, if $\mu'_\ell(\widetilde{\psi}.)$ is larger than $(\gamma+\hat{\delta}_1)$, μ''_ℓ is larger than $a(\gamma+\hat{\delta}_1/2)$ (for ε_{13} and $O_{\bar{\varepsilon}_0}(b)$ small enough). We then have:

$$J_\ell \subset \left\{ t \in J/|b|(t) > a\left(\gamma + \frac{\hat{\delta}_1}{4}\right)\right\} \cup T_\ell \subset I \cup T_\ell \tag{337}$$

and

$$a\frac{\hat{\delta}_1}{4}|T_\ell| \leq \int_{T_\ell} |\,|b| - \mu''_\ell| \leq \int_{J_\ell} |\,|b| - \mu''_\ell| = o(1). \tag{338}$$

Thus

$$C\left(1 + \int_{J_\ell} b^2\right)|J_\ell| \leq C\left(1 + \int_I b^2 + C_{41}^2|T_\ell|\right)(|I| + |T_\ell|). \tag{339}$$

Using (338) and the fact that $C(1+\int_I b^2)|I| = \varepsilon_0/800$, we derive that J_ℓ cannot be of type 2) for $O_{\varepsilon_0}(b)$ small enough ($\hat{\delta}_1$ is given), a contradiction. Thus

$$(\gamma - \tilde{\theta}_1) + o(1) < \mu_\ell(\widetilde{\psi}.) < (\gamma + \hat{\delta}_1) \tag{340}$$

$\tilde{\theta}_1, \hat{\delta}_1, o(1)$ can be taken, with $O_{\bar{\varepsilon}_0}(b)$, to be as small as we wish. The Lemma follows. We then have:

Lemma 24. *Assume that we are given a flow-line of the cancellation flow $b(s,t)$, $s \in [s_1, s_2]$. Assume that $O_{\varepsilon_0}(b)(s)$ remains small for $s \in [s_1, s_2]$. Let $\mu'_1(s)$ be the characteristic values of functions $\widetilde{\psi}. = \widetilde{\psi}.(s)$. Assume that $\int_0^1 |b(s,t)|dt \leq C_{100}$ $\forall s \in [s_1, s_2]$.*

a) There exists an increasing continuous positive function χ from $\mathbb{R}^+$ to $\mathbb{R}^+$, $\chi(0) = 0$, depending only on the geometric data of the problem, on C_{100} and on a_0 an upper bound on a and a constant $\bar{\bar{c}}$ depending only on a_0, C_{41} and the geometric data of the problem such that, for any $s' \geq s$, $s', s \in [s_1, s_2]$, for any $\mu'_1(s')$ characteristic value of $\widetilde{\psi}.(s')$ satisfying furthermore (319):

$$\chi(\inf_j |\mu'_i(s') - \mu'_j(s)|) \leq \bar{\bar{c}}/\varepsilon_{10}(a(s) - a(s')) + o(1)$$

where $o(1)$ depends only on $\sup_{s\in[s_1,s_2]} O_{\varepsilon_0}(b)(s)$ and on ε_{13}.

b) Let γ be defined for $b(s,t)$ such as in (325)-(326). If $a(s_1) - a(s_2)$ is less than $\varepsilon_{10}\bar{\bar{c}}^{-1}\chi\left(\frac{C_{41}}{16m}\right)$ then, for any $s \in [s_1, s_2]$, any oscillation on an interval J having $\bar{\bar{\mu}} \geq a(\gamma + d\gamma - \delta a(s))$ and $\underset{\sim}{\underline{\mu}} \leq a(\gamma - d\gamma + \delta a)$ is such that $C\left(|I| + \int_I b^2\right)|I| \neq \frac{\varepsilon_0}{800}$ (I is the I of the definition of θ_γ see Definition 8).

c) If $a(s_1) - a(s_2) < \varepsilon_{10}\tilde{\bar{c}}^{-1}\chi\left(\frac{C_{41}}{16m}\right)$, and if no large oscillation of b (i.e of type 2)) subdivides then, $\int_{\theta_\gamma(s)}(|b|(s,t) - a\gamma)^+ dt$ is well defined and its discontinuities are only jumps downwards when s increases from s_1 to s_2, if δ is chosen small enough.

Furthermore, if c is taken less than $\frac{\tilde{\delta}}{1+(C_{41}+1)^3}\int_0^1 b\eta_{12}$, $\tilde{\delta} > 0$ given arbitrary and the cancellation flow is built according to some other rules, then:

$$\int_{\theta_\gamma(s)}(|b|(s,t) - a\gamma)^+ dt \leq \int_{\theta_\gamma(s_1)}(|b|(s_1,t) - a\gamma)^+ + \tilde{\delta}(a(s_1) - a(s))$$

for any $s \in [s_1, s_2]$.

Proof of Lemma 24. Proof of a): When b tends to b_1 in L^1 and a tends to a_1, $|b|_\infty, |b_1|_\infty \leq C_{41} + 1$, with $\mathcal{O}_{\tilde{\varepsilon}_0}(b)$ tending to zero, then b/a is equal to $\psi. + o(1)$, b_1/a_1 to $\psi^1_. + o(1)$, $\psi.$ and $\psi^1_.$ such as in Figure R. $\psi.$ and $\psi^1_.$ are also bounded by $C_{41} + 1$ and $\psi. - \psi^1_.$ tends to zero in L^1. If $\mu'_i(\psi.)$ is a characteristic value for $\psi.$ satisfying (319), then an easy measure argument shows, when combined with the monotonicity properties of $\psi^1_.$, that $\psi^1_.$ has a characteristic value $\mu'_j(\psi^1_.)$ satisfying the less constraining (322), very close to $\mu'_i(\psi.)$. Thus:

$$\inf_j |\mu'_i(s') - \mu'_j(s)| \to 0 \text{ if } \int_0^1 |b(s,t) - b(s',t)|dt + |a(s) - a(s')| + o(1) \to 0$$

where $o(1)$ tends to zero when $\sup_{s\in[s_1,s_2]} \mathcal{O}_{\varepsilon_0}(b)(s) + \varepsilon_{13}$ tends to zero. A better statement is, with obvious notations

$$\inf_j |\mu'_i(s') - \mu'_j(s)| \to 0 \text{ if } \int_0^1 |\widetilde{\psi}.(s,t) - \widetilde{\psi}.(s',t)|dt + o(1) + |a(s) - a(s')| \to 0 \quad (341)$$

$\widetilde{\psi}.(s,t)$ and $\widetilde{\psi}.(s',t)$ are functions such as in Figure R, with a finite number (depending on C_{100}) of flat or decreasing or increasing pieces. They are all bounded by $\frac{1+C_{41}}{\underline{a}}$ where $\underline{a}$ lower bounds a. Hence, weak and strong L^2-topology or L^1-topology are equivalent for such functions. In particular, $\int_0^1 |\widetilde{\psi}.(s,t) - \widetilde{\psi}.(s',t)|dt$ tends to zero if and only if $\int_0^1 dt\left|\int_0^1 d\tau_1 \int_{\tau_1}^t dx \int_0^1 d\tau \int_\tau^x (\widetilde{\psi}.(s,z) - \widetilde{\psi}.(s',z))dz\right|$ tends to zero.

Thus, if $\int_0^1 |\widetilde{\psi}.(s,t) - \widetilde{\psi}.(s',t)|dt$ were lower bounded away from zero, then the second expression is lower bounded away from zero. Therefore, as $\mathcal{O}_{\tilde{\varepsilon}_0}(b)$ tends to zero,

$$\int_0^1 dt\left|\int_0^1 d\tau_1 \int_{\tau_1}^t dx \int_0^1 d\tau \int_{\tilde{\tau}}^x (b(s,z) - b(s',z))dz\right|$$

would be lower bounded away from zero. On the other hand, we have:

$$(1) = \int_0^1 dt \left| \int_0^1 d\tau_1 \int_{\tau_1}^t dx \int_0^1 d\tilde{\tau} \int_{\tilde{\tau}}^x (b(s,z) - b(s',z))dz \right| \tag{342}$$
$$= \int_0^1 dt \left| \int_0^1 d\tau_1 \int_{\tau_1}^t dx \int_0^1 d\tilde{\tau} \int_{\tilde{\tau}}^x dz \int_{s'}^s \left(\frac{\dot{\overline{\eta + \lambda b}}}{a} + a\eta\tau - b\eta\bar{\mu}_\xi \right) \right|$$
$$= \int_0^1 dt \left| \int_{s'}^s \int_0^1 d\tau_1 \int_{\tau_1}^t dx \frac{\dot{\eta} + \lambda b}{a}(x) \right|$$
$$+ O\left(\int_{s'}^s \int_0^1 (|\lambda + \bar{\mu}\eta|\,|b| + |a|\,|\eta| + |b|\,|\eta|)dt \right)$$
$$= \int_0^1 dt \left| \int_{s'}^s \eta \right| + O\left(C_{41} \int_{s'}^s \int_0^1 (|b| + |a|)|\eta| \right)$$
$$\le C_{41} O\left(\int_{s'}^s (|b| + |a| + 1)|\eta| \right).$$

We observe now that

$$\eta = \overbrace{\sum \alpha_{i,j} \left(\sum \psi^i_{\theta_j,\varepsilon_j}, \text{ sgn } b\eta_{0,i,j} \right)}^{=\eta_{12}} + cb \tag{343}$$

where the $\alpha_{i,j}$ are nonnegative, $\sum \alpha_{ij} = 1$, where the supports of $\eta_{0,i,j}$ is contained in $\{|b| \ge \varepsilon_{10}\}$ and where $c < \frac{\int_0^1 b\eta_{12}}{1+|b|^s_\infty}$ thus:

$$\int_0^1 |\eta| \le \frac{1}{\varepsilon_{10}} \int_0^1 b\eta_{12} + c\int_0^1 |b| \le \left(\frac{1}{\varepsilon_{10}} + 1 \right) \int_0^1 b\eta_{12}. \tag{344}$$

Thus

$$(1) \le C(C_{41}^2 + 1)(1 + a_0)\left(1 + \frac{1}{\varepsilon_{10}} \right) \times -\frac{\partial a}{\partial s} \tag{345}$$

a) follows.
Proof of b): We know that

$$\inf |\gamma - \mu_j(s_1)| \ge \frac{C_{41}}{8m}. \tag{346}$$

Thus, if at any time $s \in [s_1, s_2]$, an oscillation develops above an interval J having $\tilde{\bar{\mu}} \ge a(\gamma + d\gamma - \delta a(s))$, $\underset{\sim}{\underline{\mu}} \le a(\gamma - d\gamma + \delta a(s))$ and $C\left(|I| + \int_I b^2\right)|I| = \frac{\varepsilon_0}{800}$, then, by Lemma 23, there exists i_0 such that

$$|\gamma - \mu_{i_0}(s)| = o(1) \tag{347}$$

where $o(1)$ tends to zero with ε_{13} and $O_{\tilde{\varepsilon}_0}(b)$. Thus, if ε_{13} and $O_{\tilde{\varepsilon}_0}(b)$ are small enough:

$$|\mu_{i_0}(s) - \mu_j(s_1)| \geq \frac{C_{41}}{15m} \quad \text{for any} \quad j. \tag{348}$$

b) follows. The result extends under (328), with $\frac{\delta_{10}\gamma}{180m}$ instead of $\frac{C_{41}}{15m}$.
Proof of c): For the definition of $\theta_\gamma(s)$, it suffices to consider the intervals where $(|b| - a(\gamma - d\gamma + \delta a(s))^+$ is positive. On each such an interval, we compute $\tilde{\bar{\mu}}$. If it is above $a(\gamma + d\gamma - \delta a(s))$, we keep the interval; otherwise, we throw it away. In this way, we see that $\theta_\gamma(s)$ is well defined and that $\int_{\theta_\gamma(s)}(|b|(s,t) - a\gamma)^+ dt$ can be computed.

Let us first assume that we are given an interval J such that $\tilde{\bar{\mu}} > a(\gamma + d\gamma - \delta a)$, $\underset{\sim}{\bar{\mu}} < a(\gamma - d\gamma + \delta a)$, for $s = s_0$. Assume that $J \in \theta_\gamma(s)$ for s close enough to s_0 or for a sequence s_k tending to s_0. Let $I^* = \{t \in J/b(s_0, t) \geq a\gamma\}$. By arguments similar to the ones used for b), we never can have $C\big(|I^*| + \int_{I^*} b^2\big)|I^*| \geq \frac{\varepsilon_0}{800}$ if $a(s_1) - a(s_2) < \varepsilon_{10}\tilde{\bar{c}}^{-1}\chi\big(\frac{C_{41}}{16m}\big)$, which we assume.

Indeed, would this happen at $s = s_0$, then picking $\tilde{o}(1)$ small, positive, given, for s close enough to s_0, we would have

$$C\left(|I_1| + \int_{I_1} b^2\right)|I_1| > \frac{\varepsilon_0}{1600},$$

where $I_1 = \{t \in J/b(s,t) > a\gamma - \tilde{o}(1)\}$.

Since $J \in \theta_\gamma(s)$ for s close enough to s_0 or for a sequence tending to s_0, $C\big(|I_2| + \int_{I_2} b^2\big)|I_2| < \frac{\varepsilon_0}{800}$, where $I_2 = \{t \in J/b(s,t) > a\gamma\}$ (for such values of s).

Using the arguments of Lemma 23, this forces the existence of a characteristic value close $o(1) + \tilde{o}(1)$ to γ. Using then the assumption on $a(s_1) - a(s_2)$, we derive a contradiction.

Thus, since $C\big(|I^*| + \int_{I^*} b^2\big)|I^*| < \frac{\varepsilon_0}{800}$, if $J \in \theta_\gamma(s_k)$, s_k tending to s_0, the constraint $C\big(|I| + \int_I b^2\big)|I| < \frac{\varepsilon_0}{800}$ is never saturated and does not produce discontinuities, unless the intervals J of $\theta_\gamma(s)$ change suddenly because $\tilde{\bar{\mu}}$ or $\underset{\sim}{\bar{\mu}}$ do not satisfy the strict inequalities required anymore.

The oscillations which we are considering have size at least $1/2$ if δ is chosen small enough. Therefore, they cannot appear or disappear suddenly, as oscillations of some size. They can however subdivide or coalesce, but this is a continuous movement.

However, we have two more constraints, namely

$$\underset{\sim}{\bar{\mu}} \leq a(\gamma - d\gamma + \delta a), \qquad \tilde{\bar{\mu}} \geq a(\gamma + d\gamma - \delta a). \tag{349}$$

We claim that, if c is chosen small enough, then the saturation of these constraints will only bring drops in $\theta_\gamma(s)$ as s increases.

Indeed, if $\tilde{\bar{\mu}}$ is equal to $a(\gamma + d\gamma - \delta a(s_0))$, at some time s_0, then, assuming $\tilde{\bar{\mu}} = b(s_0, t_0)$, we have:

$$\frac{\partial}{\partial s}\frac{b}{a}(s,t_0)|_{s=s_0} = \frac{1}{a}\left(\frac{\overline{\dot{\eta} + \lambda b}}{a} + a\eta\tau - b\eta\bar{\mu}_\xi + \frac{b}{a}\int_0^1 b\eta\right)\Bigg|_{(s_0,t_0)} \tag{350}$$
$$= \frac{1}{a}\left(\frac{\overline{\dot{\eta} - \bar{\mu}\eta b}}{a} + \frac{b^2\eta}{a} + a\eta\tau - b\eta\bar{\mu}_\xi\right)\Bigg|_{(s_0,t_0)}$$
$$= \frac{1}{a}\left(\frac{\overline{\dot{\eta}_{12} - \bar{\mu}\eta_{12}b}}{a} + \frac{b^2\eta_{12}}{a} + a\eta_{12}\tau - b\eta_{12}\bar{\mu}_\xi|_{(s_0,t_0)} + \right.$$
$$\left. + c\left(\frac{\overline{\dot{b} - \bar{\mu}b^2}}{a} + \frac{b^3}{a} + ab\tau - b^2\bar{\mu}_\xi\right)\right)\Bigg|_{(s_0,t_0)}$$

We argue as if $\eta_{12} = \psi_{\theta,\varepsilon}\eta_0$; the argument extends easily to the general case. We have, using (350), (10) and the fact that $\dot{b}(s_0, t_0) = 0$:

$$\overline{\overline{\dot{\psi_{\theta,\varepsilon}\eta_0}} - \bar{\mu}\psi_{\theta,\varepsilon}\eta_0 b}\Big|_{(s_0,t_0)} \tag{351}$$
$$= -\ddot{\theta}\eta_0 - 2\dot{\theta}\dot{\eta}_0 + \frac{\partial\psi_{\theta,\varepsilon}}{\partial b}\ddot{b}\eta_0 + O(1+b^2)\psi_{\theta,\varepsilon}\eta_0 + O(\dot{\theta}b\eta_0) +$$
$$+ \psi_{\theta,\varepsilon}\left(-\frac{2}{(x_2^- - x_1^-)^2}e^{-\int_{t_1^-}^{t} -\bar{\mu}bd\tau} - C(1+b^2)\eta_0 e^{-2\int_{t_1^-}^{t} -\bar{\mu}bd\tau}\right)\Bigg|_{(s_0,t_0)}$$

$\ddot{b}(s_0, t_0)$ is negative. $\ddot{\theta}$ and $\dot{\theta}$ are very small. We can absorb all these terms in $\frac{\tilde{\delta}}{200}\int_0^1 b\eta_{12}$.

Thus, if C is large enough:

$$\overline{\overline{\dot{\psi_{\theta,\varepsilon}\eta_0}} - \bar{\mu}\psi_{\theta,\varepsilon}\eta_0 b}\Big|_{(s_0,t_0)} \leq \frac{\tilde{\delta}}{200}\int_0^1 b\eta_{12}. \tag{352}$$

We can use also $C(1+b^2)\psi_{\theta,\varepsilon}\eta_0$ to absorb $\frac{b^2\eta_{12}}{a} + a\eta_{12}\tau - b\eta_{12}\bar{\mu}_\xi$. C can be taken uniform, depending on a_0, an upper bound on the energy level.

Thus, using the fact that $\ddot{b}(s_0, t_0) \leq 0$ and $\dot{b}(s_0, t_0) = 0$:

$$\frac{\partial}{\partial s}\frac{b}{a}(s,t_0)|_{s=s_0} \leq cO(1 + |b|_\infty^3) + \frac{\tilde{\delta}}{200}\int_0^1 b\eta_{12}. \tag{353}$$

Thus, if we ask that

$$cO(1 + |b|_\infty^3) < \frac{\tilde{\delta}}{200} \int_0^1 b\eta_{12} \tag{354}$$

we derive:

$$\frac{\partial}{\partial s} \frac{b}{a}(s, t_0)|_{s=s_0} \leq -\frac{\tilde{\delta}}{100} \frac{\partial a}{\partial s}. \tag{355}$$

Therefore

$$\frac{\partial}{\partial s}\left((\gamma + d\gamma - \delta a(s)) - \frac{\tilde{\bar{\mu}}}{a}\right) \geq -\frac{\tilde{\delta}}{2} \frac{\partial a}{\partial s} > 0. \tag{356}$$

Thus, if $\tilde{\bar{\mu}}$ is even equal to $a(\gamma + d\gamma - \delta a(s))$, then $a(\gamma + d\gamma - \delta a(s))$ will move above $\tilde{\bar{\mu}}$ later, hence the oscillation will not be in $\theta_\gamma(s)$ anymore later.

For $\frac{\underset{\sim}{\bar{\mu}}}{a}$, the argument is more involved. We are seeking now a lower bound. $\ddot{b}$ is positive, $\dot{b}$ is zero (otherwise, the constraint on $\bar{\mu}$ is not saturated), $\frac{\partial \psi_{\theta,\varepsilon}}{\partial b} \ddot{b} \eta_0$ is positive: we can throw all these terms away in a lower bound. This takes care of all the terms multiplied by c. We can also easily take care of all the terms multiplied by $\ddot{\theta}$ or $\dot{\theta}$. The terms which are left are all multiplied by $\psi_{\theta,\varepsilon}$. If $\psi_{\theta,\varepsilon}(\underset{\sim}{\bar{\mu}})$ is nonzero, then, since $a\gamma$ is larger, for δ small enough, then $\underset{\sim}{\bar{\mu}} + \frac{a}{2}$ ($\underset{\sim}{\bar{\mu}} = a(\gamma - d\gamma + \delta a(s))$), we know that ($|\psi_{\theta,\varepsilon}(x) - (x - \theta)^+| \leq \varepsilon\theta$, which is very small)

$$\psi_{\theta,\varepsilon}(a\gamma) \geq \frac{a}{4}. \tag{357}$$

Denoting J the support of η_0 and

$$\tilde{I} = \{t \in J \text{ s.t. } |b|(s,t) \geq a\gamma\} \tag{358}$$

J has a $\tilde{\bar{\mu}}$ larger than $a(\gamma + d\gamma - \delta a(s))$ and since $\psi_{\theta,\varepsilon}(\underset{\sim}{\bar{\mu}})$ is positive and $\underset{\sim}{\bar{\mu}}$ is equal to $a(\gamma - d\gamma + \delta a(s))$, b, at the boundaries of J takes a value $\underset{\sim}{\bar{\mu}}(J)$ less than $a(\gamma - d\gamma + \delta a(s))$. This requires a little argument: If $\psi_{\theta,\varepsilon}(\underset{\sim}{\bar{\mu}})(\frac{2}{(x_1^- - x_2^-)^2} + C(1 + \underset{\sim}{\bar{\mu}}^2))$ can be upper bounded by $\frac{\tilde{\delta}}{200} \int_0^1 b\eta_{12}$, we are done. Otherwise, we have

$$\psi_{\theta,\varepsilon}(\underset{\sim}{\bar{\mu}})\left(\frac{2}{(x_1^- - x_2^-)^2} + C(1 + \underset{\sim}{\bar{\mu}}^2)\right) \geq \frac{\tilde{\delta}}{200} \int_{t_1^-}^{t_2^-} b\psi_{\theta,\varepsilon}\eta_0. \tag{359}$$

As $\mu_1 - \mu_1^-, \mu_2 - \mu_2^-, \mu_1 - \mu_2$ tends to zero in our construction of $\psi_{0,\varepsilon}\eta_0$, the interval $[t_1^-, t_2^-]$ does not collapse and the above lower bound does not tend to zero.

Thus, $\underset{\sim}{\bar{\mu}}$ (not the $\underset{\sim}{\bar{\mu}}(J)$ but the $\underset{\sim}{\bar{\mu}}$ which is crossing $a(\gamma - d\gamma + \delta a(s))$) can be made, in our construction, larger than $\mu_1, \mu_2, \mu_1^-, \mu_2^-$. This is in fact, included in the construction somewhat. However, we can, for the sake of the completion, ask that $\mu_1, \mu_2, \mu_1^-, \mu_2^-$ are so close that any $\underset{\sim}{\bar{\mu}}$ satisfying (359) will be above all these values.

$\tilde{\delta}$ will be chosen small later, depending on the compact which we are deforming at a given step. The rate of decrease is not primarily related to $\tilde{\delta}$ but rather to the oscillatory part $\sum \alpha_i \int b\psi_{\theta_i \varepsilon_i} \eta_{0,i}$. Therefore, the change in $\tilde{\delta}$ will not affect the amount of decease. $\tilde{\delta}$ can change at each step. The arguments are coherent. The picture is, then:

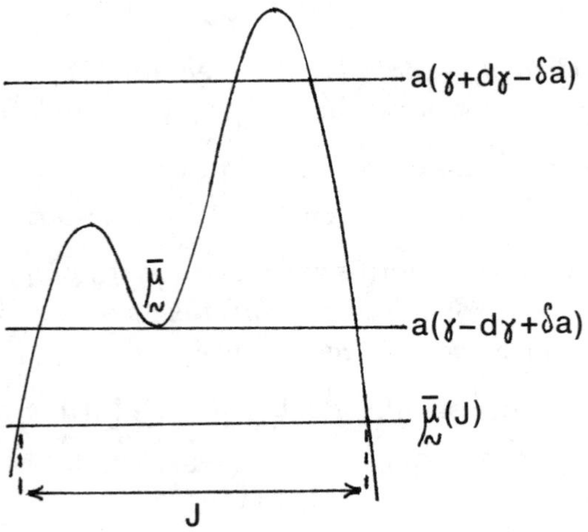

(359′)

Since J is an interval such that $|b|$ on ∂J is less than $a(\gamma - d\gamma + \delta a(s))$, any interval where $(|b| - a(\gamma - d\gamma + \delta a(s)))^+$ is positive which intersects J is contained in J. J jails any discontinuous phenomenon. There could be a problem and a jump before and after s_0 for $\theta_\gamma(s)$ only if:

$$C\left(|\tilde{I}| + \int_{\tilde{I}} b^2\right)|\tilde{I}| \geq \frac{\varepsilon_0}{800}. \tag{360}$$

Otherwise, the strict inequality:

$$C\left(|\tilde{I}| + \int b^2 1_{\tilde{I}(s_0)}\right)|\tilde{I}|(s_0) < \frac{\varepsilon_0}{800} \tag{361}$$

is preserved in a neighborhood of s_0. Since $\tilde{I}$ contains $I = \{t \in J \text{ s.t. } |b|(s,t) > a\gamma\}$, no discontinuity can take place:

Indeed, taking $\tilde{\tilde{I}}_\varepsilon$ an open set containing $\tilde{I}(s_0)$ in J, $|b|(s_0,t)$ is less than $a\gamma - \varepsilon$ on $\tilde{\tilde{I}}{}^c_\varepsilon$ (the complement is taken in J), where ε is positive. Thus, $\tilde{I}(s)$ is included into $\tilde{\tilde{I}}_\varepsilon$ for s close enough to s_0. Thus, $|\tilde{I}(s)|$ tends to $|\tilde{I}(s_0)|$ when s tends to s_0. (361) holds for $s = s_0$, hence for $I(s)$ in lieu of $\tilde{I}(s)$.

Therefore, $\theta_\gamma(s)$ is not discontinuous through s_0. (The actual description of the oscillations of $\theta_\gamma(s)$, if we cut them piece by piece, might change discontinuously, but not the union of all the intervals because there is the larger oscillation corresponding to J which is contained in $\theta_\gamma(s)$).

Thus, if there is any discontinuity, (360) must hold at $s = s_0$. Then, we have

$$|\widetilde{I}(s_0)| \geq \theta(\varepsilon_0, C_{41}) > 0 \tag{362}$$

and

$$\int\limits_J b\psi_{\theta,\varepsilon}\eta_0 \geq \psi_{\theta,\varepsilon}(a\gamma)a\gamma \int\limits_{\widetilde{I}} \eta_0 \geq \frac{a^2 d\gamma}{8} \int\limits_{\widetilde{I}} \eta_0 . \tag{363}$$

From (362), we derive that J has a length larger than $\theta(\varepsilon_0, C_{41})$. $\widetilde{I}(s_0)$ is a subset of measure at least $\theta(\varepsilon_0, C_{41})$. Thus

$$\int_J b\psi_{\theta,\varepsilon}\eta_0 \geq \frac{a^2}{8} d\gamma\theta'(\varepsilon_0, C_{41}) \geq \frac{\underline{a}^2}{8} d\gamma\theta'(\varepsilon_0, C_{41}) \tag{364}$$

where $\underline{a} > 0$ lower bounds the energy a on our flow-line (which we consider above a field value $\underline{a}$).

From (364), we derive that $\int_0^1 b\eta_{12}$ is lower bounded. We do not know how to prove here that this is impossible. However, $|b|$ has clearly over J a large oscillation, which is subdividing at $s = s_0$. Since we started with stretched curves, we need to go, at least for the first time that this subdivision occurs, through a subdivision of a type 2) oscillation. We call them large oscillation in what follows. If we exclude this, $\theta_\gamma(s)$ deforms continuously, its intervals can coalesce or subdivide, but in a continuous way if some $\underset{\sim}{\bar{\mu}}$ is equal to $a(\gamma - 1 + \delta a)$. At discontinuity points, $\int_{\theta_\gamma(s)}(|b| - a\gamma)^+(s)$ decreases. At a point of continuity, we can compute:

$$\frac{\partial}{\partial s} \int\limits_{\theta_\gamma(s)} (|b| - a\gamma)^+ = \int\limits_{I_\gamma(s)} \left(\frac{\dot{\overline{\eta + \lambda b}}}{a} + a\eta\tau - b\eta\bar{\mu}_\xi \right) \operatorname{sgn} b \tag{365}$$

where $I_\gamma(s) = \{t \in \theta_\gamma(s) \text{ such that } |b|(s,t) > a\gamma\}$. We again use here the fact that $\frac{b}{a}$ takes the value γ at a finite number of points. Proposition A1, in Appendix 1, allows to extend this particular case to the general case.

We then have, again computing as if $\eta = \psi_{\theta,\varepsilon}\eta_0 + cb$ and as if b were positive

$$\frac{\partial}{\partial s} \int\limits_{\theta_\gamma(s)} (|b| - a\gamma)^+ = \frac{\dot{\eta} + \lambda b}{a}\bigg|_{\partial I_\gamma(s)} + \int\limits_{I_\gamma(s)} (a\eta\tau - b\eta\bar{\mu}_\xi) \tag{366}$$

$$= \int\limits_{I_\gamma(s)} \left(\frac{\dot{\overline{\eta - \bar{\mu}\eta b}}}{a} + a\eta\tau - b\eta\bar{\mu}_\xi \right) + (\lambda + \bar{\mu}\eta)\gamma\bigg|_{\partial I_\gamma(s)} .$$

Observe that, since $\dot{b}$ is negative on $\partial I_\gamma(s)$:

$$c\left(\int\limits_{I_\gamma(s)} \overline{\dot{b}-\bar{\mu}b^2}\right) \leq \tilde{\delta}\int_0^1 b\eta_{12} \tag{367}$$

and, in fact, we can include all the terms multiplied by c in such a lower bound. We are thus left with (observe that $\overline{\lambda_{12}+\bar{\mu}\eta_{12}}^{\,\cdot} = b\eta_{12} - \int_0^1 b\eta_{12}$):

$$(1) = \int\limits_{I_\gamma(s)} \frac{\overline{\overline{\psi_{\theta,\varepsilon}\eta_0}^{\,\cdot} - \bar{\mu}\psi_{\theta,\varepsilon}\eta_0}^{\,\cdot}}{a} + \psi_{\theta,\varepsilon}\eta_0[(a\tau - b\bar{\mu}_\xi)+\gamma b] - \gamma\left(\int_0^1 b\psi_{\theta,\varepsilon}\eta_0\right)|I_\gamma(s)|. \tag{368}$$

Assume that we have

$$\psi_{\theta,\varepsilon}(a\gamma) \text{ is independent of } t, \text{ larger than or equal to } \frac{a\gamma}{800m} = \frac{ad\gamma}{2}. \tag{369}$$

Then ($\frac{ad\gamma}{2}$ in the general case)

$$\begin{aligned}(1) = {} & \frac{\psi_{\theta,\varepsilon}(a\gamma)}{a}\int\limits_{I_\gamma(s)} \overline{\dot{\eta}_0 - \bar{\mu}\eta_0}^{\,\cdot} + \frac{1}{a}\left(\frac{\partial\psi_{\theta,\varepsilon}}{\partial b}\dot{b} - \dot{\theta}\right)\eta_0\Bigg|_{\partial I_\gamma(s)} + \\ & + \int\limits_{I_\gamma(s)} \psi_{\theta,\varepsilon}\eta_0[(a\tau - b\bar{\mu}_\xi) + \gamma b] - \gamma\left(\int_0^1 b\psi_{\theta,\varepsilon}\eta_0\right)|I_\gamma(s)|.\end{aligned} \tag{370}$$

Observe that $\eta_0\left(\frac{\partial\psi_{\theta,\varepsilon}}{\partial b}\dot{b}\right)\big|_{\partial I_\gamma(s)}$ is negative and that

$$-\dot{\theta}\eta_0|_{\partial I_\gamma(s)} \leq C\left(\int_{t_1^-}^{t_2^-}|\ddot{\theta}| + \frac{|\dot{\theta}|_\infty}{|t_1^- - t_2^-|}\right) < \frac{\tilde{\delta}}{200}\int_0^1 b\eta_{12}. \tag{371}$$

Thus, using (370):

$$\begin{aligned}(1) \leq {} & \int\limits_{I_\gamma(s)}\left\{\frac{\psi_{\theta,\varepsilon}(a\gamma)}{a}\left(-\frac{2}{|x_1^- - x_2^-|^2}e^{-\int_{t_1^-}^t \bar{\mu}b} - C(1+b^2)\eta_0 e^{-2\int_{t_1^-}^t \bar{\mu}b}\right) + \right. \\ & \left. + \psi_{\theta,\varepsilon}\eta_0(a\tau - b\bar{\mu}_\xi + \gamma b)\right\} - \gamma\left(\int_0^1 b\psi_{\theta,\varepsilon}\eta_0\right)|I_\gamma(s)| + \frac{\tilde{\delta}}{200}\int_0^1 b\eta_{12}\end{aligned} \tag{372}$$

$\frac{\psi_{\theta,\varepsilon}(a\gamma)}{a}$ is larger than $\frac{d\gamma}{2}$. Thus, since $\psi_{\theta,\varepsilon}(b) \leq b$. (We are assuming that b is positive on the oscillation, to set the framework)

$$-C\frac{\psi_{\theta,\varepsilon}(a\gamma)}{a}(1+b^2)\eta_0 e^{-2\int_{t_1^-}^{t}\bar{\mu}b} + \psi_{\theta,\varepsilon}\eta_0(a\tau - b\bar{\mu}_\xi + \gamma b) \tag{373}$$
$$\leq -C\frac{\gamma(1+b^2)}{800m}\eta_0 e^{-2\int_{t_1^-}^{t}\bar{\mu}b} + b\eta_0(a\tau - b\bar{\mu}_\xi + \gamma b)$$

($\frac{\gamma}{800m}$ is to be replaced by $\frac{ad\gamma}{2}$ in the general case) m is bounded in function of $\int_0^1 |b|$ and ε_0 and C see (324) ($m \leq 16 \times 10^{20} C_{100}\sqrt{\frac{C}{\varepsilon_0}}$. These bounds are given once and for all; they depend only on the initial compact which we are deforming and we keep them throughout our construction. The bound on $\int_0^1 |b|$ stems from (ii) of Lemma 3; we pointed out in this statement that this bound was independent of C; later, as we create oscillations and we cancel them, the bound is ultimately preserved. It is independent of C in the definition of cases 1 and 2 and (10), because C is multiplied by a negative term $-(1+b^2)\psi_{\theta,\varepsilon}\eta_0 e^{-2\int_{t_1^-}^{t}\bar{\mu}b}$, which, therefore, does not appear in the upper bound. In fact, we take C large to absorb the other terms. And we can do it here also:

Choosing $\frac{C}{m}e^{-2\int_0^1|\bar{\mu}||b|}$, which is larger than $e^{-2\,\mathrm{Sup}|\bar{\mu}|C_{100}} \times \frac{\sqrt{C\varepsilon_0}}{16\times10^{20}\times C_{100}}$, large enough, i.e. C large enough, we can absorb all the other terms. The argument can be repeated for other values of $d\gamma$ in (331) i.e. for other oscillations. However, the value of C can be taken uniform only if δ_{10} in (328) (the size of the oscillation) is lower bounded or given once and for all. We will discuss this point later. The result follows under (369), with a C depending on $d\gamma$.

Another way to proceed, which does not involve a change of C each time we want to control the oscillations of order larger than $2\delta_0'$, when δ_0' changes, is as follows:

We observe that, under (369), if ($\bar{a}$ is an upper bound on a)

$$|x_1^- - x_2^-| < \beta(C_{41}, C_{100}, \bar{a}, d\gamma) \tag{374}$$

where β is an increasing function of $d\gamma$, then, we do have:

$$(1) \leq \frac{\tilde{\delta}}{200}\int_0^1 b\eta_{12} \tag{375}$$

since $-\frac{2}{|x_1^- - x_2^-|^2}e^{-\int_{t_1^-}^{t}\bar{\mu}b}$ can play the role that $C(1+b^2)$ was playing in the previous argument.

However, in order to have (374), we have to be a little bit careful in the construction of $[t_1^-, t_2^-]$, η_0 etc. We will also discuss this later.

At this point, we need (369). (369) follows from a maximality principle which we need to apply in the construction of the $\sum \tilde{\psi}_{\tilde{\theta}_i,\tilde{\varepsilon}_i}\tilde{\eta}_{0,i}$ pieces of our cancellation flow. These pieces are convex-combined to derive the global cancellation flow.

In each of these pieces, $\tilde{\eta}_{0,i}$ is related to one oscillation and the $\tilde{\eta}_{0,i}$'s have disjoint supports. Let us concentrate on η_0. Since b has an oscillation around $a\gamma$ of size at least $\frac{2}{3}ad\gamma$, upwards and downwards, (369) follows basically from the construction of $\psi_{\theta,\varepsilon}$ if μ_1 and μ_2 are less than $a\gamma - \frac{2}{3}ad\gamma$. Observe that, by construction of θ_γ, each interval of I_γ, $\widetilde{I}'_\gamma$, is jailed, maybe with others of the same type, into an interval J^1_γ of θ_γ thus an interval such that $|b|_{\partial I^1_\gamma}$ is less than $a\gamma - \frac{2}{3}ad\gamma$. We can restrict J^1_γ into an interval J_γ such that J_γ contains $\widetilde{I}'_\gamma$ and $|b|_{|\partial J_\gamma}$ is precisely $a\gamma - \frac{2}{3}ad\gamma$ and $|b| > a\gamma - \frac{2}{3}$ $ad\gamma$ in $\overset{\circ}{J}_\gamma$. J_γ cannot be of type 2), because, if J_γ is of type 2), $a\widetilde{\psi}.$ must have, by the arguments of Lemma 23, if $o(1)$ is small enough ($b = a\widetilde{\psi}. + o(1)$, a characteristic value $a\mu'_j(\widetilde{\psi}.)$ above $a\gamma - \frac{2}{3}ad\gamma$ or within ε_{13} of a $\gamma - \frac{2}{3}ad\gamma$, satisfying (319), with support in J_γ. Since $d\gamma = \frac{\gamma}{400m}$ for example is less than $\inf_i |\gamma - \mu'_i(\psi.(s_1))|$, would this characteristic value be between $a\gamma - \frac{2}{3}ad\gamma$ and $a\gamma + \frac{2}{3}ad\gamma$, then:

$$\frac{\gamma}{1200m} = \frac{1}{3}d\gamma \leq |\mu'_j(\widetilde{\psi}.) - \mu'_i(\psi.(s_1))| \tag{376}$$

for any i.

$\psi.(s_1)$ is a $\widetilde{\psi}.(s_1), \widetilde{\psi}.$ is $\widetilde{\psi}.(s)$, $s \in [s_1, s_2]$. Thus, applying b), $a(s_1) - a(s_2)$ is large, a contradiction. Thus:

$$\mu'_i(\widetilde{\psi}.) \geq \gamma + \frac{2}{3}d\gamma \tag{377}$$

b is very close L^1 to $a\widetilde{\psi}.$. This nearly flat piece of $\widetilde{\psi}.$ is lying in J_γ. Its support is contained in J_γ by construction and satisfies (319). As $o(1)$ tends to zero, we must thus have:

$$C\left(|\widetilde{I}_\gamma| + \int_{\widetilde{J}_\gamma} b^2\right)|\widetilde{J}_\gamma| \geq \frac{\varepsilon_0}{40} \tag{378}$$

where $\widetilde{J}_\gamma = \{t \in J_\gamma \text{ s.t. } b| \geq a\gamma + \frac{5}{8}ad\gamma\}$. However, since J_γ is contained in θ_γ, we have:

$$C\left(|I| + \int_I b^2\right)|I| < \frac{\varepsilon_0}{800} \tag{379}$$

for $I = \{t \in J_\gamma \text{ s.t. } |b|(t)| > a\gamma\}$ a contradiction. The claim follows: J_γ is not of type 2), an $\bar{\eta}_0$ and $\overline{\psi}_{\bar{\theta},\bar{\varepsilon}}$ *could* be built on J_γ and $\overline{\psi}_{\bar{\theta},\bar{\varepsilon}}$ *would* satisfy (369).

However, η_0 and $\psi_{\theta,\varepsilon}$ are given i.e. the $\eta_{0,i}$, $\psi_{\theta_i,\varepsilon_i}$'s are given, with disjoint supports. We are focusing on the $\eta_{0,i}$'s which have some support in $\widetilde{I}_\gamma$. If J_i is the interval supporting $\eta_{0,i}$, then J_i intersects $\widetilde{I}_\gamma$, hence intersects the interval J_γ. Thus, either

1) $J_i \subset J_\gamma$

2) $J_\gamma \subset J_i$
3) a boundary of J_γ is contained in J_i.

In the second and third cases, we must have:

$$\mu_1 \text{ or } \mu_2 \text{ is less than or equal to } a\gamma - \frac{2}{3}ad\gamma. \tag{380}$$

Thus, since μ_1 and μ_2 are very close, we can assume that both are less than $a\gamma - \frac{1}{2}ad\gamma$. (369) follows then, for $\eta_{0,i}$.

Would μ_1 be equal to μ_2 and μ_1^- to μ_2^- for all the oscillations involved, our argument and the related choices would then be transparent: Indeed, then, since $\mu_1 = \mu_2$ is less than or equal to $a\gamma - \frac{2}{3}ad\gamma$, a boundary point of J_i cannot lie in $\{t \in J_\gamma \text{ s.t. } |b(t)| > a\gamma - \frac{2}{3}ad\gamma\}$, which is the interior of J_γ. Thus, the third case is impossible.

Either J_γ is contained in J_i. There is then a single $\eta_{0,i}$ and it satisfies our claim. Or J_i is contained in J_γ and all the J_ℓ's which intersect J_γ are then all contained in J_γ, with disjoint supports.

Thus, if $\mu_1 = \mu_2$, the oscillations which we are considering satisfy 1) or 2) i.e. if we consider any two intervals J and K such that

i) $|b|(t) > |b|_{|\partial J} \quad \forall t \in \overset{\circ}{J}$
ii) $|b|$ is constant on ∂J

with the same properties for K, then either $J \subset K$ or $K \subset J$ if $\overset{\circ}{K} \cap \overset{\circ}{J} \neq \emptyset$. We can then pick up, given such a nested family of oscillations the smallest interval containing all of those of type 1). This smallest interval L might not be itself of type 1). However, given $\delta_{10} > 0$, we can find $L_1 \subset L$, in the same family, of type 1), such that

$$|b(t)|_{|\partial L_1} \leq |b(t)|_{\partial L} + \frac{a\delta_{10}}{800m} \times \frac{\varepsilon_{10}}{1+\varepsilon_{10}}. \tag{381}$$

The oscillation $\psi_{\theta,\varepsilon}\eta_0$ will be built, for the family, on L_1, if the family is picked up in the construction. If we apply this rule, which is simple, in our construction, then, since $J_\gamma \subset L$, any J_i involved must be such that:

$$|b(t)|_{\partial J_i} \leq |b(t)|_{\partial J_\gamma} + \frac{a\delta_{10}}{800m} \times \frac{\varepsilon_{10}}{1+\varepsilon_{10}}. \tag{382}$$

Thus

$$\mu_1, \mu_2 \leq a\gamma - \frac{2}{3}ad\gamma + \frac{a\delta_{10}}{800m} \times \frac{\varepsilon_{10}}{1+\varepsilon_{10}}. \tag{383}$$

γ is larger than ε_{10}. $d\gamma$ is lower bounded also in function of ε_{10} (in our present case, $\gamma \in \left[\frac{C_{41}}{2}, \frac{3}{4}C_{41}\right]$, in later applications, the value of γ and $d\gamma$ might change, but they are always lower bounded). Taking δ_{10} very small, chosen once and for all, observing that this construction is coherent with the requirement after (137), since once (383) holds, with $\delta_{10}, 1/k$ very small, we can still fulfill this requirement, (369) follows, if we apply this rule.

The situation is rendered a little bit more complicated by the fact that μ_1 is different from μ_2. However, if we complete this construction and we find these families with $\mu_1 = \mu_2$, and pick up these intervals, we can always, using the fact that all these intervals are such that $|b|(t) > |b|_{|\partial J} \; \forall t \in \overset{\circ}{J}$ and that b is continuous, build some values μ_1 and μ_2, μ_1^-, μ_2^- also satisfying Lemma 1, with (t_1^-, t_2^-) very close to J. It suffices to make μ_1 and μ_2 tend to $|b|_{|\partial J}$ and to complete, at the same time, the construction of Lemma 1, which depends only on the fact that b increases a little bit near μ_1 and μ_2, a condition satisfied by $|b|$ on ∂J because $|b|(t) > |b|_{|\partial J} \; \forall t \in \overset{\circ}{J}$.

Once (t_1^-, t_2^-) is found, of type 1), very close to L, our claim again holds. The proof of Lemma 24 is thereby complete. We observe that the addition of this rule does not change the proof of all the previous Lemmas which can be checked to hold after this new condition. This observation allows to keep our former construction unchanged.

When we localize the cancellation flow to $[t_1^+ + \delta, t_2^- - \delta]$, the discussion which we completed above about the maximality rule for the choice of the intervals of oscillations has obviously to be changed because of the boundaries: if an oscillation has its maximal related interval (of type 1)) in $[t_1^+ + \delta, t_2^- - \delta]$, the choices simply extend. However, if we touch the boundaries, the choice which we set before does not extend. The maximal interval has to stop, at $t_1^+ + \delta$, for example. With respect to the fulfillment of (369), there is no change if $|b|(t_1^+ + \delta)$ is less than $a\gamma - \frac{ad\gamma}{4}$ for example, only the value of few constants has to change. If $|b|(t_1^+ + \delta)$ is larger than $a\gamma - \frac{ad\gamma}{4}$, we have to include the related oscillation into another family $\theta_{\gamma'}$ of oscillations, with $\gamma' = \gamma + \frac{d\gamma}{4}$. The same arguments used for ν work for γ', with the change of some constants. The choice of $d\gamma$ small enough allows to keep γ' away from the characteristic values. Thus, the basic arguments extend with the use of two constants γ and γ', instead of one only.

The previous construction can be changed in another direction as follows: Let L be the smallest interval containing all of the intervals of type 1) corresponding to an oscillation.

Let us assume that this oscillation has a size at least $\frac{a\delta_{10}}{20m}$, where δ_{10} has been introduced in (381). Let

$$\tilde{L} = \left\{ t \in L \text{ s.t. } |b(t)| > |b|_{\partial L} + \frac{a\delta_{10}}{160m} \right\} \tag{384}$$

$\tilde{L}$ is strictly contained in L. Between any interval of $\tilde{L}$ and L, there exists an interval of type 1), this follows from the construction of L as the union of all the nested family of intervals of type 1) corresponding to one oscillation. Therefore, the intervals of $\tilde{L}$

are of type 1). Let

$$\tilde{\tilde{L}} = \left\{ t \in \tilde{L} \text{ s.t. } |b(t)| > |b|_{\partial \tilde{L}} + \frac{a\delta_{10}}{160m} = |b|_{\partial L} + \frac{a\delta_{10}}{80m} \right\}. \tag{385}$$

We then have

Lemma 25. $|\tilde{\tilde{L}}|$ *tends to zero when* $\mathcal{O}_{\bar{\varepsilon}_0}(b)$ *tends to zero, for* δ_{10} *given.*

Proof. We apply Lemma 4 to $I = \tilde{L}$, which does not contain any interval of type 2). Clearly:

$$\frac{a\delta_{10}}{80m} |\tilde{\tilde{L}}| \le \int_{\tilde{L}} (|b| - |b|_{\partial \tilde{L}})^+ < \mathcal{O}_{\bar{\varepsilon}_0}(b). \tag{386}$$

Lemma 25 follows.

Thus, if $\mathcal{O}_{\bar{\varepsilon}_0}(b)$ is small enough, we can have (374). We can have (369) at the same moment since, on any interval J of $\tilde{\tilde{L}}$, we have:

$$|b|_{\partial J} = |b|_{\partial L} + \frac{a\delta_{10}}{80m}. \tag{387}$$

and if δ_{10} is small enough, (369) follows, this without changing C, but under the role assumption that $\mathcal{O}_{\bar{\varepsilon}_0}(b)$ is small enough. Extending then our former construction, we can take values γ in the middle part of any interval of length $4\delta_{10}$ i.e.

$$\gamma_j \in ((k-1)\delta_{10}, (k+1)\delta_{10}) \qquad j \in \mathbb{Z} \tag{388}$$

$$|\gamma_j - \mu_i(\psi.)| \ge \frac{\delta_{10}}{40m} \qquad \forall i \tag{389}$$

where m is given in function of C_{100} and C. *C is given once and for all.*

We thus construct a family of values γ_j, which is finite, because we ask that:

$$|j+1|\delta_{10} \le 1 + C_{41}. \tag{390}$$

They all can be used as γ's of Lemma 24. a), b), c) of this Lemma hold, with decreases in $a(s_1) - a(s_2)$ now depending on δ_{10}. C is given, but we can use Lemma 25 and assume $\mathcal{O}_{\bar{\varepsilon}_0}(b)$ small enough so that c) will hold, if we complete our construction of $\psi_{\theta,\varepsilon}\eta_0$ on the $\tilde{\tilde{L}}$'s provided by Lemma 25. We need, for this, to have oscillations of size at least $8\delta_{10}$. We can then pick up a value among the $a\gamma_j$'s which is by at least $2\delta_{10}$ above $\underset{\sim}{\bar{\mu}}$, the infimum of the oscillation. If we construct then $\tilde{\tilde{L}}$ as in Lemma 25, $|b(t)|_{\partial \tilde{\tilde{L}}}$ will be at least $a\delta_{10}$ below $a\gamma_j$. Thus

$$\psi_{\theta,\varepsilon}(a\gamma_j) \ge a\delta_{10}. \tag{391}$$

Thus, for oscillations of size $8\delta_{10}$, we can find a related γ_j and (391) holds. For smaller oscillations e.g. those of $\theta_{\gamma_j}(b)$, taking, after (389), $d\gamma_j$ to be $\frac{\delta_{10}}{40m}$ and using (383), we see that

$$\psi_{\theta,\varepsilon}(a\gamma_j) \geq \frac{a\delta_{10}}{240m} \tag{391'}$$

Coming back to the computations of $\frac{\partial}{\partial s} \int\limits_{\theta_{\gamma_j}} (|b| - a\gamma_j)^+$, in particular to (372), we see that if $O_{\bar{\varepsilon}_0}(b)$ is small enough (in function of δ_{10}), $|\tilde{\tilde{L}}|$ is small enough, the argument developed in (374)–(375) extends. Thus, Lemma 24 extends to this framework.

We now have two approaches, one involving a dependence on δ_{10} of C, with γ_j's satisfying (326) (C depends also on C_{100}, a global bound on $\int_0^1 |b|$, given once and for all), the other one involving $\tilde{L}$ and $\tilde{\tilde{L}}$ in (385).

When we refer to the dependence of C on δ_{10}, we assume that the γ_j's satisfy (326) and, instead of (388)-(390), we build the γ_j's as follows: We consider intervals $[M_j, M_j + \delta_{10}M_j]$, with $M_j \in [\varepsilon_{10}, C_{41}+1]$, $M_j = q\delta_{10} + r\delta_{10}^2$, $q, r \geq 1$, $q, r \in \mathbb{N}$ and we construct values γ_j such that:

$$\gamma_j \in [M_j, M_j + \delta_{10}M_j] \tag{388'}$$

$$|\gamma_j - \mu_i(\psi.)| \geq \frac{\delta_{10}\gamma_j}{200m}. \tag{389'}$$

Using the arguments above, see (376)-(384) and what follows, we then have, for every oscillation of $\theta_{\gamma_j}(b)$:

$$\psi_{\theta,\varepsilon}(a\gamma_j) \geq \frac{a\delta_{10}\gamma_j}{800m}. \tag{391'}$$

We are then brought back to (373). We have seen that, considering (324) and taking C large enough in function of δ_{10}, we can derive Lemma 24, including 24 i).

The two approaches compete and one might think that the one involving the construction of $\tilde{\tilde{L}}$ is better. However, it has a flaw: the rate of decrease, in the associated deformation process, is related to a quantity smaller than $O_{\varepsilon_0}(b)$, because we restrain to the $\tilde{\tilde{L}}$'s the domain of construction of the η_{12}'s. This yields other problems, in the deformation process, which do not occur if we use the first approach, involving (388)'-(391)' and C depending on δ_{10}. δ_{10} will be, in the sequel, a fixed, given, small constant.

We now have:

Lemma 26. *C is given once and for all, large enough, independent of the parameters tending to zero. Assume that $O_{\bar{\varepsilon}_0}(b)$ is small enough and that $\sum\limits_j \int\limits_{\theta_{\gamma_j}(b)} (|b| - \gamma_j)^+$ is small enough ($< \varepsilon_{18}$, given, depending also on C_{41}).*

Then, there exists a function $\psi_.$ behaving as in Figs. G, H, I and J with flat pieces in between where $\psi_.$ takes the value μ_ℓ on intervals J_ℓ such that $C(|J_\ell|+\mu_\ell^2|J_\ell|)|J_\ell| \geq \varepsilon_0/2$ and another function ε which is $O(\delta_{10})$, such that:

$$|b - a\psi_. - \varepsilon a\psi_.|_{L^1} < \sum_j \int_{\theta_{\gamma_j}(b)} (|b| - a\gamma_j)^+.$$

Proof. $\psi_.$ is found through an iterative process, similar to the one described in (200), where oscillations of the family $\theta_{\gamma_j}(b)$ are removed; $\theta_{\gamma_j}(b)$ is the family of oscillations around γ_j of size at least $\frac{\delta_{10}\gamma_j}{40m}$, on intervals I satisfying:

$$C\left(|I| + \int_I b^2\right)|I| < \frac{\varepsilon_0}{800}. \tag{392}$$

Each time there is an oscillation of size larger than $4\delta_{10} \times M_j$ over some value M_j, we can squeeze a value γ_k in between and test whether this oscillation belongs to the family θ_{γ_j} or not. We observe there four key facts: first, the supports of the oscillations will satisfy 1) or 2) of Lemma 24 above or will be disjoint. Therefore, our iterative process is exclusive, the oscillations which we remove are independent.

Second, the proof of Lemma 23 shows that, given an interval J with an oscillation of size at least $\frac{\delta_{10}}{40m} \times \gamma_j$ for b, around any value γ_j, then, setting $I = \{t \in J \text{ s.t. } |b(t)| > a\gamma_j\}$ and (I_ℓ) the intervals of I: either

$$C\left(|I_\ell| + \int_{I_\ell} b^2\right)|I_\ell| > \varepsilon_0/3 \tag{393}$$

or

$$C\left(|I_\ell| + \int_{I_\ell} b^2\right)|I_\ell| < \frac{\varepsilon_0}{800} \tag{394}$$

if $O_{\bar{\varepsilon}_0}(b)$ is small enough (and γ_k is sizably away from the characteristic values of b, which we assume).

Third, assuming that removing oscillations of the families $\theta_{\gamma_1}(b)$, we find a first function $\psi_.^1$; then, we restart the same process with $\psi_.^1$ instead of b, and we build $\psi_.^2$ etc. we then build a sequence $\psi_.^\ell$. Removing an oscillation is replacing b on the related interval by γ_j; the process is very similar to the one described in the proof of Lemma 12.

It is clear that any $\delta_{10} \times \gamma_j/40m$ or more – oscillation of $\psi_.^\ell$ is a $\delta_{10} \times \gamma_j/40m$ or more oscillation of b around γ_j. (Indeed, and our arguments below elaborate about it extensively, any jump between two constant values of $\psi_.^\ell$ yields a larger jump for b; on any interval where $\psi_.^\ell$ is not constant, b is equal to $\psi_.^\ell$.) The question, which brings us to the fourth fact, is whether an oscillation which is in $\theta_{\gamma_j}(\psi_.^\ell)$ is in $\theta_{\gamma_j}(b)$

or not and this is tied to the bound on $b - a\psi_\cdot^{\ell+1} + O(\delta_{10}) \times a\psi_\cdot^{\ell+1}$, i.e. we claim that we have by induction – an induction which is obvious at its first step for $\psi_\cdot^1$–:

$$\text{Given any oscillation of size at least } \frac{\delta_{10} \times \gamma_j}{40m} \text{ of } \psi_\cdot^\ell \tag{395}$$

on an interval J around any value γ_j, denoting

$I = \{t \in J \text{ s.t. } |\psi_\cdot^\ell(t)| > a\gamma_j\}, (I_{\ell'})_{\ell'}$ the intervals of I, we have:

$|b(t)| = |\psi_\cdot^\ell(t)| = a\gamma_j$ on ∂I and we have the following equivalence:

For any given ℓ', $C(|I_{\ell'}| + \int_{I_{\ell'}} \psi_\cdot^{\ell_2})|I_{\ell'}| < \varepsilon_0/3 \Leftrightarrow$

$$C\left(|I_{\ell'}| + \int_{I_{\ell'}} b^2\right)|I_{\ell'}| < \frac{2\varepsilon_0}{4} \Leftrightarrow C\left(|I_{\ell'}| + \int_{I_{\ell'}} \psi_\cdot^{\ell_2}\right)|I_{\ell'}| < \frac{\varepsilon_0}{1600}$$

$$\Leftrightarrow C\left(|I_{\ell'}| + \int_{I_{\ell'}} \psi_\cdot^{\ell_2}\right)|I_{\ell'}| < \varepsilon_0/3.$$

Furthermore, we have:

$$\int (|b| - \psi_\cdot^\ell)^+ < \sum_j \int_{\theta_j(b)} (|b| - \gamma_j)^+. \tag{396}$$

Using (395) and (396), we can quietly continue our iterative process, removing oscillations around any value γ_j of size $\pm\delta_{10} \times \gamma_j/40m$. In the end, we find a function $\psi_\cdot^\infty$ which cannot have any such oscillation (these ones are in fact in finite number, because $\int |\dot{b}|$ is bounded) $\psi_\cdot^\infty$ behaves then, up to $2a\delta_{10} \times \gamma_j/40m$ as a function of Figures G, H, I and J. The intermediate pieces $J_{\ell'}$ must be such that $|\psi_\cdot^\infty|$ is identically equal to $a\gamma_{j(\ell')}$ on each of them (up to $\frac{2a\delta_{10}}{40m}$ again) and

$$C(1 + a^2\gamma_{k(\ell')}^2)|J_{\ell'}|^2 \geq \varepsilon_0/3$$

Setting $\psi_\cdot = \frac{\psi_\cdot^\infty}{a}$, we derive our claim. We thus only need to prove (395) and (396). Observe that, as pointed out in the proof of Lemma 12, $|\psi_\cdot^\ell(t)| = |b(t)|$ at any t such that $|\psi_\cdot^\ell(t)|$ is not equal to a value $a\gamma_j$, since we obtain $\psi_\cdot^\ell$ from b by replacing pieces with other pieces where $\psi_\cdot^\ell$ is equal to $a\gamma_j$, for a certain j. Near ∂I, where I is as in (395), $\psi_\cdot^\ell$ changes values infinitely many times, because it is a continuous function which has to tend to $a\gamma_j$ and is strictly larger than $a\gamma_j$ inside I. Therefore, there are infinitely many values of t, in any neighborhood of any point of ∂I, where $|b(t)| = |\psi_\cdot^\ell|(t)$. Thus, $(\psi_\cdot^\ell(t)| = |b(t)| = a\gamma_j$ on ∂I as claimed and we have: since $|\psi_\cdot^\ell| \leq |b|$

$$\begin{cases} I = \{t \in J \text{ s.t. } |\psi_\cdot^\ell(t)| > a\gamma_j\} \subset \\ \quad \subset \{t \in J \text{ s.t. } |b(t)| > a\gamma_j\} = I^0 \\ I \text{ is equal to the union of some of the} \\ \text{intervals which build } I^0. \end{cases} \tag{397}$$

Assume now that we have established (396). Then, $C(|I_{\ell'}| + \int_{I_L} \psi_\cdot^{\ell_2})|I_{\ell'}| < \varepsilon_0/3$ implies that $C(|I_{\ell'}| + \int_{I_{L'}} b^2)|I_{\ell'}|^2 < 2\varepsilon_0/5$ if $\sum_j \int_{\theta_j(b)} (|b| - \gamma_j)^+$ is small enough. Using then the arguments of Lemma 23, this implies that – I_ℓ is one the intervals of $\{t \in J$ s.t $|b(t)| > a\gamma_j\} = I^0$; by the above argument, J is the support of an oscillation of $\psi_\cdot^\ell$, hence of b, since $|b|_{|\partial J} = |\psi_\cdot^\ell|_{|\partial J}$, because $\psi_\cdot^\ell$ change values infinitely many times near ∂J –

$$C\left(|I_{\ell'}| + \int_{I_{L'}} b^2\right)|I_{\ell'}| < \frac{\varepsilon_0}{16000}. \tag{398}$$

Using (396) again, this implies that

$$C\left(|I_{\ell'}| + \int_{I_{L'}} \psi^{\ell^2}\right)|I_{\ell'}| < \frac{\varepsilon_0}{1600} \tag{399}$$

which implies that

$$C\left(|I_{\ell'}| + \int_{I_{L'}} \psi^{\ell_2}\right)|I_{\ell'}| < \frac{\varepsilon_0}{3} \tag{400}$$

which, using (396) again, implies that

$$C\left(|I_{\ell'}| + \int_{I_{L'}} b^2\right)|I_{\ell'}| < \frac{2\varepsilon_0}{5}, \tag{401}$$

which implies that

$$C\left(|I_{\ell'}| + \int_{I_{L'}} \psi_\cdot^{\ell_2}\right)|I_{\ell'}| < \varepsilon_0/3. \tag{402}$$

Thus, (395) follows. We are left with (396). Clearly, arguing inductively, $\psi_\cdot^\ell$ is obtained from b by replacing b by constants $a\gamma_j$ on intervals $I_{\ell'}$. These are disjoint intervals, all contained in some interval J where $\psi_\cdot^\ell$, hence b oscillates of at least $\pm\frac{a\delta_{10}\gamma_j}{40m}$ around $a\gamma_j$. Therefore,

$$\int (|b| - \psi_\cdot^\ell)^+ < \sum \int_{\theta_j(b)} (|b| - a\gamma_j)^+ \tag{403}$$

as claimed. Lemma 26 follows.

A direct way to reach the ν or $\tilde{\nu}$-stretched curves, arguments for the regularization

The arguments above are based on the assumption that $\mathcal{O}_{\varepsilon_0}(b)$ is small and that no subdivision is taking place in the large oscillations. However, we can trace back where the real problem lies in the extension of this argument when $\mathcal{O}_{\varepsilon_0}(b)$ is not small anymore.

Let us consider an oscillation $\tilde{\theta}_\gamma(s)$, around a value $a\delta$, of size $\pm ad\gamma + \delta a$, but without any requirement on $\{t \in \tilde{\theta}_\gamma(s)\, s.t\, |b(s,t)| > a\gamma\}$. Such an oscillation $\tilde{\theta}_\gamma(s)$, by the above arguments, cannot appear suddenly. It has to preexist. The only difficulty we have met above, with the oscillations of $\theta_\gamma(s)$, is with the requirement on $\{t \in \tilde{\theta}_\gamma(s)\, s.t\, |b(s,t)| > a\gamma\}$, not the other ones. Thus, we can trace back $\tilde{\theta}_\gamma(s)$ for reverse time and we can compute, as above,

$$\frac{\partial}{\partial s}\int_{\tilde{\theta}_\gamma(s)}(|b| - a\gamma)^+.$$

The same results hold. It also holds for $\frac{\partial}{\partial s}\int_{\tilde{\theta}_\gamma(s)}(|b| - a\gamma + a\frac{d\gamma}{2})^+$, since $\psi_{\theta,\varepsilon}(a\gamma - a\frac{d\gamma}{2})$ can also be assumed to be lowerbounded by a constant similar to the one used for $\psi_{\theta,\varepsilon}(a\gamma)$, in (369).

Thus,

$$\int_{\tilde{\theta}_\gamma(s)}\left(|b| - a\gamma + a\frac{d\gamma}{2}\right)^+(s)$$

is tiny if, in the past, it was tiny at any time, which would imply that the measure $\{t\, s.t\, |b|(s,t) > a\gamma\}$ is also tiny, hence that $\tilde{\theta}_\gamma(s)$ is an oscillation of $\theta_\gamma(s)$. Thus, $\tilde{\theta}_\gamma(s)$ is not in $\theta_\gamma(s)$ only if we could not make it tiny at an earlier time. We can choose then an earlier time where $\mathcal{O}_{\varepsilon_0}(b)$ is small, as small as we may wish since we know that we can assume that this happens on a non-compact flow-line (or a "nearly" non compact one). $\int_{\tilde{\theta}_\gamma}(|b| - a\gamma + a\frac{d\delta}{2})^+$ is then tiny, unless there is a sizable piece of a characteristic value - interval in $\tilde{\theta}_\gamma$, above $a\gamma$. The size of this piece, i.e. how large $C(|I| + \int_I b^2)|I|$ is, where $I \subset \tilde{\theta}_\gamma$ is the set where this characteristic value essentially lies, is measured only by ε_0. It is certainly larger than $\varepsilon_0/4$, for example. (Otherwise, using the fact that γ is non-characteristic, the oscillation of b is large).

Thus, these "bothering" oscillations can be traced back to a finite number of sizable oscillations. The assumption on $\mathcal{O}_{\varepsilon_0}(b)$ is not really needed. We can feel that it should rather follow from the fact that this oscillation was small in the past and from regularizing properties of the flow.

A direct way to arrive to the ν or $\tilde{\nu}$-stretched curves

At this point, we develop another argument which brings us to the $\tilde{\nu}$-stretched curves. We will later come back to the assumption on $\mathcal{O}_{\varepsilon_0}(b)$. This argument is as follows:

In order to stretch the curves, we would like to use the flow if *ii*) and *iii*) of Definition 2. The problem which we face is that the curve has oscillations which might jump from $|b|_\infty$ to $\tilde{\nu}$ or below $\tilde{\nu}$. Introducing an $\omega = \eta$ on such pieces might destroy either our control on $|b|_\infty$ or the one on $\int(|b| - \tilde{\nu})^+$. We want to avoid this.

We first observe that if $|b|$ goes below the level $\tilde{\nu} + \delta_{10}$ and, after a time span of ε_{19} and before $2\varepsilon_{19}$ again goes below this level after moving above $\tilde{\nu} + 2\delta_{10}$, then, we can use the cancellation flow on the related oscillation, which has strength, if there is a sizable set where $|b|$ is above $\tilde{\nu} + 2\delta_{10}$. We can make ε_{19} small, also in function of $|b|_\infty$, so that such an oscillation is of type 1).

Thus, we assume that this does not happen. $|b|$ remains above $\tilde{\nu} + \delta_{10}$ for time spans of the order of $\varepsilon_{19}/2$ in the sinks, if we are in the domain where $|b|$ is above $\tilde{\nu} + 2\delta_{10}$. We then introduce $\omega = \eta$, on these time intervals, corresponding to *ii)* and *iii)* of Definition 2, to which we add η_2 of the cancellation flow multiplied by a large constant M. We also require that every oscillation of size at least $2\delta_{10}$ is contributing an η_{12} to η_2, if it is of type 1). There could be a problem in this requirement if there were infinitely many of such oscillations. However, Lemma D', i.e the proof of the existence of a differential equation allowing to control $\int_0^1 \dot{b}^+$, extends after the introduction of ω and the multiplication with a fixed M. There is an added term in the right hand side, which is $\mathcal{O}(\varepsilon_{19}, |b|_\infty)(1 + \int_0^1 \dot{b}^+)$. All other terms are multiplied by M. Therefore, the introduction of ω and M does not yield an infinite number, at each time s, of oscillations of size at least $2\delta_{10}$. If such an infinite number develops, it is on an infinite period of time. Our argument can be built into the construction of our flows, using larger and larger balls for the norm $W^{1,1}$ (on b) in order to build η_1 and η_2.

If M is large enough (in function of ε_{19} and $|b|_\infty$ only), the control on $|b|_\infty$ is then kept, because the cancellation flow, i.e η_2, has the effect of decreasing $|b|_\infty$ substantially, see (24), on the oscillation where it acts and ω, i.e η_1, increases it in a way controlled by a function of ε_{19} and $|b|_\infty$. The introduction of such an M does not destroy the control on $\int \dot{b}^+$, since M is controlled by a function of ε_{19} and $|b|_\infty$. (see Lemma D').

ω does not destroy the control on $\int(|b| - \tilde{\nu})^+$, because its support does not touch the points where $|b|$ is less than $\tilde{\nu}$. Thus, ω can be introduced, with a companion $M\eta_2$, where M depends only on ε_{19}. As long as the companion oscillations on the support of ω are of type 1) with a small support and are of size at least δ_{10}, we can even use it on regions where $|b|$ assumes the value $\tilde{\nu}$ and keep control on $\int(|b| - \tilde{\nu})^+$, if these oscillations have $\underset{\sim}{\bar{\mu}}$ less than $\tilde{\nu} - \delta_{10}$, so that $\psi_{\theta,\varepsilon}(\tilde{\nu})$ is larger than δ_{10} and the computation in (365) - (375) holds. Then, replacing in (365) - (375) $a\gamma$ by $\tilde{\nu}$, observing that the introduction of ω perturbs our upperbounds by $\int_{I_{\tilde{\nu}}(s)} O(\varepsilon_{17}, |b|)$, observing finally that the cancellation flow provides a negative term in the upperbound of the type $-\int_{I_{\tilde{\nu}}(s)} \frac{\delta_{10}}{|x_1 - x_2|^2}$ which absorbs all the other terms because $|x_1 - x_2|$ is assumed to be small, we can take M large enough, in function of $\delta_{10}, \varepsilon_{19}$ and $|b|_\infty$, so that all our computations hold. C can be kept constant all along.

This allows to get rid of large pieces where $|b|$ is essentially less than $\tilde{\nu} - \delta_{10}$, because then $|b|$ cannot rise to $\tilde{\nu}$ on a large size interval. The result is that our curves have to become $\tilde{\nu}$-stretched with, maybe, lots of oscillations above their limit profile, of a very small L^1_- contribution. The regularizing flow takes care of them, see also point 6) below. In this way, we cannot cancel these oscillations on the whole unstable manifold. Our result, as it stands, is therefore weaker, but still leads us to the ν or $\tilde{\nu}$-stretched curves. In addition, using the partial argument which we built, we can easily get rid of the oscillations above the characteristic values, which can, all of them, be assumed to be equal to $|b|_\infty$. Indeed, any oscillation above the largest characteristic value can be tackled using $a\gamma$ larger than this characteristic value. For such $a\gamma$, we do not have any problem with our argument, as explained above. It is then clear, using the flow $\omega + M\eta_2$ which we just built, that we can bring our characteristic values to be nearly equal to $|b|_\infty$. We are left with oscillations above the $\tilde{\nu}$-pieces, or away from these characteristic values, in the sinks.

We can refine the construction of our flow as follows: either $\mathcal{O}_{\bar{\varepsilon}_0}(b)$ is small in function of a previous decrease in a, so small that the regularizing flow will tame b keeping control on $|b|_\infty(s)$ in function of $|b|_\infty(0) - c\Delta a$. Then b can be written as $\tilde{\psi}. + g$, where $\tilde{\psi}$ behaves like in Fig. G, -, Fig. J, g is small in norm L^1 and g has the same sign than $\tilde{\psi}..$ If $\tilde{\psi}.$ is not $\tilde{\nu}$-stretched (up to ε_{19} or a function of ε_{19}), we can introduce an $\omega + M\eta_2$.

Or $\mathcal{O}_{\bar{\varepsilon}_0}(b)$ is not small in the above sense i.e $\int b\eta_{12}$ is not small also, with a similar estimate. Then, instead of $\omega + M\eta_2$, we use $c_1\tilde{\delta}\omega + (1+M)\eta_2$, where c_1 is a small fixed constant which depends on the lowerbound on $\int b\eta_{12}$. $\tilde{\delta}$ is the constant used in Lemma 24 and in the definition of θ_γ, hence is given at each step. Since $\int b\eta_{12}$ is not small, we can choose c_1 so small so that the computation in (357)-(359) extend. Thus, for this part of the flow, the behavior of $\theta_\gamma(s)$ is similar to the one which we describe for the pure cancellation flow.

Because $c_1\tilde{\delta}$ is a fixed constant, our flow will bring the curves to be $\tilde{\nu}$-stretched (how much depends on ε_{19}), with characteristic values equal to $|b|_\infty$ up to δ_{10} and oscillations small in the L^1-sense above $\psi.$ for b. A finite number of large oscillations, made of nearly flat pieces, occur because of sharp oscillations downwards (below $|b|_\infty$ or $\tilde{\nu}$), even when $\mathcal{O}_{\bar{\varepsilon}_0}(b)$ is small.

A heuristic argument showing why large oscillations of type 2) cannot subdivide or shrink at infinity

We analyze what happens to $\theta_\gamma(s)$ in backwards time: If M is chosen large enough, also in function of $d\gamma$, the maxima of the oscillations of $\theta_\gamma(s)$ only move upwards for $\omega + M\eta_2$ and $\tilde{\delta}\omega + (1+M)\eta_2$. The minima, on the edges of an oscillation, for $c_1\tilde{\delta}\omega + (1+M)\eta_2$, always remain below $a(\delta - d\gamma) + \tilde{\delta}a^2$ for $c_1\tilde{\delta}\omega + (1+M)\eta_2$ if the oscillation had $\underset{\sim}{\bar{\mu}}$ less than $a(\gamma - d\gamma) + \tilde{\delta}a^2$. For $\omega + M\eta_2$, they might move upwards, but $\mathcal{O}_{\bar{\varepsilon}_0}(b)$ is extremely small already. We need only to analyze, on a given trajectory, what happens when, out of the regions where $O_{\tilde{\varepsilon}_0}(b)$ is extremely small, we use the flow corresponding to $c_1\tilde{\delta}\omega + (1+M)\eta_2$.

Then, an oscillation having $\underset{\sim}{\bar{\mu}}$ larger than $a(\gamma + d\gamma) - \tilde{\delta}a^2$ and $\underset{\sim}{\bar{\mu}}$ smaller than $a(\gamma - d\gamma) + \tilde{\delta}a^2$ can be traced back to an oscillation of the same type, but with $O_{\tilde{\varepsilon}_0}(b)$ extremely small. The computation in (365)-(375) extends to such oscillations, even after the introduction of ω and M, using $-\int_{I_\gamma} \frac{\psi_{\theta,\varepsilon}(a\delta)}{|x_1-x_2|^2}$ to absorb the contribution of ω which is $\int_{I_\gamma} O(\varepsilon_{19}, |b|)$.

Either, when $O_{\tilde{\theta}_0}(b)$ was extremely small, the integral of $(|b| - a\gamma)^+$ on this oscillation was tiny. Such oscillations, even taken globally, cannot change the estimates on $\int_{\theta_\gamma(s)}(|b| - a\gamma)^+$, because they were, then, in $\theta_\gamma(s)$. This works as long as there is no subdivision in the large oscillations (of type 2)).

These oscillations are very well controlled by the same estimates found in Lemma 24 c).

Or, when $O_{\tilde{\varepsilon}_0}(b)$ was extremely small, this oscillation was large of type 2). It might change $\theta_\gamma(s)$ only if it shrinks considerably or subdivides so as to fit one of its pieces into $\theta_\gamma(s)$. We conjecture that this does not happen and develop below a heuristic argument showing why: Coming back to the proof of Lemma D, we rewrite the computation of Δ_j. If there is an oscillation with ω which interferes with Δ_j and $|x_1 - x_2|$, for this oscillation, is small, in function of $|b|_\infty$ and ε_{19}, one can easily check that Δ_j is negative, even after the introduction of ω and M. Hence, no sharp downwards oscillation can develop in the oscillations supporting a $\psi_{\theta,\varepsilon}\eta_0$ and having $|x_1 - x_2|$ less that $c(|b|_\infty, \varepsilon_{19})$. If $|x_1 - x_2|$ is larger than $c(|b|_\infty, \varepsilon_{19})$, or if there is no η_2 then, this argument does not work. But ω, by itself, induces a speed of decrease in a larger than $c'(|b|_\infty, \varepsilon_{19}) > 0$.

On the other hand, on such oscillations, Δ_j is bounded by $O(\varepsilon_{19}, |b|_\infty)$. Thus, any sizable increase in $\int_{s_1}^{s} \Delta_j$ corresponds to a sizable decrease in a. Then, given an oscillation (upwards or downwards) and a local minimum on this oscillation, one can define the largest difference between a maximum to the right and this minimum and the largest difference between a maximum to the left and this minimum. One then takes the infimum of both these quantities, call it $\tilde{L}_j$. Then, one maximizes $\tilde{L}_j$ when this local minimum varies over the oscillation. this defines a quantity $\tilde{L}$, related to the oscillation, equal to zero if there is no local minimum. The $\tilde{L}_j's$ might not be continuous, because local minima can degenerate, but $\tilde{L}$ is continuous. Furthermore, using the estimates above described on Δ_j, one can derive similar estimates on $\frac{\partial}{\partial s}\tilde{L}$. Thus, $\tilde{L}$ cannot increase sizably without a corresponding sizable decrease in a. Thus, if there is not such a decrease, $\tilde{L}$ was larger than $\frac{3ad\gamma}{2}$ from the beginning. If a large oscillation, starting from the situation where $O_{\tilde{\varepsilon}_0}(b)$ is extremely small, is to subdivide and perturb θ_γ before shrinking, it had to have a large $\tilde{L}$. However, initially, when $O_{\tilde{\varepsilon}_0}(b)$ was extremely small, all oscillations above the nearly flat $|b|_\infty$-pieces were at most of amplitude $2\delta_{10}$, but for the ones corresponding or near to the sharp edges of the large oscillations. Thus, the only possibility on these pieces is to have oscillations from these sharp edges to travel or to pick up some strength on one side, in order to decrease the size of the large oscillation. If they stay near the sharp edges, since they are small in L^1-norm and since the other oscillations are small, they cannot induce a sizable subdivision. This corresponds to a shrinking phenomenon. On the nearly $\tilde{\nu}$-pieces, the oscillations, away from the edges of the large oscillation, were above $\tilde{\nu} - \delta_{10}$. Above $\tilde{\nu} + \delta_{10}$, every oscillation is in θ_γ. Hence, these ones are controlled and are globally very small. The control on them is given by Lemma 24 c). This extends for the nearly zero-pieces. Below $\tilde{\nu} - \delta_{10}$, the oscillations have to travel downwards to reach the level of $a\gamma$, which we can take to be less than $\tilde{\nu} - 5\delta_{10}$. This has to happen sizably inside the large oscillation, not near the edges, otherwise θ_γ will not be perturbed. Thus, the pieces of η_{12} near the edges do not interfere with the evolution of this minimum. Since this is almost the global minimum on the oscillation ($|b|$ is falling down considerably in the middle of the interval of the oscillation), we can also assume that the other η_{12}'s, which correspond to the small oscillations inside the large oscillation, vanish at this minimum. Thus, this minimum will decrease of δ_{10} under the sole influence of ω and the η_{12} of the large oscillation, which has an $|x_1 - x_2|$ lowerbounded. Thus, the rate of decrease of this minimum is $O(\varepsilon_{19}, |b|_\infty)$. The rate of decrease of a is at least $c'(\varepsilon_{19}, |b|_\infty)$. The conclusion follows: θ_γ cannot change by subdivision of the large oscillations. The only possibility is that one of the oscillations shrinks. Then oscillations were quite large, so large that their contribution to $-\frac{\partial a}{\partial s}$ was small, because $C(|I| + \int_I b^2)|I|$ was close to $\frac{\varepsilon_0}{2}$ on their support, hence we were starting not to include them in the construction of η_{12} by putting a small α_i in front of them. The oscillations interfere later with $\theta_\gamma(s)$ if their strength shrinks considerably, i.e $C(|I| + \int_I b^2)|I|$ has to become less than $\varepsilon_0/_{800}$. We claim that this cannot happen without a substantial decrease in a (in function of C_{41}, also). Indeed, on such oscillations, $|b|$ is almost constant, equal to $|b|_\infty$ or to $\tilde{\nu}$. The value of this constant changes very little on the s-intervals of time which we are considering, by Lemma 24 a), which applies; i.e $|b|_\infty$ changes very little, b remains essentially equal to $|b|_\infty$ or $\tilde{\nu}$ on such intervals, this if the decrease in a is small (even with respect to C_{41}). If such an oscillation shrinks substantially, the measure of the interval where it takes place has to shrink substantially: $\varepsilon_0/_2$ to $\varepsilon_0/_{800}$, it has to be divided by 200 to the least. However, these oscillations are due to sharp oscillations downwards of $|b|$ in its nearly flat pieces. Hence, there is a companion complement oscillation which has then to expand. This complement oscillation might have several nearly flat pieces, which stay essentially the same in the L^1-sense, but for one flat-piece which expands sizably. Denoting J the interval of this oscillation, $\int_J (|b| - a\gamma)^+$ has then increased substantially in time due to the expansion of this flat piece. The computation of (365)-(375) extends to J. If we can follow J in time, this substantial increase cannot occur without a substantial decrease in a. Thus, these oscillations cannot change nature without a large drop in a. $\theta_\gamma(s)$ remains stable or decreases. The only increases come from oscillations which were always previously in θ_γ, with $\frac{d\gamma}{2}$ instead of $d\gamma$. The above arguments are not rigorous because we could have exchanges of mass, through a complicated process, between two oscillations, thereby not allowing to split $\int_{\theta_\gamma(s)}(|b| - a\gamma)^+$ into two packs, corresponding to an oscillation and its complement oscillation. However, we feel that this indicates what happens.

A possible other way to regularize the whole process

Another way to proceed, now that we know how to reach ν or $\tilde{\nu}$-stretched curves without worrying about huge sharp oscillations, in order to get rid of them, is probably the following: we can, corresponding a previous decrease δa, add to our flow

$$\frac{\delta a}{1+s^2}\frac{b}{1+|b|_\infty^3}(s \text{ is the time on the flow-line})$$

i.e incoporate in it the regularizing process. The estimates on $|b|_\infty, \int_0^1 \dot{b}^2, \int_0^1 \dot{b}^+$ etc. are unchanged. The existence time become infinite because c is now lowerbounded by $\frac{\delta a}{1+s^2} \times \frac{1}{1+|b|_\infty^3}$. Each maximum of a sharp oscillation decreases fast as we have seen. Thus, since the time is infinite, these maxima should collapse. However, again, these oscillations could exchange mass through a complicated process. This is a point which deserves further study.

Proof of Lemma 20. We now complete the proof of Lemma 20:

1) First, given K, with the bound C_{41} on $|b|_\infty$ for every $x \in K$, we choose all the parameters of the cancellation flow $\varepsilon_{11}, \frac{1}{k}, \frac{1}{q}, \varepsilon_{12}$ so small and the parameters $\varepsilon_{16}, \varepsilon_{17}$ of the flow of ii) or iii) of Definition 2 so small, that applying the first flow and then after the second flow, any x of K decreases a tiny bit, if not a large amount, in energy. Any curve which does not decrease by the cancellation flow, even with tiny parameters must be ν-stretched (see above). Then, the second flow, of ii) or iii) of Definition 2 will make it decrease a tiny bit because the component b of the curve takes some positive time to go from a characteristic value μ_j to ν

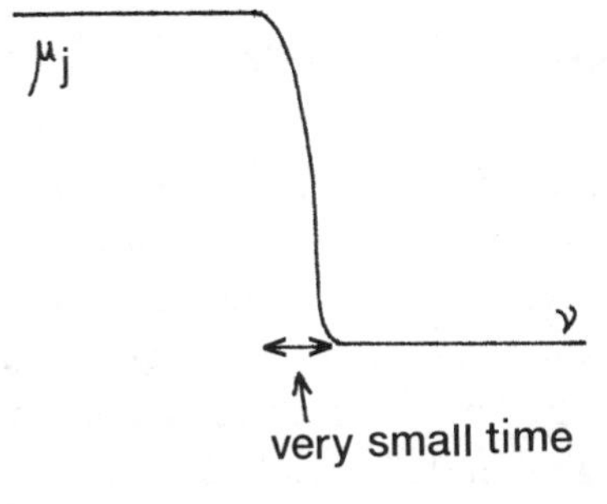

FIGURE T

During this small time, we can squeeze the flow of ii) or iii) of Definition 2 with a tiny ε_{17}. We thus, for any compact K, can decrease a a tiny bit, of $\delta(K)$, with full control on $|b|_\infty$, $\int_0^1 |b|$ etc.
2) We now use the cancellation flow with

$$\eta = \sum_i \alpha_i \left(\sum_j \widetilde{\psi}_{\tilde{\theta}_i^j, \tilde{\varepsilon}_i^j} \tilde{\eta}_{0,i,j} \right) + cb$$

c is less than $\frac{1}{(C_{41}+1)^3+1} \times \frac{\underline{\tilde{\delta}}(K)}{20} \times \frac{\delta_{10}}{C_{41}+1}$, where $\underline{\tilde{\delta}}(K) > 0$ is related to the $\delta(K)$ which has been found in the previous step. We will clarify how they are related below. δ_{10} arises in (381). Some rules, easy to follow, are to be followed, once c is chosen, for the construction of the $\widetilde{\psi}_{\tilde{\theta},\tilde{\varepsilon}}$, such as the rule related to (359). The constants in the deformation statements are however uniform, with these requirements fulfilled, as is easy to see.

The parameters of the cancellation flow are so tiny that its zero set lies in

$$\{x \text{ s.t. } a \leq a_0, |b|_\infty \leq 1 + C_{41}, b \text{ has } \mathcal{O}_{\tilde{\varepsilon}_0}(b) < \varepsilon\} = \widetilde{W}^*_\varepsilon(a_0, C_{41}) \tag{404}$$

$\varepsilon > 0$ is a given parameter. We observe that, if ε is small enough, then, for any family (γ_j) such a_0 is (388)–(389), with $\delta_{10} > 0$ given small, we have:

$$\sum_j \int_{\theta_{\gamma_j}} (|b| - \gamma_j)^+ < \frac{\underline{\tilde{\delta}}(K)}{2(1 + (1 + C_{41})^3}. \tag{405}$$

Thus, clearly, for one value γ_j, can be completed if $\mathcal{O}_{\tilde{\varepsilon}_0}(b)$ is small enough. (We come back to the definition of θ_{γ_j} and $\mathcal{O}_{\tilde{\varepsilon}_0}(b)$ to see this.) But, there are finitely many such γ_j's, for $\delta_{10} > 0$ given, in $[0, 1 + C_{41}]$. Thus, taking ε very small, we can achieve (405).

We now claim that, with this flow, we have for any $0 < c_2 < c_1$. the following:

$$J_{c_1} \cap K \text{ deforms onto a subset of } W_u(\widetilde{W}^*_\varepsilon(a_0, C_{41}) \cap J_{c_1}) \cup J_{c_2} \tag{406}$$

where W_u is the unstable manifold (of $\widetilde{W}^*_\varepsilon(a_0, C_{41}) \cap J_{c_1}$). This claim stems from a general property of pseudo-gradient and even more general flows. For pseudo-gradient flows, of Fredholm type (i.e. satisfying the Sard-Smale lemma), satisfying the Palais-Smale condition and having nondegenerate critical points or having critical points at infinity around which a Morse Lemma is available, we have established such a result in [9]. The result of [9] extends to a situation where the pseudo-gradient has a zero set which is stratified and is of Fredholm type near this zero set. The proof is involved, we skip it here.

Another approach to this result, which we will take here, derives from C. Conley and Easton's work on invariant sets (here, the zero set of the cancellation flow is the invariant set, $\widetilde{W}^*_\varepsilon(a_0, C_{41})$ is an isolating neighborhood for this invariant set), containing in an isolating neighborhood. Conley and Easton [23] build an isolating block for this situation. This isolating block $\mathcal{B}$ yields a hyperbolic neighborhood where this invariant set is the maximal invariant set S. $(\mathcal{B}, \partial\mathcal{B}^+)$ can then be continuously deformed into $(W_u(S), W_u(S) \cap \partial\mathcal{B})$ where $\partial\mathcal{B}^+$ stands for the part of $\partial\mathcal{B}$ where the flow points outwards, see Conley and Easton [23]. As pointed out above, the extension to this infinite dimensional framework is rendered possible by the viscosity coefficient $\tilde{\delta}$ which makes the exponential of the flow have nice Fredholm properties. We have to keep a J_{c_2} in the formula, with $c_2 > 0$, to avoid approaching curves having a tending to zero.

Next, we claim that, given $\varepsilon_1 > 0$ which will be chosen so that Lemma 24 applies, there exists $\bar{\delta}_2 = \delta_2(\varepsilon_1, c_1, C_{41})$ such that:

$$J_{c_1} \cap K \text{ deforms onto a subset } \mathcal{A} \text{ of } W_u(\widetilde{W}^*_\varepsilon(a_0, C_{41}) \cap J_{c_1}) \cup J_{c_1 - \bar{\delta}_2} \tag{407}$$

with

$$\begin{aligned}(\mathcal{A} - J_{c_1-\bar{\delta}_2}) \subset \{x \text{ s.t. } \int |b - \psi_\cdot| < \varepsilon_1 \text{ for a function } \psi_\cdot \text{ which behaves like}\\ \text{the functions of Fig. } 1), 2), 3), 3)' \text{ and is furthermore } \nu \text{ or } \tilde{\nu}\text{-stretched}\}.\end{aligned} \tag{408}$$

We also claim that, given ε_1 and C_{41}, there exists a constant $\bar{\bar{\delta}}_2$ independent of ε such that:

$$\bar{\delta}_2 \geq \bar{\bar{\delta}}_2 \text{ if } c_1 \text{ remains larger than any prescribed } \bar{c}_1 > 0. \tag{409}$$

This is a straightforward consequence of the arguments of Lemma 21: the tip above $J_{c_1-\bar{\delta}_2}$, for $\bar{\delta}_2$ small, must be in $\widetilde{W}^*_{\varepsilon_1}(a_0, C_{41})$, for ε_1 given. This tip is $\mathcal{A} - J_{c_1-\bar{\delta}_2}$, hence (408). (409) is straightforward, it suffices to come back to the proof of Lemma 24. Observe that

$$\mathcal{A} - J_{c_1-\bar{\delta}_2} \subset W_u(\widetilde{W}^*_\varepsilon(a_0, C_{41}) \cap J_{c_1}) - J_{c_1-\bar{\delta}_2}. \tag{410}$$

Step 3) and Step 4) below would not be needed if the addition of $\frac{\delta a}{1+s^2}\frac{b}{1+|b|^3_\infty}$ *regularized the whole process, which we do not know for the moment.*

3) Applying b) of Lemma 24 to the flow-lines of $\mathcal{A} - J_{c_1-\bar{\delta}_2}$, we pick one of them originating in x_0 with b_0 having an oscillation $\mathcal{O}_{\bar{\varepsilon}_0}(b)$ less than ε and having $|b_0|_\infty \leq C_{41}$. We can pick up values (γ_j) for b_0 away from the characteristic values of the related $\psi_\cdot$ by an amount of the size of $\delta_{10} \times \gamma_j/m$, m is an upper bound on the number of oscillations of type 2), which follows from a bound on $\int_0^1 |b|$.

c) then tells us that all of $\mathcal{A} - J_{c_1-\bar{\delta}_2}$ is contained in $\{x$ s.t. $\int|b-\psi.| < \varepsilon_1$ for a function $\psi.$ which behaves like the functions of Figures G, H, I and J and is furthermore ν or $\tilde{\nu}$-stretched $\}$. According to what we conjectured above, it is in fact contained in $\{x$ s.t.$\mathcal{O}_{\bar{\varepsilon}_0}(b) < \varepsilon_1\}$ and no large oscillation subdivides or shrinks in this set. Thus, all of Lemma 24 holds if ε_1 is well chosen, once and for all. Either $x(s,t)$ has an energy a which decreased of an amount $\varepsilon(\delta_{10}, m, C_{41})$ fixed or, combining with Lemma 26,

$$\int |b - a\psi. - O(\delta_{10}a\psi.)|(s) < \sum_j \int_{\theta_{\gamma_j}(b)} (|b| - a\gamma_j)^+(s) < \tag{411}$$

$$< \sum_j \int_{\theta_j(b_0)} (|b| - a\gamma_j)^+(0) + \frac{1+C_{41}}{\delta_{10}}(\tilde{\delta}(a(0) - a(s))$$

$$\left(\tilde{\delta} < \frac{\tilde{\delta}(K)}{20} \times \frac{\delta_{10}}{C_{41}+1} \times \frac{1}{1+(C_{41}+1)^3}\right)$$

where $\psi.$ is of the special type described in Figures G, H, I and J with flat intermediate pieces large enough, $\psi.\nu$ or $\tilde{\nu}$-stretched.

Thus, taking $\bar{\delta}_2'$ smaller than $\bar{\delta}_2$ if necessary, we derive that

$$J_{c_1} \cap K \quad \text{deforms onto some subset} \quad \mathcal{A}' \cup J_{c_1-\bar{\delta}_2'}$$

where $\mathcal{A}' \subset \Big\{x$ s.t. $|b|_\infty \le C_{41}+1$ and there exists $\psi.$ as described above s.t. $\int |b - a\psi. - O(\delta_{10}a\psi.)| < \frac{\tilde{\delta}(K)}{1+(1+C_{41})^3}\Big\}$

4) We then apply the regularizing flow to $\mathcal{A}'$, in fact to a small neighborhood $\mathcal{A}''$ of $\mathcal{A}'$, still contained $\{x$ s.t. $|b|_\infty \le C_{41}+1$ and there exists $\psi.$ as described above s.t. $\int |b - a\psi. - O(\delta_{10})| < \frac{\tilde{\delta}(K)}{1+(1+C_{41})^3}\Big\}$. The reason for taking a small neighborhood is that we need to apply the regularizing flow in $\mathcal{A}'$ during a uniform time s and progressively decrease, as we approach $\partial\mathcal{A}''$, s to zero so that the deformation extends by the identity outside of $\mathcal{A}''$. Since $\int |b - a\psi. - O(\delta_{10}a\psi.)| < \frac{\tilde{\delta}(K)}{1+(1+C_{41})^3}$ on $\mathcal{A}'$, a deformation during a small time s of b along the regularizing flow will make the oscillations of b collapse and b will basically look like $a\psi. + O(\delta_{10})$, see Lemma 13 and its estimates. In fact, using the various bounds on $\psi.$, its behavior, we can assume that, if we want s to be as small as we wish, to have taken $\underline{\tilde{\delta}}(K)$ so tiny that such an s will bring the oscillations of b to collapse. Also, playing with the smallness of $\underline{\tilde{\delta}}(K)$, we can make sure that the increase of $|b|_\infty$ in this process is controlled by $\delta(K)$, the initial decrease in a. Thus, during this process:

$$|b|_\infty(s) \le |b|_\infty(0) - \Delta a \tag{412}$$

where Δa is the total decrease since we started in K. From A' to A'', we use a time less than or equal to s. A'' is very close to A', so that (412) holds on A''. It holds outside A'' because the time is then zero. Thus, (412) holds globally, the control on $|b|_\infty$ is kept. The result is now that $J_{c_1} \cap K$ has deformed onto a subset of $\tilde{A}' \cup J_{c_1 - \bar{\delta}'_2}$, $\bar{\delta}'_2$ depending only on δ_{10}, C_{100}, C_{41}, with

$$\tilde{A}' \subset J_{c_1} \cap \{x \text{ s.t. } |b|_\infty \leq C_{41} + 1 \text{ and } b = a\psi_{\cdot} + O(\delta_{10})\psi_{\cdot}, \text{ where } \psi_{\cdot} \text{ is as above}\} \tag{413}$$

5) We can then apply the flow of ii) and iii) of Definition 2 to $b \in \tilde{A}'$. If $\psi_{\cdot}$ is not $\tilde{\nu}$-stretched, in all the intervals of size larger than ε_{17} or ε_{16} where $|b|$ is strictly less than $|b|_\infty/2$ and larger than $\tilde{\nu} + 2\delta_{10}$, or less than $\tilde{\nu} - 2\delta_{10}$, we can construct such a flow. This flow, once our curves are stretched $\tilde{\nu}$-curves, is followed by the flow of Definition 5 and the one defined by $Z_\delta(x, t_1, t_2)$. This will induce a new substantial decrease in a, unless the curves lie in $\widetilde{W}(a_0)$, described in (290). We make precise the way to proceed in order to keep ν which is involved in $\tilde{\nu} = \frac{\nu}{L}$ to be a universal constant. The process of lowering and stretching back is completed up to $0(\delta_{10})$. We can then use other smaller values δ_{11} instead of δ_{10}, with related values of $C = C(\delta_{11})$. This will allow to have smaller and smaller inessential oscillations. Ultimately, it also allows to refine more and more the neighborhoods $\widetilde{W}(a_0)$ found. It might seem that the deformation, in particular the rate of decrease and the number of essential oscillations depends on δ_{11}. But, using repeatedly, after each δ_{11}-deformation (δ_{11} tending to zero), the δ_{10}-fixed one, we can keep $\bar{\delta}'_2$, in $J_{c_1 - \bar{\delta}'_2}$, independent of δ_{11} and bring at the same time the top curves of our composition of flows, those where the flow-lines of the unstable manifold(s) originate, to have smaller and smaller inessential oscillations and ultimately to be in smaller and smaller $\widetilde{W}(a_0)$'s. The size of the jumps in b with $C = C(\delta_{10})$ depends also on C_{100}, i.e. on a bound on $\int_0^1 |b|$, which depends on ν. However, at the end of the process involving $C = C(\delta_{10})$, $C = C(\delta_{11})$, since the oscillations are now controlled, we can reapply a similar flow with $C = C_0$, independent of δ_{10}, C_{100}, etc. This new flow will keep only the nearly flat $|b|_\infty$-pieces of size at least δ_0; any other nearly flat piece, by all our previous arguments will yield a substantial decrease in a (we proceed as in Lemma 4, an $\varepsilon_1 > 0$ below the characteristic interval I associated to $C = C(\delta_{10})$, on which an η_0 corresponding to $C = C_0$ can be built, $\int_I b\eta_{12}$ is then lower bounded by $\varepsilon_1\gamma(C_{100})$, where γ is a positive function of C_{100}. If this cannot be done, then there is a characteristic value of b for $C = C_0$, which is at most ε_1 below the one corresponding to $C = C(\delta_{10})$, etc).

Composing these two flows, iteratively if necessary, we cancel the nearly flat $|b|_\infty$-pieces which do not have a δ_0-size at least. This allows to have δ_0 independent of C_{100}, hence ν, which is to be compared to δ_0, see the proof of Lemma 18, in particular ii), to be a universal constant. The flow of Definition 5 Z_δ with $\nu = \tilde{\nu}$ and of the convergence theorems (if, in particular, we use (H) is then used. Lemmas 16–18 and Lemma N apply. It is only now that we end up in the neighborhoods $\widetilde{W}(a_0)$. Our process is thereby complete. (In fact, we use $\tilde{\nu} = \frac{\nu}{L}, L$ large; hence, this problems disappears). We have established our lemma. We also now have a new compact K_1, the result of the previous deformation to which we can reapply the same process with different values of $\delta(K_1)$, $\underline{\tilde{\delta}}(K_1)$ etc. Nevertheless, $\bar{\delta}_2$ does not depend on K, as long as $|b|_\infty$ is less than $C_{41} + 1$ and $\int_0^1 |b|$ is bounded. Both properties hold because, globally on the former process, and we took very much care of this in our construction, $|b|_\infty(s)$ is controlled by $C_{41} - \Delta a$ and $\int_0^1 |b|$ is bounded. Thus, as long as $-\Delta a$ is less than 1, we can repeat our construction. Once $-\Delta a$ has reached the value 1, we have to change it but this will be at most $[a_0] + 1$ times. This justifies the construction and claims of Lemma 20. As we pointed out earlier, the lines of the deformations can be perturbed to become flow-lines of a pseudo-gradient, following the construction of [7] see (A7–A41) in particular. Our deformation process is now complete.

6) We observe furthermore that, even if the problem contained in our heuristic argument is not overcome, we still come very close to the same deformation statement: Indeed, we described above a direct way to reach the $\tilde{\nu}$-stretched curves. Combining with I.4, we progressively reach the curves which are made of ξ and $\pm v$-pieces, up to large oscillations of $|b|$ along the ξ-pieces which are L^1-small. For these oscillations $\mathcal{O}$, if they are of amplitude larger $2\delta_{10}$, we can easily show that $\int_{\mathcal{O}}(|b| - \frac{3\delta_{10}}{2})^+$ is decreasing. It suffices to notice that $|b|$, on the flat zero-pieces, has to come back fast near zero if it raised above δ_{10}. Thus, computing as in (365), we have a negative term $-\frac{\delta_{10}}{2}\int_{I_{\frac{3\delta_{10}}{2}}(s)} \frac{1}{|x_1 - x_2|^2}$ which absorbs all the other terms. Furthermore, as long as these oscillations are L^1-small, so shrinking on subdivision of oscillations of type 2) can occur among them, since they are all of type 1). The arguments of Lemma 24 c) etc. would apply if we can keep them away from the oscillations of type 2) i.e from the v-pieces. We could then regularize a whole piece of flow-line. In order to keep them away, the natural flow to use is the flow $Z_\delta(t_1, t_2, x)$, the flow used for Lemma 18 and section I.4. This flow has the required properties.

As we discussed earlier, Z_δ can be used in lieu of the cancellation / stretching flow each time all the v-pieces are larger than a given length. When they become small, then, the cancellation / stretching has an additional inwards normal which is concentrated along this small v-piece. Several of these small pieces can agglomerate to form a large v-piece. If not, the inwards normal decreases a a sizable amount and we do need any regularization.

At any rate, we can regularize all the deformation process out of curves having $\pm v$-pieces large enough. For the other ones, we need to improve the existing local analysis. This bears no difficulty and will be completed in the future.

II. The variational problem at infinity

II.1. Quantic manifolds, the second variation at infinity

The variational problem at infinity

We are repeating in what follows, in a certain way, Chapter 11 of [1], with several modifications and one key improvement which allows to give a rigorous proof to several intuitions, hints and conjectures started in [1]. In fact, the key advantage of our present paper with respect to [1] resides in three facts:

a) the deformation that we have built is locally Fredholm, while the one of [1] was not. The flow decreases also the number of zeros of b. This allows us to introduce more specific restrictions on b into our underlying space of variations, therefore brings very grounded hope to discover existence mechanisms for the periodic orbits: These, contrary to what was happening in [1], see the introduction, do now introduce, for our deformation, a difference of topology in the level sets of the functional on the space $C_\beta^k = C_\beta \cap \{b \text{ has at most } k \text{ zeros}\}$, where k is given number.

b) The variational problem at infinity, as we will prove it, is very closely tied to the full variational problem. This is the aim of the present section. We will introduce a natural set of manifolds at infinity Γ_{2k}, associated variational problems on them, and prove that the difference of the Morse indices at two critical points at infinity on Γ_k is tied in a very natural way to the same difference in the full space of variations $C_\beta^+ = \mathcal{L}_\beta \cap \{a \geq 0\}$. We will then derive the formula for the difference of topology at infinity, under no assumption if we use our flow (the one we have built earlier), under some reasonable assumptions otherwise. We will also introduce other "characteristic" manifolds $\sum_k$, related to the dynamics of α along v and show some beautiful properties linking the difference of the Morse indices of the critical points at infinity in $\sum_k$ with the same difference for the full Morse indices in an appropriate "canonical" subbundle of TC_β^+ near $\sum_k$ (the word "canonical" refers to α) and involving also the elliptic or hyperbolic (we will give a meaning to these notions for the critical points at infinity) nature of the critical points at infinity in $\sum_k$.

c) With respect to [1], we have removed several hypotheses on the dynamics of v along α, if not all. Our present paper presents a general phenomenon, which should turn out, as we described in the introduction, to explain several features of the spinorial dynamics of elementary particles in quantum theory (our framework is in fact much more general than the one of quantum theory), probably also tie these features (which mainly are expressed in the critical points at infinity) to the periodic classical motions.

We thus achieve in this section part b) of this program.

We introduce, given k an integer, two stratified sets of dimension $2k$ and k, related to ξ and v:

Definition 8'. *Γ_{2k} is a space of continuous curves with $2k$ pieces, alternatively a piece of $\pm v$-orbit (without change of sign on a given orbit) and a piece of ξ-orbit. The piece of ξ-orbits are described at a constant speed a, uniform on all the curve, while the pieces of $\pm v$-orbit are described at an infinite (constant) speed i.e. any curve of Γ_{2k} is approximated by curves of C_β such that there are k-pieces where b tends to $\pm|b|_\infty$. This parametrization will be changed later, when we will consider a curve x of Γ_k as a curve of $C_\beta^+ = \mathcal{L}_\beta \cap \{\alpha(\dot{x}) \geq 0\}$. We will also use the notation Γ_k for the same curves parametrized by arc-length or any increasing diffeomorphism of [0,1].*

Definition 9. *$\sum_k$ is the subset of Γ_{2k} built with the curve x such that any piece of $\pm v$-orbit ends at a coincidence point of the initial point it started from i.e. if the piece of v-orbit is*

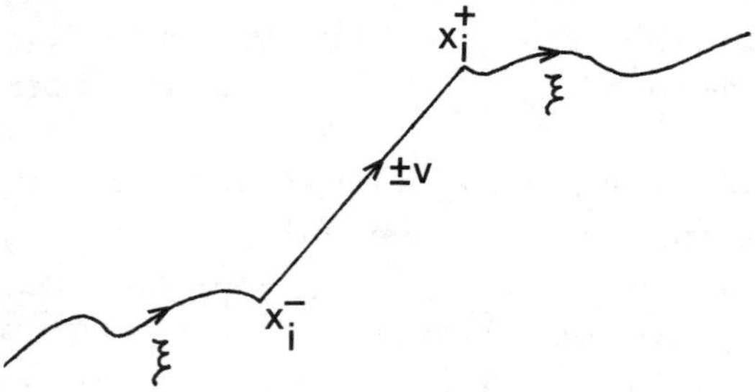

FIGURE U

then, between x_i^- and x_i^+, α has turned along the v-orbit, in the transport along v, an even number of times, as a field of planes and came back onto itself as a form, up to a multiplicative factor. Coincidence and conjugate points have been introduced and studied extensively in Chapter 1 of [1]. We refer the reader to [1] for a full and thorough study of this notion.

If ϕ_s is the one-parameter group generated by ξ, then

$$(\phi_{s_i}^*\alpha)_{x_i^-} = \lambda_i \alpha_{x_i^-}, \lambda_i > 0, \quad \text{where} \quad x_i^+ = \phi_{s_i}(x_i^-).$$

Definition 9'. *A critical point at infinity in Γ_{2k} or $\sum_k$ is a curve of Γ_{2k} such that x_i^+ is a conjugate point of x_i^- (see [1]) i.e. a coincidence point such that $\lambda_i = 1$, for each $i = 1, \ldots, k$.*

We then have:

Proposition 27. *Generically on v, i.e. up to a possible slight perturbation of v in C^∞, Γ_{2k} and $\sum_k$ are stratified sets of dimensions $2k$ and k respectively. In fact, $\Gamma_{2k} - \Gamma_{2(k-1)}$ is a manifold of* $\dim 2k$ *and $\sum_k - \sum_{k-1}$ is a manifold of dimension k.*

In the above statement, Γ_{2k} and Γ_k are parametrized by arc-length, for example.

Proof. We can start with sets $\widetilde{\Gamma}_{2k}$ and $\widetilde{\sum}_k$ of open curves, bearing similar jumps. Since these curves are not closed, their dimension is obviously $2k+3$ and $k+3$ respectively, if we do not allow degeneracies. We want now to prove that if we ask that these curves are closed, we remove precisely 3 dimensions, generically on v. Using transversality theory, we can reduce the problem to a simpler transversality result, to establish on the "Poincaré-return" map at a given closed curve of Γ_{2k} or $\sum_k$. Thus, we start with the definition of these Poincaré-return maps:

Definition of Poincaré-return maps: (Definition 10). Considering such a curve

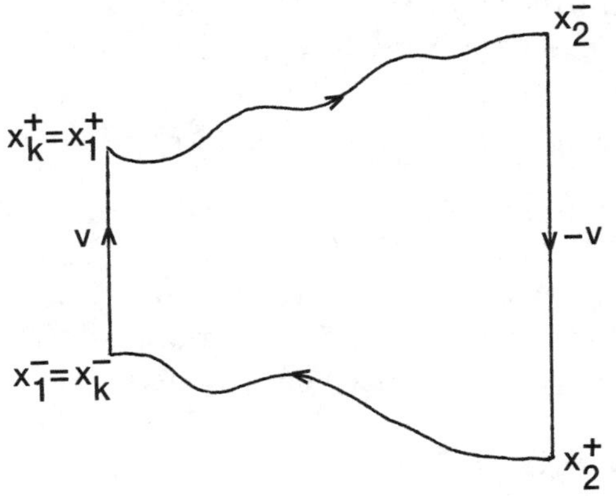

FIGURE V

we can set, for sake of simplicity, the origin of times to be at x_1^+. The Poincaré return map would then be defined to be the following map, when x is in Γ_{2k}:

We consider the $\pm v$-lengths on the curve x, denoted s_i, ϕ_s is the one parameter group of v. a_i are the ξ-lengths of the curve x, ψ_s is the one-parameter group of ξ. Then:

$$\begin{aligned} \ell_x : \quad & M^3 \longrightarrow M^3 \\ & z \longmapsto \psi_{a_k} \circ \phi_{s_k} \circ \cdots \circ \psi_{a_1} \circ \phi_{s_1} \end{aligned}$$

ℓ_x maps a section to v and ξ at x_1^+ onto itself, since $\ell_x(x_1^+) = x_1^+$.

If x is in $\sum_{2k}$, then the jumps along v are governed by relative integers m_i, where m_i designates the number of times the plane $\ker\alpha$ has rotated onto itself in the v-transport between x_i^- and x_i^+. Let θ_{m_i} be the associated map. Then, ℓ_x is defined to be:

$$\ell_x = \theta_{m_k} \circ \psi_k \circ \theta_{m_{k-1}} \circ \psi_{a_{k-1}} \circ \theta_{m_{k-2}} \circ \cdots \circ \psi_{a_1}$$

The maps θ_{m_i} are functions of the one-parameter group ϕ_s, where s depends on m_i and x, $s = s_i(x)$.

These definitions extend to open curves v of $\widetilde{\Gamma}_{2k}$ and $\widetilde{\sum}_k$, with the only difference that dl_k maps then a section to ξ and v at x_1^+ into a section of ξ and v at $\tilde{x}_1^+$, the end point of x. We then have:

Lemma 28. *Let $x \in \widetilde{\Gamma}_{2k}$ or $\widetilde{\sum}_k$. Keeping the parameters which define x unchanged (the a_i's, s_i's, x_1^+ if $x \in \widetilde{\Gamma}_{2k}$, the a_i's, m_i's, x_1^+ if $x \in \widetilde{\sum}_k$), we can perturb (C^∞) v into $\tilde{v}$, among the vector-fields in $\ker\alpha$ so that $\overline{dl}_{\bar{x}}$, where $\bar{x}$ is the new curve corresponding to the same parameters values, takes any prescribed value in a small enough neighborhood of the initial dl_x, in the bundle over $M^3 \times M^3$ whose fiber is GL_3 in the case of $\widetilde{\Gamma}_{2k}$, in the bundle over $M^3 \times M^3$ whose fiber is the subspace of GL_3 made of the linear invertible maps mapping $\ker\alpha$ at x_1^+, the first point of the product $M^3 \times M^3$ into $\ker\alpha$ at $\tilde{x}_1^+$, the second point.*

In addition, we can freely prescribe a variation $\delta\tilde{x}_1^+$ of the end point and fulfill it.

The proof of the above statement relies on the following construction and Propositions.

Let $x_0 \in M$ and let x_s be the v-orbit through x_0. Let σ_0 and σ_ε be sections to v and x_0 and at x_ε and let g: $\sigma_0 \longmapsto \sigma_\varepsilon$, $g(x_0) = x_\varepsilon$ be a local diffeomorphism.

Of special interest to us are the maps g such that

$$g = \phi + \delta g \tag{414}$$

where δg is small in the Lipschitz sense (how small depends also on ε) and ϕ: $\sigma_0 \to \sigma_\varepsilon$ is the map associated to v.

We then claim that:

Proposition 29. *v can be perturbed, in $\ker\alpha$, into a new, nonsingular vector-field $\tilde{v}$ which has also σ_0 and σ_ε as sections and x_s as an orbit. Furthermore, the associated map*

$$\tilde{\phi}: U_\varepsilon(x_0) \subset \sigma_0 \to \sigma_\varepsilon \qquad (U_\varepsilon(x_0) \text{ is a neighborhood of } x_0 \text{ in } \sigma_0)$$

coincides with g. When g is written in the form (414), with δg small, $\tilde{v}$ converges to v and $\tilde{\phi}$ to ϕ, the related map associated to v.

Proof of Proposition 29. We consider x_s, $0 \le s \le \varepsilon$, this little piece tangent to v, as a curve in $\tilde{\mathcal{L}}_\alpha$, where

$$\tilde{\mathcal{L}}_\alpha = \{x\colon\ [0,\varepsilon] \to M \quad \text{s.t.} \quad x \in H^1 \text{ and } \alpha_x(\dot{x}) \equiv 0\} \tag{415}$$

$\tilde{\mathcal{L}}_\alpha$ is a Hilbert manifold. The exponential map identifies a neighborhood of x_s in $\tilde{\mathcal{L}}_\alpha$ with the tangent space to $\tilde{\mathcal{L}}_\alpha$ to x_s. This tangent space is defined by the equation:

$$\frac{d\alpha(z)}{dt} = d\alpha(\dot{x}_s, z) = d\alpha(v, z). \tag{416}$$

Setting

$$v = \lambda\xi + \mu v + \eta w \tag{417}$$

we have

$$\dot{\overline{\lambda + \bar{\mu}\eta}} = \eta \quad \text{on} \quad [0,\varepsilon] \tag{418}$$

μ is the coordinate along v. ξ and w are transverse to v, as σ_0 and σ_ε are. We can therefore take the coordinates along $\xi(x_0)$ and $w(x_0)$ for coordinates along σ_0, near x_0 and the ones along $\xi(x_\varepsilon)$ and $w(x_\varepsilon)$ for σ_ε. Again, what we are really doing is identifying neighborhoods of x_0 and x_ε in M to neighborhoods of zero in the respective tangent spaces, and then projecting onto the (ξ, w) plane. Since σ_0 and σ_ε are transverse to v, these maps, restricted to σ_0 and to σ_ε, are invertible.

Operating in this way, the map g can be read in the (λ, η) coordinates as a diffeomorphism:

$$(\lambda_\varepsilon, \eta_\varepsilon) = g(\lambda_0, \eta_0) \tag{419}$$

or better:

$$(\lambda_\varepsilon + \bar{\mu}\eta_\varepsilon, \eta_\varepsilon) = g(\lambda_0 + \bar{\mu}\eta_0, \eta_0). \tag{420}$$

We want to solve (420) with the boundary conditions:

$$(\lambda + \bar{\mu}\eta, \eta)(0) = (\lambda_0 + \bar{\mu}\eta_0, \eta_0) \qquad (\lambda + \bar{\mu}\eta, \eta)(\varepsilon) = (\lambda_\varepsilon + \bar{\mu}\eta_\varepsilon, \eta_\varepsilon). \tag{421}$$

We will provide a solution such that

$$(\lambda + \bar{\mu}\eta, \eta)(0) \longmapsto (\lambda + \bar{\mu}\eta, \eta)(s) \tag{422}$$

is a diffeomorphism for any $s \in [0,\varepsilon]$. Let

$$x_s(\lambda_0, \eta_0) \tag{423}$$

be the associated curve in $\tilde{\mathcal{L}}_\alpha$, starting at a point near x_0 of coordinates $(\lambda_0 + \bar{\mu}\eta_0, \eta_0)$ in σ_0. $x_s(\lambda_0, \eta_0)$ is a function of s obtained through (422) and the exponential map. Clearly, (422) implies that, for ε small enough, σ_0 and σ_ε close enough, any point close to x_0 can be uniquely written as $x_s(\lambda_0, \eta_0)$ and (s, λ_0, η_0) is a local system of coordinates in a neighborhood of x_0 where σ_0 and σ_ε are included. Thus, (422) – we may always assume that v is a constant, nonzero vector-field, corresponding to the coordinate s in this neighborhood, with $s = 0$ on σ_0, $s = \varepsilon$ on σ_ε – implies that

$$\frac{\partial}{\partial s} x_s(\lambda_0, \eta_0) \tag{424}$$

defines a nonzero vector-field $\tilde{v}$ at (s, λ_0, η_0), which lies in $\ker \alpha$. Furthermore:

$$\begin{aligned} \tilde{\phi}\colon \sigma_0 &\longrightarrow \sigma_\varepsilon \\ (\lambda_0, \eta_0) &\longmapsto (\lambda_\varepsilon, \eta_\varepsilon) = \text{coordinates of } x_\varepsilon(\lambda_0, \eta_0) \text{ on } (\xi(x_\varepsilon), w(x_\varepsilon)) \end{aligned} \tag{425}$$

is precisely the map g, as claimed.

We thus only need to solve (418) together with (420) and (421).

There is an explicit solution of (418), which satisfies the boundary conditions

$$\begin{cases} \eta(s) = \eta_0 + \dfrac{s}{\varepsilon}(\eta_\varepsilon - \eta_0) + s(\varepsilon - s)\gamma \\ (\lambda + \bar{\mu}\eta)(s) = \lambda_0 + \bar{\mu}\eta_0 + \eta_0 s + \dfrac{s^2}{2\varepsilon}(\eta_\varepsilon - \eta_0) + \left(\dfrac{\varepsilon s^2}{2} - \dfrac{s^3}{s}\right)\gamma \\ \text{where} \\ \gamma = \dfrac{6}{\varepsilon^3}((\lambda_\varepsilon + \bar{\mu}\eta_\varepsilon) - (\lambda_0 + \bar{\mu}\eta_0) - \eta_0\varepsilon - \dfrac{\varepsilon}{2}(\eta_\varepsilon - \eta_0)) \end{cases} \tag{426}$$

Observe that

$$\begin{aligned} &|(\lambda + \bar{\mu}\eta, \eta)(s) - (\lambda' + \bar{\mu}\eta', \eta')(s) \le C\{|(\lambda_0 + \bar{\mu}\eta_0) - (\lambda'_0 + \bar{\mu}\eta'_0)| + \\ &+|(\lambda_\varepsilon + \bar{\mu}\eta_\varepsilon) - (\lambda'_\varepsilon + \bar{\mu}\eta'_\varepsilon)| + |\eta_0 - \eta'_0| + |\eta_\varepsilon - \eta'_\varepsilon|\}\left(1 + \frac{1}{\varepsilon}\right) \end{aligned} \tag{427}$$

where C is a universal constant.

Since $|Dg|_\infty$ is bounded, we have:

$$\begin{aligned} &|(\lambda_\varepsilon + \bar{\mu}\eta_\varepsilon) - (\lambda'_\varepsilon + \bar{\mu}\eta'_\varepsilon)| + |\eta_\varepsilon - \eta'_\varepsilon| \\ \le &C(g)|(\lambda_0 + \bar{\mu}\eta_0) - (\lambda'_0 + \bar{\mu}\eta'_0)| + |\eta_0 - \eta'_0|. \end{aligned} \tag{428}$$

Thus

$$\begin{aligned} &|(\lambda + \bar{\mu}\eta, \eta)(s) - (\lambda' + \bar{\mu}\eta', \eta')(s)| \\ \le &C(1 + C(g))(|(\lambda_0 + \bar{\mu}\eta_0) - (\lambda'_0 + \bar{\mu}\eta'_0)| + |\eta_0 - \eta'_0|)\left(1 + \frac{1}{\varepsilon}\right). \end{aligned} \tag{429}$$

This shows the continuity of the map. The differentiability is easy to prove, in a similar way.

In order to prove the invertibility of this map, we first observe that, writing $g = \phi + \delta g$, then, for ε tending to zero, ϕ tends to the identity, thus g is a perturbation of the identity. Thus, all the differences:

$$(\eta_\varepsilon - \eta_0), \qquad (\lambda_\varepsilon + \bar{\mu}\eta_\varepsilon) - (\lambda_0 + \bar{\mu}\eta_0) \tag{430}$$

are small, in the C^1-sense.

Therefore, the map $\tilde{\phi}$ is very close in the C^1-sense to:

$$(\lambda_0 + \bar{\mu}\eta_0, \eta_0) \longmapsto (\lambda_0 + \bar{\mu}\eta_0 + O(\varepsilon)\eta_0, \eta_0(1 + O(\varepsilon))) \tag{431}$$

provided g is close to the identity in the C^1-sense. How close g should be to the identity depends on ε. Clearly, the above map is invertible. So is $\tilde{\phi}$. (422) follows.

Clearly, we can modify slightly x_ε into $x_\varepsilon + \delta x_\varepsilon$. The argument is basically unchanged. We thus have:

Proposition 30. *Given $g = \phi + \delta g$, where $|\delta g|_{C^1}$ is small but $g(x_0)$ is not necessarily x_ε anymore, Proposition 29 extends.*

Going back to the proof of Proposition 27, we now prove, for example, that $\sum_k - \sum_{k-1}$ is a (k)-dimensional manifold since this is the more involved case.

Let V^∞ be the set of C^∞-vector-fields of $\ker \alpha$ which are nonsingular. Another parameter set is $a_1, \dots, a_k$. On a given open or closed curve x, which behaves as follows:

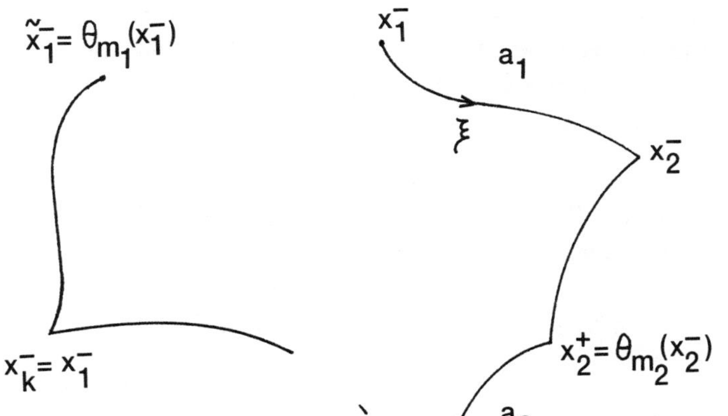

FIGURE W

we can define at $\tilde{x}_1^-$ k vectors $\xi_1, \dots, \xi_k$ which are the images, respectively, of ξ at x_{i+1}^- through the natural differential of the map associated to x from x_{i+1}^- to $\tilde{x}_1^-$: from x_2^- to $\tilde{x}_1^-$, the map is $\theta_{m_1} \circ \cdots \circ \psi_{a_2} \circ \theta_{m_2}$ etc. These maps are denoted dl_z^i. Assume that $\tilde{x}_1^-$ is close to x_1^+. In a small enough neighborhood of any point z of M^3, we can define, for y in this neighborhood, a C^∞ map

$$B_{y,z}\colon T_zM \to T_yM$$

which maps $\ker \alpha_z$ onto $\ker \alpha_y$ and satisfies the equation:

$$B_{zz} = Id.$$

Let

$$\Gamma_z = \{\text{linear maps from } z \text{ to } z \text{ mapping } \ker \alpha_z \text{ into } \ker \alpha_z\}$$

Γ_z varies when z varies but can be brought back to a single Γ, through a trivialization.

Let $\mathbb{R}_*^+$ be the set of positive numbers and let L be the following map:

$$V^\infty \times (\mathbb{R}_*^+)^k \times \mathbb{R}^k \times TM \xrightarrow{L} M \times \Gamma \times TM \times (\mathbb{R}_*^+)^k \times \mathbb{R}^k$$

$$\begin{aligned}&(v, a_1, \dots, a_k, \delta a_1, \dots, \delta a_k, z, u_z)\\ \longrightarrow &(z', B_{z',z'}^{-1} \circ dl_z, z, u_z, a_1, \dots, a_k, \delta a_1, \dots, \delta a_k)\end{aligned}$$

where dl_z is the differential of the Poincaré-map starting at $x_1^+ = z$ with the values a_i and the rotation numbers m_i and $\tilde{x}_1^- = z'$ is the endpoint of the associated curve. (We are assuming that $\tilde{x}_1^- = z'$ is close to z, restricting if necessary V^∞ and $a_1, \dots, a_k$ so that this property holds.) Our previous lemma tells us that dL is onto. Let us, then, consider the subset Ω of $M \times \Gamma \times TM \times (\mathbb{R}_*^+)^k \times \mathbb{R}^k$ defined by the equation:

$$z' = z$$

Ω is clearly a submanifold of $M \times \Gamma \times TM \times (\mathbb{R}_*^+)^k \times \mathbb{R}^k$. Since dL is onto, we can choose, by the transversality theorem, v in V^∞ so that the evaluation map:

$$L_v\colon (\mathbb{R}_*^+)^k \times \mathbb{R}^k \times TM \to M \times \Gamma \times TM \times (\mathbb{R}_*^+)^k \times \mathbb{R}^k$$

is transverse to Ω.

Thus, $L_v^{-1}(\Omega)$ is a submanifold of $(\mathbb{R}_*^+)^k \times \mathbb{R}^k \times TM$. It is defined by the equation $z' = z$. Therefore, this is precisely $\sum_k - \sum_{k-1}$. The result follows.

The above map L, which has an onto differential dL at every point, can be used in several other ways, a much weaker map is needed in order to prove that $\sum_k - \sum_{k-1}$ is a manifold. However, L and L_v are useful for other genericity statements, below.

Propositions 28 and 29 are very strong. They tell us that, along the v-pieces, any differential map g can basically be achieved. This allows the largest possible freedom for the critical curves, the second variation etc., as well as other situations. (We will encounter later what we call quantic, stable and "metastable" periodic orbits and we will claim results on the second variation along such curves. They will be derived in a similar way.) For example, we claim below that the critical points at infinity are nondegenerate. These initial points at infinity are defined by the equations:

$$\lambda_i = 1. \tag{432}$$

The second variation along a tangent vector z is built out of the variation of λ_i along $z(x_i^-)$. Clearly, using Propositions 28 and 29, using also the rotation property of $\ker\alpha$ along any vector-field v of $\ker\alpha$, see [1] Proposition 0.3, we can have for $\{x \mid \lambda_i(x) = 1\}$ any hypersurface in a neighborhood of a given one: i.e. given v and $\tilde{\tilde{\Gamma}}_i = \{x \mid \lambda_i(x) = 1\}$ for v and given $\tilde{\tilde{\Gamma}}_i^\varepsilon$ close enough to Γ_i, v can be perturbed into $\tilde{v}$ having $\tilde{\tilde{\Gamma}} = \tilde{\tilde{\Gamma}}_i^\varepsilon$. $d\lambda_i$, the differential of λ_i is also a free parameter, by Propositions 28 and 29.

Of course, the tangent space to $\Gamma_{2k}, \sum_k$ etc. changes when v changes, thus, λ_i, $d\lambda_i$ is free, but $z(x_i^-)$ etc. is also changing. However, the choice of g in Propositions 28 and 29 is completely free, provided $g = \phi + \delta g$, δg small in an appropriate sense. Thus, we can keep $z(x_i^-)$ unchanged, perturb the v-piece out of x_i^- to x_i^+, keeping x_i^-, x_i^+ unchanged, so that the new tangent vector $\tilde{z}$, corresponding in a "natural" way to z has still $\tilde{z}(x_i) = z(x_i^-)$ and even $\tilde{z}(x_i^+) = z(x_i^+)$ ("natural" means that $\tilde{z}$ is defined by the same $z(x_1^-)$ and the same δa_i's) but that $d\tilde{\lambda}_i(\tilde{z}(x_i^+))$, the corresponding variation of λ_i which we have to take along $\tilde{z}(x_i^+)$, the image of $z(x_i^-)$ though $d\theta_{m_i}(x_i^-)$, changes freely.

Thus, what we are doing is keeping $x_i^-, x_i^+, a_i, \delta a_i, z(x_i^-), z(x_i^+)$ unchanged, but changing $d\lambda_{i_0}(z(x_{i_0}^+))$. Since $d\lambda_i$ can be changed by perturbing the second variation of g (the tangent space to $\tilde{\tilde{\Gamma}}_i$ at a given point depends on the second variation of g. We can perturb it as we please, using this freedom), while $z(x_i^-)$, $z(x_i^+)$ etc. depend only on the first variation of the maps ψ_{a_i}, θ_{a_i} along the curve, we have this freedom.

This kind of argument, which we quote below, allows to prove that the critical points at infinity are locally isolated and that they are nondegenerate.

We quote them below without entering into lengthy technical proofs.

Variation and appropriate bundles

On Γ_{2k} as well as $\sum_k$, a natural functional is defined by the formula $\sum_1^k a_i$. This functional is equal to

$$J(x) = \int_0^1 \alpha_x(\dot{x})dt \quad \text{on} \quad C_\beta^+ = \{x \in \mathcal{L}_\beta \text{ s.t. } \alpha_x(\dot{x}) \geq 0\}. \tag{433}$$

We, therefore, will simply call it $J_{|\Gamma_{2k}}$ or $J_{|\sum_k}$, although, strictly speaking, J has been previously defined only on C_β.

Γ_{2k} and $\sum_k$ are in C_β^+, hence C_β comes very close to them. As we have seen previously, our critical points at infinity do not converge in C_β, but rather in C_β^+ to a curve of Γ_{2k} or $\sum_k$, for some integer k.

It is therefore very reasonable to try to understand the behavior of J near Γ_{2k} or $\sum_k$ in C_β or C_β^+ as well as to try to compare the difference of the Morse indices at two critical points at infinity in C_β^+ and in Γ_{2k} or $\sum_k$. We will complete the first part of this program later. We focus now on the second part, that is on the Morse indices.

In order to define them, we need to introduce bundles over Γ_{2k} and $\sum_k$ which are "tangent" to C_β^+ and compute the second differential of J along variations parallel to this bundle.

We will complete this program for Γ_{2k}, first, since it is easier. A curve x of Γ_{2k} is defined by $(x_1^+, a_1, \dots, a_k, s_1, \dots, s_k)$:

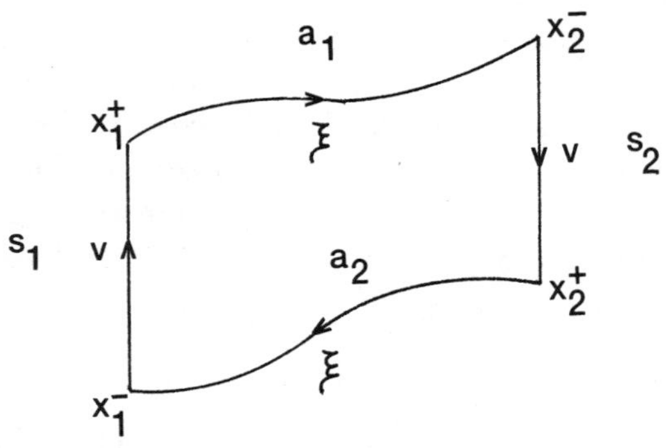

FIGURE X

Along the pieces $[x_i^+, x_{i+1}^-]$, which are tangent to ξ, the variations are free, under the condition that they stay in C_β^+, i.e. that $\dot{x}$ splits on ξ and v. This translates, on these time-intervals in the restriction:

$$\dot{\eta} = \mu a - \lambda b = \mu a \quad \text{on} \quad [x_i^+, x_{i+1}^-].$$

Also, for us, the curves of Γ_{2k} are limit-curves of C_β^+, i.e. they are curves which should be described with a constant speed a along ξ. Since the component along v, on the $[x_i^-, x_i^+]$ pieces, is infinite, a cancels, but not on the $[x_i^+, x_{i+1}^-]$. Therefore, we should add to the above equation, the equation:

$$\dot{\overline{\lambda + \bar{\mu}\eta}} = b\eta - C_i = -C_i \quad \text{on} \quad [x_i^+, x_{i+1}^-]$$

C_i depends on $[x_i^+, x_{i+1}^-]$, because there is a hidden contribution of the intervals $[x_i^-, x_i^+]$, in a time zero, which forbids to work now with a uniform constant. This is only formal and will be discussed later. Hence,

$$\begin{cases} \dot{\eta} = \mu a \\ \qquad\qquad\qquad\qquad \text{on } [x_i^+, x_{i+1}^-] \\ \overline{\lambda + \bar{\mu}\eta}^{\,\cdot} = -C_i \end{cases} \tag{434}$$

In practice, as we will see, the second equation plays no role. We could work with the functional $J(x) = \int_0^1 a\,dt$ on C_β^+, and then the second condition disappears. We keep it for the moment and we will discuss it more later.

Along the v-pieces, $[x_i^-, x_i^+]$, two types of variations are possible: the ones which are inwards C_β^+ or C_β and, therefore, will bring $\alpha_x(\dot{x})$ to become positive (or non-identically zero and nonnegative) along this portion; they are the ones we will be studying later; the other ones are tangent to the "boundary" of C_β^+ or C_β, i.e. $\alpha_x(\dot{x})$ remains zero on these pieces, the curve remains tangent to v, with possibly a time change, from s_i to $s_i + \delta s_i$. These variations obey the differential equation:

$$\begin{cases} \dot{\eta} = -\lambda b \\ \overline{\lambda + \bar{\mu}\eta}^{\,\cdot} = b\eta \end{cases}$$

Since b is infinite (on our critical points at infinity, b tends to infinity on these pieces like a constant) and is constant, we can divide by b and assume, for sake of simplicity, that the curve is directed by v (or $-v$, v for simplicity) on these pieces. If b is or were not constant, we could reparametrize by arc-length along these pieces. The equation then becomes:

$$\begin{cases} \dot{\eta} = -\lambda \\ \qquad\qquad\qquad\qquad \text{on } [x_i^-, x_i^+] \\ \overline{\lambda + \bar{\mu}\eta}^{\,\cdot} = \eta \end{cases} \tag{435}$$

The time-derivative, now, is along $\pm v$. In addition, there is a freedom on the variations of s_i along the pieces, which we can incorporate to (435) in various ways, depending on our choice, would we want to keep b constant on these pieces or leave it free to vary among all functions, which is more adequate to fill the variations of C_β^+ and C_β. Thus:

$$\dot{\mu} + a\eta\tau - b\eta\bar{\mu}_\xi = \dot{\mu} - b\eta\bar{\mu}_\xi \text{ arbitrary}$$

Dividing by b, and using then $\pm v$, we have:

$$\dot{\mu} - \eta\bar{\mu}_\xi \text{ arbitrary.} \tag{436}$$

Combining (434), (435) and (436) and asking that (λ, μ, η) are H^1-functions, which are periodic on $[0, 1]$ (after proper parametrization), we thus define a "vector-bundle" over Γ_{2k}.

$$\mathcal{H}_x \to \mathcal{H} \xrightarrow{p} \Gamma_{2k}. \tag{437}$$

We will complete a similar construction for $\sum_k$ later. There are two remarkable subbundles of $\mathcal{H}$: One is $T\Gamma_{2k}$ and is defined with the additional equation:

$$\dot{\mu} + a\eta\tau = 0 \quad \text{on the } [x_i^+, x_{i+1}^-] \text{ pieces.} \tag{438}$$

We have already seen that, after possibly perturbing v, this defines a vector-bundle of dimension $4k$ ($2k$ is the dimension of Γ_{2k}) if no s_i and a_i is zero.

The second one is denoted H: one first notices that μ is arbitrary (with the regularity H^1) on the $[x_i^+, x_{i+1}^-]$ pieces. The only constraints on the values of μ in such intervals occur at the boundaries, where it has to fit the boundary values for μ on the $[x_i^-, x_i^+]$ pieces; furthermore, μ, even on the $[x_i^-, x_i^+]$ pieces, is untied from (435), i.e. from $(\lambda + \bar{\mu}\eta, \eta)$. It is like a free variable. We define H by asking that

$$\mu_{|[x_i^+, x_{i+1}^-]} \in H_0^1[x_i^+, x_{i+1}^-] = \{\text{space of } H^1\text{-functions} \tag{439}$$

with Dirichlet boundary conditions}.

We then claim:

Proposition 30′. *The critical points of J on $\Gamma_{2k} - \Gamma_{2k-2}$ which induce a difference of topology in the level sets of J are among the critical points at infinity of J which belong to Γ_{2k}. Furthermore, possibly at the expense of perturbing slightly v, we have, at each such curve x_∞ a splitting of $\mathcal{H}_{x_\infty} = H_{x_\infty} \oplus T_{x_\infty}\Gamma_{2k}$.*

Unfortunately, such a splitting cannot be extended globally to $\Gamma_{2k} - \Gamma_{2k-2}$, at least without further study. Some subsets of Γ_{2k} seem to forbid the splitting to globalize. Since H_x is quite remarkable, it is natural to try to find a global supplement to H. This is completed as follows:

For each x_i^-, x_i^+, we can construct a μ such that

$\mu = \mu_i^\pm = 1$ at $x_i^\pm$

$(\lambda + \bar{\mu}\eta, \eta)$ are zero on the $[x_j^-, x_j^+]$ pieces and

also on all $[x_j^+, x_{j+1}^-]$ pieces for $j \neq i, \lambda + \bar{\mu}\eta$ is zero

also on $[x_i^+, x_{i+1}^-]$ (or $[x_i^+, x_i^-]$). Since $\dot{\eta} = \mu a, \eta$ cannot be zero on such

an interval ($\mu = \mu_i^\pm = 1$ at $x_i^\pm$), but is is built so that it

cancels μ and η on these intervals after a small bump.

$\eta_i = \eta$ is chosen once and for all on these intervals, for $\mu_i^\pm = 1 \cdot \eta(x_i^+)$

is zero as well as $\eta(x_{i+1}^-)$.

Finally, μ on $[x_j^-, x_j^+]$ is zero but on $[x_i^-, x_i^+]$ where

it is chosen to fit the boundary values of μ, once and for all again

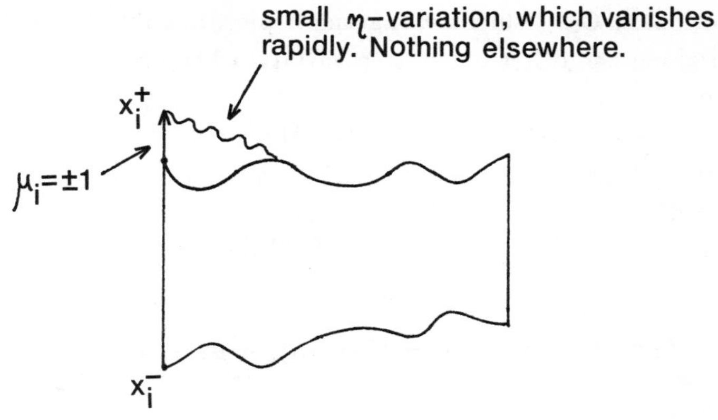

(440)

This builds a $2k$-dimensional subbundle of $\mathcal{H}$ which we denote

$$V_{2k}. \tag{441}$$

We then have:

Proposition 31.

$$H \oplus V_{2k} = \mathcal{H} \quad on \quad \Gamma_{2k} - \Gamma_{2k-2}.$$

We will prove these two propositions later. Let us now proceed with the computation of the second variation of J along $\mathcal{H}$. We have already completed this computation in [1], Chapter 11. Some details are changed here (the function $b_0(s)$ is shown to be zero. A sign mistake is corrected.). We repeat the whole argument: Let x be a curve of C_β^+, made of k pieces tangent to $\pm v$ and k other pieces where a is positive

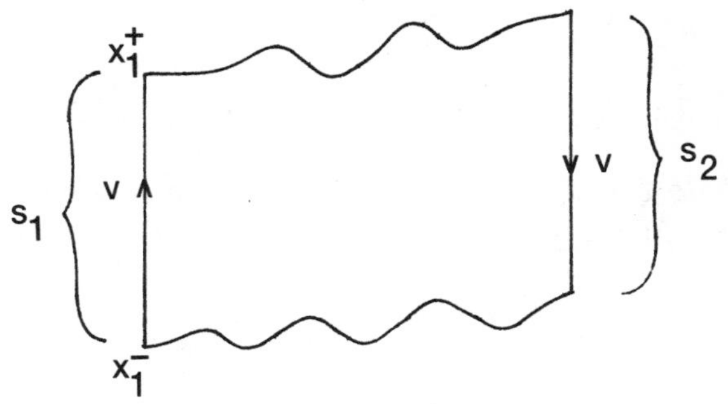

FIGURE Y

Let z be a variation in C_β^+, which, along the v-pieces, only shortens or lengthens these pieces and along the other pieces is a variation keeping $\dot{x}$ in the (ξ, v) plane.

We thus have:

$$z(x_i^+) = D\phi_{s_i}(z(x_i^-)) + \delta s_i v. \tag{442}$$

Since

$$J(x) = \int_0^1 a\,dt = \int_0^1 \alpha_x(\dot{x})dt,$$

we have:

$$(z = \lambda\xi + \mu v + \eta w = (\lambda + \bar{\mu}\eta)\xi + \mu v - \eta[\xi, v])$$

$$\partial J(x) \cdot z = \int_0^1 \left(\frac{d\alpha(z)}{dt} - d\alpha(\dot{x}, z)\right) dt \tag{443}$$

$$= \sum((\lambda + \bar{\mu}\eta)(x_i^-) - (\lambda + \bar{\mu}\eta)(x_i^+)) - \sum \int_{[x_i^-, x_{i+1}^-]} b\eta.$$

Indeed, since z preserves the v-pieces, up to contraction or dilation, $\dot{\overline{\lambda + \bar{\mu}\eta}} = b\eta$, i.e. $\frac{d\alpha(z)}{dt} - d\alpha(\dot{x}, z) = 0$ on these pieces, which yields (443).

$(\lambda + \bar{\mu}\eta)(x_i^+)$ can be derived, using (435): One introduces the coordinates $\begin{bmatrix} A(s) \\ B(s) \\ C(s) \end{bmatrix}$ of a transported vector, by v, in the $(\xi, v, -[\xi, v])$ frame. We have, using the time s along v:

$$\frac{d}{ds}\alpha(z(s)) = d\alpha(v, z(s)) \qquad (d\alpha(v, -[\xi, v]) = \beta(w) = 1) \tag{444}$$

i.e. $\dot{A} = C(= \eta)$. (There was a sign mistake, here, in [1], Chapter 11.)

$$\frac{d}{ds}\beta(z(s)) = d\beta(v, z(s)) \qquad (d\beta(v, \xi) = -\beta([v, \xi]) = d\alpha(v, [\xi, v]) = -1, \tag{445}$$

i.e.

$$\dot{C} = -A + \bar{\mu}C \quad d\beta(v, -[\xi, v]) = -d\alpha(v, [v, -[\xi, v]]) = \bar{\mu})$$

N.B. (Since $\alpha \wedge d\alpha[\xi, v, [\xi, v]) = \beta \wedge d\beta(\xi, v, [\xi, v])$, we see that $d\alpha(v, [\xi, v]) = \beta([\xi, v])d\beta(\xi, v) = -\beta([\xi, v])^2 = -d\alpha(v, [\xi, v])^2$. Thus, $d\alpha(v, [\xi, v]) = -1)$

$$\frac{d}{ds}\gamma(z(s)) = d\gamma(v, z(s)) \tag{446}$$

i.e.

$$\dot{B} = +C\bar{\mu}_\xi.$$

Thus

$$\overline{\begin{pmatrix} A \\ B \\ C \end{pmatrix}}^{\cdot} = \Gamma \begin{pmatrix} A \\ B \\ C \end{pmatrix} = \begin{pmatrix} 0 & 0 & 1 \\ 0 & 0 & -\bar{\mu}_\xi \\ -1 & 0 & \bar{\mu} \end{pmatrix} \begin{pmatrix} A \\ B \\ C \end{pmatrix}. \tag{447}$$

Let V be the resolvent matrix of (447):

$$\dot{V} = \Gamma V \qquad V(0) = Id. \tag{448}$$

V is a matrix dependent on s and on the initial point y. Thus

$$V = V(s,y) = \begin{pmatrix} a_0(s,y) & b_0(s,y) & c_0(s,y) \\ a_1(s,y) & b_1(s,y) & c_1(s,y) \\ a_2(s,y) & b_2(s,y) & c_2(s,y) \end{pmatrix} \tag{449}$$

and

$$\begin{pmatrix} A \\ B \\ C \end{pmatrix}(s) = V(s, x_i^-) \begin{pmatrix} A \\ B \\ C \end{pmatrix}(0) \tag{450}$$

Hence, in matricial writing:

$$z(x_i^+) = V(s_i, x_i^-) z(x_i^-) + \delta s_i \begin{bmatrix} 0 \\ 1 \\ 0 \end{bmatrix}. \tag{451}$$

Thus:

$$(\lambda + \bar{\mu}\eta)(x_i^+) = a_0(s_i, x_i^-)(\lambda + \bar{\mu}\eta)(x_i^-) + b_0(s_i, x_i^-)\mu(x_i^-) + c_0(s_i, x_i^-)\eta(x_i^-). \tag{452}$$

Observe that, by (448):

$$\begin{cases} \dot{b}_0 = b_2 & b_0(0) = b_2(0) = 0 \\ \dot{b}_2 = -b_0 + \bar{\mu} b_2 & \end{cases} \tag{453}$$

Thus,

$$b_0 = b_2 = 0. \tag{454}$$

(This is an improvement with respect to [1]) and

$$\begin{aligned} &-((\lambda + \bar{\mu}\eta)(x_i^+) - (\lambda + \bar{\mu}\eta)(x_i^-)) \\ &= -((a_0(s_i, x_i^-) - 1)(\lambda + \bar{\mu}\eta)(x_i^-) + c_0(s_i, x_i^-)\eta(x_i^-)). \end{aligned} \tag{455}$$

Clearly, the second variation of $-\int_{[x_i^+, x_{i+1}^-]} b\eta$ should be:

$$-\int_{[x_i^+, x_{i+1}^-]} (\dot{\mu} + a\eta\tau - b\eta\bar{\mu}_\xi)\eta \tag{456}$$

since the variation of b along these pieces is

$$(\dot{\mu} + a\eta\tau - b\eta\bar{\mu}_\xi). \tag{457}$$

There is another term equal to $-\int_{[x_i^+, x_{i+1}^-]} bz \cdot \eta$, representing the variation of z along itself (where we extend v). It is not intrinsic, but cancels on the curves on Γ_{2k}, since b is then zero on these pieces. Thus, on the curves of Γ_{2k}, the second variation along the $[x_i^+, x_{i+1}^-]$ pieces is:

$$-\int_{[x_i^+, x_{i+1}^-]} (\dot{\mu} + a\eta\tau)\eta. \tag{458}$$

On the $[x_i^-, x_i^+]$-pieces, we use (452) and we introduce the following natural quadratic form:

$$\begin{aligned} &-\left(\left(\frac{\partial a_0}{\partial s}\Big|_{(s_i, x_i^-)}(\lambda + \dot{\mu}\eta)(x_i^-) + \frac{\partial c_0}{\partial s}\Big|_{(s_i, x_i^-)}\eta(x_i^-)\right)\delta s_i + \right. \\ &\left. + \left(\frac{\partial a_0}{\partial x_i}\Big|_{(s_i, x_i^-)}(\lambda + \bar{\mu}\eta(x_i^-)) + \frac{\partial c_0}{\partial x_i^-}\Big|_{(s_i, x_i^-)}\eta(x_i^-)\right) \cdot z(x_i^-)\right). \end{aligned} \tag{459}$$

Again, there would naturally be another contribution along the self-derivative of z with z, but at a critical point at infinity, $a_0(s_i, x_i^-)$ is 1 and $c_0(s_i, x_i^-)$ is zero. Hence, this term disappears.

Observe that $a_0(s, x_i^-)(\lambda + \bar{\mu}\eta)(x_i^-) + c_0(s)\eta(x_i^-) = A(s)$. Thus, by (444):

$$\frac{\partial}{\partial s} a_0(s, x_i^-)(\lambda + \bar{\mu}\eta)(x_i^-) + \frac{\partial c_0}{\partial s}\eta(x_i^-) = \eta(s) \tag{460}$$

(460) will be used later. We have established that, at a critical point at infinity x_∞

in Γ_{2k}, the second variation is (along z)

$$\partial^2 J(x_\infty)\cdot z\cdot z = \tag{461}$$
$$= -\sum_i \left(\left(\frac{\partial a_0}{\partial s}(\lambda+\bar{\mu}\eta)(x_i^-) + \frac{\partial c_0}{\partial s}\eta(x_i^-) \right) \delta s_i + \right.$$
$$\left. + \left(\frac{\partial a_0}{\partial x_i^-}(\lambda+\bar{\mu}\eta)(x_i^-) + \frac{\partial c_0}{\partial x_i^-}\eta(x_i^-) \right) \cdot z(x_i^-) \right) \Big|_{(s_i, x_i^-)} \Bigg) -$$
$$-\sum_i \int\limits_{[x_i^+, x_{i+1}^-]} (\dot{\mu} + a\eta\tau)\eta$$
$$= -\sum_i \left(\eta(s_i^+)\delta s_i + \left(\frac{\partial a_0}{\partial x_i^-}(\lambda+\bar{\mu}\eta)(x_i^-) + \right.\right.$$
$$\left. + \frac{\partial c_0}{\partial x_i^-}\eta(x_i^-) \right) \cdot ((\lambda+\bar{\mu}\eta)(x_i^-)\xi(x_i^-) - \eta[\xi, v](x_i^-)) \Bigg) + \mu\eta \Bigg|_{x_i^+}^{x_{i+1}^-} +$$
$$+ \mu(x_i^-)v(x_i^-)\cdot(a_0(\lambda+\bar{\mu}\eta)(x_i^-) + c_0\eta(x_i^-) \Bigg) +$$
$$+ \sum_i a \int\limits_{[x_i^+, x_{i+1}^-]} (\dot{\eta}^2 - a^2\eta^2\tau).$$

Observe now that:

$$v(x_i^-)\cdot(a_0(\lambda+\bar{\mu}\eta)(x_i^-) + c_0(\eta(x_i^-)) \tag{462}$$
$$= v(x_i^-)\cdot \alpha_{\phi(s_i, x_i^-)}(D\phi_{s_i}(\lambda+\bar{\mu}\eta)(x_i^-)\xi - \eta(x_i^-)[\xi, v]))$$

We know that for any form θ and for any $\tilde{v}, \tilde{w}$:

$$\tilde{v}\cdot\theta(\tilde{w}) = \tilde{w}\cdot\theta(\tilde{v}) + \alpha([\tilde{v}, \tilde{w}]) + d\alpha(\tilde{v}, \tilde{w}). \tag{463}$$

Thus, since $\alpha(v) = 0$:

$$v(x_i^-)\cdot\phi_{s_i}^*\alpha((\lambda+\bar{\mu}\eta)(x_i^-)\xi - \eta(x_i^-)[\xi, v]) = \tag{464}$$
$$= \alpha(\phi_{s_i^*}[v, (\lambda+\bar{\mu}\eta)(x_i^-)\xi - \eta(x_i^-)[\xi, v]]) + d\alpha(D\phi_{s_i}(v(x_i^-)),$$
$$D\phi_{s_i}((\lambda+\bar{\mu}\eta)(x_i^-)\xi - \eta(x_i^-)[\xi, v])).$$

Since x_∞ is a critical point at infinity:

$$\phi_{s_i}^*\alpha = \alpha. \tag{465}$$

We also know that:

$$D\phi_{s_i}(v(x_i^-)) = v(x_i^+) \qquad D\phi_{s_i}((\lambda+\bar{\mu}\eta)(x_i^-)\xi(x_i^-) - \eta(x_i^-)[\xi,v](x_i^-)) = \tag{466}$$
$$= (\lambda+\bar{\mu}\eta)(x_i^+)\xi(x_i^+) - \eta(x_i^+)[\xi,v](x_i^+) + \theta v(x_i^+).$$

Thus

$$\begin{aligned} & v(x_i^-)\cdot\phi_{s_i}^*\alpha((\lambda+\bar{\mu}\eta)(x_i^-)\xi - \eta(x_i^-)[\xi,v]) = \\ &= -\eta(x_i^-)\alpha([v,[\xi,v]]) + d\alpha(v, -\eta(x_i^+)[\xi,v]) = \\ &= -\eta(x_i^-) + \eta(x_i^+) + (\lambda+\bar{\mu}\eta)(x_i^-)\alpha(\phi_{s_i^*}([v,\xi])) = \eta(x_i^+) - \eta(x_i^-). \end{aligned} \tag{467}$$

Thus:

$$\begin{aligned} & \partial^2 J(x_\infty)\cdot z\cdot z \\ =& -\sum_i \left(\eta(s_i^+)(\delta s_i + \mu(x_i^-) - \mu(x_i^+)) + \left(\frac{\partial a_0}{\partial x_i^-}(\lambda+\bar{\mu}\eta)(x_i^-) + \frac{\partial c_0}{\partial x_i^-}\eta(x_i^-)\right)\right. \\ & \left.((\lambda+\bar{\mu}\eta)(x_i^-)\xi(x_i^-) - \eta(x_i^-)[\xi,v](x_i^-))\right) + \\ & +\sum_i a \int_{[x_i^+, x_{i+1}^-]} (\dot{\eta}^2 - a^2\eta^2\tau). \end{aligned} \tag{468}$$

Observe now that:

$$\delta s_i = \int_{x_i^-}^{x_i^+} \dot{\mu} - \eta\bar{\mu}_\xi; \text{ thus } \delta s_i + \mu(x_i^-) - \mu(x_i^+) = -\int_{x_i^-}^{x_i^+} \eta\bar{\mu}_\xi. \tag{469}$$

Thus:

$$\begin{aligned} & \partial^2 J(x_\infty)\cdot z\cdot z \\ =& -\sum_i \left(-\eta(s_i^+)\int_{x_i^-}^{x_i^+} \eta\bar{\mu}_\xi + \left(\frac{\partial a_0}{\partial x_i^-}(\lambda+\bar{\mu}\eta)(x_i^-) + \frac{\partial c_0}{\partial x_i^-}\eta(x_i^-)\right)\cdot\right. \\ & \left.\cdot(\lambda+\bar{\mu}\eta)(x_i^-)\xi(x_i^-) - \eta(x_i^-)[\xi,v](x_i^-)\right) + \\ & +\sum_i a \int_{[x_i^+, x_{i+1}^-]} (\dot{\eta}^2 - a^2\eta^2\tau). \end{aligned} \tag{470}$$

This definition extends to any $x \in \Gamma_{2k}$ as follows:

Definition 11. *The index form on $\mathcal{H}$ is*

$$Q_x(z,z) = -\sum_i \Bigg(\Bigg(\frac{\partial a_0}{\partial x_i^-}(\lambda+\bar{\mu}\eta)(x_i^-) + \frac{\partial c_0}{\partial x_i^-}\eta(x_i^-)\Bigg)\cdot$$

$$\cdot\big(\xi(x_i^-)(\lambda+\bar{\mu}\eta)(x_i^-) - [\xi,v](x_i^-)\eta(x_i^-)\big)\Bigg)\Bigg|_{(s_i^-,x_i^-)}$$

$$-\eta(s_i^+)\int_{x_i^-}^{x_i^+}\eta\bar{\mu}_\xi ds\Bigg) +$$

$$+\sum_i a \int_{[x_i^+,x_{i+1}^-]} (\dot{\eta}^2 - a^2\eta^2\tau)$$

and we have

Proposition 32. *Q is a Fredholm quadratic form on the (λ,η) components of z in $\mathcal{H}$ (i.e. we can forget the μ-component and build an ad hoc Fredholm operator components of z in $\mathcal{H}$. It is of finite Morse index in $\mathcal{H}$ representing Q in a suitable (λ,η) space covering all the corresponding variations in $\mathcal{H}$) and the difference of the Morse indices of Q at two critical points at infinity belonging to the same connected component of Γ_{2k} is equal to the same differences but computed on $T\Gamma_{2k}$ to which we add the difference of the Morse indices of $\sum_i \int_{[x_i^+,x_{i+1}^-]} (\dot{\eta}^2 - a^2\eta^2\tau)$ computed on $\bigoplus_i H_0^1([x_i^+,x_{i+1}^-]$ at the two curves. In fact, $T_x\Gamma_{2k}$ and $\bigoplus_i H_0^1([x_i^+,x_{i+1}^-])$ are in an appropriate sense Q_x-orthogonal at any x of Γ_{2k}:*

Proof of Proposition 32. Using functions η with compact support in $[x_i^+,x_{i+1}^-]$, with $\lambda+\bar{\mu}\eta$ identically zero, as well as μ and η, but only outside of $[x_i^+,x_{i+1}^-]$, which is possible see (434), (435) and (436), we easily see that any $z\in\ker Q_x$, in $\mathcal{H}_x$, has to satisfy:

$$\dot{\mu} + a\eta\tau = 0 \quad \text{on} \quad [x_0^+,x_{i+1}^-], \qquad \forall i. \tag{471}$$

As we pointed out earlier, the additional equation (471) means that z is in $T_x\Gamma_{2k}$. Thus, $\ker Q_x \subset \ker Q_{x|T_x\Gamma_{2k}}$. We now analyze the other degeneracies, i.e. the difference between $\ker Q_{x|T_x\Gamma_{2k}}$ and $\ker Q_x$. We observe that, with $\eta\in\bigoplus_i H_0^1([x_i^+,x_{i+1}^-])$ and $C_i = 0$ in (434), we can build a subbundle $\widetilde{\bigoplus_i H_0^1([x_i^+,x_{i+1}^-])}$ of $\mathcal{H}$. Since $(\lambda+\bar{\mu}\eta,\eta) = 0$ for z in such a subbundle on $[x_i^-,x_i^+]$ and since $\dot{\mu}'+a\eta'\tau = 0$ on $[x_i^+,x_{i+1}^-]$ for $z'\in T_x\Gamma_{2k}$, it is clear that $T_x\Gamma_{2k}$ and $\widetilde{\bigoplus_i H_0^1([x_i^+,x_{i+1}^-])}$ are Q_x-orthogonal. When they intersect (i.e. their intersection is not equal, to $\{0\}$), $Q_{x|T_x\Gamma_{2k}}$

degenerate, but not Q_x necessarily: indeed, Q_x would degenerate only if an additional condition is satisfied. This is extensively explained below, in the proof of Theorem 40. We bypass this discussion here and we proceed as follows: At any rate, when $T_x\Gamma_{2k}$ and $\widetilde{\bigoplus_i H_0^1([x_i^+, x_{i+1}^-])}$ have a nonzero intersection, (438) together with $\dot{\eta} = \mu a$, for a certain index i_0, must have a nontrivial solution, under Dirichlet boundary conditions on η, i.e. the problem

$$\begin{cases} \dot{\eta} = \mu a \\ \\ \dot{\mu} + a\eta\tau = 0 \end{cases} \qquad \eta(x_{i_0}^+) = \eta(x_{i_0+1}^-) = 0$$

must have a nontrivial solution. This will clearly happen – because the Dirichlet problem on the ξ-pieces is "non-degenerate" i.e. varies as if τ was a constant when the length along ξ changes – only for characteristic values of a_{i_0}, hence on a stratified subset of codimension 1 of Γ_{2k}.

Outside of this stratified subset, $T_x\Gamma_{2k}$ and $\widetilde{\bigoplus_i H_0^1[(x_i^+, x_{i+1}^-])}$ are complement spaces in $\mathcal{H}$.

In fact, outside of this stratified set of codimension 1, they are supplement spaces: it suffices, in order to prove this, given a vector z of $\mathcal{H}$, to be able to solve

$$\begin{cases} \dot{\underline{\eta}} = \underline{\mu} a \\ \\ \dot{\underline{\mu}} + a\underline{\eta}\tau = 0 \end{cases} \qquad \underline{\eta}(x_i^+) = \eta_i^+ \qquad \underline{\eta}(x_{i+1}^-) = \eta_{i+1}^-$$

on each $[x_i^+, x_{i+1}^-]$ ξ-piece, with η_i^+, η_{i+1}^- being the values of the η-component of z at x_i^+, x_{i+1}^-. We set $\eta = \underline{\eta}$ outside on the $[x_i^-, x_{i+1}^+]$ pieces. Coupling $(\underline{\eta}, \underline{\mu})$ with $\underline{\lambda} + \bar{\mu}\underline{\eta} = \lambda + \bar{\mu}\eta$ (observe that $(\lambda + \bar{\mu}\eta, \eta)$ satisfies (435) on $[x_i^-, x_i^+]$), we construct a vector $\underline{z}$ in $T_x\Gamma_{2k}$. Clearly, $z - \underline{z}$ is in $\widetilde{\bigoplus_i H_0^1[x_i^-, x_{i+1}^+]}$. We thus have

$$\mathcal{H}_x = T_x\Gamma_{2k} \oplus \left(\widetilde{\bigoplus_i H_0^1[x_i^-, x_{i+1}^+]} \right)_x$$

outside of this stratified set and the decomposition is Q_x-orthogonal. In the proof of Theorem 40, we show that Q_x is Fredholm on $\mathcal{H}$ (actually $\overline{\mathcal{H}}$, the closure of $\mathcal{H}$ under the H^1-norm on η and λ). Using Propositions 29 and 30, we can prove that the critical points at infinity are, generically on v, not in the stratified set where the decomposition above does not hold. The claims of Proposition 32 follow.

We now study $\sum_k$, $J_{|\sum_k}$ and the second variation of J on $\sum_k$.

A curve of $\sum_k$ is made of pieces of ξ-curves and pieces of v-curves. The v-pieces run between a point x_i^- and a conjugate point of x_i^-, x_i^+, corresponding to the rotation m_i. Therefore, $\sum_k$ can be denoted

$$\sum_k = M^k(m_1, \dots, m_k) \tag{472}$$

$\overset{\circ}{M}{}^k$ will be its interior. The notation of M^k or $\sum\limits_k$ will be used indifferently in the sequel.

$$\theta_i \text{ is the collinearity coefficient} \tag{473}$$

between $\alpha_{x_i^-}$ and the pull-back along v of $\alpha_{x_i^+}$. The bundle $\mathcal{H}$ has been defined along Γ_{2k}, hence along M^k.

We define a subbundle $\mathcal{B}_x$ of $\mathcal{H}_x$ by requiring that:

$$z(x_i^+) = D\theta_{m_i}(z(x_i^-)) \qquad \forall i. \tag{473'}$$

We define a subbundle $\mathcal{F}_x$ of $\mathcal{B}_x$ as follows:

Definition 12. $\mathcal{F}_x = \{z \in \mathcal{B}_x \text{ s.t. } \alpha_{x_i^\pm}(z(x_i^\pm)) = 0\ \forall i\}$. *Observe that, if* $x \in \overset{\circ}{M}{}^k$, *then*

$$\mathcal{F}_x = \{z \in \mathcal{B}_x \text{ s.t. } \alpha_{x_i^-}(z(x_i^-)) = 0 \qquad \forall i\}.$$

We then have:

Proposition 33. *Let x be a critical point at infinity. Generically on v, we have:*

$$\mathcal{B}_x = T_x \overset{\circ}{M}{}^k \oplus \mathcal{F}_x.$$

Proof. The statement on $\mathcal{B}_x = T_x \overset{\circ}{M}{}^k \oplus \mathcal{F}_x$ amounts to prove that the $\alpha_{x_i^-}(z(x_i))$'s are free parameters on $T_x \overset{\circ}{M}{}^k$.

This is obvious when we want to solve (435). The value of $(\lambda + \bar{\mu}\eta, \eta)(x_i^+)$ is derived from the value of $(\lambda + \bar{\mu}\eta, \eta)(x_i^-)$; we must furthermore have:

$$z(x_i^+) = D\theta_{m_i}(z(x_i^-)). \tag{474}$$

Then, $(\lambda + \bar{\mu}\eta, \eta)$ is given at every x_i^-, x_i^+. The various values of $\lambda + \bar{\mu}\eta$ can be glued up, using the constants C_i in (434). We are thus left with extending η and μ to the ξ-pieces i.e. with solving $\begin{cases} \dot{\mu} + a\eta\tau = 0 \\ \dot{\eta} = \mu a \end{cases}$ $\eta(x_i^+)$ given, $\mu(x_i^+)$ given. This initial value problem can be solved step by step. The values derived of $\eta(x_{i+1}^-), \mu(x_{i+1}^-)$ and the one presribed $\lambda(x_{i+1}^-)$ allow to continue the process. There is a compatibility condition: Solving inductively, ξ-piece by ξ-piece, then going through the v-branches, we are left with an equation of the type: $dl(u) - u = A$, where A is a fixed vector in $\ker\alpha$. This is identical to the study of $T_x \sum_k$ which we completed earlier. A is in $\ker\alpha$ because the curve is critical, hence $\theta_i = 1 \quad \forall i$. This equation needs to be solved in u. There could be a problem if $dl - Id|_{\ker\alpha}$ was not invertible. Propositions 29 and 30 allow to rule it out on critical points at infinity. Proposition 33 follows.

Proposition 33 fails at other points on curves x of $\overset{\circ}{M}{}^k(\sum_k)$ and $\mathcal{B}_x$ is not necessarily the direct sum of $T_x \overset{\circ}{M}{}^k$ and $\mathcal{F}_x$. We therefore introduce an abstract bundle $\mathcal{H}'$ as follows:

Definition 13. *The fiber $\mathcal{B}'_x$ at x is the direct sum of $T_x \overset{\circ}{M}{}^k$ and $\mathcal{F}_x$. Both spaces are equipped with the H^1-norm inherited from $\mathcal{B}_x$, hence from $\mathcal{H}_x$.*

Then $\mathcal{B}'$ is a new vector-bundle over $\overset{\circ}{M}{}^k$, whose fiber can be thought to coincide with the fiber of $\mathcal{B}$ in some open subset of $\overset{\circ}{M}{}^k$, which contains the critical points at infinity of $J_{|\overset{\circ}{M}{}^k}$.

We now define a quadratic form on $\mathcal{H}$ as follows: Let $z, z' \in \mathcal{H}_x$; $\gamma(z)$ is the component of z along v.

Definition 14. *The incomplete index form $\mathcal{Q}_x$ at $x \in \overset{\circ}{M}{}^k$ is:*

$$\mathcal{Q}_x(z,z') = \sum_{i=1}^{k}\prod_{j=1}^{i}\theta_j \left(- \int_{[x_i^+, x_{i+1}^-]} \left(\frac{d\gamma(z)}{dt} - a_i d\gamma(\xi, z)\right)\beta(z')d\tau - \right.$$
$$- \int_{[x_i^-, x_{i+1}^+]} \left(\frac{d\gamma(z')}{dt} - a d\gamma(\xi, z')\right)\beta(z)d\tau + \alpha_{x_i}(z(x_i^-))\frac{d\theta_i}{\theta_i^2}(z'(x_i^-)) +$$
$$\left. + \alpha_{x_i^-}(z'(x_i^-))\frac{d\theta_i}{\theta_i^2}(z(x_i^-)) \right).$$

Observe that $\mathcal{Q}_x(z,z') = \mathcal{Q}_x(z+\lambda\xi+\mu v, z') = \mathcal{Q}_x(z, z'+\lambda\xi+\mu v)$, where λ is any H^1-function vanishing on the v-pieces and where μ is any H^1-function vanishing on the ξ-pieces. Indeed, $\gamma(\xi)$ is zero since ξ is characteristic for $d\alpha$ and $\beta(v)$ is zero since $\beta = d\alpha(v,\cdot)$. We could therefore consider that $\mathcal{Q}_x$ is defined on $\mathcal{B}/\sim$ where $\sim$ is the following equivalence relation:

$$z \sim z_1 \quad z, z_1 \in \mathcal{B}_k \quad \textit{if and only if} \quad z - z_1 = \lambda\xi + \mu v, \tag{475}$$

λ vanishing on the v-pieces, μ vanishing on the ξ-pieces (λ and μ are both H^1-function).

However, even on $\mathcal{B}/\sim$, $\mathcal{Q}_x$ is not a Fredholm quadratic form i.e. the associated linear symmetric operator is not Fredholm and it seems difficult to change the norm of $\mathcal{B}$ and take the closure of $\mathcal{B}$ in the new norm so that $\mathcal{Q}_x$ becomes Fredholm on the new space.

This change can be completed on $\mathcal{B}'$, provided we slightly modify $\mathcal{Q}_x$, as follows:

Definition 15. *The complete index form* $\widetilde{Q}_x$ *at* $x \in \overset{\circ}{M}{}^k$ *is a quadratic form defined on* $\mathcal{B}'_k$ *as follows.*

Let $\tilde{z} = \bar{z} + z$, $\bar{z} \in T_x \overset{\circ}{M}{}^k$, $z \in \mathcal{F}_x$, be an element of $\mathcal{B}'_x$. Then:

$$\widetilde{Q}_x(\tilde{z},\tilde{z}) = Q_x(\bar{z},\bar{z}) + Q_x(z,z) - 2\left(1 - \prod_{j=1}^{k} \theta_j\right) \gamma(z(x_1^-))\beta(z(x_1^-)).$$

In order to make our construction more transparent, we prove the following Lemma:

Lemma 34. *(i)* $T_x \overset{\circ}{M}{}^k$ *and* $\mathcal{F}_x$*, as subspaces of* $\mathcal{B}_x$*, are* Q_x*-orthogonal at any curve* $x \in \overset{\circ}{M}{}^k$ *such that* $\prod_{j=1}^{k} \theta_j = 1$.

(ii) $\widetilde{Q}_{x'} Q_x$ *and* Q_x *coincide at any critical point at infinity x (observe that* $\mathcal{B}'_x = \mathcal{B}_x$ *at such a point).*

(iii) There exist kC^∞ *functions* $w_1(x), \dots, w_k(x)$*, such that for any* $z \in \mathcal{F}_x \subset \mathcal{B}'_x$ *(*$x \in \overset{\circ}{M}{}^k$*, arbitrary),*

$$\widetilde{Q}_x(z,z) = 2\sum_{i=0}^{k}\left\{\prod_{j=1}^{i}\theta_j\left(\left(\frac{d\beta(z)}{dt}\right)^2 dt + \int_{[x_i^-,x_i^+]} a_i\, d\gamma(\xi,z)\beta(z)dt\right) + \right.$$
$$\left. + \,\omega_i(x_i^-)\beta(z(x_i^-))^2\right\}.$$

We will prove Lemma 34 later. At this point, we recall the following identity from [1] (see [1] pp. 2–3, see also Proposition A2 of this paper)

$$d\gamma(\xi,\cdot) = -\tau\beta(\cdot) \qquad (\tau \in C^\infty(M,\mathbb{R})) \tag{476}$$
$$d\beta(\xi,\cdot) = \gamma(\cdot). \tag{477}$$

From these identities and from (iii) of Lemma 34, we derive that if $z \in \mathcal{F}_x$, $\widetilde{Q}_x(z,z)$ reads as follows:

$$\widetilde{Q}_x(z,z) = 2\sum_{i=1}^{k}\left\{\prod_{j=1}^{i}\theta_j\left(\int_{[x_i^+,x_{i+1}^-]} \frac{1}{a_i}\left(\frac{d\beta(z)}{dt}\right)^2 dt - \int_{[x_i^+,x_{i+1}^-]} \tau\, a_i\beta(z)^2 dt\right) + \right.$$
$$\left. + \,\omega_i(x_i^-)\beta(z(x_i^-))^2\right\} \tag{478}$$

Writing $\widetilde{Q}_x$ under this form, we are very much tempted to consider a new space, different from $\mathcal{F}_x$, where the variable would be $\beta(z)$, $\beta(z)$ belonging to this space $\oplus H^1([x_i^+, x_{i+1}^-]$, i.e. $\beta(z)$ is H^1 on each on the pieces $[x_i^+, x_{i+1}^-]$. We would, however, keep track of our original problem by imposing boundary conditions on $\beta(z)$ at each of the x_i^+, x_{i+1}^-. Observe that, if $z \in \mathcal{F}_x$, we can write, since $\alpha(z(x_i^\pm)) = 0$, $z(x_i^\pm)$ as follows:

$$z(x_i^\pm) = \mu_i^\pm v(x_i) - \eta_i^\pm[\xi, v](x_i). \tag{479}$$

Indeed, v and $-[\xi, v]$ span $\ker\alpha$ since $d\alpha(v, [\xi, v]) = -1$ (see [1] pp. 2–3 and Proposition A2). We then have:

$$\beta(z(x_i^\pm)) = d\alpha(v, z(x_i^\pm)) = \eta_i^\pm \tag{480}$$

$$\gamma(z(x_i^\pm)) = \alpha(z, w) = d\alpha(z, -[\xi, v] + \bar{\mu}\xi) = \mu_i^\pm. \tag{481}$$

Since $D\theta_{m_i}(z(x_i^-)) = z(x_i^+)$, we have:

$$D\theta_{m_i}(\gamma(z(x_i^-))v - \beta(z(x_i^-))[\xi, v]) = \gamma(z(x_i^+))v - \beta(z(x_i^+))[\xi, v]. \tag{482}$$

Observing then that, by (476)

$$\mu_i^\pm = \gamma(z(x_i^\pm)) = d\beta(\xi, z(x_i^\pm)) = \frac{1}{a_i^\pm}\frac{d}{dt}\beta(z)(x_i^\pm), \quad (a_i^- = a_{i-1}, a_i^+ = a_i) \tag{483}$$

we can rewrite (482) as:

$$D\theta_{m_i}\left(\frac{1}{a_{i-1}}\frac{d}{dt}\beta(z)(x_i^-)v - \beta(z(x_i^-))[\xi, v]\right) = \frac{1}{a_i}\frac{d}{dt}\beta(z)(x_i^+)v - \beta(z(x_i^+))[\xi, v]. \tag{484}$$

When i runs from 1 to k, (484) provides a family of boundary conditions and it is natural to introduce the space:

Definition 16. *Let* $\mathcal{F}_x(\eta) = \left\{\eta \in \bigoplus_{i=1}^{k} H^2([x_i^+, x_{i+1}^-])\right.$ *s.t.* $(\eta(x_i^-), \quad \dot\eta(x_i^-),$ $\eta(x_{i+1}^+), \dot\eta(x_{i=1}^+))$ *satisfy, for each* $i = 1, \ldots, k$, *the condition:*

$$\left. D\theta_{m_i}\left(\frac{1}{a_{i-1}}\dot\eta(x_i^-)v - \eta(x_i^-)[\xi, v]\right) = \frac{1}{a_i}\dot\eta(x_{i+1}^+)v - \eta(x_{i+1}^+)[\xi, v]\right\}.$$

We denote $\overline{\mathcal{F}}_x(\eta)$ *the closure of* $\mathcal{F}_x(\eta)$ *under the norm on* η: $|\eta|_{H^1}^2 = \sum_{i=1}^{k} \int_{[x_i^+, x_{i+1}^-]} (\dot\eta^2 + \eta^2)dt$. *It is easy to see that the matrix of* $D\theta_{m_i}|_{\ker\alpha}$ *is triangular inferior from the basis* $(-[\xi, v], v)(x_i^-)$ *to* $(-[\xi, v], v)(x_i^+)$, *since* $D\theta_{m_i}(v)$ *is collinear to* v. *Thus,*

$$D\theta_{m_i}|_{\ker\alpha} = \begin{bmatrix} \tilde\gamma & 0 \\ \tilde\mu & \tilde\delta \end{bmatrix}.$$

Starting from any function η in $H^2(S^1)$, i.e. H^2 and periodic, we can build, by simple changes, a function $\tilde{\eta}$ which will between 0^- and 0^+ undergo a jump governed by $D\theta_{m_i}$. Indeed, if we keep η for $t < 0$ and replace η by $\tilde{\eta} = \varphi(t)\eta(\gamma_1 t)$ for $t > 0$, then:

$$\begin{pmatrix} \tilde{\eta}(0) \\ \dot{\tilde{\eta}}(0) \end{pmatrix} = \begin{bmatrix} \varphi(0) & 0 \\ \varphi'(0) & \gamma_1\varphi(0) \end{bmatrix} \begin{bmatrix} \eta(0) \\ \dot{\eta}(0) \end{bmatrix}.$$

Thus, the changes of $\mathcal{F}_x(\eta)$, the jumps, can be completed through appropriate changes of variables $\eta \mapsto \tilde{\eta}$. Thus, our spaces are well defined, it suffices to start from $\eta \in H^2$. The closure $\overline{\mathcal{F}_x(\eta)}$ is isomorphic to $H^1(S^1)$. On $\overline{\mathcal{F}}_x(\eta)$, we introduce the quadratic form:

Definition 17.

$$Q_x(\eta, \eta) = 2\sum_{i=1}^{k}\left\{\prod_{j=1}^{i}\theta_j\left(\int_{[x_i^+, x_{i+1}^-]} \frac{1}{a_i}\dot{\eta}^2\,dt - \int_{[x_i^+, x_{i+1}^-]} a_i\tau\eta^2\,dt\right) + \omega_i(x_i^-)\eta(x_i^-)^2\right\}$$

We then have:

Lemma 35. *Let x be a curve of $\overset{\circ}{M}{}^k$. (i) the map N: $\mathcal{F}_x \longrightarrow \mathcal{F}_x(\eta)$*

$$z \longmapsto \beta(z)$$

is a linear continuous map. Furthermore, $N(\mathcal{F}_x)$ is dense in $\overline{\mathcal{F}}_x(\eta)$, for the H^1-topology on η.

(ii) Q_x is a Fredholm quadratic form on $\overline{\mathcal{F}}_x(\eta)$. If $\prod_{j=1}^{k}\theta_j = 1$, $\ker Q_x = N(\mathcal{F}_x \cap T_x\overset{\circ}{M}{}^k)$. (The intersection $\mathcal{F}_x \cap T_x\overset{\circ}{M}{}^k$ is taken in $\mathcal{B}_x$).

(iii) Let y_0 and y_1 be two distinct critical points at infinity in $\overset{\circ}{M}{}^k$, s.t. $\pi\theta_i = 1$, which belong to the same connected component of $\overset{\circ}{M}{}^k \cap \left\{\prod_{i=1}^{k}\theta_i = 1\right\}$; then, generically on v, Q_{y_0} and Q_{y_1} have the same number of negative eigenvalues if y_0 and y_1 are hyperbolic. If one is hyperbolic and the other is elliptic, there is a jump of $+1$ or -1.

(iv) Generically on v, we have $\mathcal{F}_x \cap T_x\overset{\circ}{M}{}^k = \{0\}$ at any critical point at infinity x s.t. $\pi\theta_i = 1$.

(v) Let x be a critical point at infinity of J in $\overset{\circ}{M}{}^k$ s.t. $\pi\theta_i = 1$; let $\bar{z} \in \mathcal{B}_x$, $\bar{z} = \bar{z} + z$, $\bar{z} \in T_x\overset{\circ}{M}{}^k$, $z \in \mathcal{F}_x$. Then, $\mathcal{Q}_x(z + \bar{z}, z + \bar{z}) = \mathcal{Q}_x(\bar{z}, \bar{z}) + Q_x(\beta(z), \beta(z))$. In particular, let G_x be a subspace of $\mathcal{B}_x$ of maximal dimension, in which $\mathcal{Q}_x$ is definite negative. Then, $\dim G_x < +\infty$ and $\dim G_{y_0} - \dim G_{y_1} =$ Morse index$_{y_0}(J|_{\overset{\circ}{M}{}^k})$ - Morse index$_{y_1}(J|_{\overset{\circ}{M}{}^k})$ for any two critical points at infinity y_0 and y_1 such that

$\pi\theta_i = 1$, in the same connected component of $\overset{\circ}{M}{}^k \cap \{\pi\theta_i = 1\}$, if y_0 and y_1 are both hyperbolic, or both elliptic. Otherwise, there is a jump of 1 in the difference of Morse indices.

Proof of Lemma 34. Proof of (i): Let $\bar{z} \in T_x \overset{\circ}{M}{}^k$ and $z \in \mathcal{F}_x$. Then

$$\mathcal{Q}_x(z, \bar{z}) = \sum_1^k \prod_{j=1}^i \theta_j \left(- \int_{[x_i^+, x_{i+1}^-]} \left(\frac{d\gamma(z)}{dt} - a_i d\gamma(\xi, z) \right) \beta(\bar{z}) dt - \right. \tag{485}$$

$$\left. - \int_{[x_i^+, x_{i+1}^-]} \left(\frac{d\gamma(\bar{z})}{dt} - a_i d\gamma(\xi, \bar{z}) \right) \beta(z) dt + \alpha_{x_i^-}(\bar{z}(x_i^-)) \frac{d\theta_i}{\theta_i^2}(z(x_i^-)) \right)$$

z and $\bar{z}$ satisfy the following equations on $[x_i^+, x_{i+1}^-]$:

$$\frac{d\beta(z)}{dt} = a_i d\beta(\xi, z) = a_i \gamma(z) \tag{486}$$

$$\frac{d\beta(\bar{z})}{dt} = a_i d\beta(\xi, \bar{z}) = a_i \gamma(\bar{z}) \tag{487}$$

$$\text{on } [x_i^+, x_{i+1}^-]$$

We thus have:

$$- \int_{[x_i^-, x_{i+1}^-]} \frac{d\gamma(z)}{dt} \beta(\bar{z}) dt = -\gamma(z)\beta(\bar{z}) \Big|_{x_i^+}^{x_{i+1}^-} + \int_{x_i^+}^{x_{i+1}^-} \gamma(z) \frac{d}{dt} \beta(\bar{z}) dt =$$

$$= - \gamma(z)\beta(\bar{z}) \Big|_{x_i^+}^{x_{i+1}^-} + \int_{[x_i^+, x_{i+1}^-]} \frac{d}{dt} \beta(z) \gamma(\bar{z}) dt \tag{488}$$

$$= \beta(z)\gamma(\bar{z}) - \gamma(z)\beta(\bar{z}) \Big|_{x_i^+}^{x_{i+1}^-} - \int_{[x_i^+, x_{i+1}^-]} \beta(z) \frac{d}{dt} \gamma(\bar{z}) dt$$

Using (488), we derive from (485):

$$\mathcal{Q}_x(z, \bar{z}) = \sum_{i=1}^k \prod_{j=1}^i \theta_j \left(-2 \int_{[x_i^+, x_{i+1}^-]} \beta(z) \left(\frac{d\gamma(\bar{z})}{dt} + \tau a_i \beta(\bar{z}) \right) dt + \right. \tag{489}$$

$$\left. + \beta(z)\gamma(\bar{z}) - \gamma(z)\beta(\bar{z}) \Big|_{x_i^+}^{x_{i+1}^-} + \alpha_{x_i}(\bar{z}(x_i^-)) \frac{d\theta_i}{\theta_i^2}(z(x_i^-)). \right.$$

Observe that, since $d\alpha(v,w)=1$ and $d\alpha(\xi,\cdot)=0$, we have:

$$\beta(z)\gamma(\bar{z})-\gamma(z)\beta(\bar{z})=-d\alpha(z,\bar{z}). \tag{490}$$

Thus,

$$\begin{aligned}\mathcal{Q}_x(z,\bar{z})=\sum_{i=1}^{k}\prod_{j=1}^{i}\theta_j\Bigg\{-2\int\limits_{[x_i^-,x_{i+1}^+]}\left(\frac{d\gamma(\bar{z})}{dt}+\tau a_{i+1}\beta(\bar{z})\right)\beta(z)dt-\\ -d\alpha(z,\bar{z})\Big|_{x_i^+}^{x_{i+1}^-}+\alpha_{x_i^-}(\bar{z}(x_i^-))\frac{d\theta_i}{\theta_i^2}(z(x_i^-)).\end{aligned} \tag{491}$$

Since $\bar{z}\in T_x\overset{\circ}{M}{}^k$, the first variation, along $\bar{z}$, of the v-component of $\dot{x}$, the tangent vector to x, is zero. This component is $\gamma(\dot{x})$ and its first variation is therefore $\frac{d\gamma(\bar{z})}{dt}-a_i d\gamma(\xi,\bar{z})$. Thus, we have:

$$\frac{d}{dt}\gamma(\bar{z})+\tau a_i\beta(\bar{z})=0. \tag{492}$$

Therefore:

$$\mathcal{Q}_x(z,\bar{z})=\sum_{i=1}^{k}\prod_{j=1}^{i}\theta_j\left\{-d\alpha(z,\bar{z})\Big|_{x_i^+}^{x_{i+1}^-}+\alpha_{x_i^-}(\bar{z}(x_i^-))\frac{d\theta_i}{\theta_i^2}(z(x_i^-))\right\}. \tag{493}$$

Since z and $\bar{z}\in\mathcal{B}_x$, $z(x_i^+)$, $\bar{z}(x_i^-)$ are respectively equal to $D\theta_{m_i}(z(x_i^-))$, $D\theta_{m_i}(\bar{z}(x_i^-))$. On the other hand:

$$\theta_{m_i}^*\alpha=\frac{\alpha}{\theta_i}. \tag{494}$$

Thus

$$\theta_{m_i}^*\,d\alpha=-\frac{d\theta_i}{\theta_i^2}\wedge\alpha+\frac{1}{\theta_i}d\alpha. \tag{495}$$

Hence, using (495) and the fact that $\alpha(z(x_i^-))=0$,

$$d\alpha(z(x_i^+),\bar{z}(x_i^+))=\frac{1}{\theta_i}d\alpha(z(x_i^-),\bar{z}(x_i^-))-\frac{d\theta_i}{\theta_i^2}(z(x_i^-))\alpha(\bar{z}(x_i^-)). \tag{496}$$

Using (496) in (493) and recalling that we are assuming that $\prod_{j=1}^{k}\theta_j=1$, we derive that $\mathcal{Q}_x(z,\bar{z})$ is identically zero. Hence, (i) of Lemma 34.

Proof of (ii): At a critical point at infinity, the θ_i's are equal to 1. Therefore, $\widetilde{\mathcal{Q}}_x(\tilde{z},\tilde{z})$ is the sum of $\mathcal{Q}_x(\bar{z},\bar{z})$ and $\mathcal{Q}_x(z,z)$, where $\tilde{z}=\bar{z}+z$, $\bar{z}\in T_x\overset{\circ}{M}{}^k$, $z\in\mathcal{F}_x$.

On the other hand, since $\pi\theta_j = 1$ at such a point, we have just established that $\mathcal{F}_x$ and $T_x \overset{\circ}{M}{}^k$ are $\mathcal{Q}_x$-orthogonal. Hence, $\mathcal{Q}_x(\tilde{z}, \tilde{z})$ is also the sum of $\mathcal{Q}_x(\bar{z}, \bar{z})$ and $\mathcal{Q}_x(z, z)$. Thus, $\widetilde{\mathcal{Q}}_x(\tilde{z}, \tilde{z}) = \mathcal{Q}_x(\tilde{z}, \tilde{z})$. (ii) of Lemma 35 is established, up to the equality with Q_x. Observe that $a_0(s_i, x_i^-)(\lambda + \bar{\mu}\eta)(x_i^-) + c_0(s_i, x_i^-)\eta(x_i^-)$ is equal to $\theta_i(\lambda + \bar{\mu}\eta)(x_i^-)$. Indeed, $z(x_i^+) = D\theta_{m_i}(z(x_i^-))$ for $z \in \mathcal{B}$, thus $\theta_{m_i}^* \alpha = \theta_i \alpha$ and our claim holds. Furthermore, $\theta_i = 1$ at a critical point at infinity. Thus, the computations of (459)–(470), using the arguments of (495)–(496), can be carried out on $\mathcal{Q}_x$, since $\pi\theta_j = 1$. The two formulae of $\mathcal{Q}_x$ and Q_x coincide, then, at a critical point at infinity: For $z \in \mathcal{B}$, $a_0(\lambda + \bar{\mu}\eta) + c_0\eta$ is $\theta_i(\lambda + \bar{\mu}\eta)$; $D\theta_{m_i}(z)(x_i^-)) = z(x_i^+)$. Thus, z preserves the fact that the v-branch is between two coincidence points with m_i revolutions of $\ker \alpha$ between them. Thus, the derivative along z of $a_0(\lambda + \bar{\mu}\eta) + c_0\eta$ is the same than the one of $\theta_i(\lambda + \bar{\mu}\eta)$. Since the θ_i's are equal to 1, $\mathcal{Q}_x$ is equal to Q_x on $\mathcal{B}_x$.

Proof of (iii): Let z be in $\mathcal{F}_x$. Using (486) and (477), we have:

$$\mathcal{Q}_x(z, z) = -2 \sum_{i=1}^{k} \prod_{j=1}^{i} \theta_j \left(\int\limits_{[x_i^+, x_{i+1}^-]} \left(\frac{d\gamma(z)}{dt} + a_i d\gamma(\xi, x) \right) \beta(z) dt = \right. \tag{497}$$

$$= -2 \sum_{i=1}^{k} \prod_{j=1}^{i} \theta_j \left(- \int\limits_{[x_i^+, x_{i+1}^-]} \frac{1}{a_i} \left(\frac{d}{dt} \beta(z) \right)^2 dt + \right.$$

$$\left. + \int\limits_{[x_i^+, x_{i+1}^-]} a_i d\gamma(\xi, z) \beta(z) dt + \gamma(z)\beta(z) \Big|_{x_i^+}^{x_{i+1}^-} \right).$$

Thus, reordering the terms, we have:

$$\widetilde{\mathcal{Q}}_x(z, z) = -2 \sum_{i=1}^{k} \prod_{j=1}^{i} \theta_j \left(\int\limits_{[x_i^+, x_{i+1}^-]} - \left(\frac{1}{a_i} \left(\frac{d}{dt} \beta(z) \right)^2 + a_i \, d\gamma(\xi, z) \beta(z) \right) \right) dt + \tag{498}$$

$$+ \left(\frac{\gamma(z)(x_i^-))\beta(z(x_i^-))}{\theta_i} - \gamma(z(x_i^+)\beta(z(x_i^+)) \right).$$

As one can notice, the addition of the term $\left(1 - \prod_{j=1}^{k} \theta_j \right) \gamma(z(x_1^-))\beta(z(x_1^-))$ in $\widetilde{\mathcal{Q}}_x$ was meant to allow such a reordering.

Let us turn back to the notations of (479)–(484). We have:

$$z(x_i^-) = \mu_i^- v(x_i^-) - \eta_i^-[\xi, v](x_i^-) = \gamma(z(x_i^-))v(x_i^-) - \beta(z(x_i^-))[\xi, v](x_i^-) \tag{499}$$

$$z(x_i^+) = \mu_i^+ v(x_i^+) - \eta_i^+[\xi, v](x_i^+) = \gamma(z(x_i^+))v(x_i^+) - \beta(z(x_i^+))[\xi, v](x_i^+) \tag{500}$$

$$D\theta_{m_i}(z(x_i^-)) = z(x_i^+). \tag{501}$$

Clearly, $D\theta_{m_i}(v)$ is collinear to v. We denote $\bar{\omega}_i$ this collinearity coefficient:

$$D\theta_{m_i}(v) = \bar{\omega}_i v \tag{502}$$

$D\theta_{m_i}(-[\xi, v](x_i^-))$ splits on $v(x_i^+)$, $-[\xi, v](x_i^+)$. We denote $\bar{\omega}_i'$, γ_i these two components:

$$D\theta_{m_i}(-[\xi, v](x_i^-)) = \bar{\omega}_i' v - \gamma_i[\xi, v]. \tag{503}$$

Using the identity:

$$d\alpha(D\theta_{m_i}(v), -D\theta_{m_i}([\xi, v])) = \left(\frac{1}{\theta_i} d\alpha(\cdot,\cdot) - \frac{d\theta_i}{\theta_i^2} \wedge \alpha\right)(v, -[\xi, v]), \tag{504}$$

we derive (recall that $d\alpha(v, -[\xi, v]) = d\alpha(v, w) = 1$ and $\alpha(v) = \alpha([\xi, v]) = 0$):

$$\bar{\omega}_i \gamma_i = \frac{1}{\theta_i}. \tag{505}$$

Using (501), we also have:

$$\mu_i^+ = \mu_i^- \bar{\omega}_i + \eta_i^- \bar{\omega}_i' \tag{506}$$

$$\eta_i^+ = \eta_i^- \gamma_i. \tag{507}$$

Therefore,

$$\frac{\mu_i^- \eta_i^-}{\theta_i} = \mu_i^- \eta_i^- \gamma_i \bar{\omega}_i = \mu_i^+ \eta_i^+ - \bar{\omega}_i' \gamma_i \eta_i^{-2}. \tag{508}$$

Hence, we derive from (495) and (508):

$$\widetilde{Q}_x(z, z) = -2 \sum_{i=1}^{k} \prod_{j=1}^{i} \theta_j \left(\int\limits_{[x_i^+, x_{i+1}^-]} -\left(\frac{1}{a_i}\left(\frac{d\beta(z)}{dt}\right)^2 - a_i d\gamma(\xi, z)\beta(z)\right) dt - \right. \tag{509}$$

$$\left. - \bar{\omega}_i' \gamma_i \beta(z(x_i^-))^2 \right).$$

Observe that the function $\omega_i(y) = \bar{\omega}_i'(y)\gamma_i(y)$ is defined and is $C^\infty(M, \mathbb{R})$. (iii) of Lemma 34 follows then from (509). The proof of Lemma 34 is thereby complete.

Proof of Lemma 35. Proof of (i): $\mathcal{F}_x$ is equipped with the H^1-topology on $z = \lambda\xi + \mu v + \eta w$, i.e. λ, μ, η are H^1 in each piece $[x_i^+, x_{i+1}^-]$. We have:

$$\dot{\eta} = \mu a_i \quad \text{on} \quad [x_i^+, x_{i+1}^-]. \tag{510}$$

Thus, η is in fact H^2 on each of these pieces. Hence, $\mathcal{N}$ is well defined and continuous. $\mathcal{N}(\mathcal{F}_x)$ is dense in $\overline{\mathcal{F}_x(\eta)}$ since $\mathcal{N}(\mathcal{F}_x) = \mathcal{F}_x(\eta)$. Indeed, if η belongs to $\mathcal{F}_x(\eta)$, then the vector $z = \eta w + \frac{\dot{\eta}}{a_i}v$ is H^1 on each piece $[x_i^+, x_{i+1}^-]$. Furthermore, $D\theta_{m_i}(z(x_i^-)) = z(x_i^+)$ and $\frac{d\beta(z)}{dt} = \dot{\eta} = a_i d\beta(\xi, z)$ on each piece $[x_i^+, x_{i+1}^-]$. Thus, z is well defined on the pieces $[x_i^+, x_{i+1}^-]$ and satisfies the constraints. We can extend z to the pieces $[x_i^+, x_{i+1}^-]$, which are v-oriented, as follows: $\theta_{m_i}(y)$ can be uniquely written as $\phi_{s_i(y)}(y)$, where ϕ_s is the one-parameter group generated by v and s_i is a C^∞-function. Any point z on the piece $[x_i^-, x_i^+]$ can uniquely written as $z = \phi_{s_i(x_i^-)t}(x_i^-)$, for $t \in [0, 1]$.

We then set

$$z(\phi_{s_i(x_i^-)t}(x_i^-)) = D_y(\phi_{s_i(x_i^-)t}(y))_{x_i^-}(z(x_i^-)). \tag{511}$$

Since $\phi_{s_i(y)}$ is θ_{m_i}, this coincides, for $t = 1$, with $D\theta_{m_i}(z(x_i^-)) = z(x_i^+)$. z is thus H^1 on all the curve x and satisfies all the requirements to be in $\mathcal{B}$. Hence, (i) of Lemma 35.

Proof of (ii): Q_x is obviously a Fredholm quadratic form on $\overline{\mathcal{F}_x(\eta)}$. Let us now study $\ker Q_x$. Clearly, if $\bar{\eta}$ belongs to $\ker Q_x$, then, $\bar{\eta} \in \bigoplus_{i=1}^{k} H^2([x_i^+, x_{i+1}^-])$ and $\bar{\eta}$ satisfies:

$$\ddot{\bar{\eta}} = -\tau a_i^2 \bar{\eta} \quad \text{on the piece} \quad [x_i^+, x_{i+1}^-] \tag{512}$$

which we rewrite as:

$$\left(\frac{\dot{\bar{\eta}}}{a_i}\right)^{\cdot} = -\tau a_i \bar{\eta} \quad \text{on} \quad [x_i^+, x_{i+1}^-]. \tag{513}$$

Let η_1 be a function in $\mathcal{F}_x(\eta)$. Using (513), we have:

$$\begin{aligned} 0 = Q_x(\bar{\eta}, \eta_1) &= 2\sum_{i=1}^{k}\prod_{j=1}^{i}\theta_j\left(\frac{1}{a_i}\dot{\bar{\eta}}\eta_1\Big|_{x_i^+}^{x_{i+1}^-} + \omega_i(x_i^-)\bar{\eta}(x_i^-)\eta_1(x_i^-)\right) \\ &= 2\sum_{i=1}^{k}\prod_{j=1}^{i}\theta_j(\omega_i(x_i^-)\bar{\eta}(x_i^-)\eta_1(x_i^-) - \frac{1}{a_i}\dot{\bar{\eta}}(x_i^+)\eta_1(x_i^+) + \\ &\quad + \frac{1}{\theta_i}\frac{\dot{\bar{\eta}}(x_i^-)}{a_i}\eta_1(x_i^-)) - 2\left(1 - \prod_{j=1}^{k}\theta_j\right)\frac{\dot{\bar{\eta}}(x_k^-)\eta_1(x_k^-)}{a_{k-1}}. \end{aligned} \tag{514}$$

Using (507), we can replace $\eta_1(x_{i+1}^+)$ by $\gamma_i\eta_1(x_i^-)$ in (514). We then have:

$$-2\left(1-\prod_{j=1}^{k}\theta_j\right)\frac{\dot{\eta}(x_k^-)}{a_{k-1}}\eta_1(x_k^-)+2\sum_{i=1}^{k}\prod_{j=1}^{i}\theta_j(\omega_i(x_i^-)\bar{\eta}(x_i^-)- \tag{515}$$
$$\left.-\frac{1}{a_i}\dot{\bar{\eta}}(x_i^+)\gamma_i+\frac{1}{\theta_i}\frac{\dot{\bar{\eta}}(x_i^-)}{a_{i-1}}\right)\eta_1(x_i^-)=0.$$

Since $\prod_{j=1}^{k}\theta_j=1$, (515) reduces to:

$$\sum_{i=1}^{k}\prod_{j=1}^{i}\theta_j(\omega_i(x_i)\bar{\eta}(x_i)-\frac{1}{a_i}\dot{\bar{\eta}}(x_{i+1}^+)\gamma_i+\frac{1}{\theta_i}\frac{\dot{\bar{\eta}}(x_i^-))}{a_{i-1}}\eta_1(x_i^-)=0. \tag{516}$$

The values $\eta_1(x_i^-)$ are free parameters when η_1 runs in $\mathcal{F}_x(\eta)$. We therefore derive from (516):

$$\omega_i(x_i^-)\bar{\eta}(x_i^-)+\frac{1}{\theta_i}\frac{\dot{\bar{\eta}}(x_i^-)}{a_{i-1}}=\frac{\dot{\bar{\eta}}(x_i^+)\gamma_i}{a_i}. \tag{517}$$

Since $\omega_i(x_i^-)=\bar{\omega}_i'(x_i^-)\gamma_i(x_i^-)$ and $\frac{1}{\theta_i\gamma_i}=\bar{\omega}_i$, we have:

$$\frac{\dot{\bar{\eta}}(x_i^+)}{a_i}=\frac{\dot{\bar{\eta}}(x_i^-)}{a_{i-1}}\bar{\omega}_i+\bar{\omega}_i'\bar{\eta}(x_i^-). \tag{518}$$

On the other hand, (507) implies that

$$\eta(x_i^+)=\gamma_i\eta(x_i^-). \tag{519}$$

Since $\overline{\mathcal{F}_x(\eta)}$ is the H^1-closure of $\mathcal{F}_x(\eta)$, (519) holds also on $\overline{\mathcal{F}_x(\eta)}$ and we thus have:

$$\bar{\eta}(x_i^+)=\gamma_i\bar{\eta}(x_i^-). \tag{520}$$

Comparing (517)–(519) to (514)–(515), we see that, if we denote

$$\mu_i^+=\frac{1}{a_i}\dot{\bar{\eta}}(x_i^+),\quad \mu_i^-=\frac{1}{a_{i-1}}\dot{\bar{\eta}}(x_i^-),\quad \eta_i^+=\eta(x_i^+),\eta_i^-=\eta(x_i^-),$$

(506)–(507) is satisfied, hence

$$D\theta_{m_i}\left(\frac{\dot{\bar{\eta}}}{a_{i-1}}(x_i^-)v-\bar{\eta}(x_i^-)[\xi,v]\right)=\frac{\dot{\bar{\eta}}}{a_i}(x_i^+)v-\bar{\eta}(x_i^+)[\xi,v]. \tag{521}$$

Therefore, the vector $z = -\bar{\eta}[\xi, v] + \frac{\dot{\bar{\eta}}}{a_i} v$ on $[x_i^+, x_{i+1}^-]$ is in $\mathcal{F}_x$ and $\bar{\eta}$ is equal to $\mathcal{N}(z)$. Since z satisfies also (514), the first variation of $\gamma(\dot{x})$ along z is zero along the ξ-piece $[x_{2i+1}, x_{2i+2}]$, and z is therefore in $T_x \overset{\circ}{M}{}^k$.

The proof of (ii) of Lemma 35 is thereby complete.

Proof of (iii): In view of (ii), we need to understand, once a degeneracy occurs, how the index behaves nearby.

When a degeneracy occurs at x_0, then we can find $u \in \ker\alpha \neq 0$ such that

$$dl_{x_0}(u) = u. \tag{522}$$

We can assume that the subset S_3 of $\overset{\circ}{M}{}^k$ defined by

$$S_3 = \{x \in \overset{\circ}{M}{}^k \cap \{\pi\theta_i = 1\} / \det(dl_x - Id_{|\ker\alpha}) = 0\} \tag{523}$$

is a manifold near x_0. Since $\pi\theta_i = 1$ near x_0, the eigenvalue 1 is double at x_0 and along S_3 for $dl_{x_0}|_{\ker\alpha}$.

Indeed,

$$\ell^*\alpha = \pi\theta_i\alpha. \tag{524}$$

Thus

$$\ell^*(\alpha \wedge d\alpha) = (\pi\theta_i)^2 \alpha \wedge d\alpha. \tag{525}$$

By (524), $\pi\theta_i$ is an eigenvalue for dl^t, with eigenvector α. Therefore, $\det dl_x|_{\ker\alpha} = \pi\theta_i$ and if 1 is an eigenvalue of $dl_x|_{\ker\alpha}$, then it has to be a double eigenvalue.

Near x_0, S_3 divides the set $\overset{\circ}{M}{}^k \cap \{\pi\theta_i = 1\}$ into two regions: $dl_x|_{\ker\alpha}$ can be reduced at x_0 to the form

$$\begin{pmatrix} 1 & c \\ 0 & 1 \end{pmatrix} \qquad c \neq 0 \quad (x_0 \text{ generic}). \tag{526}$$

The first factor is u. The second can be assumed to be any other vector in $\ker\alpha$, for example v. If v and u are dependent at x_0, then by genericity, they are not at a nearby point. In fact, these points x_0 where u and v are dependent are on a codimension one set in S_3 (see Propositions 29 and 30). Therefore, paths transverse to S_3 can be assumed to avoid them. Nearby, $dl_x|_{\ker\alpha}$ can be denoted:

$$\begin{pmatrix} 1+a_1 & c+a_3 \\ a_2 & 1+a_4 \end{pmatrix} \quad (1+a_1)(1+a_4) - a_2(c+a_3) = 1 (\text{because } \pi\theta_i = 1) \tag{527}$$

in a C^∞ extension (u_x, v_x) of the basis $(u_{x_0}, v_{x_0}) \cdot a_1, a_2, a_3, a_4$ are C^∞ functions which vanish at x_0.

On one side of S_3, denoted S_3^+, the eigenvalues of (527) are real and their sum is larger than 2. They are λ and $\frac{1}{\lambda}$.

$$\text{on } S_3^+ \colon \; a_1 + a_4 > 0. \tag{528}$$

On S_3^-, the other side, the eigenvalues are λ and $\bar{\lambda}$ and

$$\text{on } S_3^- \colon \; a_1 + a_4 < 0. \tag{529}$$

Indeed, the characteristic polynomial is, taking into account the fact that $(1 + a_1)(1 + a_4) - (c + a_3)a_2 = 1$

$$((1 + a_1) - x)(1 + a_4 - x) - a_2(c + a_3) = x^2 - x(2 + a_1 + a_4) + 1. \tag{530}$$

Hence, its discriminant is $(2 + a_1 + a_4)^2 - 4 = 4(a_1 + a_4) + (a_1 + a_4)^2$, hence the above discussion.

Let $w_\lambda, w_{\bar{\lambda}} = \bar{w}_\lambda$ be the eigenvectors of $dl_x|_{\ker\alpha}$ on S_3^-. We can write

$$w_\lambda = u + dv, \quad w_{\bar{\lambda}} = u + \bar{d}v, \qquad d(x_0) = 0. \tag{531}$$

We then have:

$$\begin{aligned} dl(w_\lambda + w_{\bar{\lambda}}) &= \lambda w_\lambda + \bar{\lambda} w_{\bar{\lambda}} = w_\lambda + w_{\bar{\lambda}} + (\lambda - 1)w_\lambda + (\bar{\lambda} - 1)w_{\bar{\lambda}} \\ &= w_\lambda + w_{\bar{\lambda}} + (\lambda + \bar{\lambda} - 2)u + ((\lambda - 1)d + (\bar{\lambda} - 1)\bar{d})v \\ &= w_\lambda + w_{\bar{\lambda}} + (a_1 + a_4)u + ((\lambda - 1)d + (\bar{\lambda} - 1)\bar{d})v \\ &= w_\lambda + w_{\bar{\lambda}} + (a_1 + a_4)\frac{(dl(v) - v)}{c} + ((\lambda - 1)d + (\bar{\lambda} - 1)\bar{d})v + O(d(x, S_3)^2). \end{aligned} \tag{532}$$

Observe that $dl(w_\lambda) = \lambda w_\lambda$. Thus:

$$1 + a_1 + d(c + a_3) = \lambda \quad a_2 + d(1 + a_4) = d(1 + a_1 + d(c + a_3)). \tag{533}$$

Now, by (530),

$$\begin{aligned} \lambda &= \frac{2 + a_1 + a_4}{2} + i\sqrt{-(a_1 + a_4) - \frac{(a_1 + a_4)^2}{4}} \\ &(a_1 + a_4 < 0 \text{ on } S_3^- \text{ close to zero}). \end{aligned} \tag{534}$$

Thus, near x_0,

$$\lambda - 1 \sim i\sqrt{-(a_1 + a_4)}. \tag{535}$$

Hence, by (533), since a_1 and a_3 are differentiable and zero at x_0:

$$cd \sim \lambda - 1 \sim i\sqrt{-(a_1 + a_4)} \tag{536}$$

and

$$d+\bar{d}=\frac{\lambda+\bar{\lambda}-2-2a_1}{c+a_3}=\frac{a_1+a_4-2a_1}{c+a_3} \tag{537}$$

which is differentiable.

Thus $w_\lambda + w_{\bar{\lambda}}$ is differentiable. Hence, (532) can be rewritten as:

$$dl(\psi)-\psi=((\lambda-1)d+(\bar{\lambda}-1)\bar{d})v+O(d(x,S_3)^2) \tag{538}$$

where ψ is differentiable on $\overline{S_3^-}$, is in $\ker\alpha$, and $\psi(x_0)=2u$. Observe that $(\lambda-1)d+(\bar{\lambda}-1)\bar{d}$, by (536) is equivalent to $\frac{2}{c}\ \mathrm{Re}(\lambda-1)^2 \sim \frac{2(a_1+a_4)}{c}$, which is also differentiable on $\overline{S_3^-}$, since a_1+a_4 extend to $S_3^-\cup S_3^+\cup S_3$. We can include, of course, $O(d(x,S_3)^2)$ into $dl(\psi)-\psi$, since $\det(dl-Id|_{\ker\alpha})=(\lambda-1)(\bar{\lambda}-1)=-(a_1+a_4)\geq c_1 d(x,S_3)$, $c_1>0$ for $x\in S_3^-\cup S_3$.

We thus have:

$$dl(\widetilde{\psi})-\widetilde{\psi}=((\lambda-1)d+(\bar{\lambda}-1)\bar{d})v,\quad \widetilde{\psi}\in\ker\alpha,\ \text{differentiable},\ \widetilde{\psi}(x_0)=2u \tag{539}$$

$(\lambda-1)d+(\bar{\lambda}-1)\bar{d}$ is a C^1-function on $\overline{S_3^-}$. We then conclude that (539) defines a vector-field on S_3^-, which lies on $\overline{\mathcal{F}_x}$, or else that the component of this vector-field on $-[\xi,v]$, denoted η, belongs to $\overline{\mathcal{F}_x(\eta)}$.

Indeed, since $dl(\widetilde{\psi})-\widetilde{\psi}$ differs by a component on v, all the conditions set in Definition 16 are satisfied, except for the first one, which is satisfied not using $\frac{\dot{\eta}(\text{at the initial point})}{a_0}$, but $\frac{\dot{\eta}(\text{at the initial point})}{a_0}-((\lambda-1)d+(\bar{\lambda}-1)\bar{d})$.

The function η is still in $\overline{\mathcal{F}_x(\eta)}$, since the closure is taken in the H^1-sense, hence values of $\dot{\eta}$ at a point are irrelevant. Furthermore, this vector-field at x_0 is $2u$. Thus, from the sign of the quadratic form Q_x on this extension, we can understand how the degeneracy evolves. Calling $\bar{\eta}$ the component on $-[\xi,v]$ of the vector-field, we have, using the fact that the vector-field is ξ-transported along the $[x_i^+,x_{i+1}^-]$-pieces:

$$\ddot{\bar{\eta}}=-\tau a_i^2\bar{\eta}\quad\text{on}\quad [x_i^+,x_{i+1}^-]. \tag{540}$$

Thus, using (540), we have:

$$Q_x(\bar{\eta},\bar{\eta})=2\sum_{i=1}^{k}\prod_{j=1}^{i}\theta_j\left(\frac{1}{a_i}\dot{\bar{\eta}}\bar{\eta}\Big|_{x_i^+}^{x_{i+1}^-}+w_i(x_i^-)\bar{\eta}(x_i^-)^2\right) \tag{541}$$

$$=2\sum_{i=1}^{k}\prod_{j=1}^{i}\theta_j\left(w_i(x_i^-)\bar{\eta}(x_i^-)^2-\frac{1}{a_i}\dot{\bar{\eta}}(x_i^+)\bar{\eta}(x_i^+)+\frac{1}{\theta_i}\frac{\dot{\bar{\eta}}(x_i^-)}{a_{i-1}}\bar{\eta}(x_i^-)\right)$$

$$-2\left(1-\prod_{j=1}^{k}\theta_j\right)\frac{\dot{\bar{\eta}}(x_k^-)\bar{\eta}(x_k^-)}{a_{k-1}}=_{(\pi\theta_j=1)}2\sum_{i=1}^{k}\prod_{j=1}^{i}\theta_j(\omega_i(x_i^-)\bar{\eta}(x_i^-)-$$

$$-\frac{1}{a_i}\dot{\bar{\eta}}(x_i^+)\bar{\eta}(x_i^+)+\frac{1}{\theta_i}\frac{\dot{\bar{\eta}}(x_i^-)}{a_{i-1}}\bar{\eta}(x_i^-)\bigg).$$

Our vector-field satisfies (501) by construction, but with

$$\mu_1^- = \frac{\dot{\eta}(x_1^-)}{a_{k-1}} - ((\lambda - 1)d + (\bar{\lambda} - 1)\bar{d}) \tag{542}$$

x_1^- is the initial point.

Indeed, our vector-field is built by taking $\widetilde{\psi}$ and its images through $D\theta_{m_i}$ and the ξ-transport along $[x_i^+, x_{i+1}^-]$ until, once the whole curve has been described, we return to

$$\widetilde{\psi} + ((\lambda - 1)d - (\bar{\lambda} - 1)\bar{d})v.$$

Thus, $\frac{\dot{\eta}(x_1^-)}{a_{k-1}} = \mu_1^- + (\lambda - 1)d + (\bar{\lambda} - 1)\bar{d}$. We also have:

$$\mu_j^{\pm} = \frac{\dot{\eta}(x_j^{\pm})}{a_{\delta(j)}} \qquad \delta(j) = j - 1 \text{ for } x_j^-, \delta(j) = j \text{ for } x_j^+. \tag{543}$$

For $j = 1$, (542) holds. Since (501) holds, (506) and (508) hold. Thus:

$$\begin{aligned} Q_x(\bar{\eta}, \bar{\eta}) = 2 \sum_{i=2}^{k} \prod_{j=1}^{i} \theta_j (\omega_i(x_i^-)\bar{\eta}(x_i^-) - \mu_i^+ \gamma_i + \frac{1}{\theta_i}\mu_i^-)\bar{\eta}(x_i^-) + \\ + 2\theta_1 \left(\omega_1(x_1^-)\bar{\eta}(x_1^-) - \mu_1^+\gamma_1 + \frac{\mu_1^- + (\lambda - 1)d + (\bar{\lambda} - 1)\bar{d}}{\theta_1} \right) \bar{\eta}(x_1^-). \end{aligned} \tag{544}$$

Observe that, by (504), $\frac{1}{\theta_i} = \bar{\omega}_i\gamma_i$ and by (506) and (508):

$$\omega_i(x_i^-) = \bar{\omega}_i'\gamma_i \tag{545}$$

$$\mu_i^+\gamma_i = \mu_i^-\bar{\omega}_i\gamma_i + \bar{\eta}_i^+\bar{\omega}_i'\gamma_i = \frac{\mu_i^-}{\theta_i} + \bar{\eta}_i^-\omega_i(x_i^-). \tag{546}$$

Thus

$$Q_x(\bar{\eta}, \bar{\eta}) = 2((\lambda - 1)d + (\bar{\lambda} - 1)\bar{d})\bar{\eta}(x_1^-). \tag{547}$$

We have seen that $(\lambda - 1)d + (\bar{\lambda} - 1)\bar{d}$ is equivalent to $\frac{2(a_1+a_4)}{c}$. Thus

$$Q_x(\bar{\eta}, \bar{\eta}) \sim 4(a_1 + a_4)\frac{\bar{\eta}(x_1^-)}{c}. \tag{548}$$

Since $a_1 + a_4$ is negative, this has the same sign as $-\bar{\eta}(x_1^-)c$. Recall now that:

$$cu = dl(v) - v \tag{549}$$

$$\widetilde{\psi}(x_1^-) = 2u(x_1^-) \text{ (at the curve } x_0 \text{ on } S_3). \tag{550}$$

Thus,

$$d\alpha(dl(v),v) = d\alpha(dl(v) - v, v) = c\, dl(u,v) \qquad (551)$$
$$= -c(-[\xi, v]\text{-component of } u).$$

The $-[\xi, v]$-component of u is nonzero, since $u(x_1^-)$ and $v(x_1^-)$ are independent. Because of (550), its sign is the same, in a neighborhood of the curve x_0, than the one of $\bar{\eta}(x_1^-)$. Thus, $Q_x(\bar{\eta}, \bar{\eta})$ has the same sign than $d\alpha(dl(v), v)$.

We thus have established, that at the crossing of a degeneracy, the change of index, when going into the region where the eigenvalues are complex, is provided by $d\alpha(dl(v), v)$.

Observe now that $d\alpha(dl(v), v)$ does not vanish when the eigenvalues are complex (not real). Indeed, if $d\alpha(dl(v), v)$ is zero, then $dl(v)$ is collinear to v, hence $dl|_{\ker\alpha}$ has a real eigenvalue.

Thus any crossing to a region where the eigenvalues are real, starting from the same region where the eigenvalues are complex, induces the same change of index in Q_x on $\overline{\mathcal{F}_x(\eta)}$. Thus, at the crossing of a region with complex eigenvalues, the index might change of $+1$ or -1, but when we come back to a hyperbolic region, this change is cancelled. Outside these crossings, the index is preserved. (iii) follows (iv) follows from Proposition 33. The first statement of (v) follows from (i) of Lemma 34. The second one follows from (iii) of Lemma 35. The proof of Lemma 35 is thereby complete.

II.2. A refined deformation Lemma

Before starting the next section, we state here a deformation result which follows from the construction of our flow in section I and from some additional natural hypotheses. We claim the following holds:

Deformation Lemma. *Assume that either*

a) $\sum_{i=1}^{k} a_i$ *satisfies the Palais-Smale condition on* Γ_{2k}, $\forall k$ *or*

b) Every point of M *has at least one coincidence point distinct from itself (i.e.* α *turns well along* v*, see Proposition A3 or Proposition 9 of [1]).*

Then, there are, generically on v*, isolated critical points at infinity* $(x_j^\infty)_{j=1}^\infty$ *(false and true ones, as we will see in the next section) and a suitable flow which does not increase the number of zeros of* b *such that, for any* $a_2 > a_1 > 0$, J_{a_2} *retracts by deformation onto*

$$J_{a_1} \cup \left(\bigcup_{a_1 \le J(x_j^\infty) \le a_2} W_u(x_j^\infty) \right) \cup \left(\bigcup_{\substack{\bar{x}\ \xi-\text{periodic} \\ \text{orbit},\ a_1 \le J(\bar{x}) \le a_2}} W_u(\bar{x}) \right).$$

Proof. Under b), $\sum_{\kappa}^{m_1,\dots,m_k} \cap J_{a_2}$ is compact, where $\sum_{\kappa}^{m_1,\dots,m_k}$ is the subset of Γ_{2k} defined by the rotation numbers m_i between x_i^- and x_i^+. Thus, Γ_{2k} can be cut in slices by the $\sum_{\kappa}^{m_1,\dots,m_k}$'s and each slice, when intersected with J_{a_2} is a compact stratified set. We can then define a pseudo-gradient flow for $\sum a_i$ on Γ_{2k} which preserves the slices, for example we can prescribe that this flow would preserve each $\sum_{\kappa}^{m_1,\dots,m_k}$. We then obtain a pseudo-gradient flow which will satisfy the Palais-Smale condition on $\bigcup_k \Gamma_{2k}$. Using our previous transversality results, we can assume that this flow has only nondegenerate, hence isolated critical points $(x_j^\infty)_{j=1}^\infty$.

Thus, under b), a) holds and we need to prove the same deformation result in both cases. Our previous deformation result stated that J_{a_2} could be deformed onto $J_{a_1} \cup \left(\left(\bigcup_k W_u(\Gamma_{2k}) \right) \cap J_{a_2} \right) \cup \left(\bigcup_{\bar{x}} W_u(\bar{x}) \right)$. W_u is here the unstable manifold for the iterative process which we built in section I.

We thus have two "flows", now: the first one is the result of our former iterative process. It has this special feature that if a flow-line in J_{a_2} is in a small enough neighborhood of Γ_{2k}, then, it will stay in a little bit larger neighborhood of Γ_{2k} unless it decreases a substantially.

The second flow is defined at infinity, on $\bigcup_k \Gamma_{2k}$, and can be easily extended to a tubular neighborhood of $\bigcup_k \Gamma_{2k}$ in C_β^+. Since it satisfies the Palais-Smale condition, for each k and a_2, we can find $C(a_2, k)$ such that, denoting s the time along the flow-lines of this flow, we have:

$$\int_0^1 (|b| - \nu)(s) \leq \int_0^1 (|b| - \nu)(0) + C(a_2, k)(a(0) - a(s))$$

($a(0) = \sum a_i(0), a(s) = \sum a_i(s)$, $\int_0^1 (|b| - \nu)(\tau) = \sum |b_i| =$ sum of the lengths of the v-pieces).

The extension of this flow to the tubular neighborhood can easily be made to satisfy the same type of inequality. Furthermore, because the initial flow is on $\bigcup_k \Gamma_{2k}$, i.e. preserves infinity, there is no problem for controlling $|b|_\infty(s)$ in the extension. Thus, for each slice $\Gamma_{2k} \cap J_{a_2}$, the extension of this second flow enjoys the same properties than our first sequence of flows of section I.

Combining these two flows, we find the following problem: the constant involved in the control of $\int (|b| - \nu)^+(s)$ for the second flow, $C(a_2, k)$, depends on k. Therefore, if we would have an infinite sequence of jumps from Γ_{2k} to $\Gamma_{2k'}$, etc., we could loose control of $\int (|b| - \nu)^+$. However, this does not happen with the second flow, since it preserves $\bigcup_k \Gamma_{2k}$, hence can be made to preserve any prescribed small neighborhood of $\bigcup_k \Gamma_{2k}$. The flow-lines in this neighborhood will mimic the ones on $\bigcup_k \Gamma_{2k}$, thus will stay near Γ_{2k} if they started nearby.

If it happens for the first flow, then we have to move away from a neighborhood of infinity and as we pointed out, this would require a substantial decrease in a.

Thus, the combination of the two flows (one after the other, to avoid having to prove estimate on $\frac{\partial}{\partial s}\int_0^1 \dot{b}^+$ for the second flow. Thus, instead of convex-combining them as vector-fields, we use the first one, and then the second one we refer to Appendix 4 for complete justifications.) The first global one and the second near infinity, can be completed. It will drive down $\bigcup_k W_u(\Gamma_{2k}) \cap J_{a_2}$ onto $\cup W_u(x_j^\infty) \cap J_{a_2}$. The result follows.

II.3. False and true critical points at infinity; their index

Let

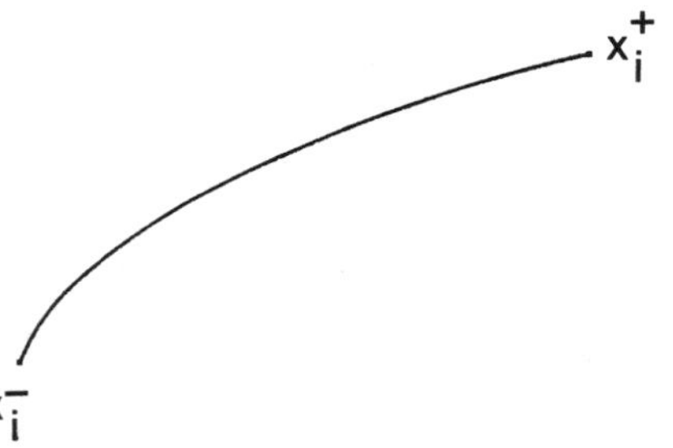

(552)

be a piece of v-orbit between two conjugate points. We will assume that it is parameterized by s on the time-interval $[\alpha, \beta]$. Let $x_1, \dots, x_k$ be the coincidence points of x_i^-, hence of x_i^+, in the segment $[x_i^-, x_i^+]$. Let s_j be the time needed to travel from x_i^- to x_j and let

$$\phi_{s_j}^* \alpha = \alpha_{x_j}(D\phi_{s_j}(\cdot)) = \lambda_j \alpha_{x_i}. \tag{553}$$

We then have:

Lemma 36. *i) Let us consider, for $\varphi \geq 0$ on $[\alpha, \beta)$, the differential equation*

$$(\varphi)\begin{cases} \dot{\overline{\lambda + \bar{\mu}\eta}} = \eta + \varphi \\ \dot{\eta} = -\lambda \end{cases}$$

Then, $\int_\alpha^\beta \eta$ is nonpositive if and only if $\int_\alpha^\beta (\alpha_{x_i^-}(D\phi_{-s}(\xi)) - 1)\varphi$ is nonpositive.

ii) Any solution, with $\varphi \geq 0$, of (φ) has $\int_\alpha^\beta \eta$ nonpositive if and only if $\lambda_j > 1$, for any $j = 1, \dots, k$.

Proof of Lemma 36. We have:

$$\frac{\partial}{\partial s}(\alpha_{x_i^-}(D\phi_{-s}(\xi))) = \alpha_{x_i^-}(D\phi_{-s}([v,\xi])), \tag{554}$$

since $D\phi_{-s}$ transports $\frac{\partial}{\partial s}$ which acts like v on ξ; and:

$$\frac{\partial}{\partial s}\alpha_{x_i^-}(D\phi_{-s}([v,\xi])) = \alpha_{x_i^-}(D\phi_{-s}([v,[v,\xi]])). \tag{555}$$

Since $d\alpha(v,[v,\xi])$ is 1, the component of $[v,[v,\xi]]$ on ξ is -1. Since $\bar{\mu} = d\alpha(v,[v,[\xi,v]])$ (see [1], also Proposition A2 of this paper), the component of $[v,[\xi,v]]$ along $-[\xi,v]$ is $\bar{\mu}$ and $[v,[v,,\xi]] = -\xi + \bar{\mu}([\xi,v]) + zv$. Thus

$$\begin{aligned}\frac{\partial}{\partial s}\alpha_{x_i^-}(D\phi_{-s}([v,\xi])) &= -\alpha_{x_i^-}(D\phi_{-s}(\xi)) + \bar{\mu}\alpha_{x_i^-}(D\phi_{-s}([\xi,v])) \\ &= -\alpha_{x_i^-}(D\phi_{-s}(\xi)) - \bar{\mu}\alpha_{x_i^-}(D\phi_{-s}([v,\xi])).\end{aligned} \tag{556}$$

Thus; setting

$$\psi = \alpha_{x_i^-}(D\phi_{-s}(\xi)) - 1 \tag{557}$$

we have:

$$-\left(\frac{\partial^2\psi}{\partial s^2} + \psi + \bar{\mu}\frac{\partial\psi}{\partial s}\right) = 1. \tag{558}$$

Let, now, $(\eta, \lambda + \bar{\mu}\eta)$ be a solution of (φ). Thus, η satisfies:

$$-\ddot{\eta} + \dot{\overline{\bar{\mu}\eta}} - \eta = \varphi. \tag{559}$$

We multiply (559) by ψ and integrate by parts. We derive:

$$\int_\alpha^\beta \eta + \dot{\psi}\eta - \psi\dot{\eta}\Big|_\alpha^\beta = \int_\alpha^\beta \varphi\psi. \tag{560}$$

Since x_i^+ is conjugate to x_i^-:

$$\psi(\alpha) = \psi(\beta) = 0. \tag{561}$$

Thus

$$\int_\alpha^\beta \eta + \dot{\psi}\eta\Big|_\alpha^\beta = \int_\alpha^\beta \varphi\psi. \tag{562}$$

Observe that we have also:

$$\dot{\psi}(\alpha) = \dot{\psi}(\beta) = 0. \tag{563}$$

This is easily derived from (554) and from the fact that x_i^+ is conjugate to x_i^-, hence $[\xi, v]$ is transported into $\ker\alpha$ between x_i^- and x_i^+. Thus

$$\int_\alpha^\beta \eta = \int_\alpha^\beta \varphi\psi \tag{564}$$

(i) follows.

In order to prove (ii), we observe that if $\int_\alpha^\beta \eta$ is nonpositive for every $\varphi \geq 0$, then ψ has to be nonpositive. Since ψ vanishes at α and β, we consider the points where ψ attains its maxima between α and β. Using (554), these are points $\tilde{s}$ such that:

$$\alpha_{x_i^-}(D\phi_{-\tilde{s}}([v,\xi])) = 0. \tag{565}$$

Thus, $\ker\alpha$ has turned and come back on itself as a plane between 0 and $\tilde{s}$. These are the coincidence points of x_i^-. Those which have the opposite orientation obviously have ψ negative. At the other ones, which are one of the x_j's, $i = 1, \dots, k$, $\alpha_{x_i^-}(D\phi_{-\tilde{s}}(\xi))$ is equal to $\frac{1}{\lambda}$. Thus, ii) holds if and only if

$$\frac{1}{\lambda_j} < 1 \Leftrightarrow \lambda_j > 1 \qquad \forall j = 1, \dots, k \qquad \text{q.e.d.} \tag{566}$$

The flow that we have built is a little bit more special and does not allow general φ's. Indeed, along the pieces where b is extremely large and where Diracs develop, one can check that b is almost constant equal to $|b|_\infty$. Since a is constant, with respect to the limit curve in C_β^+, the φ's with arise are constant: they are basically equal, if starting from a curve of C_β^+ we use (φ) to push inwards in C_β^+ during the time ε, to a_ε, the ξ-component of $\dot{x}_\varepsilon$ (x_ε is the curve of C_β^+ obtained after the time ε) along these time-intervals, i.e. we have:

$$a_\varepsilon / b_\varepsilon \simeq \varepsilon\varphi \tag{567}$$

b_ε is almost constant equal to $|b|_\infty$ (which tends to $+\infty$). Thus:

$$a_\varepsilon / |b|_\infty \simeq \varepsilon\varphi. \tag{568}$$

Since our curves are in fact in C_β, a_ε has to be constant. Thus, φ has to be constant, the same constant along all these pieces. We are thus led to study the problem:

$$\begin{cases} \dot{\overline{\lambda_i + \bar{\mu}\eta_i}} = \eta_i + 1 \\ & \text{on } [\alpha_i, \beta_i] \\ \dot{\eta}_i = -\lambda_i \end{cases} \tag{569}$$

Let ψ_i be the solution of the dual problem (558). We then have:

Lemma 37. *Assume that we are at a critical point at infinity*

$$\sum \int_{\alpha_i}^{\beta_i} \eta_i \text{ is critical if and only if } \sum \int_{\alpha_i}^{\beta_i} \psi_i \text{ is nonpositive.}$$

Lemma 37 is obvious. As we establish rigorously below, introducing a parametrization of C_β near Γ_{2k}, the condition on $\sum \int_{\alpha_i}^{\beta_i} \psi_i$ is the right condition for us. However, it can, at any time, by a modification of the flow – see below – be replaced by the condition of Lemma 36 ($\psi_i \leq 0 \forall i$). The exchange of conditions cannot change the differences of topology which occur. There are therefore some hidden identities in the differences of topology as expressed using our flow or the flow as we will modify it to adjust to the conditions of Lemma 36. In addition, our flow requires $\eta_i \geq 0$. The above analysis holds if we bring in the conjugate point flow, which we use under (H) or if α turns well along v. The requirement $\eta_i \geq 0$ has to be analyzed separately when we do not use these conditions. This can be done with a more refined Morse Lemma, which will be completed elsewhere.

We then have the following straightforward generalization of Lemmas 36 and 37:

Lemma 38. *Let x be a curve of $\Gamma_{2k} \cdot [x_i^-, x_i^+]$ a piece of v-orbit of x, described on the interval $[\alpha_i, \beta_i]$.*

i) the problem (φ) has no solution, for $\varphi \geq 0$, having $\int_{\alpha_i}^{\beta_i} \eta$ positive if and only if ψ is negative on $[\alpha_i, \beta_i]$.

ii) Let (λ_i, η_i) be a solution of (569) on $[x_i^-, x_i^+]$. Then

$$\sum \int_{\alpha_i}^{\beta_i} \eta_i = \sum \int_{\alpha_i}^{\beta_i} \psi_i - \sum (\dot{\psi}(\beta_i)\eta_i(\beta_i) - \psi(\beta_i)\dot{\eta}_i(\beta_i)).$$

Proof. Observe that $\psi_i(\alpha_i) = \dot{\psi}_i(\alpha_i) = 0$. Then (ii) follows immediately from (560) and (569). For (i), we observe that we can ask in (φ) that $\eta_i(\beta_i) = \dot{\eta}_i(\beta_i) = 0$. The boundary term then vanishes in (560) and we have:

$$\int_{\alpha_i}^{\beta_i} \eta = \int_{\alpha_i}^{\beta_i} \varphi\psi. \tag{570}$$

The conclusion follows.

We now prove the following:

Lemma 39. *Let $\bar{x}$ or $\bar{x}_\infty$ be a critical point or a critical point at infinity of J in C_β, of critical value c. For sake of simplicity; we assume that $\bar{x}$ or $\bar{x}_\infty$ is the unique critical point at the level c. We also assume that, for a time t_0 where $\bar{x}$ or $\bar{x}_\infty$ is tangent to ξ (to simplify, our two previous lemmas take care of the situation where $\bar{x}$ or $\bar{x}_\infty$ is tangent to v at t), the function $\psi_t(s) = \alpha_{\bar{x}(t)}(D\phi_{-s}(\xi)) - 1$ (or*

$\alpha_{\bar{x}_\infty(t)}(D\phi_{-s}(\xi)) - 1)$ *is positive for some time* s_0 *on the* v*-orbit through* $\bar{x}(t)$ *or* $\bar{x}_\infty(t)$. *There exists then a* $\delta_0 > 0$ *such that* $J_{c+\delta_0}$ *retracts by deformation onto* $J_{c-\delta_0}$.

Proof. First, using our flow and the related deformation lemmas, we can deform $J_{c+\delta_0}$, for δ_0 small enough, onto $W_u(V) \cup J_{c-\delta_0}$, where V is any presented $W^{1,1}$ neighborhood of $\bar{x}$ or $\bar{x}_\infty$ in C_β^+.

Since, given δ_0, V is as small as we wish, we can choose V such that

$$c - \delta = \operatorname*{Min}_V J > c - \delta_0. \tag{571}$$

Choosing a fixed neighborhood V_1 and letting V be smaller and smaller, we claim that we will have eventually:

$$W_u(V) \cap \partial V_1 \subset J_{c-\delta_1} \tag{572}$$

where δ_1 is a fixed positive number. We can take δ_1 to be δ_0 in the sequel.

(572) holds because a trajectory starting from V will not reach ∂V_1, using our flow which has strong compactifying properties and does not allow sharp oscillations to develop, unless J has sufficiently decreased (see the deformation lemmas, also Lemma 24). Thus, $J_{c+\delta_1}$ retracts by deformation onto some set $\widetilde{B}$ included into $(J_{c+\delta_1} \cap \overline{V}_1) \cup J_{c-\delta_1}$ and which intersects ∂V_1 along a subset of $J_{c-\delta_1}$. Let $B = \widetilde{B} \cap J_{c+\delta_1} \cap \overline{V}_1$. It suffices therefore to prove that the injection

$$(B, B \cap \partial V_1) \longrightarrow (J_{c+\delta_1}, J_{c-\delta_1})$$

is homotopic to zero. We will build a retraction of this map to zero, which will stay constant, equal to the injection of $B \cap \partial V_1$ into $J_{c-\delta_1}$, on the second factor. Observe that δ_1 can be made smaller, once V_1 is given. (572) still holds.

The idea of this deformation is quite simple and making it to work practically is, after the construction of our flow in the first sections of this paper, quite standard. We will only outline the construction, the details bear no real difficulty:

First, we observe, as we already did in [1] and have recalled in the introduction of the present work, that if, at $t = t_0$, and for some $s_0 > 0$,

$$\alpha_{\bar{x}(t_0)}(D\phi_{-s_0}(\xi)) > 1 \tag{573}$$

then, we have some unusual variation of the curve $\bar{x}$ decreasing it level below the level c.

We take the piece of curve $\bar{x}$ tangent to ξ around $\bar{x}(t_0)$

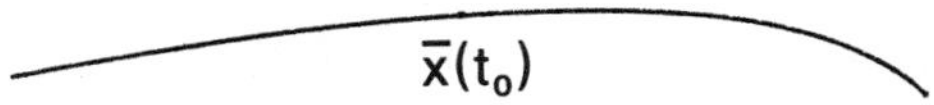

FIGURE Z

In a first step, we build a v-oscillation progressively, which takes the form of a combination of one positive and one negative Dirac masses z at $\bar{x}(t_0)$, which move progressively, along the v-orbit through $\bar{x}(t_0)$ from $\bar{x}(t_0)$ to $\phi_{s_0}(\bar{x}(t_0))$:

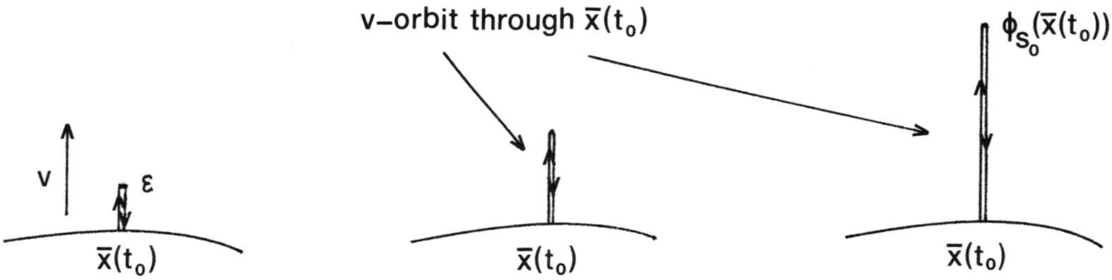

FIGURE AA

This is the first stage of our deformation in C_β^+. Once we have arrived at $\phi_{s_0}(\bar{x}(t_0))$, we open up the curve at $\phi_{s_0}(\bar{x}(t_0))$, creating an $\varepsilon\xi$ piece:

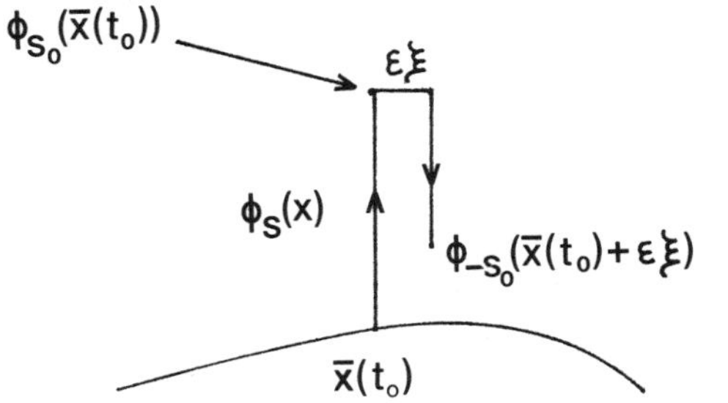

FIGURE BB

The vector $\frac{\phi_{-s_0}(\bar{x}(t_0)+\varepsilon\xi)-\bar{x}(t_0)}{\varepsilon} = D\phi_{-s_0}(\xi) + o(1)$ is the direction of opening downstairs, $\phi_{-s_0}(\bar{x}(t_0)+\varepsilon\xi)$ does not lie on the curve $\bar{x}(t)$. We need to close our variation i.e. to define, for $t \in [t_0, t_0 + \delta]$, λ, η such that:

$$\begin{cases} \dot{\eta} = \mu a \qquad \dot{\overline{\lambda + \bar{\mu}\eta}} = \gamma(t) \\ \lambda\xi(\bar{x}(t_0)) + \mu(t_0)v(\bar{x}(t_0)) - \eta(t_0)[\xi, v](\bar{x}(t_0)) = D\phi_{-s_0}(\xi) \\ \lambda\xi(\bar{x}(t_0 + \delta)) + \mu(t_0 + \delta)v(\bar{x}(t_0 + \delta)) - \eta(t_0 + \delta)[\xi, v](\bar{x}(t_0 + \xi)) = 0. \end{cases} \tag{574}$$

The above system of equations has of course a solution, which defines a variation of $\bar{x}(t)$, which keeps $\dot{\tilde{x}}(t)$ as $a\xi + bv$, $a > 0$ ($\tilde{x}$ is the curve after deformation), opens it at $\bar{x}(t_0)$ with the direction $D\phi_{-s_0}(\xi)$ and dies at the time $t_0 + \delta$. Rigorously speaking, this system defines a tangent vector to the open curve $\bar{x}$ (opened at t_0) in the space of open curves whose tangent vector is in $\ker \beta$. Using the associated natural exponential map, we can define this deformation for $\varepsilon > 0$, using a piece of curve above tangent to $\varepsilon\xi$:

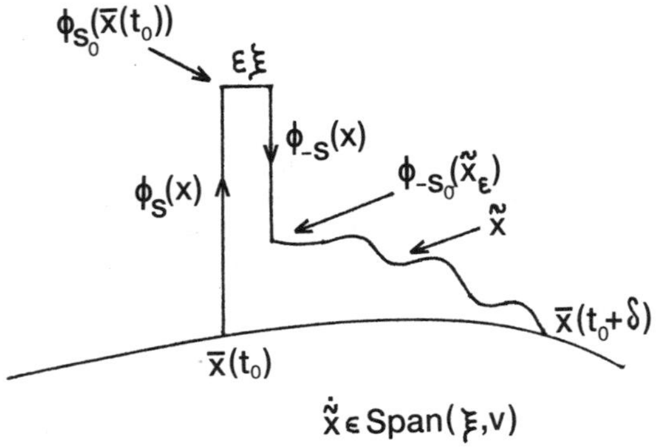

FIGURE CC

This procedure can be easily applied, in a continuous way to all the curves x in a $W^{1,1}$-neighborhood of $\bar{x}$, building therefore, when we combine (AA), (BB) and (CC) the following deformation, in a continuous way, in a $W^{1,1}$-neighborhood of $\bar{x}$:

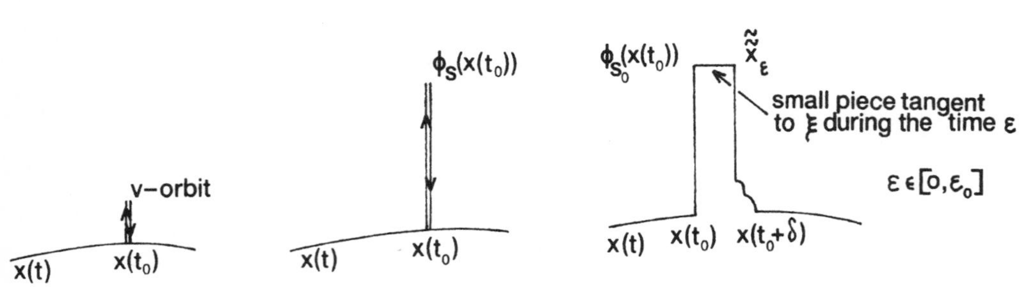

FIGURE DD

All this process takes place in C_β^+.

We can achieve this process progressively from ∂V_1, using inwards, so that the curves of $B \cap \partial V_1$ are kept unchanged, while, as we move slightly inwards, we end with the third phase of our deformation.

We can compute the variation of $J(x)$ between the initial time and the third step of our deformation; at first order in ε, this variation is:

$$\varepsilon\left(1+\int_{t_0}^{t_0+\delta}(\overline{\dot{\lambda+\bar{\mu}\eta}}-b\eta)\right)=\varepsilon\left(1-(\lambda+\bar{\mu}\eta)(t_0)-\int_{t_0}^{t_0+\delta}b\eta\right)$$

$$=\varepsilon\left(1-\alpha(D\phi_{-s_0}(\xi))-\int_{t_0}^{t_0+\delta}b\eta\right)=\varepsilon(1-\alpha(D\phi_{-s_0}(\xi))+\varepsilon|\eta|_\infty\int_{t_0}^{t_0+\delta}|b|$$

η has only to satisfy (574). It is easy to see that therefore, we can ask:

$$|\eta|_\infty \leq C(|D\phi_{-s_0}(\xi)|+1) \tag{575}$$

where C is a universal constant.

$\int_{t_0}^{t_0+\delta}|b|$ is zero on $\bar{x}$, hence can be made as small as we wish if we restrict the $W^{1,1}$-neighborhood of $\bar{x}$ where we pick up x. Thus, we derive that the variation of J will be larger in absolute value, than:

$$\frac{\varepsilon}{2}|1-\alpha(D\phi_{-s_0}(s))| \tag{576}$$

and negative, for any $\varepsilon \in [0,\varepsilon_0]$.

We thus have decreased at least of

$$\frac{\varepsilon_0}{2}(1-\alpha(D\phi_{-s_0}(\xi)))$$

which is a fixed amount.

Choosing δ_1 such that:

$$\delta_1 < \frac{\varepsilon_0}{4}|1-\alpha(D\phi_{-s_0}(\xi))|$$

we can easily time our deformation so that it induces the desired result on the injection $(B, B\cap\partial V_1) \hookrightarrow (J_{c+\delta_1}, J_{c-\delta_1})$. Therefore, our result holds for C_β^+. We would like to prove it for C_β and this requires a little bit more work.

First, we slightly modify the first step of our deformation, when we create v-Diracs. Instead of creating a simple v-Dirac, back and forth, which we open later, we open it from the beginning, creating a very tiny ξ-piece:

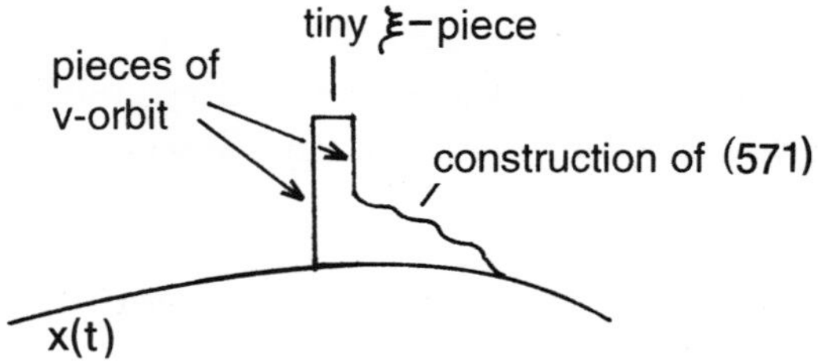

FIGURE EE

The modification which we have completed with respect to (I) is simply that a v-oscillation of this type:

FIGURE FF

has been transformed into

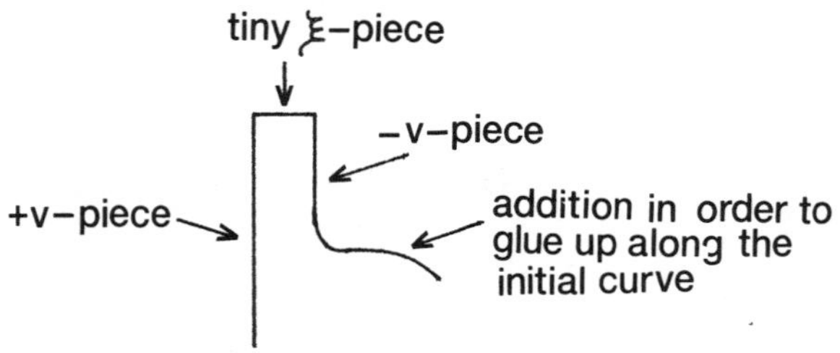

FIGURE GG

The tiny ξ-piece cancels as the oscillation cancels. It grows and becomes a piece of ξ-orbit during the time ε to ε_0 when s becomes s_0. Before the time s_0, it is extremely tiny. In this process, our former deformation has been modified very slightly. Therefore, the deformation statement is unchanged, the energy levels remaining essentially the same.

Our task is now to modify a little bit this construction as that a becomes positive on this deformation, yielding after reparametrization a deformation in C_β.

Observe that we face the following problem. We have a piece:

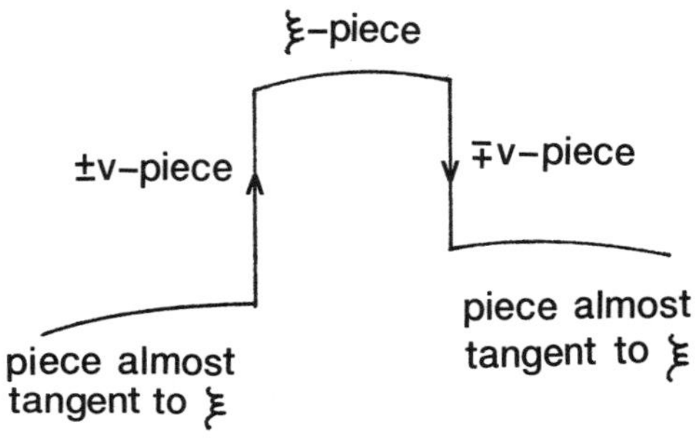

(577)

and we wish to push it into $a > 0$. The last piece is almost tangent to ξ because, in (574), $|\mu|_{C^2}$ depends only δ and $|D\phi_{-s_0}(\xi)|$ i.e.

$$|\dot{\mu} + a\eta\tau - b\eta\bar{\mu}_\xi| \leq \frac{C}{\delta^2}|D\phi_{-s_0}(\xi)| \tag{578}$$

on this deformation

$$|\dot{\overline{\lambda + \bar{\mu}\eta}} - b\eta| \leq C|D\phi_{-s_0}(\xi)|(1 + |b|). \tag{579}$$

We start with curves better than $W^{1,1}$ close to $\bar{x}$. Then x's which we obtain using our flow are, near $\bar{x}(t_0)$, close $W^{1,\infty}$ to $\bar{x}$. Thus, we can assume that $|b|$ is locally less than 1 and

$$|\dot{\mu} + a\eta\tau - b\eta\bar{\mu}_\xi| \leq \frac{C}{\delta^2}|D\phi_{-s_0}(\xi)| \tag{580}$$

$$|\dot{\overline{\lambda + \bar{\mu}\eta}} - b\eta| \leq 2C|D\phi_{-s_0}(\xi)|. \tag{581}$$

During the time ε_0, we have ($a(0)$ is the value of a at the time zero of the deformation, $a(\varepsilon_0)$ is any value of $a(t)$ at the time ε_0 of the deformation)

$$a(\varepsilon_0)(t) \geq a(0)(t) - 2C\varepsilon_0|D\phi_{-s_0}(\xi)| \tag{582}$$

$$|b(\varepsilon_0)|(t) \leq |b(0)|(t) + \frac{C\varepsilon_0}{\delta^2}|D\phi_{-s_0}(\xi)|.$$

Using our flow, $\frac{|b(0)|(t)}{|a(0)|(t)}$ is small.

Taking ε_0 small enough, for δ given, $\frac{|b(\varepsilon_0)|(t)}{a(\varepsilon_0)(t)}$ remains small as claimed. As we already explained earlier, V_1 being given, δ_1 can be chosen smaller, in function of ε_0, if necessary. δ is given, once and for all, and depends only on the curve $\bar{x}$.

We need, thus, to push the curves drawn in (577) into $\{a > 0\}$.

Let us introduce the differential equation, on the $\pm v$-piece of (577):

$$\begin{cases} \dot{\overline{\lambda + \bar{\mu}\eta}} = b\eta + \varphi & \varphi > 0 \ b \text{ a constant, equal to } +1 \text{ or } -1 \\ \dot{\eta} = -\lambda b & \\ \begin{cases} \lambda + \bar{\mu}\eta(0) = -1 \\ \eta(0) = 0 \\ \mu(0) = -b(t_0)/a \end{cases} & \begin{cases} (\lambda + \bar{\mu}\eta)(s_1) = \gamma \quad \gamma > 0 \\ \eta(s_1) = 0 \\ \mu(s_1) = 0 \end{cases} \end{cases} \tag{583}$$

where s_1 is the time spent on this piece

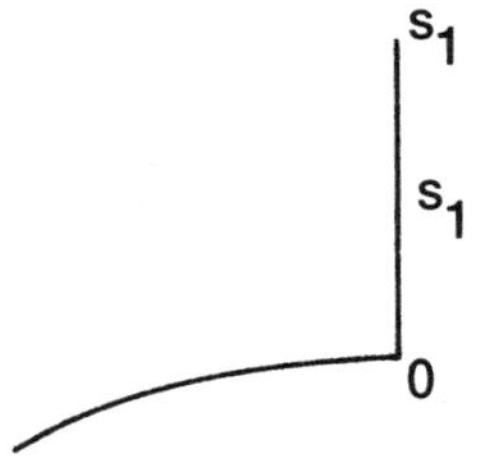

FIGURE HH

and $b(t_0), a$ are the components of $\dot{x}$ at the time t_0. If we are able to find a $\varphi > 0$ for which we can solve (583), then we can push this piece into $\{a > 0\}$. Observe that the initial point along this v-orbit, which is $x(t_0)$, will be pushed backwards into $x(t_0 - \varepsilon)$, since $z(0) = -\xi - \frac{b(t_0)}{a} v$ is parallel to $-\dot{x}(t_0)$, while the final point will be pushed along the ξ-piece we have created a little bit forward, since $z(s_1) = \gamma\xi$. Thus, we would have:

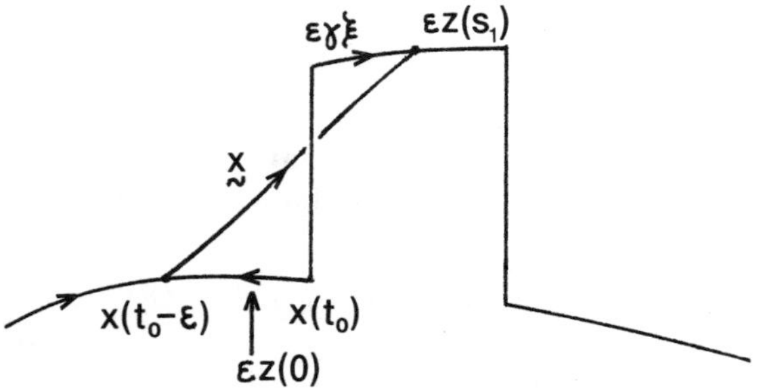

(584)

The new $\underset{\sim}{x}$, because $\dot{\overline{\lambda + \bar{\mu}\eta - b\eta}} = \varphi > 0$ will be in $a > 0$. (Again, we are replacing vector-fields by deformations, using exponential maps.) We can complete the same operation on the other side by solving the same kind of equation. (583) reverses fully, we obtain the same equation. We can time ε on one side and ε' on the other side so that they coincide above, yielding a curve $\underset{\sim}{x}$:

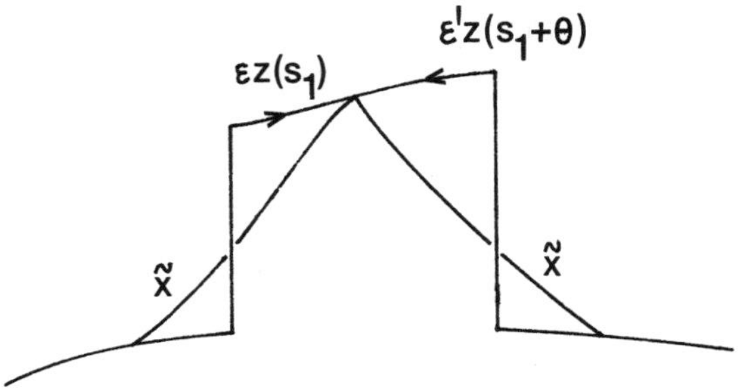

FIGURE II

which is closed and is contained in $\{a > 0\}$.

Again, the variation in $J(x)$ can be made as small as we wish, so that our deformation statement still holds. Thus, we are left with (583), where we can forget the requirements on $\mu(0)$ and $\mu(s_1)$ which are very easy to fulfill.

A remarkable fact is that we can also drop the sign requirement on γ in (583). If γ is negative instead of positive, we would rather describe:

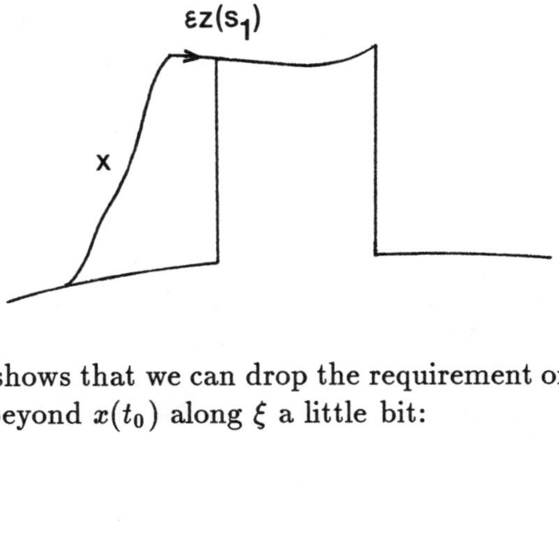

(585)

A similar argument shows that we can drop the requirement on $(\lambda + \bar{\mu}\eta)(0)$, allowing the curve to travel beyond $x(t_0)$ along ξ a little bit:

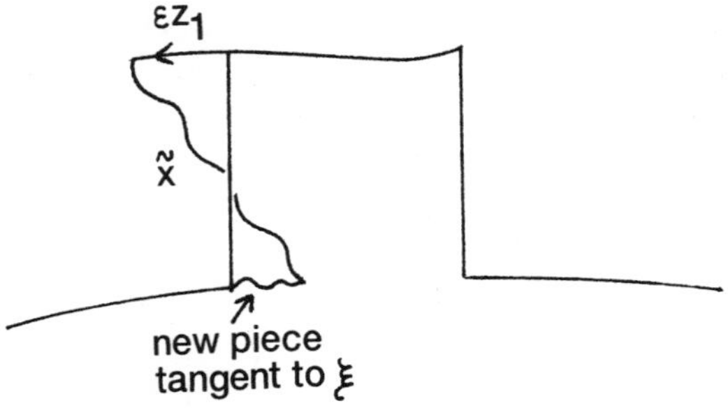

(586)

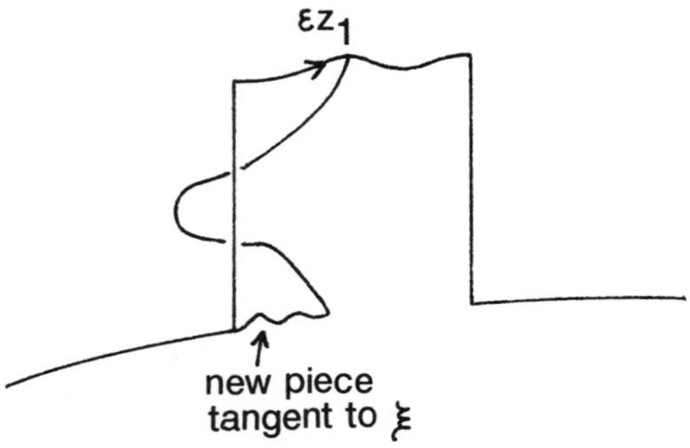

(587)

and, in fact, we can also drop the requirement on $\eta(0)$ to be zero, i.e. η at the time t_0. We would then extend η before t_0, on $[t_0 - \delta, t_0]$ so that it vanishes at $t_0 - \delta$ and satisfies:

$$\dot{\eta} = \mu a \quad \text{on} \quad [t_0 - \delta, t_0] \tag{588}$$

$\lambda + \bar{\mu}\eta$ would also be extended so as to vanish before $t_0 - \delta$. The curves would still be in $\{a > 0\}$.

The exponential map allows to transform these local vector-fields into deformations. We only need the continuity of the process, which is equivalent to the existence and continuity of such a vector-field, which, now, has only to satisfy, after our previous remarks:

$$\begin{cases} \dot{\overline{\lambda + \bar{\mu}\eta}} = b\eta + \varphi \quad \varphi > 0 & b = +1 \text{ or } -1 \\ \dot{\eta} = -\lambda b & \eta(s_1) = 0 \end{cases} \tag{589}$$

Such a couple $(\lambda + \bar{\mu}\eta, \eta)$ is very easy to define continuously on $[0, s_1]$. The extension of $\eta, \lambda + \bar{\mu}\eta$, η satisfying (588), (λ, η) vanishing before $t_0 - \delta$ is also very easy to complete in a continuous way.

A similar construction can be completed for the other side of the oscillation.

Combining all our arguments and using the exponential map, we construct a deformation in $\{a > 0\}$, which satisfies the requirements of our lemma. Reparametrizing, we can ask that this deformation takes place in C_β. Our argument is now complete.

We introduce the following "canonical" subset of $\sum_k$, which will turn out to be a manifold with very special properties.

Definition 18. *We define $\widetilde{\sum}_{k-1}$, the quantic manifold of α of order k to be the subset of*

$$\sum_k \text{ defined by the relation } \prod_{i=1}^{k} \theta_i = 1$$

and we have the following:

Theorem 40.

i) All the critical points of $\sum a_i$, on Γ_{2k}, which are not critical points at infinity (i.e. those which are not described in the introduction) do not induce, generically on v, any difference of topology in the level sets of J.

ii) Generically on v, $\widetilde{\sum}_{k-1}$ is a manifold of dimension $k-1$. The only critical points of $\sum a_i$ on $\widetilde{\sum}_{k-1}$, which are also critical in the sense of Lemma 36, are the critical points at infinity (the quantic periodic orbits). Furthermore, the difference of the Morse indices of $J_{|\widetilde{\sum}_{k-1}}$, in a same connected component of $\widetilde{\sum}_{k-1}$, for two critical points at infinity, equals the difference when computed on $\mathcal{B}$. Finally, $\mathcal{B}_x = \mathcal{H}_x$ at any such critical point at infinity.

iii) The deformation of the levels sets of J near a critical curve of $\sum a_i$ which is not critical at infinity in the sense of the introduction (i.e. some of the v-jumps are not between conjugate points) can be made into a Fredholm deformation.

Proof. Let us first prove that the difference of indices in $\widetilde{\sum}_{k-1}$ equals the one in $\mathcal{B}$. Coming back to Lemma 35, we know that, on $\sum_k$, we have equality between these differences, up to ± 1, depending on the elliptic or hyperbolic nature of the critical point at infinity. This is an immediate corollary of the behavior of Q_x on $\mathcal{F}_x$ and of the orthogonality of $T_x \sum_x$ and $\mathcal{F}_x$. The ± 1 comes from the crossing of a hypersurface in $\widetilde{\sum}_{k-1}$ (see the path of Lemma 35) where $dl - Id|_{\ker \alpha}$ is noninvertible i.e. there exists $u \in \ker \alpha$ such that

$$dl(u) = u \qquad \alpha(u) = 0. \tag{590}$$

Observe that this u defines a vector-tangent to $\sum_k$.

However, this vector is not tangent to $\widetilde{\sum}_{k-1}$ unless it satisfies one more equation, namely:

$$\sum \frac{d\theta_i}{\theta_i}(u_i^-) = 0 \tag{591}$$

where u_i^- is the value of this vector at x_i^-. (591) defines one more restriction on u, independent of (590). Propositions 29 and 30 imply that the set of curves x of $\widetilde{\sum}_{k-1}$ satisfying (590) and (591) is of codim 2, hence can be avoided along a path.

Thus, the degeneracy does not occur in $T_x\widetilde{\sum}_{k-1}$ at such a crossing. It does not occur also in the full space, i.e. on $\mathcal{B}_x$. Indeed, computing as in (491)–(496), but with $\alpha(z(x_i^-))$ not necessarily zero anymore, we obtain: ($\tilde{u}$ is the degenerate direction in $T_x\sum_k$)

$$\mathcal{Q}_x(z,\tilde{u}) = 2\sum\prod_{j=1}^{i}\theta_j\alpha(z(x_i^-))\frac{d\theta_i}{\theta_i^2}(u_i) \tag{592}$$

for any $z \in \mathcal{B}_x$. On $\mathcal{B}_x, \alpha(z(x_i^-))$ is a free parameter. Since (591) is not satisfied, we can build z in $\mathcal{B}_x$ such that $\mathcal{Q}_x$ in span$(z,\tilde{u})$ does not degenerate.

z can be extended before and after the degeneracy occurs along the path in $\widetilde{\sum}_{k-1}$, as well as $\tilde{x}$ which we did extend in (539). We then can come back to our usual $T_x\sum_k \oplus\mathcal{F}_x$ decomposition, which yields a Fredholm $\mathcal{Q}_x$ before and after the degeneracy (in $\overline{\mathcal{F}}_x$, the closure of $\mathcal{F}_x$ under the H^1-norm on η and λ). We can split, before and after, $T_x\sum_k \oplus\mathcal{F}_x$ into Span$(z,\tilde{u})$ (after extension) and its $\mathcal{Q}_x$-orthogonal. The index in Span$(z,\tilde{u})$ is the same, since the determinant of $\mathcal{Q}_x|_{\text{Span}(z,\tilde{u})}$ does not vanish, before and after the degeneracy. Its $\mathcal{Q}_x$-orthogonal does not contain $\tilde{u}$ at the degeneracy, which is assumed to be simple in $\mathcal{F}_x$. Thus, would $\mathcal{Q}_x$ be Fredholm on $\overline{\mathcal{B}}_x$ (again the closure of B_x under the H^1-norm on η and λ), our conclusion would hold, since the total index would be equal to the index on Span$(z,\tilde{u})$ which is unchanged, plus the index on its $\mathcal{Q}_x$-orthogonal which is again unchanged. Observe that $\overline{\mathcal{B}}_x$ is a Hilbert space equal to $\overline{\mathcal{H}}_x$: the boundary conditions at $x_i^\pm$ are irrelevant, for the v-component in the H^1-closure, only the components on ξ and $[\xi,v]$ play a role in the closure. Thus, in the H^1-closure, $\overline{\mathcal{H}}_x = \overline{\mathcal{B}}_x$.

$\overline{\mathcal{F}}_x$ is a finite-codimension subspace of $\overline{\mathcal{H}}_x = \overline{\mathcal{B}}_x$ where $\mathcal{Q}_x$ is Fredholm. $\overline{\mathcal{F}}_x$ is H^1-closed (see the discussion about $\overline{\mathcal{F}_x(\eta)}$ after Definition 16, for example. $\overline{\mathcal{F}} = \cup\overline{\mathcal{F}}_y \to \sum_k$ defines a bundle via $\overline{\mathcal{F}_y(\eta)}$. $\overline{\mathcal{H}}$ also defines a bundle and $\overline{\mathcal{F}}$ is a closed subbundle of $\overline{\mathcal{H}}$ of finite codimension, equal to k (the $\alpha(z(x_i^-))$'s are zero on $\mathcal{F}_x$, while they are free parameters on $\overline{\mathcal{H}}_x = \overline{\mathcal{B}}_x$) $\mathcal{Q}$ is Fredholm on $\overline{\mathcal{F}}$. It is easy to construct a supplement G of $\overline{\mathcal{F}}$ in $\overline{\mathcal{B}}$ of dimension k such that $\mathcal{Q}$ is obviously continuous on $\overline{\mathcal{F}}\oplus G = \overline{\mathcal{B}}$. Indeed, we can pick up values $\alpha(z(x_{i_0}^-)) = 1, \alpha(z(x_j^-)) = 0$ for $j \neq i_0$ and build a vector f_{i_0} in $\mathcal{B}$ satisfying these conditions. $\operatorname{Span}_{i=1}^{k}(f_i)$ is G. Clearly, $\overline{\mathcal{B}} = \overline{\mathcal{F}}\oplus G$. Computing, for $z = f+g$, $f \in \mathcal{F}_x$, $g \in G_x$, we find

$$\mathcal{Q}_x(f+g,f+g) = \mathcal{Q}_x(f,f) + \mathcal{Q}_x(g,g) + 2\mathcal{Q}_x(f,g). \tag{593}$$

Arguing as in (491)–(496) ($\dot{\mu}_g + a\eta_g\tau$ is not zero, here, contrary to what happening with $\bar{z}$ in (491)–(496)), we have

$$\mathcal{Q}_x(f,g) = \sum \prod_{j=1}^{i} \theta_j \left(-2 \int (\dot{\mu}_g + a\eta_g \tau)\eta_f \right). \tag{594}$$

This is clearly a continuous linear form, for each g, on $\overline{\mathcal{F}_x(\eta)}$. The conclusion follows.

We thus are left with proving that the other degeneracies occur in $T_x\widetilde{\sum}_{k-1}$, a more precise statement than $T_x \sum_k$. At any other x, we have:

$$\mathcal{Q}_x(z,z') = \sum \prod_{j=1}^{i} \theta_j \left(\alpha_{x_i^-}(z_{x_i^-}) \frac{d\theta_i}{\theta_i^2}(z'(x_i^-)) + \right. \tag{595}$$

$$\left. + \alpha_{x_i^-}(z'(x_i^-)) \frac{d\theta_i}{\theta_i^2}(z(x_i^-)) \right) \text{ for } z, z' \in T_x \sum\nolimits_k .$$

We also have: (these translate the v and ξ-transport equations on z and z')

$$\left(d\left(\frac{1}{\theta_j}\right) \wedge \alpha + \frac{1}{\theta_j} d\alpha \right) (z(x_j^-), z'(x_j^-)) = d\alpha(z(x_j^+), z'(x_j^+)) \tag{596}$$

$$d\alpha(z(x_j^+), z'(x_j^+)) = d\alpha(z(x_{j+1}^-), z'(x_{j+1}^-)). \tag{597}$$

Since $\pi\theta_j = 1$, we have:

$$\sum \prod_{j=1}^{i} \theta_j \left(d\alpha(z(x_i^+), z'(x_i^+)) - \frac{1}{\theta_i} d\alpha(z(x_i^-), z'(x_i^-)) \right) \tag{598}$$

$$= \sum \prod_{j=1}^{i} \theta_j \left(d\alpha(z(x_{i+1}^-), z'(x_{i+1}^-)) - \frac{1}{\theta_i} d\alpha(z(x_i^-), z'(x_i^-) \right)$$

$$= \sum \prod_{j=1}^{i} \theta_j d\alpha(z(x_{i+1}^-), z'(x_{i+1}^-)) - \sum \prod_{j=1}^{i-1} \theta_j d\alpha(z(x_i^-), z'(x_i^-))$$

$$= 0.$$

Thus,

$$\mathcal{Q}_x(z,z') = 2 \sum \prod_{j=1}^{i} \theta_j \alpha_{x_i^-}(z(x_i^-)) \frac{d\theta_i}{\theta_i^2}(z'(x_i^-)). \tag{599}$$

On the other hand, since, at such an x, $dl - Id|_{\ker\alpha}$ is invertible, since $\ell_*(\alpha \wedge d\alpha) = \alpha \wedge d\alpha$, the equation of the tangent space to $\sum_k$ at $x \in \widetilde{\sum}^k$:

$$dl(u) - u = \sum \delta a_i \xi_i \tag{600}$$

is easy to satisfy. Given δa_i, $\sum \delta a_i \alpha(\xi_i)$ is equal to $\sum \delta a_i \prod_{j=i}^{k} \frac{1}{\theta_j}$. Since $\pi\theta_j = 1$, $\alpha(dl(u))$ is equal to $\alpha(u)$. The only constraint on $\sum \delta a_i \xi_i$ is:

$$\sum \delta a_i \prod_{j=i}^{k} \frac{1}{\theta_j} = 0. \tag{601}$$

Then, one can solve (600), setting:

$$u = ((dl - Id)|_{\ker \alpha})^{-1} \left(\sum \delta a_i \xi_i\right). \tag{602}$$

On the other hand, we then have:

$$\delta a_i = \alpha(z(x_{i+1}^-)) - \alpha(z(x_i^+)) = \alpha(z(x_{i+1}^-)) - \frac{1}{\theta_i}\alpha(z(x_i^-)). \tag{603}$$

With these notations, since $\pi\theta_j = 1$, (601) is automatically satisfied, i.e. the $\alpha(z(x_1^-))$, $\dots, \alpha(z(x_k^-))$ are then free parameters at such an x (since $\prod \theta_j = 1, dl - Id$ has a kernel, which is not in $\ker \alpha$. This direction allows to have $\alpha(z(x_1^-))$ as a free parameter). We derive a tangent vector by setting $\delta a_i = \alpha(z(x_{i+1}^-)) - \frac{1}{\theta_i}\alpha(z(x_i^-))$ and solving (602). Thus, if $z' \in \ker S_x|_{T_x \sum_k}$, with $dl - Id|_{\ker\alpha}$ invertible, $x \in \widetilde{\sum}_{k-1}$, then, since the $\alpha_{x_i^-}(z(x_i^-))$ are free parameters, we must have:

$$d\theta_i(z'(x_i^-)) = 0. \tag{604}$$

Thus

$$\sum \frac{d\theta_i}{\theta_i}(z'(x_i^-)) = 0. \tag{605}$$

Thus, $z' \in T_x\widetilde{\sum}_{k-1}$.

The statement about the difference of indices follows.

Observe that, as far as the evolution of the index on $\overline{\mathcal{B}}$ or $\overline{\mathcal{H}}$ is concerned, a similar phenomenon occurs for the quadratic form Q_x of Definition 11 on $\overline{\mathcal{H}}$, when there is a crossing of a submanifold of Γ_{2k} where $T_x\Gamma_{2k}$ and $\bigoplus_i H_0^1[x_i^+, x_{i+1}^-]$ (see Proposition 32) intersect. Indeed, since these two spaces are Q_x-orthogonal, the restriction of the quadratic form to each of them degenerates. However, the quadratic form, which is Fredholm as stated, does not degenerate on $\overline{\mathcal{B}}_x$: Coming back to Proposition 32, z' being the vector in $T_x\Gamma_{2k} \cap \left(\bigoplus_i H_0^1[x_i^+, x_{i+1}^-]\right)$, we have, with Q_x

the quadratic form of Definition 11 and z a vector of $\mathcal{H}_x$:

$$Q_x(z,z') = -\sum_i \left(\eta(s_i^+) \int_{x_i^-}^{x_i^+} \eta' \bar{\mu}_\xi ds + a \int_{[x_i^+, x_{i+1}^-]} (\dot\eta \dot\eta' - a^2 \eta\eta'\tau) \right) \tag{606}$$
$$= -\sum_i \left(\eta(s_i^+) \int_{x_i^-}^{x_i^+} \eta' \bar{\mu}_\xi ds + a\eta\dot\eta' \Big|_{x_i^+}^{x_{i+1}^-} \right)$$

z', by construction, see (434)–(436), has $\eta' = 0$ on $[x_i^-, x_i^+]$: it is built with a nontrivial solution of

$$\begin{cases} \dot\mu' + a\eta'\tau = 0 \\ \dot\eta' = a\mu' \end{cases} \quad \eta'(x_{i_0}^+) = \eta'(x_{i_0+1}^-) = 0 \tag{607}$$

which one extends with $\eta' = 0$ outside of this interval ($\lambda' + \bar\mu\eta'$ is also zero).

Thus,

$$Q_x(z,z') = -\sum_i a\eta\dot\eta' \Big|_{x_i^+}^{x_{i+1}^-} = -a\eta\dot\eta' \Big|_{x_{i_0}^+}^{x_{i_0+1}^-} \tag{608}$$

$\dot\eta'$ cannot vanish at $x_{i_0}^+, x_{i_0+1}^-$, otherwise z' is identically zero (see (607)) $\eta(x_{i_0}^+)$, $\eta(x_{i_0+1}^-)$ are free parameters on $\mathcal{H}_x$. Thus, Q_x does not degenerate. A similar phenomenon occurs. This justifies our claim in the proof of Proposition 32, the proof of which we left aside earlier.

We now prove our claim about the critical points of $\sum a_i$ on Γ_{2k} and $\widetilde{\sum}_{k-1}$ which are not critical points at infinity in the sense given in the introduction.

On $\widetilde{\sum}_{k-1}$, at any of these, we have one θ_j, θ_{j_0} for example which is not 1. Since $\pi\theta_j$ is 1, there is one θ_{j_0} larger than 1 and one less than 1 (strictly). Hence the conditions of ii) of Lemma 36 are not fulfilled. We can solve the problem (φ) of Lemma 36 and have at the same time $\varphi > 0$ and $\int \eta > 0$. The critical point at infinity is a false one. Hence the result on $\widetilde{\sum}_{k-1}$.

For Γ_{2k}, we consider a critical curve for $\sum a_i$. We apply i) of Lemma 38: if this critical curve induces any difference of topology in the level sets, then the function ψ of (557) is negative along any v-jump. There are, in fact, two functions ψ depending on the initial conditions: If the jump is between A and B:

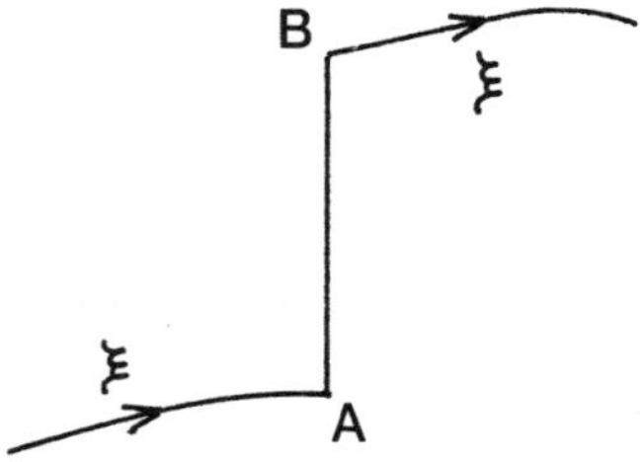

FIGURE JJ

we can ask $\psi(A) = 0$, $\dot{\psi}(A) = 0$, which yields a first function ψ_A; or ask $\psi(B) = 0$, $\dot{\psi}(B) = 0$, which yields a second function ψ_B.

If $\psi_A(B)$ is zero and $\psi_B(A)$ is zero, we have a very special jump. If $A - B$ is not a jump between conjugate points, $\psi_A(B) = \psi_B(A) = 0$ provides one more condition, which is independent of the criticality requirements on the curve x. Such curves can be assumed not to exist using a genericity argument on v (see Propositions 29 and 30). Thus, we can assume that $\psi_A(B)$, is nonzero for example, hence $\psi_A(B) < 0$. However, then, the following variation in C_β decreases the energy of the curve (i.e. decreases J):

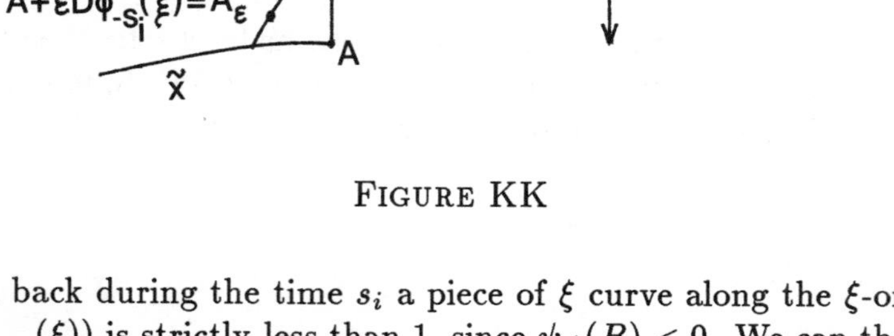

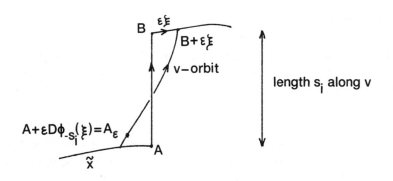

FIGURE KK

i.e. if we pull back during the time s_i a piece of ξ curve along the ξ-orbit out of B, then $\alpha_A(D\phi_{-s_i}(\xi))$ is strictly less than 1, since $\psi_A(B) < 0$. We can then extend the "vector" $\overrightarrow{AA_\varepsilon}$ to $\tilde{x}$ downstairs, before A, so as to build a deformation in C_β. The construction of this deformation is very much the same than the one completed in the proof of Lemma 39. Since $\psi_A(B) < 0$, this deformation decreases J. The result follows.

Observe that the above construction is Fredholm: it does not create any new oscillation in v, because the jump already exists. This is in sharp opposition with (574)–(589) where we had to create a v-oscillation. In our new deformation, the function b deforms only in a compact way, which depends only on the endpoints of the type of $B + \varepsilon\xi$ and A_ε. Hence iii) of Theorem 40.

We devote now our next section to the study of the behavior of these false critical points at infinity (metastable quantic orbits) and we start a thorough study of their behavior and of the behavior of the second derivative of $\sum a_i$ on these curves.

Our study, here, although rigorous, is only sketchy in the sense that several directions could be pushed much more. This will be left to a later work.

The false critical points at infinity

We could have called these false critical points at infinity using some denomination such as "Metastable quantic orbits" or another expression of this kind and the notation would make sense: it is very probable that these false critical points at infinity yield, for some modified variational problems, which might also have a meaning in Physics, true critical points at infinity. Such phenomena are seen to occur in Yamabe-type problems see [2], [7].

We thus start a study of these curves. Namely, we prove the following Proposition/description of these special curves:

Proposition 41. *i) on $\widetilde{\sum}_{k-1}$, the false critical points of infinity occur when a solution of $dl(w) = w$, $w \neq 0$, satisfies $\sum_{i=1}^{k} \frac{d\theta_i}{\theta_i}(w_i^-) = 0$, i.e. is tangent to $\widetilde{\sum}_{k-1}$. We must have: $\prod_1^i \theta_i \frac{d\theta_i}{\theta_i^2}(w_i^-) = C\left(1 - \frac{1}{\theta_i}\right)$, for every i, with an appropriate constant C. The second variation at x can be naturally decomposed into an orthogonal decomposition $T_x\widetilde{\sum}_{k-2} \oplus \mathbb{R}w_x$, where $\widetilde{\sum}_{k-2}$ is the submanifold of $\widetilde{\sum}_{k-1}$ defined by the equation $\sum_{i=1}^{k} \frac{d\theta_i}{\theta_i}(w_i^-) = 0$ (i.e. the submanifold of $\widetilde{\sum}_{k-1}$ along which $w \neq 0$, an eigenvector of eigenvalue 1 of dl, is tangent to $\widetilde{\sum}_{k-1}$). w_x is the tangent vector defined by w at x.*

ii) On Γ_{2k}, the false critical points at infinity occur when one or more of the distances a_i become characteristic for x_i^+, i.e. the Dirichlet problem on $[x_i^+, x_{i+1}^-]$ has a nontrivial solution. Any v-jump between two non-characteristic ξ_i pieces is a jump between conjugate points. Any v-jump from a non-characteristic ξ-piece to a characteristic ξ-piece has a_0 (of (449)) equal to 1 in the direction going from the non-characteristic ξ-piece to the characteristic one. Any v-jump from a characteristic ξ-piece to a non-characteristic one has a_0 equal to 1 in the related direction. The other v-jumps, between characteristic ξ-pieces, have their a_0, c_0's tied between themselves and tied to the c_0's of the v-jumps from a characteristic ξ-piece to a non-characteristic one or conversely. Each block of consecutive characteristic ξ-pieces *which is preceded by a non-characteristic ξ-piece and followed by a non-characteristic one can be considered separately. If, in a block, there are $\ell\xi$-pieces, hence $\ell+1$ v-jumps (but for $\ell = 1$, where there is only one v-jump), there are $\ell+2$ relations between the a_0's and c_0's of these jumps. Assuming for example that there is one characteristic and one non-characteristic ξ-piece, the a_0's (in the direction we described) of the v-jumps must be equal to 1. In addition, their c_0's are in a ratio equal to minus the ratio of the derivatives of the nontrivial solution of the Dirichlet problem on the non-characteristic ξ-piece.*

Finally, the second variation at such a curve has an orthogonal decomposition built as follows: Any characteristic ξ-piece yields a tangent vector z, built with the nontrivial solution of the associated Dirichlet problem. Summing up the span of these directions, we obtain a space Z_x ($\subset T_x\Gamma_{2k}$). On the other hand, if the piece $[x_i^+, x_{i+1}^-]$ is characteristic, then x belongs to the submanifold Γ^i_{2k-1}, which is defined by the fact that the piece $[y_i^+, y_{i+1}^-]$ of the curve y of Γ^i_{2k-1} should be characteristic.

Thus, if x has $[x_{i_1}^+, x_{i_1+1}^-], \dots, [x_{i_\ell}^+, x_{i_\ell+1}^-]$ as ξ-characteristic pieces, then $x \in \bigcap_{j=1}^{\ell} \Gamma^{i_j}_{2k-1} = \Gamma_{i_1,\dots,i_\ell}$. We claim that $T_x\Gamma_{2k} = Z_z \oplus T_x\Gamma_{i_1,\dots,i_\ell}$ and that this decomposition is orthogonal for the second variation of $\sum a_i$ at x.

Proof of Proposition 41. We study first the critical points of $\sum a_i$ on Γ_{2k} and $\widetilde{\sum}_{k-1}$. The equations of the tangent spaces are:

$$dl(u) - u = \sum_{i=1}^{k} \delta a_i \xi_i + \sum_{i=1}^{k} \delta s_i v_i, \tag{609}$$

where ξ_i is $dl_i(\xi_{x_i^+})$, $v_i = dl_i(v_{x_i^+})$ for Γ_{2k}, and:

$$dl(u) - u = \sum_{i=1}^{k} \delta a_i \xi_i \quad \text{for} \quad \widetilde{\sum}_{k-1}. \tag{610}$$

This defines a vector z equal to z_i^- at x_i^- and we also require:

$$\sum_{i=1}^{k} \frac{d\theta_i}{\theta_i}(z_i^-) = 0. \tag{611}$$

We need to study, under these constraints, the equation:

$$\sum_{i=1}^{k} \delta a_i = 0. \tag{612}$$

Let us assume, in the case of second system of constraints, that:

$$(dl - Id) \text{ is onto } \ker\alpha. \tag{613}$$

Observe that on $\widetilde{\sum}_{k-1}$, we have:

$$\ell^*\alpha = \alpha. \tag{614}$$

Thus, $dl - Id$ is valued in $\ker\alpha$; (613) makes sense. Under (613), (610) simply implies that $\sum_{i=1}^{k} \delta a_i \xi_i$ is in $\ker\alpha$. Conversely, assuming that $\sum_{i=1}^{k} \delta a_i \xi_i$ is in $\ker\alpha$, we can find u solving (610). Thus, we must ask

$$\sum_{i=1}^{k} \delta a_i \alpha(\xi_i) = 0 \tag{615}$$

and we can forget (610).

Since $\ell^*\alpha = \alpha$, dl^t, hence dl, has the eigenvalue 1. Let w be such that

$$dl(w) = w, \qquad w \neq 0. \tag{616}$$

If

$$\sum_{i=1}^{k} \frac{d\theta_i}{\theta_i}(w_i^-) \neq 0 \tag{617}$$

then, given δa_i satisfying (615), it is easy to construct z satisfying (610) and (611) having the δa_i's as components along ξ_i. Indeed, w yields a tangent vector $\bar{z}$ which has the δa_i's equal to zero, but $\sum_{i=1}^{k} \frac{d\theta_i}{\theta_i}(w_i^-)$ nonzero. We thus can add or subtract $\lambda\bar{z}$ at ease, without changing the δa_i's and satisfy thereby (611). Thus, at a critical point, the only constraint, under (610) and (611) is (610) and we must have $\sum \delta a_i = 0$.

This implies that:

$$\alpha(\xi_i) = C \qquad \forall i. \tag{618}$$

Since $\alpha(\xi_1) = 1, C$ is equal to 1. On the other hand,

$$\alpha(\xi_i) = \prod_{i+1}^{k} \frac{1}{\theta_j}. \tag{619}$$

Thus,

$$\theta_j \equiv 1 \qquad \forall j \tag{620}$$

and we are at a critical point at infinity, as described in [1]. Assume that (617) holds, but $w \notin \ker\alpha$. We claim that we are not then at a critical point. Indeed, we have: (θ_{m_i} is the map associated to v and to $w_i \in \mathbb{Z}$)

$$\theta^*_{m_i}\alpha_{x_i^+} = \frac{1}{\theta_i}\alpha_{x_i^-}. \tag{621}$$

Thus, given z and w, z satisfying (610) and w satisfying (616):

$$d\alpha_{x_i^+}(z_i^+, w_i^+) = \left(d\left(\frac{1}{\theta_i}\right)\wedge\alpha\right)(z_i^-, w_i^-) + \frac{1}{\theta_i}d\alpha(z_i^-, w_i^-). \tag{622}$$

We also have:

$$d\alpha_{x_i^+}(z_i^+, w_i^+) = d\alpha_{x_{i+1}^-}(z_{i+1}^-, w_{i+1}^-) \tag{623}$$

$$\alpha_{x_i^-}(w_i^-) = \alpha_{x_{i-1}^+}(w_{i-1}^+) = \frac{1}{\theta_{i-1}}\alpha_{x_{i+1}^-}(w_{i-1}^-) = \cdots = \frac{1}{\prod_{j=1}^{i-1}\theta_{i-j}}\alpha_{x_1^-}(w_1^-). \tag{624}$$

Thus

$$\begin{aligned} &d\alpha_{x_{i+1}^-}(z_{i+1}^-, w_{i+1}^-) - \frac{1}{\theta_i}d\alpha_{x_i^-}(z_i^-, w_i^-) \\ &= -\frac{d\theta_i}{\theta_i}(z_i^-)\frac{\alpha_{x_1^-}(w_i^-)}{\prod_{j=0}^{i-1}\theta_{i-j}} + d\left(\frac{1}{\theta_j}\right)(w_i^-)\alpha(z_i^-). \end{aligned} \tag{625}$$

Thus, since $\pi\theta_j = 1$.

$$\begin{aligned} 0 &= \sum\prod_{j=0}^{i-1}\theta_{i-j}\left(d\alpha_{x_{i+1}^-}(z_{i+1}^-, w_{i+1}^-) - \frac{1}{\theta_i}d\alpha_{x_i^-}(z_i^-, w_i^-)\right) \\ &= -\sum\frac{d\theta_i}{\theta_i}(z_i^-)\alpha_{x_1^-}(w_1^-) + \sum d\left(\frac{1}{\theta_i}\right)(w_i^-)\alpha(z_i^-)\prod_{j=0}^{i-1}\theta_{i-j}. \end{aligned} \tag{626}$$

Hence, since $w \notin \ker\alpha$, (611) is equivalent to:

$$\sum\prod_{j=0}^{i-1}\theta_{i-j}d\left(\frac{1}{\theta_i}\right)(w_i^-)\alpha(z_i^-) = 0. \tag{627}$$

Setting

$$\delta a_i = \alpha(z_{i+1}^-) - \frac{\alpha(z_i^-)}{\theta_i}, \tag{628}$$

(615) is immediately satisfied since

$$\alpha(\xi_i) = \prod_{i+1}^{k} \frac{1}{\theta_j} \quad \text{and} \quad \pi\theta_j = 1. \tag{629}$$

Thus, the $\alpha(z_i^-)$ have only to satisfy (627) and under this constraint, we must have:

$$\sum \delta a_i = \sum \left(\alpha(z_{i+1}^-) - \frac{\alpha(z_i^-)}{\theta_i} \right) = \sum \alpha(z_i^-) \left(1 - \frac{1}{\theta_i} \right) = 0. \tag{630}$$

Thus:

$$\prod_{j=0}^{i-1} \theta_{i-j} \frac{d\theta_i}{\theta_i^2}(w_i^-) = C \left(1 - \frac{1}{\theta_i} \right). \tag{631}$$

Hence

$$\frac{d\theta_i}{\theta_i}(w_i^-) = C \frac{1}{\prod_1^{i-1} \theta_j} \left(1 - \frac{1}{\theta_i} \right). \tag{632}$$

Thus

$$\sum \frac{d\theta_i}{\theta_i}(w_i^-) = 0 \tag{633}$$

a contradiction. Thus, (633) must hold and we have related critical points satisfying (632). If $w \in \ker \alpha$, (621)–(626) still holds. Thus, (627) holds and, using a continuity argument, (627) can be taken in place of (611). Indeed, both requirements are continuous and are equivalent on a dense subset of $\widetilde{\sum}_{k-1}$ (Another consequence of Propositions 29 and 30.) Thus, (631) and (632) hold at a critical point. We thus have, for $C \neq 0$, false critical points at infinity. Observe that, since $\pi\theta_j = 1$, one of the θ_j's is larger than 1 if one is less. Thus, we can apply to such critical points the procedures of Lemma 39. These are fake critical points at infinity for J on C_β. It is still interesting to study the behavior of the second variation of $\sum a_i$, restricted to $T_x\Gamma_{2k}$, at these critical points.

We observe here-there are more possible remarks – that there is a natural orthogonal splitting for the second variation on $T_x\Gamma_{2k}$. Namely, we introduce the hypersurface $\widetilde{\sum}'_{k-2}$ of $\widetilde{\sum}_{k-1}$ defined by the equation:

$$\sum \frac{d\theta_i}{\theta_i}(w_i^-) = 0. \tag{634}$$

The false critical points belong to $\widetilde{\sum}'_{k-2}$. At any x critical belonging to $\widetilde{\sum}^1_{k-2}$, we can split

$$T_x\Gamma_{2k} = T_x\widetilde{\sum}'_{k-2} \oplus \mathbb{R}w$$

where $\mathbb{R}w$ is the direction parallel to the vector-field defined by w. w is generically transverse to $\widetilde{\sum}'_{k-2}$ at such an x. All along $\widetilde{\sum}'_{k-2}$, w is defined. Therefore, it is defined near x and if $z' \in T_x\widetilde{\sum}^1_{k-2}$, it makes sense, after extending w from x to a neighborhood in $\widetilde{\sum}'_{k-2}$ to compute the second variation along z' and w. We have:

$$z' \cdot \left(\sum \delta a_i\right) = \sum z' \cdot \left(\alpha(w_i^-)\left(1 - \frac{1}{\theta_i}\right)\right) \tag{635}$$
$$= \sum \alpha(w_i^-)\frac{d\theta_i(z'^-_i)}{\theta_i^2} + \sum z' \cdot \alpha(w_i^-)\left(1 - \frac{1}{\theta_i}\right).$$

Along $\widetilde{\sum}'_{k-2}$, we have

$$\alpha(w_i^-) = \frac{1}{\prod_1^{j-1} \theta_j}\alpha_{x_1^-}(w_1^-). \tag{636}$$

We can assume that $\alpha_{x_1^-}(w_1^-) = 1$, the case where $w_1^- \in \ker\alpha$ is considered independently in the sequel. Thus

$$z' \cdot \alpha(w_i^-) = \alpha_{x_i^-}(w_1^-)z' \cdot \left(\frac{1}{\prod_{j=1}^{i-1} \theta_j}\right) = -\alpha(w_i^-)\sum_{j=1}^{i-1}\frac{d\theta_j(z'^-_j)}{\theta_j}. \tag{637}$$

Thus:

$$z' \cdot \left(\sum \delta a_i\right) = \sum \frac{\delta\theta_j}{\theta_j}(z'^-_j)\left(\frac{\alpha(w_i^-)}{\theta_i} - \sum_{i+1}^{k}\alpha(w_\ell^-)\left(1 - \frac{1}{\theta_\ell}\right)\right). \tag{638}$$

We know that $\alpha(w_{\ell+1}^-) = \frac{1}{\theta_\ell}\alpha(w_\ell^-)$. Thus

$$z' \cdot \left(\sum \delta a_i\right) = \sum \frac{d\theta_j}{\theta_j}(z'^-_j)\left(\frac{\alpha(w_i^-)}{\theta_i} - \alpha(w_{i+1}^-) - \frac{\alpha(w_k^-)}{\theta_k}\right) \tag{639}$$
$$= -\left(\sum \frac{d\theta_j}{\theta_j}(z'^-_j)\right)\alpha(w_1^-).$$

$\sum \frac{d\theta_j}{\theta_j}(z'^-_j)$ is zero because z' is tangent to $\widetilde{\sum}_{k-1}$. Our claim about the orthogonal splitting follows.

Some more work can be done to compute the second variation of $\sum a_i$ in the direction of w. The result depends on C. We leave the description as it is.

We are thus left with the case where:

$$w \neq 0 \quad dl(w) = w, w \in \ker\alpha, \sum_{i=1}^{k} \frac{d\theta_i}{\theta_i}(w_i^-) = 0. \tag{640}$$

On $\widetilde{\sum}_{k-1}$, (640) yields two independent conditions. (Propositions 29 and 30.) Hence, it is satisfied on a set of dimension $k-3$. The fact that we are at a critical point usually translates into $k-1$ conditions. However, w yields a tangent vector satisfying $\sum \delta a_i = 0$. We are thus left with $k-2$ conditions, which are independent, or any further dependence yields one more constraint.

Thus, we have $k-2$ conditions for $k-3$ parameters. We can assume, generically, that these phenomena do not occur. We now consider the case where

$$dl - Id \text{ is not onto } \ker\alpha. \tag{641}$$

In particular, $dl - Id|_{\ker\alpha}$ is not invertible. Also $\ell^*\alpha \wedge d\alpha = \alpha \wedge d\alpha$. The matrix of dl is, thus, in an appropriate basis, where the two last vectors span $\ker\alpha$, one of them being the eigenvector in $\ker\alpha$, and where the first vector of the basis is ξ:

$$\begin{pmatrix} 1 & 0 & 0 \\ \alpha & 1 & 0 \\ \beta & \gamma & 1 \end{pmatrix}. \tag{642}$$

Since $dl - Id$ is not onto $\ker\alpha$, $\alpha\gamma$ is zero, i.e. α or γ is zero. Assuming that (β,γ) is nonzero, $\begin{pmatrix} -\gamma \\ \beta \\ 0 \end{pmatrix}$ is another eigenvector of eigenvalue 1, besides $\begin{pmatrix} 0 \\ 0 \\ 1 \end{pmatrix}$. One of them satisfies (617), unless we have one more independent constraint.

Then (611) yields no condition on the δa_i's. We only have to satisfy (610). $R(dl - Id)$ is one dimensional. Thus, there are two constraints on the δa_i's. We have $k-1$ free parameters (again, Propositions 29 and 30). We have to satisfy the criticality conditions, i.e. $k-1$ conditions, one of them to drop because one of the eigenvectors of $dl - Id$ satisfies (611) and has $\sum \delta a_i$ equal to zero. If both satisfy (611), we drop two conditions, but add one constraint. The result is the same:

Thus, we have $k-1$ free parameters for $k-2$ conditions, plus the constraints $dl - Id|_{\ker\alpha}$ noninvertible and $\alpha\gamma = 0$. Hence, too many constraints for the number of parameters. Again, we can generically avoid such situations. The same argument repeats with $k-3$ conditions, but three constraints if both eigenvectors satisfy (611), hence the same conclusion.

Assuming now that β and γ are zero, $dl|_{\ker\alpha}$ is the identity. We have an eigenspace, corresponding to the eigenvalue 1, of dimension 2. The same arguments repeat. The only case left is $\alpha = \beta = \gamma = 0$. Then, dl is equal to the identity. This happens on a subset of $\widetilde{\sum}_{k-1}$ of codimension 4 (another consequence of Propositions 29 and 30).

There exists at least one eigenvector of dl of eigenvalue 1 such that the associated tangent vector to Γ_{2k} does not satisfy (611). Otherwise, two more constraints are added. Using this eigenvector, we can free ourselves from (611). We are left with (610), with $dl = Id$. Under (610), which defines the tangent space to $\sum_k$, we must have $\sum \delta a_i = 0$.

Thus, we have to satisfy k criticality conditions, three of them being already satisfied on the eigenspace of dl. Thus, we only have $k-3$ independent conditions, for k free parameters. To these independent conditions, we add four constraints, corresponding to $dl = Id$, in fact five: they are only four on $\widetilde{\sum}_{k-1}$. Otherwise, we have to add $\pi\theta_j = 1$. Again, too many conditions to fulfill. Our result follows also in this case. Again, if (617) is not satisfied for any eigenvector of dl, the drop in the number of criticality conditions is compensated by the additional constraints. Hence, the result for $\widetilde{\sum}_{k-1}$. We now consider Γ_{2k}, thus (609) only. Assume first that, at a curve $x \in \Gamma_{2k}$, no ξ-piece is degenerate, in the following sense:

The boundary value problem:

$$\begin{cases} \dot{\mu} + a\eta\tau = 0 \\ \eta(x_i^+) = \eta_1 \qquad \eta(x_{i+1}^-) = \eta_2 \\ \dot{\eta} = \mu a \end{cases} \tag{643}$$

can be solved, for any i.

Then, we can build a tangent vector to Γ_{2k} at x by solving, for $((\lambda + \bar{\mu}\eta)(x_i^-), \eta(x_i^-))$ given, the differential equation

$$\begin{cases} \dot{\overline{\lambda + \bar{\mu}\eta}} = \eta \\ \qquad\qquad\qquad ((\lambda + \bar{\mu}\eta)(x_i^-), \eta(x_i^-)) \text{ given} \\ \dot{\eta} = -\lambda \end{cases} \tag{644}$$

until x_i^+. Then, we obtain values $\eta(x_i^+)$. With $(\eta(x_i^+), \eta(x_{i+1}^-))$, we solve the previous problem. We glue up, with constants, $(\lambda + \bar{\mu}\eta)(x_i^+)$ and $(\lambda + \bar{\mu}\eta)(x_{i+1}^-)$ (see (434)). Thus, if no ξ-piece is degenerate, $((\lambda + \bar{\mu}\eta)(x_i^-), \eta(x_i^-))$ are free parameters. $\sum \delta a_i$ is equal, on such a tangent vector, to $-\sum((\lambda + \bar{\mu}\eta)(x_i^+) - (\lambda + \bar{\mu}\eta)(x_i^-))$. $(\lambda + \bar{\mu}\eta)(x_{i_0}^+)$ depends only on $((\lambda + \bar{\mu}\eta)(x_{i_0}^-), \eta(x_{i_0}^-))$, i.e. if we take $((\lambda + \bar{\mu}\eta)(x_i^-), \eta(x_i^-)) = 0$ for $i \neq i_0$, the variation $\sum \delta a_i$ is equal to

$$-((\lambda + \bar{\mu}\eta)(x_{i_0}^+) - (\lambda + \bar{\mu}\eta)(x_{i_0}^-)).$$

If x is critical, we derive that

$$(\lambda + \bar{\mu}\eta)(x_{i_0}^+) = (\lambda + \bar{\mu}\eta)(x_{i_0}^-) \qquad \forall i_0 \tag{645}$$

for all solutions of (644). Hence a critical point at infinity. Observe that if we have a critical point with two consecutive nondegenerate ξ-pieces, then any v-jump between them is between conjugate points. Indeed, at one end of each of these nondegenerate ξ-pieces, the end which does not follow by the other piece, we can ask that $(\lambda+\bar{\mu}\eta,\eta)$ is zero. Then, z, the tangent vector is zero outside of these two pieces and the v-jump between them. However, $(\lambda+\bar{\mu}\eta,\eta)$ remain free parameters at one end of this v-jump. The previous argument extends, yielding our claim. We will use this later.

We are left with the case where there is a degenerate ξ-piece, for example $[x_1^+, x_2^-]$.

Using this degenerate ξ-piece, we can build, by solving (643) for $i = 1$, a tangent vector z_0 which has

$$(\lambda_0 + \bar{\mu}\eta_0)(x_i^\pm) = \eta_0(x_i^\pm) = 0 \qquad \forall i = 1, \dots, k \tag{646}$$

and which is still nonzero, because η is nonzero on this $[x_1^+, x_2^-]$-piece. We will see later how the second derivative behave with respect to z_0. We compute the first and second variations of $\sum a_i$ at such a curve:

The first variation is:

$$-\sum\left[((\lambda+\bar{\mu}\eta)(x_i^+) - \lambda + \bar{\mu}\eta(x_i^-)) - \int_{x_i^+}^{x_{i+1}^-} b\eta\right]. \tag{647}$$

Therefore, the second variation is, along a tangent vector z':

$$-\sum\left[z'\cdot((\lambda+\bar{\mu}\eta)(x_i^+) - (\lambda+\bar{\mu}\eta)(x_i^-)) - \int_{x_i^+}^{x_{i+1}^-}(\dot{\mu}' + a\eta'\tau)\eta\right] - \int_{x_i^+}^{x_{i+1}^-} bz'\cdot\eta \tag{648}$$

b is zero on $[x_i^+, x_{i+1}^-]$. Thus $\int_{x_i^+}^{x_{i+1}^-} bz'\cdot\eta$ drops. $\dot{\mu}' + a\eta'\tau$ is also zero for our tangent vectors. $(\lambda+\bar{\mu}\eta)(x_i^+)$ is equal to $a_0^{i,-}(s_i, x_i^-)(\lambda+\bar{\mu}\eta)(x_i^-) + c_0^{i,-}(s_i, x_i^-)\eta(x_i^-)$. a_0 and c_0 are as in (449). The index i refers to the i-jump and $-$ to the fact that we are taking the jump starting from x_i^-. There are similar quantities $a_0^{i,+}, c_0^{i,+}$ related to the same jump, when it is considered as starting from x_i^+

$$\begin{aligned}
&z'\cdot(\lambda+\bar{\mu}\eta(x_i^+) - (\lambda+\bar{\mu}\eta)(x_i^-)) \\
=&(z'\cdot(a_0^{i,-}(s_i, x_i^-) - 1))(\lambda+\bar{\mu}\eta)(x_i^-) + \\
&+ (z'\cdot c_0^{i,-}(s_i, x_i^-))\eta(x_i^-) + (a_0^{i,-} - 1)z'\cdot(\lambda+\bar{\mu}\eta)(x_i^-) + c_0^{i,-}z'\cdot\eta(x_i^-).
\end{aligned} \tag{649}$$

The expression $\sum_i((z'\cdot(a_0^{i,-}(s_i, x_i^-) - 1))(\lambda_0+\bar{\mu}\eta_0)(x_i^-) + (z'\cdot c_0^{i,-}(s_i, x_i^-))\eta_0(x_i^-))$ is zero, for any z', because of (646). Our second variation along (z_0, z') reduces to

$$\sum_i((a_0^{i,-}, -1)z'\cdot(\lambda_0+\bar{\mu}\eta_0)(x_i^-) + c_0^{i,-}z'\cdot\eta_0(x_i^-)) \tag{650}$$

where λ_0, η_0 have been extended to a neighborhood of x. Let us first argue in the simpler case where there is only one characteristic ξ-piece, for example the one corresponding to the length a_1, on $[x_1^+, x_2^-]$. There is a hypersurface, corresponding to the equation $a_1 = a_{1,c}$, $a_{1,c}$ a characteristic value of the ξ-space, on which our curve x lies. Let $\widetilde{\Gamma}^1_{2k-1}$ be this hypersurface. Generically on v, $\widetilde{\Gamma}^1_{2k-1}$ is, near x, a manifold of dimension $2k-1$ and the vector z_0 which we built earlier, related to this ξ-piece, is transverse to $\widetilde{\Gamma}^1_{2k-1}$ at x (another statement which should follow from Propositions 29 and 30) z_0 can be built in a neighborhood of x in $\widetilde{\Gamma}^1_{2k-1}$. For y a curve in this neighborhood, we have:

$$(\lambda_0 + \bar{\mu}\eta_0)(y_i^-) = \eta_0(y_i^-) = 0 \qquad \forall i. \tag{651}$$

Thus, for any z' tangent to $\widetilde{\Gamma}^1_{2k-1}$ at x, we have:

$$\sum_i ((a_0^{i,-} - 1)z' \cdot (\lambda_0 + \bar{\mu}\eta_0)(x_i^-) + c_0^{i,-} z' \cdot \eta_0(x_i^-)) = 0. \tag{652}$$

Also, with evident notations:

$$\sum_i ((a_0^{i,\varepsilon_i} - 1)z' \cdot (\lambda_0 + \bar{\mu}\eta_0)(x_i^{\varepsilon_i}) + c_0^{i,\varepsilon_i} z' \cdot \eta_0(x_i^{\varepsilon_i})) = 0 \text{ where } \varepsilon_i = \pm \tag{653}$$

i.e. if we wish to compute the second derivative using rather x_i^+ for certain jumps, the result is unchanged for z' tangent to $\widetilde{\Gamma}_{2k-1}$ at x; it is zero.

We are left with z' equal to a direction transverse to $\widetilde{\Gamma}^1_{2k-1}$ at x, which we can take to be z_0 itself, properly extended. There are several ways to complete this extension which we do not discuss here. Clearly, our statement about the second variation along z_0 and the orthogonal decomposition between $\mathbb{R}z_0$ and $T_x\widetilde{\Gamma}^1_{2k-1}$ follows from (652). The same argument extends to the case where there are several characteristic ξ-pieces, yielding our claim about the second variation and the related orthogonal decomposition.

A deeper study of the behavior of the second derivative along Z_x (see the statement of the Proposition) should allow to lower bound the index of the second variation at such critical curves. We will not complete this here.

We now prove that these critical curves behave as described in the Proposition. We have seen that any jump between two non-characteristic ξ-pieces must be a jump between conjugate points. The problem with the characteristic ξ-pieces is that, the Dirichlet problem (643) being degenerate, it can be solved only if η satisfies:

$$\eta(x_{i+1}^-) = \overline{C}_i \eta(x_1^+), \tag{654}$$

where $\overline{C}_i$ is an appropriate constant, on the degenerate ξ-piece. Thus, if we jump from one degenerate ξ-piece to another one, the boundary values of η have to satisfy consecutive relations. At a jump, $(\lambda+\bar{\mu}\eta)(x_{i+1}^-)$ or $(\lambda+\bar{\mu}\eta)(x_{i+1}^+)$ was before a free parameter. As long as the chain of relations (654) is not closed, i.e. as long as all ξ-pieces are not characteristic, we can choose to keep this freedom on $\lambda+\bar{\mu}\eta$ and instead impose conditions on $\eta(x_{i+1}^-)$, the end of the new ξ-piece, so as to satisfy (654). Since the chain of relations (654) does not close, we end up with a non-characteristic ξ-piece where the boundary values of η are free. Writing the criticality of the curve x, we have:

$$\sum((a_0^{i,\varepsilon_i}-1)(\lambda+\bar{\mu}\eta)(x_i^{\varepsilon_i})+c_0^{i,\varepsilon_i}\eta(x_i^{\varepsilon_i}))=0. \tag{655}$$

If we consider a block of consecutive ξ-pieces which starts with a non-characteristic one and ends with a non-characteristic one, then the last $(\lambda+\bar{\mu}\eta)(x_i^-)$ and the first one, which should rather be thought of as $(\lambda+\bar{\mu}\eta)(x_i^+)$, are free, since the η which follows is free. Thus:

$$a_0^{i_0,+}=1, \qquad a_0^{j_0+1,-}=1 \tag{656}$$

where i_0 is the index we start with and j_0 the one we end with.

Besides these two restrictions and the fact that the jump between non-characteristic ξ-pieces are between conjugate points, we have the relations internal to one block. If the block contains ℓ characteristic pieces, there are $\ell+2$ relations, two are described in (656). The ℓ remaining, we will describe for $\ell=1$ for sake of simplicity. We then have a curve such as:

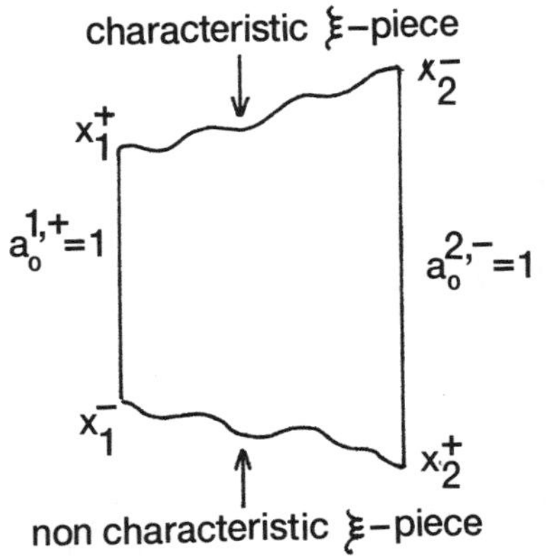

(657)

Since $[x_i^+,x_2^-]$ is characteristic, the Dirichlet boundary value problem (643) can be solved only if

$$\eta(x_2^-)=\overline{C}_2\eta(x_1^+) \tag{658}$$

where

$$\overline{C}_2=\frac{\dot{\psi}(x_2^-)}{\dot{\psi}(x_1^+)} \tag{659}$$

and

$$\left|\begin{array}{l}\ddot{\psi} + a^2\psi\tau = 0\\ \psi(x_1^+) = \psi(x_2^-) = 0, \quad \psi \neq 0.\end{array}\right. \tag{660}$$

Coming back to (643) and using the freedom in x_1^- and x_2^+ for η, due to the fact that $[x_1^-, x_2^+]$ is not characteristic we derive that:

$$c_0^{2,-}\dot{\psi}(x_1^+) = c_0^{1,+}\dot{\psi}(x_2^-) \tag{661}$$

which provides another relation on the critical curve besides (656) and makes a finite possible set of choice of such curves, generically on v.

Indeed, we have:

$$a_0^{1,+} = 1 \; a_0^{2,-} = 1, \; [x_1^+, x_2^-] \text{ is characteristic}, \; c_0^{2,-}\dot{\psi}(x_1^+) = c_0^{1,+}\dot{\psi}(x_2^-). \tag{662}$$

We are thus left with the case where all ξ-pieces are degenerate, in which case we have a family of relations (654) which start at x_1^+ and end at $x_{k+1}^- = x_1^-$. In order to fulfill this set of relations, we can keep $(\lambda + \bar{\mu}\eta)(x_i^-)$ free and derive, by transport, $\eta(x_{i+1}^+)$, but for $(\lambda + \bar{\mu}\eta)(x_k^-)$. Indeed, since $\eta(x_{k+1}^-) = \eta(x_1^-)$ is given (by transport back from x_2^+, where $((\lambda+\bar{\mu}\eta)(x_2^+), \eta(x_2^+))$ is given for example, or because we choose $((\lambda + \bar{\mu}\eta)(x_1^-), \eta(x_1^-)))$, $\eta(x_k^+)$ is derived using (654), since the ξ-piece $[x_k^+, x_1^-]$ is degenerate. Thus, $(\lambda + \bar{\mu}\eta)(x_k^-)$ cannot be free. In other words, we must have an appropriate relation

$$F_x((\lambda + \bar{\mu}\eta)(x_1^-), \ldots, (\lambda + \bar{\mu}\eta)(x_k^-), \eta(x_1^-)) = 0. \tag{663}$$

This relation is linear, nondegenerate i.e. does not cancel (this would require a proof, which should follow, in a natural way, from Propositions 29 and 30). Writing the criticality of the curve x, we have:

$$\sum((a_0^{i,-} - 1)(\lambda + \bar{\mu}\eta)(x_1^-) + c_0^{i,-}\eta(x_i^-)) = 0 \tag{664}$$

under (663). Of course

$$\eta(x_{i+1}^-) = \overline{C}_i\eta(x_i^+) \tag{665}$$

is a function of $((\lambda + \bar{\mu}\eta)(x_1^-), \eta(x_i^-))$, thus ultimately of $(\lambda + \bar{\mu}\eta(x_1^-), \ldots, (\lambda + \bar{\mu}\eta)(x_k^-), \eta(x_1^-))$. This yields k conditions on $(a_0^{i,-}, c_0^{i,-})$ which do not degenerate. Our curves are then fully determined. Any other restriction would cancel them generically. When we turn to the second, derivative, along the family of k-vectors, similar to z_0 which has been built in (646) corresponding to each ξ-piece, the argument is completely similar. Our definition and proof is by now complete.

III. The underlying variational space; the classical periodic motions

III.1. Topological study of C_β

We prove here the following theorem, which we already announced in [1]. We then continue studying some S^1-properties of C_β.

Theorem 42. *Assume α (hence β) turns well along v (see Hypotheis A1 in [1], or below). Then, the injection of C_β into $\Omega(S^1, M)$ is a homotopy equivalence.*

Proof. Since α is a contact form, there is a property of rotation of $\ker\alpha$ in the transport by v, described in Proposition 9 of [1]. Namely, the trace of α in section to v rotates monotonically in a frame transported by v. α turns well along v if this rotation is at least 2π.

We prove now that any continuous map

$$\sigma\colon\ S^k \longrightarrow \Omega(S^1, M)$$

can be deformed into a map valued in C_β. The result follows. Since we already know that $\mathcal{L}_\beta \hookrightarrow \Omega(S^1, M)$ is a homotopy equivalence, we can assume that σ is valued into $\mathcal{L}_\beta$. Using standard approximation and regularization arguments, we can assume that $\sigma(S^k)$ is made of C^∞-curves and that each zero of a or b, a and b being the components of $\dot{x}$ on ξ and v respectively, is of finite order.

We, of course, have a problem with any piece of curve $\sigma(z)$, $z \in S^k$, where a is negative. We need to deform this piece of curve so as to bring it into $a > 0$. Modifying a little bit σ by adding a tiny ξ-piece described back and forth to each curve (and regularizing it, then), we can always assume that a^+ is not identically zero, at any curve $\sigma(z)$.

Using the fact that α turns well along v, and that β is transverse to α, hence has also to turn well along v, there is a natural map

$$\begin{array}{ccc} M^3 & \xrightarrow{\varphi} & M^3 \\ & & \\ x & \longmapsto & \phi_{s(x)} \end{array}$$

such that $d\varphi(\xi) = -\theta\xi + \lambda v$, $\theta > 0$ i.e. $\phi_{s(x)}$ is the map built out of ϕ_s, the one-parameter group generated by v, mapping the direction ξ to the direction $-\xi$ transversally to v. $s(x)$ can be defined to be the first positive time along v such that this happens.

If $\sigma(z)(t)$, for $t \in [t_1, t_2]$, has $a \leq 0$, then $\varphi(\sigma(z))(t)$ has $a \geq 0$ for $t \in [t_1, t_2]$. Since a (of $\sigma(z)(t)$) is zero only at t_1 and t_2, on the interval $[t_1, t_2]$, so is a for

$\varphi(\sigma(z))(t)$. We, thus, have built another piece of curve having $a > 0$ above the piece of $\sigma(z)$ having $a < 0$:

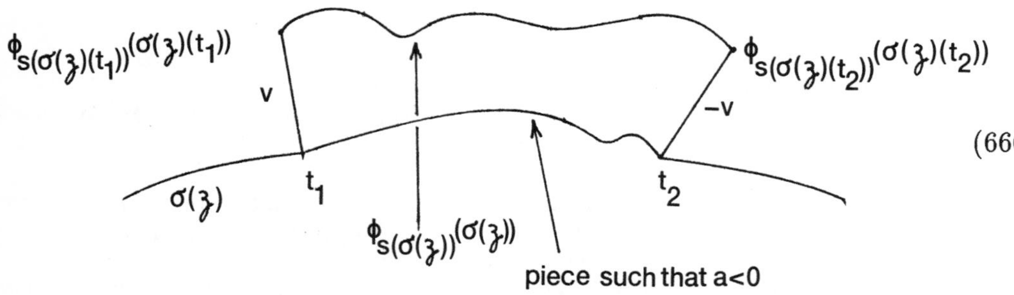

(666)

However, we have paid a price: we have created two v-jumps, one at the time t_1, going from $\sigma(z)(t_1)$ to $\varphi(\sigma(z)(t_1))$, the other one at the time t_2 going from $\varphi(\sigma(z)(t_2))$ to $\sigma(z)(t_2)$. We need to prove that we can incorporate them into a continuous deformation of σ. Once this deformation is completed continuously in $a \geq 0$, with some v-jumps, we can easily adapt the argument of Lemma 39 and push $\tilde{\sigma}$, the deformed map of σ, which we have obtained into $a > 0$, yielding thus our result.

We can, to deform σ into $\tilde{\sigma}$, start by creating a v-oscillation back and forth, at $\sigma(z)\left(\frac{t_1+t_2}{2}\right)$, which takes amplitude progressively, until it oscillates between $\sigma(z)\left(\frac{t_1+t_2}{2}\right)$ and
$\phi_{s(\sigma(z)(\frac{t_1+t_2}{2}))}\left(\sigma(z)\left(\frac{t_1+t_2}{2}\right)\right)$:

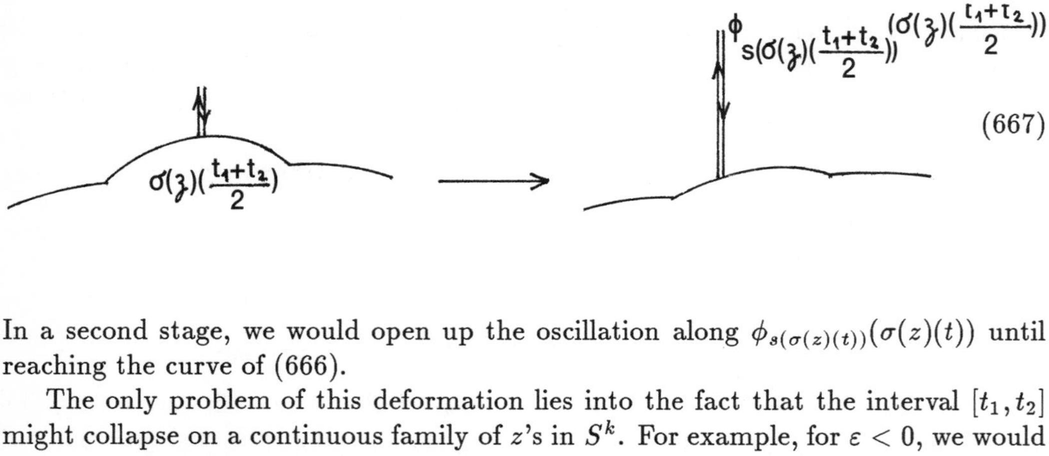

(667)

In a second stage, we would open up the oscillation along $\phi_{s(\sigma(z)(t))}(\sigma(z)(t))$ until reaching the curve of (666).

The only problem of this deformation lies into the fact that the interval $[t_1, t_2]$ might collapse on a continuous family of z's in S^k. For example, for $\varepsilon < 0$, we would have a nontrivial interval $[t_1, t_2]$. But, for $\varepsilon = 0$, $t_1 = t_2$ and no interval $[t_1, t_2]$ thereafter.

We can assume that we have changed σ, using (667) only on the pieces where $t_2 - t_1$ is "large enough" (i.e. not equal to zero, outside an arbitrary given neighborhood of zero); thus, that we have built the v-oscillations as $t_2 - t_1$ started to "grow" and have not touched anything when it was too small. We then need to modify σ on these small pieces of negativity of a.

They can, of course, cumulate, i.e. the intermediate intervals where a is positive also collapse. In order to overcome this problem, we first observe that we have the following property:

Lemma 43. *1) Let C be given. For any $\theta, \theta_1 > 0$, there exists a $\delta_c(\theta, \theta_1) > 0$ such that, for any $\tilde{x}_0 \in M$, for any C^1-curve $x = x(t)$, $t \in \Delta = [0, \bar{t}]$ having:*

$$\begin{cases} x(0) = \tilde{x}_0 & \dot{x}(t) = a(t)\xi(x(t)) + b(t)v(x(t)), \\ a(t) \geq \theta_1 & \text{for } t \in [0, \underline{t}], \underline{t} \in [0, \bar{t}], \\ \int_0^{\underline{t}} a \geq \theta, & a(t) \geq 0 \text{ for } t \in \Delta, \underset{\Delta}{Sup}(|a| + |b|) \leq C \end{cases}$$

any point y belonging to a δ-neighborhood of $x(\bar{t})$ in M, with $\delta \leq \delta_c(\theta, \theta_1)$, can be reached from $\tilde{x}_0$ with a curve $x_y(t)$ having $\dot{x}_y(t) = a_y\xi + b_y v$, with $a_y > 0$. Furthermore, $x_y(t)$ depends continuously on the curve x and on y. It is $O(\delta)$ C^1-close to $x(t)$.

2) Denoting V_δ this δ-neighborhood of $x(\bar{t})$, the same property – including the continuous dependence – holds for any $y \in \bigcup_{0 \leq s \leq s(x(\bar{t}))} \phi_s(V_\delta)$. If y now belongs to $\phi_s(V_\delta)$, then $x_y(t)$ is $O(\delta)$ C^1-close to the curve obtained from $x(t)$ by adding $\bigcup_{\tau \in [0,s]} \phi_\tau(x(\bar{t}))$.

The previous lemma, of local type, allows to build up or cancel v-oscillations progressively, replacing pieces of such curves with other pieces having $a > 0$. It also allows, if we have small intervals where a changes sign for curves of σ, which are preceded by a piece of curve having $a > 0$, $\int a \geq \theta$, or followed by such a piece, to cancel these small intervals, bringing a, on the new piece of curve, to the be positive.

The details of the proof of Theorem 42 are as follows, after Lemma 43: C of Lemma 43 upperbounds $\underset{z \in S^k}{\text{Sup}} |\dot{\overline{\sigma(z)}}|_\infty$. After completing the construction of (666), i.e. transforming the pieces where $\sigma(z)$ has $a < 0$ using $\phi_{s(x)}$, we have a continuity problem i.e. this operation can be only achieved continuously (globally) when the intervals where a is negative do not collapse; i.e. we start to develop v-oscillations back and forth when the size of such an interval is very small but not zero, but we have to come back to the curves $\sigma(z)$ when this size becomes extremely small (how small, we can choose) so as not to have discontinuities due to a v-oscillation persisting on as the interval collapses and disappears. The creation of v-oscillations, progressively between $\sigma(z)\left(\frac{t_1+t_2}{2}\right)$ and $\phi_{s(\sigma(z)(\frac{t_1+t_2}{2}))}(\sigma(z)\left(\frac{t_1+t_2}{2}\right)$ is as in (667). As the size of the intervals of negativity evolves from 2ε to ε, for example, where ε is some positive parameter, we move back from our new map $\tilde{\sigma}$ to our old map σ as follows:

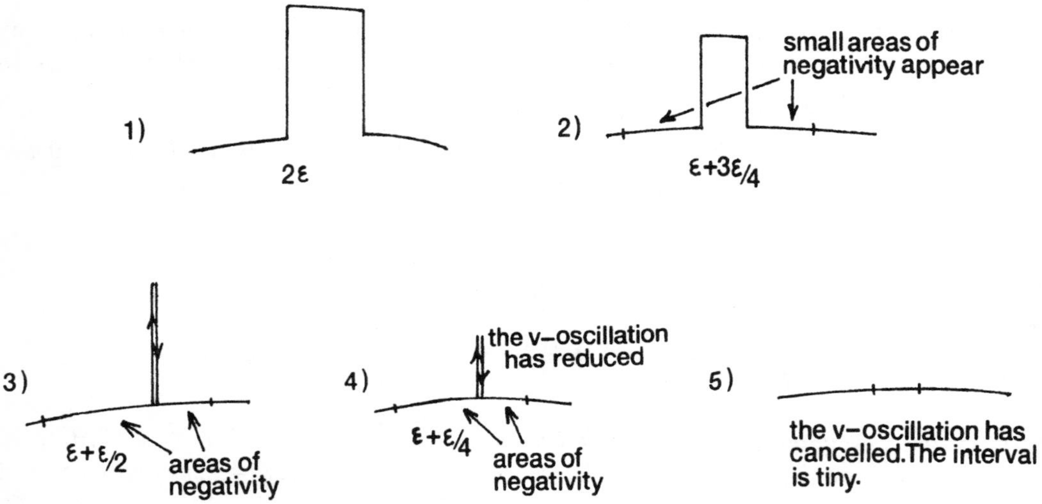

FIGURE LL

Several of these phenomena can cumulate, as several such intervals, together with the intermediate intervals of positivity collapse. Accordingly, we can define strata, a stratum being associated to a certain number of collapses and changes of sign (a stratum corresponds to a zero of a of a certain order). These strata can be assumed, by a standard transversally argument on σ, to be in finite number i.e. the zeros of a are of an upperbounded order, globally on σ. Furthermore, although we do not need it, we can assume that each stratum is a manifold. It is outside a small neighborhood, as small as we please, of the "singular" set that σ transforms into $\tilde{\sigma}$ as in 1)–5). Taking a point (a curve $\sigma(z)$) in this singular set, corresponding to one or several zeros of a of order two at least and related negativity (and positivity) intervals for a which collapse, there is one interval of positivity I^- which precedes the collapse and another one which follows, I^+, which do not collapse. They are available, in a continuous way, in a neighborhood of the curve. We can assume that $\tilde{\sigma}$ transforms into σ in the neighborhood where these intervals are defined. Observe that I^- and I^+ are also available for the curves of $\tilde{\sigma}$ in this neighborhood, as well as all along the transition process from 1) to 5). The transition process from 1) to 5), even when cumulated, we can assume to take place in intervals of negativity and positivity as small as we please, since we are near points where a collapse is occurring. Thus, the pieces of curves $\sigma(z)$ where this transition is taking place are fading away, since the related time-intervals are collapsing. Due to $\tilde{\sigma}$ and the v-oscillations, we can have spreading along v, between $\phi_{s(\sigma(z)(t))}(\sigma(z)(t))$ and $\sigma(z)(t)$, but the base in

$\sigma(z)$ is fading. Thus, the points y involved in the transition process from $\tilde{\sigma}$ to σ are very close to $\phi_{s(x^-)\tau}(x^-)$, $\tau \in [0,1]$, x^- being the right end-point of I^-.

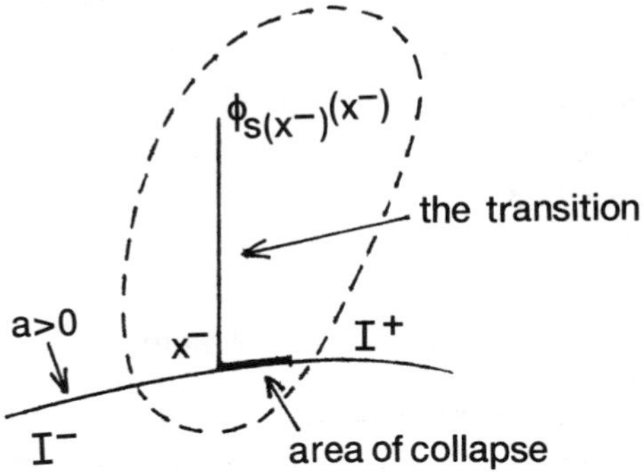

(668)

We can then use, for example, I^- as the interval Δ of Lemma 43. Applying 1) and 2) of this Lemma , we can continuously replace all the curves involved in the transition as well as those of σ, near this $\sigma(z)$, by curves having $a > 0$. We have to use now Lemma 43 repeatedly, so as to cover all the "singular" set. The only thing we have to worry about is to make sure that we only have to apply it to curves such as in (668) i.e. that a former application of this lemma does not destroy this picture. However, our lemma tells us that the new curves that we build are $O(\delta)$ C^1-close to the former ones. As explained above, δ we can choose as small as we please. The number of times we will apply Lemma 43 depends only on the "singular" set of σ i.e. on how many neighborhoods corresponding to intervals I^-, I^+ etc. we need to cover this compact "singular" set. There will be a finite number of these neighborhoods, their union will be a neighborhood of the "singular" set of σ in $\sigma(S^k)$. There will be uniform θ, θ_1, $\delta_c(\theta, \theta_1)$ corresponding to them. Asking that the transition takes place very close to the "singular" set, we will have δ as small as we wish, hence we will make sure that, even after the maximal number of applications of the Lemma which we will complete, which is provided by the number of neighborhoods needed for the covering, we still will be within $\delta_c(\theta, \theta_1)$ of the extremity x^- of I^-. Our claim follows.

We need now only to prove Lemma 43. Lemma 43 2) follows from Lemma 43 1) and from the constructions of the proof of Lemma 39. We have seen how to push such v-oscillations or jumps into $a > 0$ continuously. We are left with Lemma 43 1). Observe that the first portion of the curves under study, from 0 to $\underline{t}$, allows through the exponential map to reach any point in the neighborhood of $x(\underline{t})$. This neighborhood is a uniform δ-neighborhood because the hypotheses of 1) of Lemma 47 provide very stringent bounds on the curves on Δ, $W^{1,\infty}$ bounds. After $\underline{t}$, we can build a time-dependent vector-field through the differential equation:

$$\begin{cases} \dot{z}(t) = a(t)\xi(z(t)) + b(t)v(z(t)) \\ z(0) = y \qquad y \text{ belonging to a neighborhood of } x(\underline{t}) \end{cases}$$

$a(t)$ and $b(t)$ are the component of $\dot{x}(t)$. The above differential equation will map the neighborhood of $x(\underline{t})$ into a neighborhood of $x(\bar{t})$. This map will even be onto, if the neighborhood of $x(\bar{t})$ is small enough because a and b are controlled in $W^{1,\infty}$, hence the resolvent matrix associated to the linearization of the above differential equation is very well controlled, as well as its inverse.

The result follows with $a_y \geq 0$, $a_y \geq \theta_{1/2}$ on $[0,\underline{t}]$. Arguing as in (583)–(589) – we need an easy extension – we can push a_y into $a_y > 0$ on $[0,t]$, using the portion $[0,\underline{t}]$ where a_y is larger than $\theta_{1/2}$. The proof of Lemma 43 is complete.

A remark about the S^1-Faddell-Rabinowitz index of $\Omega(M)$ and C_β

We assume here, an assumption which is probably partly removable that

(H) $\qquad\qquad \pi_1(M)$ is finite.

Thus, M is covered by a three-dimensional homotopy sphere Γ_3

$$\Gamma_3 \longrightarrow M \tag{669}$$

M can be mapped into S^3 with a degree 1 map. Thus, we have a map:

$$\Gamma_3 \longrightarrow M \overset{\psi}{\longrightarrow} S^3 \tag{670}$$

ψ is orientation preserving, with nonzero degree. Thus, ψ is a homotopy equivalence.

We then have a map:

$$S^3 \overset{\gamma}{\longrightarrow} \Gamma_3 \tag{671}$$

which is the topological inverse of ψ. Thus, we have a map:

$$S^3 \longrightarrow M \longrightarrow S^3. \tag{672}$$

The composite is a homotopy equivalence.

Since $\pi_1(\Omega(S^3)) = 0$, thus $\pi_1(\Omega(S^3) - S^3) = 0$, any S^1-equivariant map:

$$S^1 \overset{g}{\longrightarrow} \Omega(S^3) - S^3 \tag{673}$$

is homotopic to zero, thus yields an S^1-equivariant map:

$$S^3 \overset{h}{\longrightarrow} \Omega(S^3) - S^3. \tag{674}$$

The image of g can be taken to be a tiny loop along which we shift the time as we rotate in S^1. Therefore, its homotopy to zero can be built out of a contraction of this tiny loop to zero in S^3, occurring in a small disk D. Let

$$\begin{array}{rccc} p: & \Omega(S^3) & \longrightarrow & S^3 \\ & x & \longrightarrow & x(0) \end{array} \tag{675}$$

h is built over a small contraction, using the S^1-action to make this contraction equivariant. This construction, due to Krasnoselskii for Z_2-equivariant maps ([25]) can be extended to S^1-equivariant maps ([26]). Thus, $p \circ h$ takes its values in a small disk. It is therefore homotopic to zero. Thus, since $\pi_4(S^3) = \mathbb{Z}_2$ and $\pi_3(\Omega(S^3)) = \pi_4(S^3) \oplus \pi_3(S^3)$,

$$h^2\colon S^3 \longrightarrow \Omega(S^3) \tag{676}$$

is homotopic to zero, because $p \circ h^2$, which yields the class of h^2 in $\pi_3(S^3) = \mathbb{Z}$ is zero (for the same reason than $p \circ h$) and because h^2, being a square, is zero in $\pi_4(S^3)$. h^2 can be made into an S^1-equivariant map, although the action might be different from the standard one: to obtain h^2, we can compare h with the map:

$$\begin{array}{ccc} S^3 & \longrightarrow & S^3 \\ (z_1, z_2) & \longrightarrow & \left(\dfrac{z_1^2}{1 + |z_1|^4 + |z_2|^2}, \dfrac{z_2}{1 + |z_1|^4 + |z_2|^2} \right) \end{array}$$

Since all these S^1-actions on S^2 are equivalent for the S^1-index (they are all effective and have the same action on the Chern class of the universal bundle), we can work with this S^1-equivariant form of h^2, which is homotopic to zero. Let

$$\Gamma\colon S^4 \longrightarrow \Omega(S^3) \tag{677}$$

be its contraction to zero. By general position arguments, it can be assumed to avoid $S^3 \subset \Omega(S^3)$. Thus:

$$\Gamma\colon S^4 \longrightarrow \Omega(S^3) - S^3 \tag{678}$$

contracts h^2, thus yields an S^1-equivariant map

$$S^5 \longrightarrow \Omega(S^3) - S^3. \tag{679}$$

The procedure iterates, to build S^1-equivariant maps of any order into $\Omega(S^3) - S^3$, using the fact that $\pi_{2q+1}(S^3)$ is finite for $2q + 1 > 3$.

Using then:

$$S^3 \xrightarrow{\theta} M \tag{680}$$

which yields:

$$\tilde{\theta}\colon \Omega(S^3) \longrightarrow \Omega(M), \quad S^1 - \text{ equivariant}, \tag{681}$$

we can perturb any map:

$$\Gamma\colon\ S^{2k+2} \longrightarrow \Omega(S^3) - S^3 \tag{682}$$

so that, when composed with $\tilde{\theta}$, $\tilde{\theta} \circ \Gamma$ avoids M in $\Omega(M)$. Inductively, we then build S^1-equivariant maps of any order $S^{2k+3} \longrightarrow \Omega(M) - M$, with the associated map:

$$S^{2k+3} \longrightarrow \Omega(M) \tag{683}$$

homotopic to zero, because, up to the previous slight homotopy, it factorizes through $\Omega(S^3)$. (The construction for $\Omega(S^3)$ maps into the construction for $\Omega(M)$, up to this slight perturbation of Γ, which is necessary to avoid M, but does not affect the homotopy classes of the maps we build in $\Omega(S^3)$ or $\Omega(M)$.) Thus, we have proved:

Proposition 44. *Under (H), there are S^1-equivariant maps from $S^{2k+1} \longrightarrow \Omega(M)$, for any k. Therefore, the S^1-Faddell-Rabinowitz index of $\Omega(M)$ is infinite.*

III.2. Classical periodic motions of multipolarized spin-particles

We studied, up to now, the quantic motions of particles with spin. However, we want to show, in this section, how, under reasonable hypotheses, *which are very likely removable* thanks to an extension – to be completed – of our technical tools, these quantic motions indicate the presence of classical motions and are intimately tied to them. Despite the sophisticated tools involved in the foundations of quantum mechanics, this mechanics does not escape the classical framework, it is simply a sophisticated and highly nonobvious extension of this framework. The full strength of the theory of probabilities combined with the Hilbert theory of symmetric unbounded operators did not suffice to cut completely off this tie.

In order to see how the quantic motions lead and imply the classical ones, we consider a very simple case where the first ones generate and are generated by the second ones. We then prove, under reasonable hypotheses, that we can embed this simple case as a limit phenomenon in any contact structure $\{\alpha = 0\}$. This starts as follows: Assume – which is a very strong assumption – that
$(*)$ α turns well along v and any oriented coincidence point is a conjugate point i.e. λ_i in Definition 9 is always equal to 1.

For example, if v preserves the volume form defined by $\alpha \wedge d\alpha$, $(*)$ holds.

Then, any of the quantic motions we described, i.e. any critical point at infinity, generates a periodic orbit. Indeed, by $(*)$:

$$d\theta_{m_i}(\xi) = \xi \tag{684}$$

since

$$\theta^*_{m_i}\alpha = \alpha. \tag{685}$$

Therefore, any critical point at infinity:

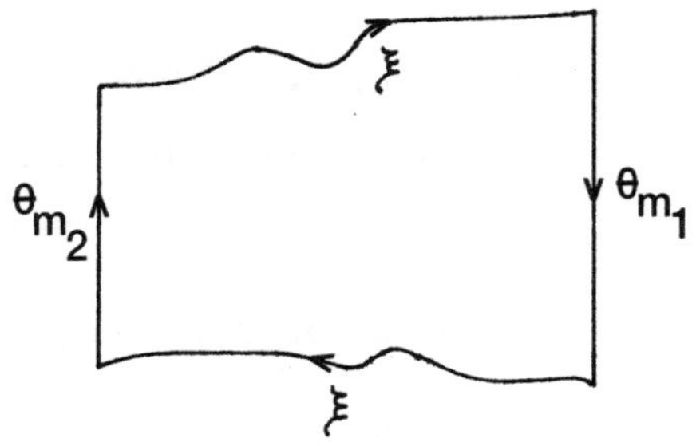

(686)

can be brought back to

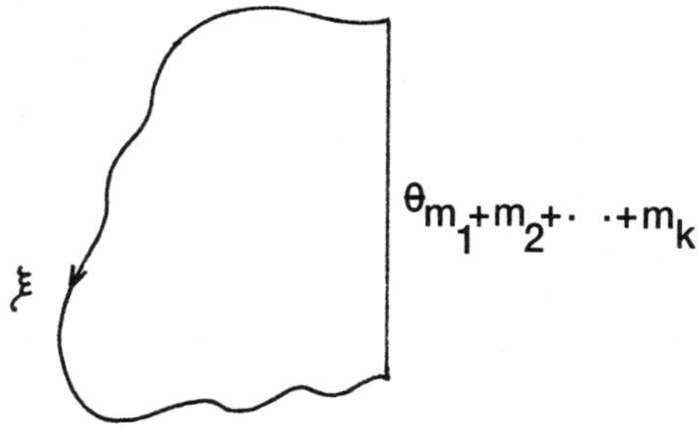

(687)

If

$$\sum m_i = 0 \tag{688}$$

this yields a periodic orbit.

If we consider the quantic manifold

$$\mathcal{M}(m_1, \ldots, m_k) \tag{689}$$

with

$$\sum m_i = 0 \tag{690}$$

$\mathcal{M}(m_1, \ldots, m_k)$ might have strata. Each stratum is a manifold. We can consider the functional $\sum a_i$ on $\mathcal{M}$ and we can minimize it.

If $\mathcal{M}(m_1, \ldots, m_k)$ is nonempty, a point which we will discuss later, we have such a critical point at infinity, hence a related periodic orbit.

Introducing $W_\ell = \{x \in C_\beta^+$ such that b has at most ℓ zeros$\}$, we know that our flow preserves W_ℓ. Let us assume:

(**) There exists an $\mathcal{M}(m_1, \dots, m_k)$ or a family of such $\mathcal{M}(m_1, \dots, m_k)$, with $\sum m_i = 0$, whose topology does not vanish in C_β^+ or in W_ℓ, for ℓ given large.

A better hypothesis than (**), in view of Theorem 40 ii), is:

(* * *) There exist $m_1, \dots, m_k$ such that $\sum m_i = 0$ and such that the topology of $\{x \in \mathcal{M}(m_1, \dots, m_k)$ such that $\prod_{i=1}^{k} \theta_i = 1\}$ or of a union of such sets does not totally cancel in C_β^+ or in W_ℓ, for ℓ given large.

When we then deform the contact form, keeping the contact structure unchanged i.e. we change:

$$\alpha^s = ((1-s)\theta + s\tilde{\theta})\alpha_0 \qquad s \in [0,1] \tag{691}$$

starting from this symmetric case and moving to other non-symmetric cases, we can keep the same v. Thus, we can introduce

$$\tilde{\beta}^s = d\alpha^s(v, \cdot). \tag{692}$$

Assuming:

($*^{\mathrm{V}}$) $\tilde{\beta}^s \wedge d\tilde{\beta}^s$ is contact form, $s \in [0,1]$, we can normalize v into v^s so that:

$$\beta^s = d\alpha^s(v^s, \cdot) \qquad \beta^s \wedge d\beta^s = \alpha^s \wedge d\alpha^s. \tag{693}$$

We can then introduce C_{β^s} and follow our variational problems

$$J_s(x^s) = \int_0^1 \alpha^s(\dot{x}^s)dt \qquad x^s \in C_{\beta^s} \tag{694}$$

when s varies from 0 to 1.

We can then follow the quantic manifolds $\mathcal{M}^s(m_1, \dots, m_k)$ and their submanifolds having $\prod_{i=1}^{k} \theta_i = 1$. We can also follow the periodic orbits and their unstable manifolds through the deformation.

It is not very difficult to see that (**) or (* * *) implies a similar property $(**)_s$ or $(* * *)_s$, provided no critical point at infinity and no periodic orbit has an energy tending to $+\infty$ when s varies from 0 to 1. Similarly, the global topological contribution of the periodic orbits does not change. It might be trivial in C_β^+, and not in W_ℓ, due to the Fredholm character of our deformation. Assuming

($*^{\mathrm{VI}}$) β^s turns well along ξ^s,
this cannot happen (the proof of all our claims will be provided elsewhere).

Thus, under (*)–($*^{\mathrm{VI}}$) (where (**) or (* * *) can be assumed, not both of them), if the topological contribution of the periodic orbits were not zero for $s = 0$, it will be not zero for $s = 1$, in a certain W_ℓ, for example. The existence of periodic orbits would follow.

We now observe that since $\mathcal{M}^0(m_1, \dots, m_k)$, i.e. the initial quantic manifolds, can be renormalized through the process described in (686)–(690) into periodic orbits, their topological contribution is directly related to the one of the periodic orbits. $(**)$ or $(***)$ implies (another claim to be established elsewhere) that the topological contribution of the periodic orbits at $s = 0$, in C_β^+ or in W_ℓ, depending on where $(**)$ or $(***)$ holds, is nonzero. $(*^{\mathrm{VI}})$ allows to continue this property when s changes from 0 to 1, yielding the existence of periodic orbits for $s = 1$.

We thus have a scheme, which central part is in the verification of $(**)$ or $(***)$, preferably in some W_ℓ, ℓ large, in order to derive the existence of periodic motions from the existence of quantic motions and their topological weight, in the case where the family of contact forms $\omega\alpha$ contains a symmetric form with respect to v, i.e. a form α satisfying $(*)$.

We claim that, under an additional reasonable hypothesis, every such family contains a family of "nearly symmetric" forms. This family opens for us the possibility to use the above described scheme.

The idea is as follows:

$(*)$ is removed. We only assume

$(*)'$ α turns well along v.

Let θ_1 be the map, from M to M, which associates to a point the next conjugate one along the v-orbit. Let

$$\theta_1^* \tag{695}$$

be the associated pull-back map, which we can iterate. We denote

$$\theta_m^* \tag{696}$$

the m-th iterate of θ_1^*.

Let k_0 be an integer. Let

$$\tilde{\alpha}_{k_0} = \frac{1}{2k_0+1} \sum_{-k_0}^{k_0} \theta_m^* \alpha \tag{697}$$

be the form α symmetrized at the order k_0.

Observe that $\theta_m^*\alpha$ is collinear to α. Thus

$$\theta_m^* \alpha = \lambda_m \alpha \tag{698}$$

Let k_1 be an index very small with respect to k_0. Then

$$\begin{aligned} \theta_{k_1}^* \tilde{\alpha}_{k_0} &= \frac{1}{2k_0+1} \sum_{-k_0}^{k_0} \theta_{m+k_1}^* \alpha = \frac{1}{2k_0+1} \sum_{-k_0+k_1}^{k_0+k_1} \theta_m^* \alpha \\ &= \tilde{\alpha}_{k_0} + \frac{1}{2k_0+1} O\left(\sum_{-k_0}^{-k_0+k_1-1} |\lambda_i| + \sum_{k_0+1}^{k_0+k_1} |\lambda_i| \right). \end{aligned} \tag{699}$$

Assuming ($*^{\mathrm{VII}}$) λ_i and $d\lambda_i$ are bounded independently of i, we see that $\tilde{\alpha}_{k_0}$ is almost k_1-symmetric. If we combine this construction with bounds on k_1, coming from the knowledge of the topological class involved in ($**$) or ($* * *$) and related bounds on $m_1, \ldots, m_k$ and on the energy (which follows from an index estimate and from the assumption ($*^{\mathrm{VI}}$)), we see how we can hope to complete the extension of our argument above.

This goes beyond the aims of the present paper, which contains a detailed analysis of the quantic motions, establishes the existence of a nearly compact Fredholm deformation of pseudo-gradient type and shows, as we just did, that there is a serious link between the periodic classical and quantic motions.

III.3. Extensions to more general situations hinted by the Stark effect

So far, we have studied a purely spinorial effect. Indeed, α is not time-dependent. In the physical experiments related to quantum mechanics, this happens in the Zeeman effect. With the Stark effect, and the introduction of an electric field, the Hamiltonian depends on the classical (the nonspinorial) motion of the particle, hence on the time. We suggested, in the introduction, a classical explanation of the Stark effect. However, one can easily foresee situations involving quantic motions where the Hamiltonian depends, as in the Stark effect, on the classical motion of the particle. We indicate here briefly why all our previous results extend to this more general framework. In the language of $\mathbb{R}^4$, the Hamiltonian is time-dependent. This is not a true statement: what happens physically, see for example Souriau [13], is that the spin component is part of a wider system of spatial/physical/spinorial coordinates for the particle. There is an inside splitting in the sense that the symplectic form and the space of motions splits into a symplectic product of a spatial/physical component and a spinorial component, living in a space S^2. This is in the weak models currently used today. Probably, this splitting will be broken in the future.

The quantization can be thought as a by-product of separate quantizations, of the spatial/physical component on one hand, of the spinorial component on the other. Related to the spinorial component, the bundle $S^3 \to S^2 (P\mathbb{R}^3 \to S^2)$ is always embedded in this construction, in the relativistic as well as the non-relativistic models (see, for example, Souriau [13]).

The contact structure on S^3 (or a quotient of S^3 by some finite group) is the same, equal to α_0, the standard one. The contact form changes due to the spinorial as well as the spatial/physical components. α is $\lambda\alpha_0$, λ depends on the spin and on the position of the particle.

When this particle is a free particle, or is submitted to a magnetic field such as in the Zeeman effect, λ depends only on the spin. In the other cases, there is also a dependence on the classical motion, i.e. the spatial motion, or better, on the position in the spatial/physical world.

The optimal result would be derived from the analysis of the full spatial/spinorial system. This should be completed in the future. We are analyzing only the spinorial component of it. In the present model which we are analyzing, the spin acts on the spatial/physical component through the spin number $|s|$ only, for example. This model occurs in the Zeeman effect as well as in the Stark effect or in the relativistic Dirac equation; in the theoretical physics books there are very few cases (I know of none) where the spin interferes with the spatial/physical Hamiltonian as a moment. It interferes only by $|s| = 0, \frac{1}{2}, 3/2$ etc.

From a theoretical point of view, it is possible to incorporate an action of the spin component on the spatial/physical variable. This would lead to study, if we isolate the spinorial variational problem and consider the spatial/physical problem as solved, variational problems of the type

$$\int_0^1 \omega(t,x)\alpha_x(\dot{x})dt + \int_0^1 g(t,x)dt \tag{700}$$

where α is a contact form, ω is positive, g is a function of t and x.

In what follows, we consider (700) with $g = 0$, i.e. cases where the spin interferes with the spatial/physical component only through the spin number. However, the presence of $\omega(t,x)$ indicates that the spatial/physical component interferes with the spin. This is reasonable: as a particle enters new materials, it *acquires* polarization, polarization is not given a priori. It has been partly the result of the spatial movement of the particle.

As we said above, a natural goal is to introduce, in the future, the function g in (700). Therefore, we have to introduce some time-dependence of the gauge of α, ω^t i.e. α instead of being a constant form should be a variable form, with $\ker\alpha$ constant. This will provide a good model for the spinorial motions of a particle in any potential field.

We claim that our results extend to this framework, with the (expected) difference that the ξ-pieces in any kind of critical points (at infinity) should be replaced by ξ_t-pieces, where ξ_t is the characteristic vector-field of α^t, the form at the time t. In order to establish our claim, we repeat, step by step, all our constructions, using the same vector-field v. Since $\ker\alpha^t$ is not changed, we can take the same v for the whole family. We define

$$\begin{cases} \alpha^t = \omega^t\alpha_0 \\ \tilde{\beta}^t = d\alpha^t(v,\cdot). \end{cases} \tag{701}$$

We assume that

$$\tilde{\beta}^t \text{ is a contact form.} \tag{702}$$

We then introduce the unique form β^t collinear to $\tilde{\beta}^t$ such that

$$\begin{cases} \beta^t \wedge d\beta^t = \alpha^t \wedge d\alpha^t, \quad \beta^t = d\alpha^t(v^t,\cdot) \\ v^t = \theta^t v \end{cases} \tag{703}$$

We then define:

$$\mathcal{L}_{\beta^t} = \{x \in H^1(S^1,M) \quad \text{s.t.} \quad \beta^t(\dot{x}) = 0\} \tag{704}$$

$\dot{x}$ can then be uniquely written as

$$\dot{x} = a\xi_t + bv^t \tag{705}$$

where ξ_t is the characteristic vector-field of α^t. We also define:

$$C_{\beta^t} = \{x \in \mathcal{L}_{\beta^t} \text{ s.t. } a \text{ is a positive constant}\}. \tag{706}$$

We claim that:

Proposition 45. *C_{β^t} is a manifold. The critical points of $J(x) = a$ on C_{β^t} are the periodic orbits of the time-dependent characteristic vector-field of α^t, ξ^t, with one additional (natural) condition.*

Denoting b the component if $\dot{x}$ on v^t, the variation of b, $\frac{\partial b}{\partial s}$, along a tangent vector z to C_{β^t}, reads as in the time-dependent case, but for the addition of a term which is $O\left(\int_0^1 b\eta\right) = O\left(-\frac{\partial a}{\partial s}\right)$ (provided $b\eta \geq 0$, which we always embed in our construction). Therefore, the whole previous theory extends; the only difference being that it is not anymore obvious that the number of zeros of b can be made not to increase for an appropriate Z which decreases a, as in the time independent case. This is due to the term

$$O\left(-\frac{\partial a}{\partial s}\right).$$

Remark. Some more work, maybe on a slightly modified function b should preserve this property.

Proof. The forms are now time dependent. The tangent space to C_{β^t} reads:

$$\begin{cases} \dfrac{d}{dt}(\alpha^t(z)) - \dfrac{\partial \alpha^t}{\partial t}(z) = d\alpha^t(\dot{x}, z) - C \\ \dfrac{d}{dt}(\beta^t(z)) - \dfrac{\partial \beta^t}{\partial t}(z) = d\beta^t(\dot{x}, z) \end{cases} \tag{707}$$

z is again:

$$z = \lambda\xi^t + \mu v^t + \eta w^t \tag{708}$$

where ξ^t is the characteristic vector-field of α^t, v^t has been defined above, w^t is the characteristic vector-field of β^t. Thus, (707) reads:

$$\begin{cases} \dot{\overline{\lambda + \bar{\mu}^t\eta}} = \dfrac{\partial \alpha^t}{\partial t}(z) + b\eta - C = \dfrac{\partial \omega^t}{\partial t} \times \dfrac{\lambda + \bar{\mu}^t\eta}{\omega^t} + b\eta - C \\ \dot{\eta} = \dfrac{\partial \beta^t}{\partial t}(z) + \mu a - \lambda b \end{cases} \tag{709}$$

The variation of b along z is computed after noticing that

$$b = \gamma^t(\dot{x}) \quad \text{where} \quad \gamma^t = d\alpha([\xi^t, v^t], \cdot). \tag{710}$$

Thus:

$$\begin{aligned} z \cdot b = \frac{\partial b}{\partial s} &= \frac{d}{dt}(\gamma^t(z)) - \frac{\partial \gamma^t}{\partial t}(z) - d\gamma^t(\dot{x}, z) \\ &= \dot{\mu} + a\eta^t - b\eta\bar{\mu}^t_{\xi^t} - \frac{\partial \gamma^t}{\partial t}(z). \end{aligned} \tag{711}$$

Using (709), we derive:

$$\begin{aligned} z \cdot b = \frac{\partial b}{\partial s} &= \underbrace{\frac{\overline{\dot{\eta} + \lambda b}}{a} + a\eta\tau^t - b\eta\bar{\mu}^t_{\xi^t}}_{(1)} - \underbrace{\frac{1}{a}\overline{\frac{\partial \beta^t}{\partial t}(z)}^{\cdot} - \frac{\partial \gamma^t}{\partial t}(z)}_{(2)} \\ &= (1) - (2) \end{aligned} \tag{712}$$

(1) is as in the time independent case. We now study (2) and discuss also, whether, as in the time-independent case, η can be chosen freely; this led to our previous construction.

Observe now that:

$$\begin{aligned} \frac{\partial \gamma^t}{\partial t}(z) &= (\lambda + \bar{\mu}\eta)\frac{\partial \gamma^t}{\partial t}(\xi^t) + \mu\frac{\partial \gamma^t}{\partial t}(v^t) + O(\eta) \\ &= (\lambda + \bar{\mu}\eta)\frac{\partial \gamma^t}{\partial t}(\xi^t) - \frac{1}{a}\frac{\partial \beta^t}{\partial t}(z)\frac{\partial \gamma^t}{\partial t}(v^t) + C(\eta, \dot{\eta}) \end{aligned} \tag{713}$$

We can incorporate $-C(\eta, \dot{\eta})$ in (1), it does not change its fundamental behavior.

On the other hand:

$$\begin{aligned} \frac{\partial \beta^t}{\partial t}(z) &= \frac{\partial}{\partial t}(d(\omega^t\alpha_0)(\theta^t v, \cdot))(z) \\ &= \frac{\partial}{\partial t}(\theta^t d(\omega^t\alpha_0))(v, z) = (\lambda + \bar{\mu}\eta)\frac{\partial}{\partial t}(\theta^t d(\omega^t\alpha_0))(v, \xi^t) + \\ &\quad + O(\eta). \end{aligned} \tag{714}$$

and, using (709):

$$\begin{aligned} \frac{1}{a}\overline{\frac{\partial}{\partial t}\beta^t(z)}^{\cdot} &= \overline{(\lambda + \bar{\mu}\eta)}^{\cdot}\frac{\partial}{\partial t}(\theta^t d(\omega^t\alpha_0))(v, \xi^t) + \\ &\quad + (\lambda + \bar{\mu}\eta)\frac{d}{dt}\left(\frac{\partial}{\partial t}(\theta^t d(\omega^t\alpha_0))(v, \xi^t)\right) + O(\eta, \dot{\eta}) \\ &= \nu(t)(\lambda + \bar{\mu}\eta) + \nu_1(t)C + O_1(\eta, \dot{\eta}). \end{aligned} \tag{715}$$

Using (713), (714) and (715), we derive that:

$$-(2) = O_2(\eta, \dot{\eta}) + \tilde{\nu}(\lambda + \bar{\mu} + \eta) + \tilde{\nu}_1 C \tag{716}$$

where $\tilde{\nu}$ and $\tilde{\nu}_1$ are appropriate functions of t; C is defined in (709).

Thus

$$z \cdot b = \frac{\partial b}{\partial s} = (1)' + \tilde{\nu}(\lambda + \bar{\mu}^t \eta) + \tilde{\nu}_1 C \tag{717}$$

where $(1)'$ behaves just as (1) does.

From (709), we derive, setting:

$$\hat{\omega}^t = \int_0^t \frac{1}{\omega^t} \frac{\partial \omega^t}{\partial t} dt \tag{718}$$

that

$$e^{-\hat{\omega}^t}(\lambda + \bar{\mu}^t \eta)(t) = (\lambda + \bar{\mu}^t \eta)(0) + \int_0^t e^{-\hat{\omega}^z}(b\eta - C)dz. \tag{719}$$

Thus,

$$(\lambda + \bar{\mu}^t \eta)(t) = e^{\hat{\omega}^t}\left((\lambda + \bar{\mu}^t \eta)(0) + \int_0^t e^{-\hat{\omega}^z}(b\eta - C)dz\right). \tag{720}$$

Setting, for example,

$$(\lambda + \bar{\mu}^t \eta)(0) = 0 \tag{721}$$

we derive that:

$$C = \int_0^1 e^{-\hat{\omega}^z} b\eta dz \bigg/ \int_0^1 e^{-\hat{\omega}^t} dz. \tag{722}$$

Clearly, we can keep $b\eta \geq 0$ and have $c > 0$, hence decrease a . Then,

$$\lambda + \bar{\mu}^t \eta = 0\left(\int_0^1 b\eta\right) = O\left(-\frac{\partial a}{\partial s}\right) \tag{723}$$

$$C = O\left(\int_0^1 b\eta\right) = O\left(-\frac{\partial a}{\partial s}\right) \tag{724}$$

as claimed. $\frac{\partial b}{\partial s}$ behaved as stated in the proposition. The critical points of $a = J(x)$ satisfies:

$$C = 0 \quad \text{for any} \quad \eta. \tag{725}$$

Thus

$$b \equiv 0. \tag{726}$$

Coming back to (720), we see that we must also have

$$\hat{\omega}^1 = \int_0^1 \frac{\partial \omega^t}{\partial t} \frac{1}{\omega t} dt = 0. \tag{727}$$

It is natural, since ω depends on the time, that there is an extra-condition of this type since not every ξ^t periodic orbit will combine with a spatial/physical periodic orbit to form a global periodic orbit. Our proof is thereby complete.

For a small: (about extending the deformation theory to curves with very small energy)
There is a problem for the control of $\int(|b|-\nu)^t$ due to the presence of $\frac{\lambda b}{a}$ (i.e. a is in the denominator). This term might ruin the estimate. However, this should behave like $\sum \nu\frac{\Delta\lambda}{a} = \nu O\left(-\frac{\frac{\partial a}{\partial s}}{a}\right)$. Hence, we should obtain:

$$\frac{\partial}{\partial s}\int(|b|-\nu)^+ \leq \frac{\partial}{\partial s}\left(C\nu \operatorname{Log}\frac{1}{a}\right) - \frac{\partial}{\partial s}(Ca)\left(1+\frac{1}{\nu}\right).$$

Thus:

$$\int(|b|-\nu)^+(s) \leq C_\nu(0) + \operatorname{Log}\frac{1}{a^{C\gamma}} - Ca\left(1+\frac{1}{\gamma}\right).$$

Thus:

$$a^{C\nu}e^{\int(|b|-\nu)^+} \leq \widetilde{C}_\nu(0).$$

Thus:

$$ae^{1/C\nu\int(|b|-\nu)^+} \leq \widetilde{\widetilde{C}}_\nu(0).$$

This should be enough in order to make sure that all the small curves are almost closed v-orbits. The ones having one or more zeros $\left(\int b = 0\right)$ are clearly of S^1-index 1. The ones having $\int b$ away from zero and bounded also, **together** with the previous ones (they are in *different* connected components, hence their total index does not add up, it is equal to the largest one), build on S^1-index 1. We are left with these having $\left|\int b\right| \to +\infty$. They are of two types: near the periodic orbits. These are again fine. The other ones are recurrent ones. We need a result here, about the existence of sections.

Appendix 1

Some technical justifications

We justify here the computations of $\frac{\partial}{\partial s}\int(|b|-\nu)^+$ and $\frac{\partial}{\partial s}\int(|b|-\gamma)^+$ in (26)–(33), of Lemma 6, of (141) and (366). A clear justification, which requires however a lengthy proof, can be obtained through the fact that $b(s,t)$ is Georey in t, a classical statement for solutions of classical parabolic evolution equations. Our evolution equation is parabolic, but nonlocal and involves also the coefficients α_i of a partition of unity. Therefore, the proof of such a claim is not straightforward. We provide therefore another argument in order to justify the computations of (26)–(33), Lemma 6, (141) and (366).

We first observe that these computations are justified if $\{t$ such that $|b(s,t)| = \nu\}$ or $\{t$ such that $|b(s,t)| = \gamma\}$ for (26)–(33) or (366), or $\{t$ such that $|b(s,t)| = \frac{m}{q}\}$ for (141) is finite, almost everywhere in s. We then think of (26)–(33), (141), (366) or Lemma 6 in an integral form i.e. we replace these inequalities by inequalities involving $\int(|b|-\nu)^+(s) - \int(|b|-\nu)^+(s')$ for example, $s > s'$, and we observe, then, that under this form, it suffices to prove that they hold for a sequence (ν_ℓ) or (γ_ℓ) or $\left(\frac{m}{q}\right)_\ell$ tending to ν or γ or $\frac{m}{q}$ or $\tilde{\nu}$. If we are able to provide such sequences so that $\{t$ such that $|b(s,t)| = \nu_\ell$ or γ_ℓ or $\left|\frac{m}{q}\right|_\ell$ or $\tilde{\nu}_\ell\}$ is finite almost everywhere in t, then the computations of $\frac{\partial}{\partial s}\int(|b|-\nu)^+$, because ν_ℓ tends to ν (or $\frac{\partial}{\partial s}\int(|b|-\gamma)^+$ because γ_ℓ tends to γ etc.) can be applied, rigorously, to ν_ℓ: now, with ν_ℓ, for example, $|b|$ assumes, almost everywhere in s, a finite number of times the value ν_ℓ. Since ν_ℓ is very close to ν, the arguments of (26)–(33) hold for ν_ℓ. The claim follows. We can even free ourselves more of the constraints: If we prove that these inequalities hold for a sequence of initial conditions b_0^ℓ tending to our initial condition $b_0 = b(0,t)$, the result will follow for $b(s,t)$. We, in fact, given a flow-line, need to prove such a result only on δ-slices of this flow-line, where δ is given, positive. Therefore, we need only to prove:

Proposition A1. *i) Given any real γ, any flow-line $b(s,t)$, $s \in [s_1, s_2]$, for the cancellation flow, there exists $\delta > 0$ such that, for any subinterval of $[s_1, s_2]$ of size $\delta, [\tilde{s}_1, \tilde{s}_1 + \delta]$, there exists a sequence (γ_ℓ) tending to γ and b_0^ℓ tending to $b_0 = b(\tilde{s}_1, t)(= b(0,t)$ for the sake of keeping simple notations) such that for every ℓ, $\{t$ such that $b^\ell(s,t) = \pm\gamma_\ell\}$ is finite almost everywhere in s, for $s \in [\tilde{s}_1, \tilde{s}_1 + \delta]$.*

ii) A similar argument justifies the computation of Lemma 6.

Proof of Proposition A1. We will work as if $[\tilde{s}_1, \tilde{s}_1 + \delta] = [\varepsilon, T - \varepsilon]$, a large interval. At the end of the proof, we will need to consider small intervals. By Lemma M, $b(s,t)$ is C^∞ in function of t for $s > 0$ a.e. in s, until the blow-up time, hence is C^∞ in (s,t) for $s > 0$. Let T denote this blow-up time.

The map

$$\begin{array}{ccc}(0,T)\times S^1 & \longrightarrow & \mathbb{R}\\ (s,t) & \longrightarrow & b(s,t)\end{array}$$

is C^∞ in t and C^∞ is s as well a.e in s: once it is C^∞ in t, it is easy to see that it is C^∞ in s, it suffices to differentiate in the distributional sense and use the t-estimates to conclude.

Therefore, Sard's Theorem applies near any s_0 where b is C^k, since $b(s,t)$ is then C^k for $s\in[s_0,s_0+\delta s_0),\delta s_0>0$. Near any value γ, we can find a sequence of regular values $(\tilde\gamma_\ell)$ tending to γ, $\tilde\gamma_\ell$ can be assumed to be positive. Since $\tilde\gamma_\ell$ is regular, the set:

$$\widetilde{\mathcal{M}}_\ell=\{(s,t)\quad\text{such that}\quad |b|(s,t)=\tilde\gamma_\ell\}$$

is a one-dimensional manifold ($|b|$ replaces b, but this requires only a minor modification of the standard argument). Taking $\varepsilon>0$ and reducing the variations of s to $[\varepsilon,T-\varepsilon]$, we can assume that $\widetilde{\mathcal{M}}_\ell$ is compact, one dimensional (with boundary). For z_ℓ close enough to $\tilde\gamma_\ell$,

$$\widetilde{\mathcal{M}}_\ell(z_\ell)=\{(s,t),s\in[\varepsilon,T-\varepsilon]\quad\text{such that}\quad |b|(s,t)=z_\ell\}$$

is then also a one-dimensional manifold diffeomorphic to $\widetilde{\mathcal{M}}_\ell$. Indeed, $\widetilde{\mathcal{M}}_\ell(z_\ell)$ is close to $\widetilde{\mathcal{M}}_\ell$. $\left|\frac{\partial b}{\partial s}\right|+\left|\frac{\partial b}{\partial t}\right|$ is nonzero on $\widetilde{\mathcal{M}}_\ell$, hence on $\widetilde{\mathcal{M}}_\ell(z_\ell)$. This will hold for $z_\ell\in(\tilde\gamma_\ell-\varepsilon_\ell,\tilde\gamma_\ell-\varepsilon_\ell)$, $\varepsilon_\ell>0$. This will also hold for $\tilde b$, if $\tilde b$ is the v-component of $\tilde x$ and $\tilde x(0,t)$ is close enough to $x(0,t)$ in C^{15} for example. (By Lemma M, $x(0,t)\longmapsto x(s,t)$ is continuous from C^{15} to C^{15}). We thus can introduce:

$$\widetilde{\mathcal{M}}_\ell(z_\ell,\tilde x_0)=\{(s,t),s\in[\varepsilon,T-\varepsilon]\quad\text{such that}\quad |\tilde b|(s,t)=z_\ell\}$$

and

$$\psi_{z_\ell,\tilde x_0}\colon\ \widetilde{\mathcal{M}}_\ell\longmapsto\widetilde{\mathcal{M}}_\ell(z_\ell,\tilde x_0)$$

the related diffeomorphism, which depends C^{15} on z_ℓ and $\tilde x_0$. We then have:

$$\begin{array}{c}V_{C^{15}}(x(0,t))\times(\tilde\gamma_\ell-\varepsilon_\ell,\tilde\gamma_\ell+\varepsilon_\ell)\times\widetilde{\mathcal{M}}_\ell\xrightarrow{A_\ell}\mathbb{R}\\ (\tilde x_0,z_\ell,(\sigma,\tau))\longrightarrow\dfrac{\partial}{\partial t}\tilde b(\psi_{z_\ell,\tilde x_0}(\sigma,\tau)).\end{array}$$

We claim that A_ℓ is C^∞ and DA_ℓ is onto at 0 i.e. that 0 is a regular value for A_ℓ. By the transversality theorem, we can then find $(\tilde x_0^\ell,z_\ell)$ such that $\widetilde{\mathcal{M}}_\ell(z_\ell,\tilde x_0^\ell)$ is a one-dimensional manifold where 0 is a regular value of $\frac{\partial\tilde b}{\partial t}$, therefore the zeros of $|\tilde b|-z_\ell$, in t, are simple or double, hence in finite number, for every s.

Thus, Proposition A1 follows from our statement on A_ℓ. We will prove this statement below. Here, we observe that we also have the same property for:

$$V_{15}(x(0,t))\times[\varepsilon,T-\varepsilon]\times S^1\xrightarrow{B_\ell}\mathbb{R}$$

$$(\tilde{x}_0, s, t) \longrightarrow \frac{\partial \tilde{b}}{\partial t}(s,t)$$

We thus can find near $x(s,t)$, as close as we wish, $\tilde{x}(s,t)$ such that the zeros of $\frac{\partial \tilde{b}}{\partial t}(s,t)$ are simple. The estimates on $\frac{\partial}{\partial s}\int \dot{b}^+$ in Lemma D follow therefore. We now prove our claims on A_ℓ and B_ℓ.

$$\begin{array}{rcl} (\tilde{\gamma}_\ell - \varepsilon_\ell, \tilde{\gamma}_\ell + \varepsilon_\ell) \times \widetilde{\mathcal{M}}_\ell & \longrightarrow & [\varepsilon, T-\varepsilon] \times S^1 \\ (z_\ell, (\sigma, \tau)) & \longmapsto & \psi_{z_\ell, \tilde{x}_0}(\sigma, \tau) \end{array}$$

is a local C^∞-diffeomorphism, for every $\tilde{x}_0$. Using the continuity of ψ with respect to z_ℓ and $\tilde{x}_0$, the statements about A_ℓ and B_ℓ are equivalent. We, therefore, just need to prove that:

$$(\tilde{x}_0, s, t) \rightarrow \frac{\partial \tilde{b}}{\partial t}(s,t) \text{ is } C^\infty \text{ and}$$

has a nonzero differential at any $(\tilde{x}_0, s_0, t_0)$ such that $\frac{\partial \tilde{b}}{\partial t}(s_0, t_0) = 0$. We compute this differential using the evolution equation of the cancellation flow. This is a fully nonlinear elliptic equation, with other nonlocal terms due to the presence of $\dot{\overline{\lambda b}}$ in $\frac{\partial b}{\partial s} \cdot \lambda + \bar{\mu}\eta$ satisfies

$$\dot{\overline{\lambda + \bar{\mu}\eta}} = b\eta - \int_0^1 b\eta, \qquad \lambda + \bar{\mu}\eta = \int_0^t b\eta - t\int_0^1 b\eta$$

η is of local type, equal to $\sum_i \alpha_i \widetilde{\psi}_{\tilde{\theta}_i, \tilde{\varepsilon}_i} \tilde{\eta}_{0,i} + cb$ and we have:

$$\left| \begin{array}{l} \dfrac{\partial b}{\partial s} \\ b(0,t) = b_0(t) \end{array} \right. = \frac{\dot{\overline{\dot{\eta} + \lambda b}}}{a} + a\eta\tau - b\eta\bar{\mu}_\xi$$

$b(t,s)$ is C^∞ for $s > 0$. The above evolution equation, when we consider it in itself, i.e. independently of $x(s,t)$, is easily seen to be C^∞ with respect to b_0, for any $s > 0$: if we introduce a variation h_0 of b_0, the related evolution equation has the form:

$$\begin{aligned} \frac{\partial h}{\partial s} &= \frac{1}{a}\left[\dot{\overline{\sum_i \alpha_i \frac{\partial \widetilde{\psi}}{\partial b}_{\tilde{\theta}_i, \tilde{\varepsilon}_i}(b)\tilde{\eta}_{0,i}\dot{h} + c\dot{h}}} \right] + K_1^s(h) + K_2^s(\dot{h}) \\ &= \mathcal{L}_s h + K_1^s(h) + K_2^s(\dot{h}) \end{aligned}$$

where K_1^s and K_2^s are continuous linear operators from L^2 into L^2, which are differentiable functions of s. This statement is almost obvious after looking at the expression of η and of $\lambda + \bar{\mu}\eta$, but for the slight dependence of $\tilde{\eta}_{0,i}$, $\tilde{\theta}_i$, $\tilde{\varepsilon}_i$ and α_i on b and their related contribution, which, we claim, can be embodied into K_1^s and K_2^s.

This is clear for $\tilde{\eta}_{0,i}$, which, for a given i, depends on b only through $e^{\int_0^t \bar{\mu}b}$, see (58)′. α_i is related to an H^1 partition of unity, it can be built with a standard family of functions, depending on one parameter θ, ω_θ. α_i will then be of the form $\omega_{\theta_i}(|b-b_i|^2_{H^1})$. θ_i depends only on $b_i \cdot \omega_{\theta_i}$ vanishes for $|b-b_i|^2_H$ larger than θ_i. Clearly, such an α_i is differentiable in b and its differential is

$$2\nabla\omega_{\theta_i}(b-b_i,h)_{H^1}.$$

Its contribution can be considered to be part of $K_2^s(\dot{h})$. $\tilde{\theta}_i$ and $\tilde{\varepsilon}_i$ can be built – see (γ) – using $b(t^-_{1,i})$, $b(t^-_{2,i})$, $b(t^+_{1,i})$, $b(t^+_{2,i})$, in a standard way: it is obvious for $\tilde{\varepsilon}_i$. For $\tilde{\theta}_i$, we observe that we need, having θ_i for b_i (b_i is the center of the ball B_i involved in this partition of unity), only to rescale a little bit θ_i, according to the new values of b at $t^\pm_{1,i}$, $t^\pm_{2,i}$ in order to derive $\tilde{\theta}_i$. This rescaling takes the form of a linear transformation of θ_i, which depends differentiably on $b(t^\pm_{1,i})$, $b(t^\pm_{2,i})$. The map

$$b \longmapsto b(t^\pm_{1,i}) \quad \text{or} \quad b(t^\pm_{2,i})$$

is linear and continuous from H^1 into $\mathbb{R}$.

Therefore, the related differential can also be thrown into $K_2^s(\dot{h})$ and $K_1^s(h)$. This justifies our claim but for an additional point: there is also a variation, when b varies of h, or x of δx, of the coefficients $\bar{\mu}$, $\bar{\mu}_\xi$ and τ which are involved in the evolution equation. Typically, we will have:

$$\delta\bar{\mu} = \nabla\bar{\mu}(\delta x).$$

We would also like to include such terms in the general form which we wrote above.

We observe that:

$$\frac{\partial x}{\partial s} = \lambda\xi + \mu v + \eta w = (\lambda + \bar{\mu}\eta)\xi + \frac{\dot{\eta} + \lambda b}{a} v - \eta[\xi, v]$$

where $\eta = \sum \alpha_i \widetilde{\psi}_{\tilde{\theta}_i,\tilde{\varepsilon}_i}\eta_{0,i} + cb$, $\lambda + \bar{\mu}\eta = \int_0^t b\eta - t\int_0^1 b\eta$. If b varies of h, $\delta\eta(h)$ can be derived easily, as well as $\delta(\lambda + \bar{\mu}\eta)(h)$. Thus:

$$(*) \qquad \begin{aligned} \frac{\partial \delta x(h)}{\partial s} &= (\lambda + \bar{\mu}\eta)D\xi(\delta x) + \frac{\dot{\eta} + \lambda b}{a} Dv(\delta x) - \eta D[\xi, v](\delta x) + \\ &\quad + \delta(\lambda + \bar{\mu}\eta)(h)\xi + \left(\frac{\dot{\overline{\delta\eta(h)}} + \delta\lambda b}{a} + \frac{\lambda h}{a}\right) v - \delta\eta(h)[\xi, v] \\ &= \Gamma^s(\delta x) + \mathcal{L}_1^s(h) + \mathcal{L}_2^s(\dot{h}) \end{aligned}$$

where Γ^s and $\mathcal{L}_1^s, \mathcal{L}_2^s$ are linear, nonlocal in t, continuous from L^2 into L^2.

Thus, the above differential equation is equivalent to:

$$(**)\qquad \delta x(h) - \int_0^s \Gamma^s(\delta x(h)) = \int_0^s (\mathcal{L}_1^\tau(h) + \mathcal{L}_2^\tau(\dot{h}))d\tau.$$

Using the linearity of Γ, for s small, the above equation can be easily solved in

$$\delta x = (Id - \theta)^{-1}\left(\int_0^s (\mathcal{L}_1^\tau(h) + \mathcal{L}_2^\tau(\dot{h}))\right)$$

where

$$\theta\colon\ y \longmapsto \int_0^s \Gamma^\tau(y)d\tau \text{ is an } L^2\text{-function of } s \text{ and } t.$$

Thus, δx is a linear transformation of $\int_0^s(\mathcal{L}_1^\tau(h) + \mathcal{L}_2^\tau(\dot{h}))$. Thus, when we write:

$$\frac{\partial h}{\partial s} = \mathcal{L}_s h + K_1^s(h) + K_2^s(\dot{h})$$

we should allow $K_1^s(h)$ and $K_2^s(\dot{h})$ to be nonlocal in s, or rather write:

$$\frac{\partial h}{\partial s} = \mathcal{L}_s h + \overline{K}_1^s(h) + \overline{K}_2^s(\dot{h}) + \overline{K}_3^s\left(\int_0^s \mathcal{L}_1^\tau h\right) + \overline{K}_4^s\left(\int_0^s \mathcal{L}_2^\tau(\dot{h})\right)$$

where $\overline{K}_i^s$ are linear continuous operators, for each s, from L^2 into L^2 (in t) involving iterated integrations in s, which arise naturally in $(Id-\theta)^{-1}$. $\mathcal{L}_1^\tau h$ an $\mathcal{L}_2^\tau \dot{h}$ are linear functions of h and $\dot{h}$, nonlocal in t because there are integrations on $[0,t]$ in order to compute $\lambda + \bar{\mu}\eta$.

$\mathcal{L}_s$ is uniformly elliptic, of second order; we will work on an s-interval which will be small. The smallness of the interval depends only on the flow-line i.e. given any flow-line $x(s,t)$, $s \in [s_1, s_2]$, we will cut it into a finite number of small pieces and we will discuss each small piece separately. On each of these pieces, a sequence b^ℓ and a related sequence x^ℓ tending to b and x will be found, satisfying Proposition A1. Thus, all our computations in Lemma 6, in (26)–(33) etc. will hold on each small piece for b^ℓ, thus for b up to $o(1)$, where $o(1)$ tends to zero when ℓ tends to $+\infty$. Since the pieces are in finite number, the total sum of the various $o(1)$, on the various intervals, tends to zero. Proposition A1 follows and our computations are justified. We thus focus on:

$$(***)\qquad \frac{\partial h}{\partial s} = \mathcal{L}_s h + \overline{K}_1^s(h) + \overline{K}_2^s(\dot{h}) + \overline{K}_3^s\left(\int_0^s \mathcal{L}_1^\tau h\right) + \overline{K}_4^s\left(\int_0^s \mathcal{L}_2^\tau \dot{h}\right).$$

We observe that, if the variation of b at the time 0, h_0, corresponds to a variation δx_0 of the curve, then, at later times $s > 0$, the computation of δx_s and $\delta b(s,t) = h$

can be brought back, by our arguments above, to the study of this single equation on h: the equation on δx_s follows, by the formula:

$$\delta x_s = (Id - \theta_s)^{-1}\left(\int_0^s (\mathcal{L}_1^T(h) + \mathcal{L}_2^T(\dot{h}))\right).$$

We will see that $(***)$ is a well-posed initial value problem, with a unique solution.

Not all variations h_0 of b_0 correspond to a variation δx_0 of the curve. Indeed, since

$$\delta x_0 = \lambda_0 \xi + \frac{\dot{\eta}_0 + \lambda_0 b_0}{a} v + \eta_0 w$$

we must have:

$$h_0 = \frac{\overset{\cdot}{\overline{\dot{\eta}_0 + \lambda_0 b}}}{a} + a\eta_0 \tau - b_0 \eta_0 \bar{\mu}_\xi.$$

We can view this equation as an equation on η_0 only, since $\overset{\cdot}{\overline{\lambda_0 + \bar{\mu}\eta_0}} = b_0\eta_0 - \int_0^1 b_0\eta_0$.

It is of the form:

$$\ddot{\eta}_0 + K_0(\eta_0) = h_0 \qquad \eta_0(0) = \eta_0(1), \dot{\eta}_0(0) = \dot{\eta}_0(1)$$

where K_0 is compact with respect to $\eta_0 \longmapsto \ddot{\eta}_0$. This equation can be solved if $\ddot{\eta}_0 + K_0(\eta_0)$ has no kernel under periodic boundary conditions. Otherwise, there is a restriction, h_0 should belong to the range of $\ddot{\eta}_0 + K_0(\eta_0)$. However, once h_0 is chosen in this way and h solves $(***)$, we can derive a $\delta x(h)$ by $(**)$, which solves $(*)$. We then claim that $h = h(s,t)$ belongs to the range of $\ddot{\eta} + K_s(\eta)$, i.e. corresponds to a variation δx. This is somewhat clear from our construction, but the following argument makes it transparent:

We can extend h_0 to a neighborhood of $x_0 = x(0,t)$ into a C^∞-vector-field $\tilde{h}_0$ just by setting

$$\tilde{h}_0 = \ddot{\eta}_0 + \widetilde{K}_0(\eta_0)$$

where η_0 is the function found for η_0 and $\widetilde{K}_0$ is the analogue of K_0 at the curve $\tilde{x}_0$ close to x_0 in H^1. We can then integrate locally the vector-field $\tilde{h}_0$, in particular we have a small segment $x(0,t,z)$, z is the time along $\tilde{h}_0$, starting at $x(0,t,0) = x(0,t) = x_0$ and tangent to $\tilde{h}_0$ when z varies, hence to h_0 at $z = 0$.

We can then solve the evolution equation of the cancellation flow with initial data $x(0,t,z)$. We derive a family of curves $x(s,t,z)$. It is then easy to see that $x(s,t,z)$ is differentiable in s and z as well as $b(s,t,z)$ and that $\frac{\partial}{\partial z}b(s,t,z)$ has to satisfy $(***)$, while $\frac{\partial}{\partial z}x(s,t,z)$ has to satisfy $(*)$. This is just derived by differentiation of the $b(s,t,z)$-evolution equation and $x(s,t,z)$ evolution equation in s. The differentiation is justified by the existence and continuity of solutions $(*)$–$(***)$.

If η_z is the component of $\frac{\partial}{\partial z}x(s,t,z)$ along w and λ_z its component along ξ, we must have:

$$\frac{\partial}{\partial z}b(s,t,z) = \frac{\overset{\cdot}{\overline{\eta_z + \lambda_z b}}}{a} + a\eta_z b - b\eta_z \bar{\mu}_\xi$$

i.e. $\frac{\partial}{\partial z}b(s,t,z)$ belongs to the range of $\ddot{\eta}+K_z(\eta)$. At $z=0$, $\frac{\partial}{\partial z}b(s,t,0)$ can be identified as $h(s,t)$ since $\frac{\partial}{\partial z}b(0,t,0)=h_0$ and $\frac{\partial}{\partial z}b(s,t,0)$ a_0 well as $h(s,t)$ solve $(***)$. Our claim follows.

We thus study in the sequel $(***)$ solely, with an initial condition $h_0=\ddot{\eta}_0+K_0(\eta_0)$.

In order to see that $(***)$ has a solution, for any H^1-initial data h_0, one repeats the arguments for the existence and continuity for the cancellation flow i.e. one regularizes $(***)$ by replacing h by $\phi_\varepsilon(h)$ in the right hand side of $(***)$, where $\phi_\varepsilon(h)$ is an ε-regularization of h. For the cancellation flow, we replaced b by $\phi_\varepsilon(b)$ in $\frac{\partial x}{\partial s}$ and $\frac{\partial b}{\partial s}$. We therefore derived a C^1-vector-field Z_ε, which we integrated locally. We then had ε tend to zero. Here, after replacing h by $\phi_\varepsilon(h)$ in $(***)$, our flow is still not the flow of a vector-field, due to the presence of terms such as $\int_0^s \mathcal{L}_1^\tau\phi_\varepsilon(h)$ or $\int_0^s \mathcal{L}_2^\tau\phi_\varepsilon(\dot{h})$. However, the fixed point method to solve differential equations works very well. We thus have existence and continuity for the ε-$(***)$ problem. To $(***)$ or ε-$(***)$, we can easily apply a t-derivative. We derive, on $\dot{h}$, an equation very similar to $(***)$ or ε-$(***)$, denoted $(*\,*\,**)$ or ε-$(*\,*\,**)$. Multiplying $(***)$ by h, $(*\,*\,**)$ by $\dot{h}$ as well as their ε-versions and integrating in t on $[0,1]$, we derive easily the desired bounds. For example, let us see why the non-local terms do not impede the usual estimates. Multiplying $(***)$ and $(*\,*\,**)$ by h and $\dot{h}$ respectively and integrating, we obtain: (observe that $\int_0^s\int_0^{s_m}\cdots\int_0^{s_1}\int_0^1 h^2ds_1\dots ds_m dt\le\int_0^s\int_0^1 h^2dt$ if $s\le 1$)

$$\frac{\partial}{\partial s}\int_0^1 h^2+c\int_0^1\dot{h}^2\le c\left(\int_0^1 h^2+\int_0^s\int_0^1 h^2+\int_0^s\int_0^1\dot{h}^2\right)$$
$$\frac{\partial}{\partial s}\int_0^1 \dot{h}^2+c\int_0^1\ddot{h}^2\le c\left(\int_0^1 \dot{h}^2+\int_0^s\int_0^1 \dot{h}^2+\int_0^s\int_0^1\ddot{h}^2\right).$$

Assuming h_0 is H^1 and integrating between 0 and s, we derive:

$$\int_0^1 h^2+c\int_0^s\int_0^1\dot{h}^2\le c\left(\int_0^s\int_0^1 h^2(1+s)+s\int_0^s\int_0^1\dot{h}^2\right)+\int_0^1 h_0^2$$
$$\int_0^1 \dot{h}^2+c\int_0^s\int_0^1\ddot{h}^2\le c\left(\int_0^s\int_0^1 \dot{h}^2(1+s)+s\int_0^s\int_0^1\ddot{h}^2\right)+\int_0^1 \dot{h}_0^2.$$

Taking Cs small enough with respect to c, we derive that $\int_0^1 h^2$, $\int_0^1\dot{h}^2$ is bounded, uniformly on s, for s small. For $\phi_\varepsilon(h)$ instead of h in the right hand side of $(***)$ and $(*\,*\,**)$, the arguments are the same.

We now start the final argument in the proof of Proposition A1: We need, for this, to prove that we can find h_0 in the range of $\ddot{\eta}_0+K_0(\eta_0)$ such that $\frac{\partial h}{\partial t}(s_0,0)$, for $s_0\in[\tilde{s}_1,\tilde{s}_1+\delta]$, $\delta>0$ uniform on the flow-line ($\tilde{s}_1\in[s_1,s_2]$), is nonzero. 0, in $\frac{\partial h}{\partial t}(s_0,0)$, can be replaced, of course, by any other value t_0.

We observe that $\dot{h} = \frac{\partial h}{\partial t}(s,t)$ satisfies an equation very much similar to $(***)$. This equation is obtained from $(***)$ by t differentiation, hence is:

$$(****) \qquad \frac{\partial \dot{h}}{\partial s} = \mathcal{L}_s(\dot{h}) + \widetilde{K}_1^s(h) + \overline{\widetilde{K}}_1^s(\dot{h}) + \widetilde{K}_2^s(\ddot{h}) + \\ + \widetilde{K}_3^s\left(\int_0^s \widetilde{\mathcal{L}}_1^\tau h\right) + \overline{\widetilde{K}}_3^s\left(\int_0^s \overline{\widetilde{\mathcal{L}}}_1^\tau \dot{h}\right) + \overline{\widetilde{K}}_4^s\left(\int_0^s \widetilde{\mathcal{L}}_2^\tau \ddot{h}\right).$$

Let us introduce the solutions of the two following problems: ($\tilde{s}_1$ is set to be zero for the sake of simplicity)

$$\widetilde{(***)} \qquad \frac{\partial \tilde{h}}{\partial s} = \mathcal{L}_s \tilde{h} \qquad \tilde{h}(0,t) = h_0(t)$$
$$\widetilde{(****)} \qquad \frac{\partial \tilde{\omega}}{\partial s} = \mathcal{L}_s \tilde{\omega} \qquad \tilde{\omega}(0,t) = \dot{h}_0(t)$$

We will prove later that $\tilde{h}$ and $\tilde{\omega}$ satisfy the following estimates, classical for elliptic equations, for $s \in [0, \delta]$:

$$\text{(A)} \qquad \begin{cases} |\tilde{h}|_{L^2} \leq C|h_0|_\infty & |\tilde{\omega}|_{L^2} \leq C|\dot{h}_0|_\infty \\ |\dot{\tilde{h}}|_{L^2} \leq \frac{C}{\sqrt{s}}|h_0|_\infty & |\dot{\tilde{\omega}}|_{L^2} \leq \frac{C}{\sqrt{s}}|\dot{h}_0|_\infty \\ |\ddot{\tilde{h}}|_{L^2} \leq \frac{C}{s}|h_0|_\infty & |\ddot{\tilde{\omega}}|_{L^2} \leq \frac{C}{s}|\dot{h}_0|_\infty. \end{cases}$$

We introduce then the function:

$$k = h - \tilde{h}, \qquad z = \dot{h} - \tilde{\omega}$$

k is zero as well as z at time $s = 0$. Furthermore, k and z satisfy:

$$(***)' \qquad \frac{\partial k}{\partial s} = \mathcal{L}_s k + \overline{K}_1^s(k) + \overline{K}_2^s(\dot{k}) + \overline{K}_3^s\left(\int_0^s \mathcal{L}_1^\tau k\right) + \overline{K}_4^s\left(\int_0^s \mathcal{L}_2^\tau \dot{k}\right) + \\ + \overline{K}_1^s(\tilde{h})\overline{K}_2^s(\dot{\tilde{h}}) + \overline{K}_3^s\left(\int_0^s \mathcal{L}_1^\tau \tilde{h}\right) + \overline{K}_4^s\left(\int_0^s \mathcal{L}_2^\tau \dot{\tilde{h}}\right)$$

$$(****)' \qquad \frac{\partial z}{\partial s} = \mathcal{L}_s(z) + \widetilde{K}_1^s(k + \tilde{h}) + \overline{\widetilde{K}}_1^s(z) + \widetilde{K}_2^s(\dot{z}) + \widetilde{K}_3^s\left(\int_0^s \widetilde{\mathcal{L}}_1^\tau(k + \tilde{h})\right) + \\ + \overline{\widetilde{K}}_3^s\left(\int_0^s \overline{\widetilde{\mathcal{L}}}_1^\tau z\right) + \overline{\widetilde{K}}_4^s\left(\int_0^s \widetilde{\mathcal{L}}_2^\tau \dot{z}\right) + \overline{\widetilde{K}}_1^s(\tilde{\omega}) + \widetilde{K}_2^s(\dot{\tilde{\omega}}) + \\ + \overline{\widetilde{K}}_3^s\left(\int_0^s \overline{\widetilde{\mathcal{L}}}_1^\tau \tilde{\omega}\right) + \overline{\widetilde{K}}_4^s\left(\int_0^s \widetilde{\mathcal{L}}_2^\tau \dot{\tilde{\omega}}\right).$$

In fact, we will need one more differentiation i.e. we need also the evolution equation on $\dot{z}$. However, studying $(***)'$ and $(****)'$, we will derive a good estimate on $|k|_\infty$. It will then be easy to derive the estimate which holds on $|z|_\infty$.

We multiply $(***)'$ by k, $(****)'$ by z and integrate between 0 and . We derive, with a suitable $c > 0$ and using (A):

$$\frac{\partial}{\partial s}\int_0^1 k^2 + c\int_0^1 \dot{k}^2 \leq C\left(\int_0^1 k^2 + \int_0^s\int_0^1 k^2 + \int_0^s\int_0^1 \dot{k}^2 + |h_0|_\infty\left(\int_0^1 k^2\right)^{1/2}\left(s+1+\frac{1}{\sqrt{s}}\right) + |h_0|_\infty \cdot \int_0^s \frac{1}{\sqrt{\tau}}\left(\int_0^1 k^2\right)^{1/2}\right)$$

$$\frac{\partial}{\partial s}\int_0^1 z^2 + c\int_0^1 \dot{z}^2 \leq C\left(\int_0^1 z^2 + \int_0^s\int_0^1 z^2 + \int_0^s\int_0^1 \dot{z}^2 + |\dot{h}_0|_\infty\left(\int_0^1 z^2\right)^{1/2}\left(s+1+\frac{1}{\sqrt{s}}\right) + |\dot{h}|_\infty\int_0^s \frac{1}{\sqrt{\tau}}\left(\int_0^1 z^2\right)^{1/2} + \int_0^1 k^2 + \int_0^s\int_0^1 k^2 + |h_0|_\infty(1+s)\left(\int_0^1 z^2\right)^{1/2}\right).$$

Integrating the first inequality, we derive:

$$\int_0^1 k^2 + \frac{c}{2}\int_0^s\int_0^1 \dot{k}^2 \leq C'\left(\int_0^s\int_0^1 k^2 + |h_0|_\infty\int_0^2 \frac{1}{\sqrt{\tau}}\left(\int_0^1 k^2\right)^{1/2}\right).$$

Setting $u = \int_0^s \frac{1}{\sqrt{\tau}}\left(\int_0^1 k^2\right)^{1/2}$, we have:

As long as $\int_0^1 k^2 \leq 2|h_0|_\infty^2$ and $s \leq 1$, the following estimate holds:

$$su'^2 \leq C^{11}|h_0|_\infty u,$$

since $\int_0^s\int_0^1 k^2$ is then less than $\int_0^s \frac{|h_0|_\infty}{\sqrt{\tau}}\left(\int_0^1 k^2\right)^{1/2}$. Thus,

$$u \leq C_3|h_0|_\infty s$$

and

$$\int_0^1 k^2 + \frac{c}{2}\int_0^s\int_0^1 \dot{k}^2 \leq C_4|h_0|_\infty^2 s.$$

If $C_4 s < 1$, $\int_0^1 k^2 \leq 2|h_0|_\infty^2$, hence the above estimate holds. It is then easy to see that, if $C_5 s < 1$:

$$\int_0^1 z^2 + \frac{c}{2}\int_0^s \int_0^1 \dot{z}^2 \leq C_5\left(|h_0|_\infty^2 + |\dot{h}_0|_\infty^2\right) s$$

z is not $\dot{k}$ because $\tilde{\omega}$ is not $\dot{\tilde{h}}$, but the difference is quite little and using the above estimates, we can easily derive that

$$\int_0^1 (\tilde{\omega} - \dot{\tilde{h}})^2 \leq C_6(|h_0|_\infty^2 + |\dot{h}_0|_\infty^2)s.$$

Thus,

$$\int_0^1 k^2 + \int_0^1 \dot{k}^2 \leq C_7(|h_0|_\infty^2 + |\dot{h}_0|_\infty^2)s.$$

Hence,

$$|k|_\infty \leq C_8(|h_0|_\infty + |\dot{h}_0|_\infty)\sqrt{s}.$$

We also have, using similar estimates:

$$|z|_\infty + |\dot{k}|_\infty \leq C_9(|h_0|_\infty + |\dot{h}_0|_\infty + |\ddot{h}_0|_\infty)\sqrt{s}.$$

Thus,

$$\dot{h} = \tilde{\omega} + O(|h_0|_\infty + |\dot{h}_0|_\infty + |\ddot{h}_0|_\infty)\sqrt{s}$$

h_0 has to be chosen under the form $\ddot{\eta}_0 + K_0(\eta_0)$, where η_0 is an arbitrary C^∞ periodic function, for example. Thus, $\dot{h}_0 = \dddot{\eta}_0 + \widetilde{K}_0(\eta_0) + K_1(\dot{\eta}_0)$ and

$$\tilde{\omega}(s,t) = \int_0^1 G_s(t,\tau)\dot{h}_0(\tau)d\tau$$

where G_s is the Green's function of $\frac{\partial}{\partial s} - \mathcal{L}_s$. Thus,

$$\begin{aligned}\dot{h} = \int_0^1 & G_s(t,\tau)(\dddot{\eta}_0 + \widetilde{K}_0(\eta_0) + K_1(\dot{\eta}_0)) + \\ & + O(|\eta_0|_\infty + |\dot{\eta}_0|_\infty + |\ddot{\eta}_0|_\infty + |\dddot{\eta}_0|_\infty + |\eta_0^{(iv)}|_\infty)\sqrt{s}.\end{aligned}$$

Taking any function such that $\eta_0^{(3)}(0)$ is nonzero, η_0 a fixed C^∞-function, and letting s be very small, we will have $\dot{h}(s,0)$ nonzero.

How small should be s depends on $\widetilde{K}_0, K_1, 0$ and $G_s(t,\tau)$. On $[s_1, s_2]$, all of these have uniform bounds (including bounds from below for $G_s(t,\tau)$, in appropriate forms). Therefore, the smallness of s can be taken uniform on this time-interval.

We are left with the (A)-estimates. The first estimate is easily derived from:

$$\frac{\partial}{\partial s}\int_0^1 \tilde{h}^2 + c\int_0^1 \dot{\tilde{h}}^2 \leq 0.$$

Thus

$$\int_0^1 \tilde{h}^2(s) + c\int_0^s\int_0^1 \dot{\tilde{h}}^2 \leq \int_0^1 \tilde{h}^2(0) \leq |h(0)|_\infty^2.$$

Differentiating the equation, we have:

$$\frac{\partial}{\partial s}\dot{\tilde{h}} = \mathcal{L}_s\dot{\tilde{h}} + \dot{\mathcal{L}}_s\tilde{h}.$$

Thus

$$\frac{\partial}{\partial s}\int_0^1 \dot{\tilde{h}}^2 + c\int_0^1 \ddot{\tilde{h}}^2 \leq \bar{\delta}\int_0^1 \dot{\tilde{h}}^2 + \frac{C}{\bar{\delta}}\int_0^1 \tilde{h}^2$$

$\bar{\delta}\int_0^1 \dot{\tilde{h}}^2$ can be absorbed by $\frac{c}{2}\int_0^1 \ddot{\tilde{h}}^2$ Thus:

$$\int_0^1 \dot{\tilde{h}}^2(s) - \int_0^1 \dot{\tilde{h}}^2(\tau) \leq \frac{C}{\bar{\delta}}|h(0)|_\infty^2(s-\tau).$$

Integrating in τ on $[0,s]$, we derive:

$$s\int_0^1 \dot{\tilde{h}}^2(s) \leq |h(0)|_\infty^2\left(1+\frac{Cs^2}{2\bar{\delta}}\right).$$

Thus,

$$|\dot{\tilde{h}}|_{L^2} \leq \frac{C|h(0)|_\infty}{\sqrt{s}}.$$

We also have, by integration:

$$\begin{aligned}\int_0^1 \dot{\tilde{h}}^2(s) + c\int_{s/2}^s \ddot{\tilde{h}}^2(\tau)d\tau &\leq \int_0^1 \dot{\tilde{h}}^2(s/2) + C|h(0)|_\infty^2 \\ &\leq \frac{C'|h(0)|_\infty^2}{s}.\end{aligned}$$

Thus,

$$\int_{s/2}^s \ddot{\tilde{h}}^2(\tau)d\tau \leq C^{11}\frac{|h(0)|_\infty^2}{s}.$$

Finally,

$$\frac{\partial}{\partial s}\ddot{\tilde{h}} = \mathcal{L}_s\ddot{\tilde{h}} + \ddot{\mathcal{L}}_s\tilde{h} + \dot{\mathcal{L}}_s\dot{\tilde{h}}.$$

Thus

$$\frac{\partial}{\partial s}\int_0^1 \ddot{\bar{h}}^2 + c\int_0^1 \dddot{\bar{h}}^2 \leq \bar{\delta}\int \dddot{\bar{h}}^2 + \frac{C}{\bar{\delta}}\int_0^1 \ddot{\bar{h}}^2 + \int_0^1 \dot{\bar{h}}^2 + \int_0^1 \bar{h}^2.$$

Thus,

$$\frac{\partial}{\partial s}\int_0^1 \ddot{\bar{h}}^2 \leq C|h(0)|_\infty^2$$

$$\int_0^1 \ddot{\bar{h}}^2(s) - \int_0^1 \ddot{\bar{h}}^2(\tau) \leq C|h(0)|_\infty^2(s-\tau).$$

We integrate between $\frac{s}{2}$ and s in τ and derive the last estimate:

$$\frac{s}{2}\int_0^1 \ddot{\bar{h}}^2(s) \leq C^{111}\frac{|h(0)|_\infty^2}{s}.$$

Thus,

$$\int_0^1 \ddot{\bar{h}}^2(s) \leq C_4\frac{|h(0)|_\infty^2}{s^2} \qquad \text{q.e.d.}$$

Appendix 2

The dynamics of α along v: basic computation
Assume that $\tilde{v}$ is given in $\ker\alpha$ such that

$$\text{(A2)} \qquad d\alpha(\tilde{v},\cdot) = \tilde{\beta} \quad \text{is a contact form}$$

Assume that

(B) There exists at least one point of v having an oriented coincidence point distinct from itself in the α-rotation along v.

Then, $\tilde{\beta} \wedge d\tilde{\beta}$ and $\alpha \wedge d\alpha$ have the same sign because $\tilde{\beta}$ and α are transverse to each other, hence homotopic as field of planes (see Proposition A3 below).

We normalize $\tilde{v}$ into $\lambda\tilde{v} = v$ so that

$$\text{(A3)} \qquad \beta = d\alpha(v,\cdot), \qquad \beta \wedge d\beta = \alpha \wedge d\alpha.$$

Let w be the contact vector-field of β.

We then have:

Proposition A2. *Let $\bar{\mu} = \alpha(w), \bar{\mu}_\xi = d\bar{\mu}(\xi)$, $\bar{\mu}_{\xi\xi} = d\bar{\mu}_\xi(\xi)$. We then have:*

$$\begin{aligned}
&w = -[\xi, v] + \bar{\mu}\xi \\
&[\xi,[\xi,v]] = -\tau v \qquad \tau \in C^\infty(M,\mathbb{R}) \\
&\bar{\mu} = d\alpha(v,[v,[\xi,v]]) \\
&\bar{\mu}_\xi = d\alpha([\xi,v],[v,[\xi,v]]) \\
&\bar{\mu}_{\xi\xi} + \tau\bar{\mu} = -d\tau(v) = -\tau_v \\
&\gamma = -d\alpha(w,\cdot) = d\beta(\xi,\cdot).
\end{aligned}$$

Proof of Proposition A2. Since $\beta \wedge d\beta = \alpha \wedge d\alpha$ and w is the contact vector-field of β,

$$\beta \wedge d\beta(\xi, v, w) = \alpha \wedge d\alpha(\xi, v, w).$$

Thus,

$$d\beta(\xi, v) = d\alpha(v, w) = 1.$$

On the other hand, since $\beta(\xi) = \beta(v) = 0$

$$d\beta(\xi, v) = -\beta([\xi, v]) = -d\alpha(v, [\xi, v]).$$

We thus have:

$$d\alpha(v, [\xi, v]) = -1$$

which we differentiate with respect to ξ. $d\alpha$ and ξ can be taken to be constant in the same chart. Thus:

$$d\alpha(v, [\xi, [\xi, v]]) = 0$$

$[\xi, v]$, $[\xi, [\xi, v]]$ are easily seen to belong to $\ker \alpha$. Thus,

$$[\xi, [\xi, v]] = -\tau v, \qquad \tau \in C^\infty(M, \mathbb{R}).$$

We also have:

$$w = -[\xi, v] + \bar{\mu}\xi + \theta v.$$

We want to prove that θ is zero. We compute

$$\begin{aligned} 0 = d\beta(\xi, w) &= -\beta([\xi, w]) = -d\alpha(v, [\xi, -[\xi, v] + \bar{\mu}\xi + \theta v]) \\ &= -d\alpha(v, \tau v + \bar{\mu}_\xi \xi + d\theta(\xi) v + \theta[\xi, v]) = \theta. \end{aligned}$$

Thus $\theta = 0$ and $w = -[\xi, v] + \bar{\mu}\xi$ as claimed.,

We then observe that:

$$\begin{aligned} 0 = d\beta(w, v) &= -\beta([w, v]) = -d\alpha(v, [w, v]) = -d\alpha(v, [-[\xi, v] + \bar{\mu}\xi, v]) \\ &= d\alpha(v, [[\xi, v], v]) - \bar{\mu} d\alpha(v, [\xi, v]). \end{aligned}$$

Thus,

$$\bar{\mu} = d\alpha(v, [v, [\xi, v]]).$$

In coordinates where ξ and $d\alpha$ are constant, we then have:

$$\bar{\mu}_\xi = d\bar{\mu}_\xi = d\alpha([\xi, v], [v, [\xi, v]]) + d\alpha(v, [\xi, [v, [\xi, v]]]).$$

Since

$$[\xi, [v, [\xi, v]]] + [[\xi, v], [\xi, v]] + [v, [[\xi, v], \xi]] = 0$$

and since $[\xi, [\xi, v]] = -\tau v$, we have:

$$\bar{\mu}_\xi = d\alpha([\xi, v], [v, [\xi, v]]).$$

Differentiating again along ξ;

$$\begin{aligned} d\bar{\mu}_\xi(\xi) = \bar{\mu}_{\xi\xi} &= d\alpha([\xi, [\xi, v]], [v, [\xi, v]]) + d\alpha([\xi, v], [\xi, [v, [\xi, v]]]) \\ &= d\alpha(-\tau v, [v, [\xi, v]]) + d\alpha([\xi, v], [v, [[\xi, v], \xi]]) \\ &= -\tau\bar{\mu} + d\alpha([\xi, v], [v, \tau v]) = -\tau\bar{\mu} + d\tau(v). \end{aligned}$$

Thus,

$$\bar{\mu}_{\xi\xi} + \tau\bar{\mu} = -\tau_v$$

as claimed.

Finally,

$$\gamma(\xi) = -d\alpha(w,\xi) = d\beta(\xi,\xi) = 0$$
$$\gamma(v) = -d\alpha(w,v) = 1 \quad \text{and}$$
$$d\beta(\xi,v) = -d\alpha(v,[\xi,v]) = 1.$$

Thus

$$\gamma(v) = d\beta(\xi,v)$$

and

$$\gamma([\xi,v]) = -d\alpha(w,[\xi,v]) = -d\alpha(-[\xi,v] + \bar{\mu}\xi,[\xi,v]) = 0$$

while

$$d\beta(\xi,[\xi,v]) = -\beta([\xi,[\xi,v]]) = -d\alpha(v,-\tau v) = 0.$$

Thus

$$\gamma = -d\alpha(w,\cdot) = d\beta(\xi,\cdot)$$

as claimed. The proof of Proposition A2 is thereby complete.

We repeat now here, for the sake of completeness, Proposition 9 of [1], with its proof. This Proposition expresses a property of rotation of a contact form α along the flow-lines of a vector-field v in its kernel. The direction of the rotation is related to the sign of $\alpha \wedge d\alpha$. Therefore, when we have two contact forms such as α and β, having a common vector-field in their kernel, they both rotate along the flow-lines of v.

If $\alpha \wedge d\alpha$ and $\beta \wedge d\beta$ have different signs, the rotations are in opposite directions. If one of them rotates "enough" at any given point i.e. if there is a point $x_0 \in M$ having an oriented coincidence point distinct x_1 from itself for α for example (along v) (such as in (B)), then, between x_0 and x_1, α has rotated for at least 2π in one direction. β has rotated in the opposite direction. Thus, $\{\beta = 0\}$ and $\{\alpha = 0\}$ should coincide at some point along the v-orbit between x_0 and x_1. Thus, if β and α are transverse, under (B), $\beta \wedge d\beta$ and $\alpha \wedge d\alpha$ must have the same sign, as we wrote in (A3).

We now prove this property of rotation of α along v: Let x_0 be a point of M and $\phi_s(x_0)$ be the integral curve of v passing through x_0.

Let $e_1(x_0)$ and $e_2(x_0)$ be two tangent vectors to M at x_0 such that $(\alpha \wedge d\alpha)_{x_0}(e_1(x_0), e_2(x_0), v(x_0))$ is negative. We denote:

$$e_1(s) = D\phi_s(e_1(x_0)), \qquad e_2(s) = D\phi_s(e_2(x_0)).$$

At $x_s = \phi_s(x_0), (e_1(s), e_2(s), v(x_s))$ is again a basis and we have:

$$(\alpha \wedge d\alpha)_{x_s}(e_1(s), e_2(s), v(x_s)) < 0.$$

Since $\alpha(v) = 0$, we can define the trace of $\alpha = 0$ in Span$(e_1(s), e_2(s))$. It is given by:

$$\alpha(e_2(s))e_1(s) - \alpha(e_1(s))e_2(s) = u(s).$$

We follow this trace when s changes.

We then have:

Proposition A3. *i) The trace of α in Span$(e_1(s), e_2(s))$ rotates from e_1 to e_2 when s grows.*

ii) Under (B), $\beta \wedge d\beta = \alpha \wedge d\alpha$.

Proof of Proposition A3. We discussed ii) above. We are left with i).

Since $\alpha \wedge d\alpha(e_1(s), e_2(s), v) < 0$, we have:

$$\alpha(e_1)d\alpha(e_2, v) - \alpha(e_2)d\alpha(e_1, v) < 0.$$

Let

$$\alpha(e_2) = \rho\cos\widetilde{\varphi} \qquad \alpha(e_1) = -\rho\sin\widetilde{\varphi}$$

so that

$$u = \rho[\cos\widetilde{\varphi}e_1 + \sin\widetilde{\varphi}e_2].$$

Let us consider a local coordinate chart near the orbit x_s where v is read as a constant vector. $\frac{\partial}{\partial s}$ and v are then the same derivation, $e_1(s)$ and $e_2(s)$ are v-transported, therefore constant. We can also assume that we have extended them near x_s and that they are constant near x_s.

We then have:

$$d\alpha(e_1, v) = e_1 \cdot \alpha(v) - v \cdot \alpha(e_1) - \alpha\left(\left[\frac{\partial}{\partial s}, e_1\right]\right)$$

$\alpha(v)$ is zero, $\left[\frac{\partial}{\partial s}, e_1\right]$ is zero and $v \cdot \alpha(e_1) = \frac{\partial}{\partial s}\alpha(e_1)$. Thus,

$$d\alpha(e_1, v) = -\frac{\partial}{\partial s}\alpha(e_1) = -\frac{\partial}{\partial s}(\rho\cos\widetilde{\varphi}).$$

Similarly,

$$d\alpha(e_2, v) = -\frac{\partial}{\partial s}\alpha(e_2) = -\frac{\partial}{\partial s}(\rho\sin\widetilde{\varphi}).$$

Thus,

$$\begin{aligned}&\alpha(e_1)d\alpha(e_2, v) - \alpha(e_2)d\alpha(e_1, v)\\ =&\rho\sin\widetilde{\varphi}\frac{\partial}{\partial s}(\rho\cos\widetilde{\varphi}) - \rho\cos\widetilde{\varphi}\frac{\partial}{\partial s}(\rho\sin\widetilde{\varphi})\\ =&-\rho^2\dot{\widetilde{\varphi}} < 0.\end{aligned}$$

Hence, $\dot{\widetilde{\varphi}}$ is positive. Proposition A3 follows.

Appendix 3

Some additional remarks about the physical interpretation, two mathematical conjectures

1. An extremality principle for the spin component, despite the possible analogies, has to be quite different from the extremality principle for a classical system: the variations of the spin are confined to S^3. There might be some analogy with the fixed energy problem for a Hamiltonian system; but this problem is an artificial mathematical construction: the classical particle is not, a priori, confined to this hypersurface, its energy level brings it there.

The mathematical treatment of both situations has to be different. Beyond this fact, it is unclear whether the same laws of physics apply to the spatial/physical world and to the world of spins. If we keep the analogy with a self-rotation, then we have to apply such laws, which, however, have to account for flips and transitions. This is tantamount to the study of the variational problem on S^3.

In addition, we only have access, experimentally to *the projection of the spin on* S^2. Hence, if we want to have any control on the S^3-behavior which takes root in our world, we need to postulate that the direction of the motion of the spin of the particle splits on the Hamiltonian direction and the direction of projection $S^3 \to S^2$.

We are then naturally led to $\dot{x} = a\xi + bv, v \in \ker \alpha$ hence to a space such as $\mathcal{L}_\beta$ (a reconstructive approach). This model offers some hope to account for some paradoxes or some phenomena still not accounted for.

2. The hypothesis that α turns well along v has to be opposed to the situation where v spans $\ker \alpha \wedge \ker \gamma$ and γ is a foliation. As shown in the last chapter of Pseudo-Orbits of contact forms, this situation may arise.

Then α cannot turn well along v.

We *conjecture* that, given a contact form, or a contact structure, there are vector-fields v of the two types (with repect to α) in $\ker \alpha$ or $\ker(\alpha + df)$.

This would have the interesting feature to allow to move continuously (through a homotopy between the two kinds of v's) from a "spin" situation (when α turns well) to a "classical" situation (α does not turn well, there are no coincidence points). This has obvious physical consequences.

3. Taking $\eta = b$ in the evolution equations which we studied, we conjecture that $b(s,t)$, at the explosion time, behave as $\sum_{i=1}^{m} c_i \delta_{t_i}$, where δ_{t_i} is the Dirac mass at t_i and $|c_i| \geq \gamma > 0, \gamma$ a fixed constant. This result would allow to shortcut the first half of the present work.

Appendix 4

How to take care of the problem of the small normals, description of the retraction on $\bigcup_k W_u(\Gamma_{2k})$

We improve in two directions our results and get rid of the problem of the small normals. Our retraction by deformation on $\bigcup_k W_u(\Gamma_{2k})$ becomes thereby neat.

First, we show how to improve some of the estimates of Figures O-Q.

We have not led, for the sake of the concision, the possible improvements to their full end.

We had:

$$Y_i = \psi_{\delta s}(x_i + (\ell_i^k - \ell_i^q)v) = x_{i+1} + (\ell_{i+1}^k - \ell_{i+1}^q)v - (\delta s' - \delta s)\xi(Y_i) +$$
$$+ 0((\delta s' - \delta s)^2) + 0\left(\frac{1}{|b|_\infty} + \frac{1}{L^2} - \Delta a_i - \Delta a_{i+1} + \frac{\delta s}{L} + \frac{\delta s'}{L}\right)$$

$0\left(\frac{1}{L^2}\right)$ is, in fact, as small as we wish, as well as $0(\Delta a_i + \Delta a_{i+1})$, if we do not use the conjugate points flow. We therefore concentrate on $0\left(\frac{\delta s}{L} + \frac{\delta s'}{L}\right)$.

These two terms are due to the fact that the almost ξ-pieces are in fact tangent to $a\xi + 0\left(\frac{\nu}{L}\right)v$.

These pieces, using the one-parameter group ψ_τ generated by ξ, can be replaced by a piece of ξ-orbit during the same time δs or $\delta s'$, plus a piece of curve tangent to $D\psi_\tau\left(0\left(\frac{\nu}{L}\right)v\right)$, during the time δs or $\delta s'$.

Hence, $0\left(\frac{\delta s}{L}\right)$ can be replaced, in local coordinates, by the contribution of a small curve z_1, ending at x_{i+1}, tangent to $D\psi_\tau\left(0\left(\frac{\nu}{L}\right)v\right)$ during the time δs. Similarly, $0\left(\frac{\delta s'}{L}\right)$ can be replaced, in local coordinates, by the contribution of a small curve z_2 starting at $\psi_{\delta s'}(x_i + (\ell_i^k - \ell_i^q)v)$, tangent to $D\psi_\tau\left(0\left(\frac{\nu}{L}\right)v\right)$, during the time $\delta s'$. If the first interval ends at the time t_{i+1}, we have:

$$z_1(t) = \psi_{-a(t-t_{i+1})}(x(t)).$$

A similar formula holds for $z_2(t)$.

We then obtain using the arguments of Figures M-Q:

$$(\delta s' - \delta s)(1 + o(1)) = 0(\delta s(\ell_i^k - \ell_i^q)^2 + \frac{1}{|b|_\infty} + \frac{1}{L^2} - \Delta a_i - \Delta a_{i+1}) +$$
$$+ 0\left(\left(\frac{\nu}{L}\right)^2 \delta s^2 + \left(\frac{\nu}{L}\right)^2 \delta s'^2 + \frac{\nu}{L}\delta s'(\ell_{i+1}^k - \ell_{i+1}^q)\right).$$

Indeed, the equation on $\delta s' - \delta s$ is obtained making use of $\alpha_{\psi_{\delta s}(x_i)}$, i.e applying it to the identity $Y_i = \psi_{\delta s}(x_i + (\ell_i^k - \ell_i^q)v) = \dots$

Of course

$$\psi_{\delta s}(x_i) = x_{i+1} + 0\left(\frac{\nu}{L}\delta s\right).$$

Thus,

$$\alpha_{\psi_{\delta s}(x_i)} = \alpha_{x_{i+1}} + 0\left(\frac{\nu}{L}\delta s\right).$$

Since $D\psi_\tau\left(0\left(\frac{\nu}{L}\right)v\right)$ is in $\ker\alpha(\psi_s$ is generated by $\xi)$,

$$\alpha_{z_j(t)}(\dot z_j(t)) = o \text{ for } j = 1, 2.$$

Also, $z_1(\delta s) = x_{i+1}$ and:

$$\alpha_{\psi_{\delta s}(x_i)}(z_1(\delta s) - z_1(o)) =$$

$$\int_0^{\delta s}\left(\alpha_{z_1(t)}(\dot z_1(t)) + 0\left(d(z_1(t), z_1(\delta s)) + 0\left(\frac{\nu}{L}\delta s\right)\right)|\dot z_1(t)|\right) =$$

$$= 0\left(\frac{\nu}{L}\right)\int_0^{\delta s}\left(\int_0^t |\dot z_1(x)|dx\right)dt + 0\left(\frac{\nu^2}{L^2}\right)\delta s^2 = 0\left(\frac{\nu^2}{L^2}\delta s^2\right).$$

Similarly, we have:

$$\psi_{\delta s'}(x_i + (\ell_i^k - \ell_i^q)v) = x_{i+1} + (\ell_{i+1}^k - \ell_{i+1}^q)v + 0\left(-\Delta a_i - \Delta a_{i+1} + \frac{1}{|b|_\infty} + \frac{\nu\delta s'}{L}\right)$$

$$= \psi_{\delta s}(x_i) + (\ell_{i+1}^k - \ell_{i+1}^k)v + 0\left(-\Delta a_i - \Delta a_{i+1} + \frac{1}{|b|_\infty} + \frac{\nu\delta s'}{L} + \frac{\nu\delta s}{L}\right).$$

Thus:

$$\alpha_{\psi_{\delta s}(x_i)}(z_2(\delta s') - z_2(o))$$

$$= 0\left(\frac{\nu}{L}\delta s'\right)\left(|\ell_{i+1}^k - \ell_{i+1}^q| - \Delta a_i - \Delta a_{i+1} + \frac{1}{|b|_\infty} + \frac{\nu\delta s'}{L} + \frac{\nu\delta s}{L}\right).$$

Our claim follows.

The equation on $\ell_i^k - \ell_i^q$ (hence by a symmetry argument, on $\ell_{i+1}^k - \ell_{i+1}^q$) also improves. The curves z_1 and z_2 have $\dot z_i$ in $\ker\alpha$, equal to $0\left(\frac{\nu}{L}\right) + 0\left(\frac{\nu}{L}(\delta s + \delta s')\right)$, since $|\tau|$ is less than $\delta s + \delta s'$.

Thus, since v is constant, the projection of $z_1(\delta s) - z_1(o)$ or $z_2(\delta s') - z_2(o)$ onto $[\xi, v](\psi_{\delta s}(x_i))$ is $0\left(\frac{\nu}{L}(\delta s + \delta s')^2\right)$.

Thus:

$$\tau\delta s(\ell_i^k - \ell_i^q) = 0(\delta s|\ell_i^k - \ell_i^q|^2 + |\delta s' - \delta s|\,|\ell_i^k - \ell_i^q| +$$

$$+ (\delta s' - \delta s)^2 + \frac{1}{|b|_\infty} + \frac{1}{L^2} - \Delta a_i - \Delta a_{i+1} + \frac{\nu}{L}(\delta s + \delta s')^2).$$

A similar equation holds for $\ell^k_{i+1} - \ell^q_{i+1}$.

Combining, we derive:

$$(\delta s' - \delta s)(1 - o(1)) = 0\left(\frac{1}{|b|_\infty} + \frac{1}{L^2} - \Delta a_i - \Delta a_{i+1}\right) + \frac{\nu^2}{L^2}\left((\delta s + \delta s')^2\right).$$

This is a much better estimate. Furthermore, $0\left(\frac{1}{|b|_\infty} + \frac{1}{L^2} - \Delta a_i - \Delta a_{i+1}\right)$ reduces to $0\left(\frac{1}{|b|_\infty}\right)$ if we do not use the conjugate - points flow.

Thus, in our general construction, when we do not use the conjugate-points flow, we have:

$$(\delta s' - \delta s)(1 + o(1)) = 0\left(\frac{1}{|b|_\infty} + \frac{\nu^2}{L^2}\delta s^2\right). \qquad \text{(E)}$$

The result extends when δs is not small anymore.

The other estimates also improve, but we do not focus on these improvements here.

A second question which we address now is how to take care of the deconcentration flow-lines out of $\bigcup_k \Gamma_{2k}$.

Namely, our flow has brought us near the $\frac{\nu}{L}$-stretched curves, and thereafter, combining the $\frac{\nu}{L}$ and $\frac{2\nu}{L}$ - flows which we have constructed previously and which displace very little the v-branches, near $\bigcup_k \Gamma_{2k}$. There are accordingly flow-lines flowing out of the $\frac{\nu}{L}$-stretched curves and ultimately of $\bigcup_k \Gamma_{2k}$.

If the v-branches of the $\frac{\nu}{L}$-stretched curves or of the curves of $\bigcup_k \Gamma_{2k}$ are "large enough", the unstable lines of these flows out of these curves do not touch these v-branches, since there are then no positive solution of:

$$\begin{cases} -\ddot{\eta}_0 - C(1+b^2)\eta_0 = -\frac{2}{|x_1^- - x_2^-|^2} e^{-\int_{t_1^-}^{t} \bar{\mu} b} \\ \eta_0(x_1^-) = 0 = \eta_0(x_2^-) \end{cases} \qquad \text{(Equation (10))}.$$

Thus, our general flow and the more special flows $Z_\delta(t_1, t_2, x)$ coincide. We can assume that we use the Z_δ's instead of our general flow near such curves and all our deformation statements and estimates hold. (the ones related to Figures M-Q in particular).

But, as soon as one of the v-branches has some critical length, the linear operator of equation (10) degenerates and a deconcentration line develops along such a v-branch, out of this curve at infinity into C_β:

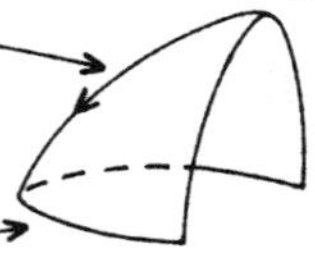

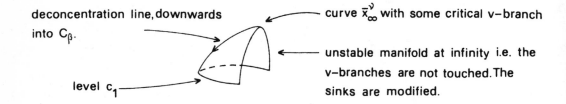

FIGURE A4.1

$\bar{x}^{\nu}_{\infty}$ could be a $\frac{\nu}{L}$ or $\frac{2\nu}{L}$-stretched curve or a curve of $\bigcup_k \Gamma_{2k}$. We have built a process to go from $\bar{x}^{\nu}_{\infty}$ to $\bar{x}^{0}_{\infty}$, a curve of $\bigcup_k \Gamma_{2k}$. This process has been built "at infinity" i.e assuming that we started with a $\frac{\nu}{L}$-stretched curve. But it has an obvious extension to a neighborhood, in $W^{1,1}$, of those curves i.e if we have nearly v-branches and sinks, we can repeat the same construction. Thus, as long as the above figure featuring the unstable manifold of $\bar{x}^{\nu}_{\infty}$, with a deconcentration line embedded in it, remains in such a neighborhood, we can move all this figure downwards. It deforms, through a continuous movement, into the unstable manifolds of various new critical points at infinity for the $\frac{\nu}{L}$ or $\frac{2\nu}{L}$ -flow, or of $\bigcup_k \Gamma_{2k}$. At the end of the process these will be curves of $\bigcup_k \Gamma_{2k}$. There can be only one of them, because of the arguments of Figures M-Q. Indeed, if there were two, their v-branches would be extemely close, because they have to be extremely close to those of $\bar{x}^{\nu}_{\infty}$ (they cannot come from inside C_β because oue process keeps the nearly v-branches essentially unchanged. Since at the end they have to be stretched, they should be stretched from the beginning). Thus, these would be curves made of the same number of v and ξ-pieces, with the v-branches essentially confounded, up to small differences in length. The arguments of Figures M-Q show that they are then close up to $0(c)$ i.e up to a small number depending only on how small we choose to take c in (15).

All these curves can be considered to be essentially one curve from the deformation point of view: the fact that there could several, if not a continuum of very close curves, is due to the fact that our flows have large zero-sets. If we perturb them, so that they have a discrete zero set, the problem disappears. If we do not, we have local family, which are essentially constant. This is a minor technical detail which we will address later. At this point, proceeding as if this very compounded family were reduced to one curve, we have the following figure:

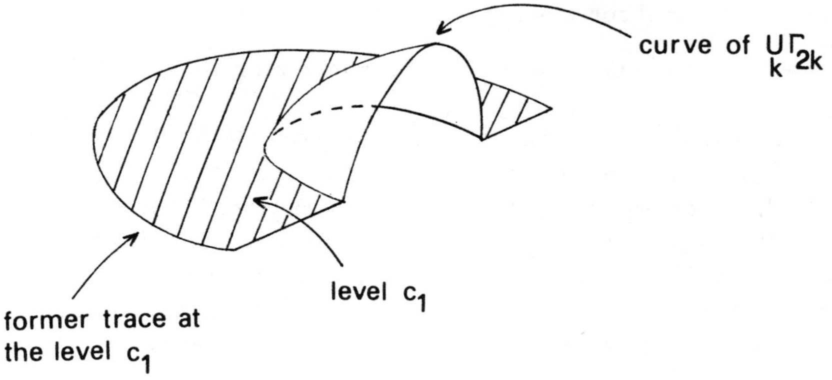

FIGURE A4.2

We can then glue up, along the former trace at the level c_1, to this figure, the continuation of the unstable manifold of $\bar{x}_\infty^\nu$ thus obtaining:

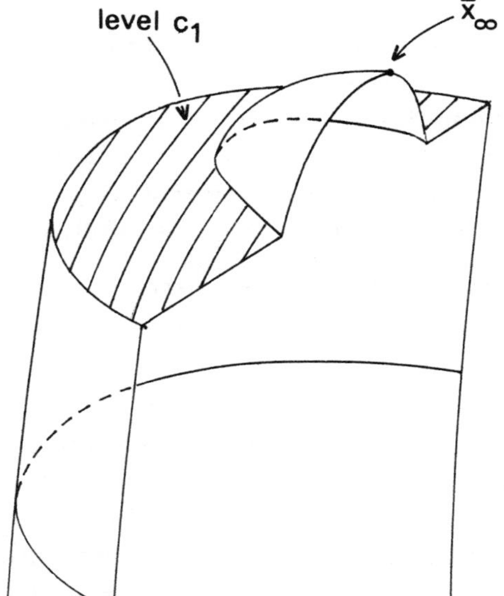

FIGURE A4.3

The level c_1 can be bent a little bit so that all this above figure can be thought of as the unstable manifold of $\bar{\bar{x}}_\infty$. All bounds and estimates hold.

Thus, the deconcentration lines fith with our deformation process, as long as the level of $\bar{\bar{x}}_\infty$, the final curve of $\bigcup_k \Gamma_{2k}$ does not lie too much below $\bar{x}^\nu_\infty$ i.e we need that the piece of unstable manifold of $\bar{x}^\nu_\infty$ above its level is in a small $W^{1,1}$-neighborhood of $\bar{x}^\nu_\infty$.

This condition drops if the difference between the level of $\bar{x}^\nu_\infty$ and c_1 is a fixed constant, which we can even take to be of the type $\frac{\varepsilon_{30}}{L}$.

Indeed, our deformation, at each stage, takes place between a level a and $a - \Delta a$ with $-L\Delta a < \varepsilon_{30}$. Thus, if the unstable manifold of $\bar{x}^\nu_\infty$ is built in such a way that it is still in this $W^{1,1}$-neighborhood when it reaches the level of $\bar{x}^\nu_\infty$ minus $\frac{\varepsilon_{30}}{L}$, we do not have to worry about the level of $\bar{\bar{x}}_\infty$. If it is above this level, the above argument works. Otherwise, it is below and the deformation reaches the level $a - \frac{\varepsilon_{30}}{L}$ to our satisfication, with L becoming $L+1$.

We are going to replace our general flow near the deconcentration set, at infinity, by another flow which will have exactly the same properties plus another one: unless the level has decreased of $\frac{\varepsilon_{30}}{L}$, the deconcentration line and the associated piece of unstable manifold as described above remain in the $W^{1,1}$-neighborhood when our iterative process $Z^{\nu/L}/Z^{2\nu/L}$ is defined. Thus, by the above arguments, we will have built a flow deforming C_β on $\bigcup_k W_u(\Gamma_{2k})$ (up to the unstable manifolds of possible periodic orbits).

This modification does not affect our former conclusions because this flow will have exactly the same properties.

First, before describing this modification which will focus on one piece of v-branch along which a deconcentration line builds and on the two nearby ξ-pieces, we show that, considering a topological class of dimension k_0 which we wish to deform, we will never encounter curves of $\bigcup_k \Gamma_{2k}$ or $\frac{\nu}{L}$ $\left(\frac{2\nu}{L}\right)$-sketches curves having more than $k_0 + 2$ such characteristic v-pieces where a deconcentration line can be built.

The reason is obviously a dimensional one. We have to make it rigorous and clear. This works as follows:

Let us call L^ν_{2m} (respectively $L^{2\nu}_{2m}, L_{2m}$) the space of curves having m v-jumps and m pieces where $\dot{x} = a\xi + bv$ when b is $\frac{\nu}{L}$ stretched (respectively $\frac{2\nu}{L}$ or zero).

In L^ν_{2m}, the subspace of curves having k_0+2 characteristic or more v-pieces is a stratified set of codimension larger than or equal to $k_0 + 2$.

Since our deformation class is assumed to be of dimension k_0 it can be assumed to avoid objects of codimension larger than k_0, unless these objects are critical points themselves, in which case the dimesnion of their stable manifold enters into play. In this case, with the flow as we built it, all curves of L^ν_{2m} are critical, including the curves we wish to avoid. This problem can be overcome as follows, one needs to introduce a suitable flow at infinity which has (nearly) all the properties of our previous flow but has a smaller zero set which will not intersect the set we want to avoid. The claim then will follow by standard arguments of general position.

This flow which we describe on L^ν_{2m}, can be extended, as we show below, to a $W^{1,1}$-neighborhood of $\left(\bigcup_m L^\nu_{2m}\right)$. It can be used in lieu of our former flow and yields the desired conclusion.

On L^ν_{2m}, we consider the functionals:

$$I(x) = \int_0^1 \left(|b| - \frac{\nu}{L}\right)^+$$

and

$$J(x) = \int_0^1 a.$$

I and J are differentiable on L^ν_{2m} (we skip here some technical details, for example, since $|b| \leq \frac{\nu}{L}$ in the sinks, I measures the length of the v-branches, which is a differentiable function on L^ν_{2m}. Also L^ν_{2m} is a stratified set ...)

The set where I' and J' are dependent is generically of dimension 1. Hence, using a general position argument, we can assume that it avoids the set of curves having $k_0 + 2$ or more characteristic pieces.

Outside of the set when I' and J' on L^ν_{2m}, are dependent, we can build a flow on L^ν_{2m} which decreases I and J, thus a flow which controls $\int_0^1 \left(|b| - \frac{\nu}{L}\right)^+$ and decreases $J(x)$. This flow has to be made zero on the set of dependence. But this is far from the set we wish to avoid along our deformation.

As we already said, this flow defined at infinity, can be extended to a small neighborhood of $\bigcup_m L^\nu_{2m}$, with all the properties of the former flow: on $\bigcup_m L^\nu_{2m}$, it already controls $\int_0^1 \left(|b| - \frac{\nu}{L}\right)^+$ and decreases $\int_0^1 a$. We will show that $\int_0^1 \dot{b}^2$ behaves well on the deformation lines of this flow as well as $|b|_\infty$. All of this is completed below.

Let us describe, when we have a characteristic v-jump, what is the deconcentration flow at infinity.

A characteristic v-jump is a piece of $\pm v$-orbit along the curve, preceded and followed by $\xi + 0\left(\frac{\nu}{L}\right)$ v-pieces:

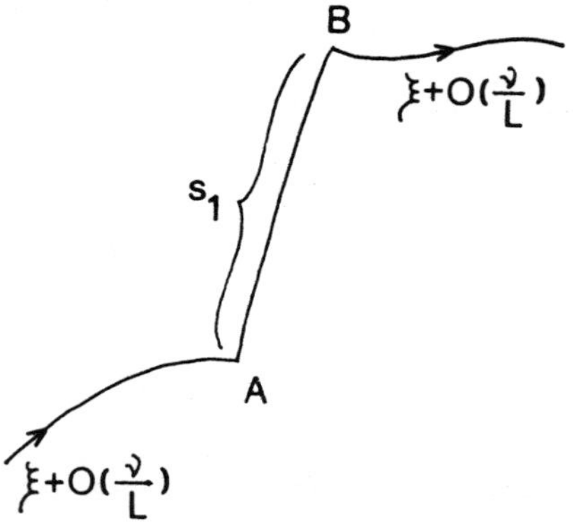

FIGURE A4.4

such that

$$-\ddot{\eta} - C(1+b^2)\eta,$$

under Dirichlet boundary conditions, degenerates between A and B (for the first eigenvalue). When c increases, the length of [AB] along v decreases. Thus, taking C large enough, we can assume that [AB] is of a given length, but small.

Let us compute and estimate

$$\alpha_B\left(D\phi_{s_1}(\xi_A)\right)$$

i.e we consider the vector ξ at A, transport it in B using the one-parameter group of v and compute α on this vector.

We have denoting $x_s = \phi_s(A)$:

$$\begin{aligned}\frac{\partial}{\partial s}\left(\alpha_{x_s}(D\phi_s(\xi))\right) &= D\phi_s(\xi_A)\cdot\alpha_{x_s}(v) - \alpha_{x_s}\left(\left[\frac{\partial}{\partial s}+v, D\phi_s(\xi_A)\right]\right) + \\ &+ d\alpha(v, D\phi_s(\xi_A)) = \\ &= d\alpha(v, D\phi_s(\xi_A)).\end{aligned}$$

At $s=o, D\phi_s(\xi_A)=\xi_A$ and $\frac{\partial}{\partial s}(\alpha_{x_s}(D\phi_s(\xi_A)))=o$.

We compute therefore $\frac{\partial^2}{\partial s^2}(\alpha_{x_s}(D\phi_s(\xi_A))) = \frac{\partial}{\partial s}(d\alpha(v, D\phi_s(\xi_A))) =$ $\frac{\partial}{\partial s}\left(\beta_{x_s}(D\phi_s(\xi_A))\right)$.

We have:

$$\begin{aligned}\frac{\partial}{\partial s}\left(\beta_{x_s}(D\phi_s(\xi_A))\right) &= D\phi_s(\xi_A)\cdot\beta_{x_s}(v) - \beta_{x_s}\left(\left[\frac{\partial}{\partial s}+v, D\phi_s(\xi_A)\right]\right) + \\ &+ d\beta_{x_s}(v, D\phi_s(\xi_A)) = d\beta_{x_s}(v, D\phi_s(\xi_A)).\end{aligned}$$

At $s = o$, this is equal to

$$d\beta(v,\xi) = -d\alpha(v,[v,\xi]) = -1.$$

Thus, if s_1 is small, which we can assume:

$$\alpha_B(D\phi_{s_1}(\xi_A)) < 1.$$

In our curves, $0\left(\frac{\nu}{L}\right)$ is $0\left(\frac{\nu}{L}\right)v$. Thus, since $D\phi_{s_1}(v) = v$, which belongs to $\ker\alpha$:

$$\alpha_B\left(D\phi_{s_1}\left(\xi_A + 0\left(\frac{\nu}{L}\right)\right)\right) < 1 = \alpha_A\left(\xi_A + 0\left(\frac{\nu}{L}\right)v\right).$$

Let us now assume that we remove from the $\xi + 0\left(\frac{\nu}{L}\right)$-piece preceding A a small piece, basically equal to $\varepsilon\left(\xi + 0\left(\frac{\nu}{L}\right)\right)$. We then obtain a point A_ε, which we map by ϕ_{s_1}, thus obtaining:

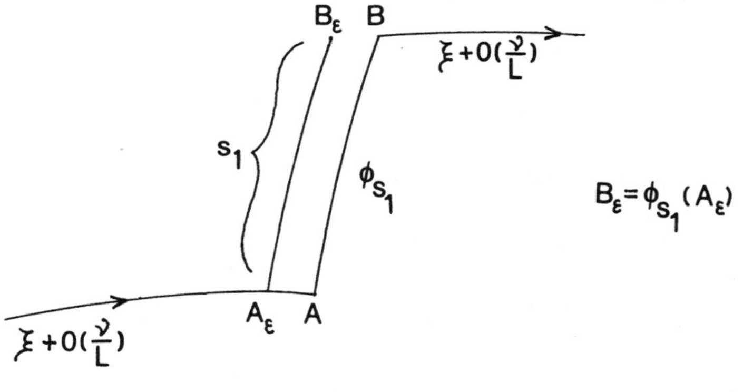

FIGURE A4.5

Assume now that we manage to close the gap between B_ε and B - we will see how - thus obtaining a curve x_ε:

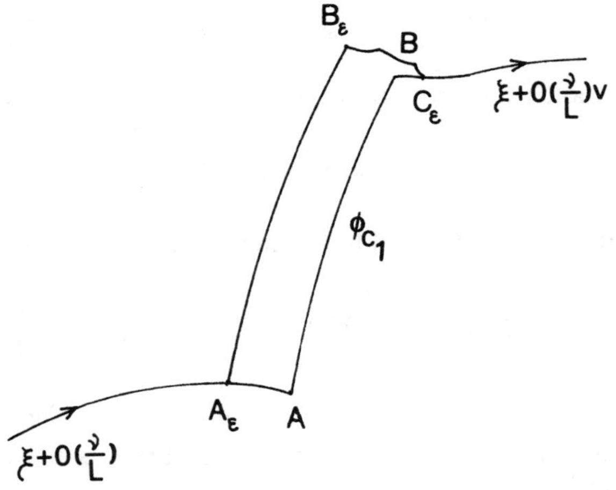

FIGURE A4.6

The variation in $\int_0^1 a$ is then, at first order, if we kept $\eta = 0(\varepsilon)$:

$$- \varepsilon \alpha_A \left(\xi + 0\left(\frac{\nu}{L}\right) v\right) - \alpha_B(\overrightarrow{BB_\varepsilon}) - \int_{[BC_\varepsilon]} b\eta =$$
$$- \varepsilon + \varepsilon\alpha_B(D\phi_{s_1}(\xi)) + \varepsilon 0\left(\frac{\nu}{L}\right) + 0(\varepsilon^2) =$$
$$= \varepsilon\left(\alpha_B(D\phi_{s_1}(\xi)) - 1 + 0\left(\frac{\nu}{L}\right)\right) + 0(\varepsilon^2).$$

Taking ν small enough, this is negative, for ε small. ($L \geq 100$).
Also, for $o \leq s \leq s_1$,

$$- \beta_{x_s}\left(D\phi_s\left(\xi_A + 0\left(\frac{\nu}{L}\right) v\right)\right) =$$
$$= -\beta_{x_s}\left(D\phi_s\left(\xi_A\right)\right) \geq o, > o \text{ if } 0 < s \leq s_1.$$

Indeed,

$$\beta_A(\xi_a) = o$$
$$- \frac{\partial}{\partial s}\left(\beta_{x_s}(D\phi_s(\xi_A))\right)\Big|_{s+o} = -d\beta_A(v, \xi_A) = 1.$$

Since s_1 can be assumed to be small, our claim follows.

After having indicated the construction of these two flows at infinity, we show how to extend them to a suitable $W^{1,1}$-neighborhood of infinity and we describe their properties. We review in the end the related deformation arguments.

In order to see how the first flow is defined and extends, we need to understand first its limit behavior at infintiy i.e. on Γ_{2k} or L^{ν}_{2m}.

Let us study Γ_{2k} for example. The extension to L^{ν}_{2m} will be easy.

In part II of this book, we study the tangent space to Γ_{2k}. Γ_{2k} is made of v and ξ-pieces:

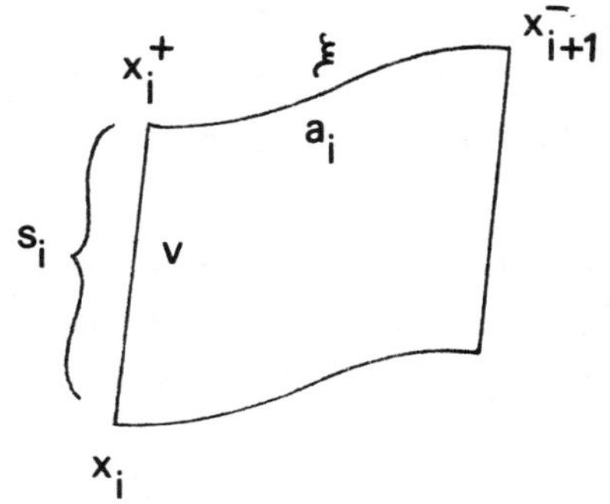

FIGURE A4.7

A tangent vector at a curve in x of Γ_{2k} is the datum of a variation of the base point x_0, of δx_0, combined with variation δs_i and δa_i.

Accordingly, the curve being parametrized by $\dot{x} = a\xi + bv$, it is given by a periodic solution of:

$$(D)\begin{cases} \dot{\eta} = \mu a - \lambda b \\ \dot{\mu} = -a\eta\tau + b\eta\bar{\mu}_\xi + \sum_{i=1}^{k} \delta s_i \delta_{x_i^+} \qquad (\eta, \mu, \lambda)(x_0) = \delta x_0 \\ \overline{\lambda + \bar{\mu}\eta}^{\,\cdot} = b\eta + \sum_{i=1}^{k} \delta a_i \delta_{x_{i+1}^-} \end{cases}$$

where δ_z is the Dirac mass at z.

(On L^{ν}_{2m}, (D) is modified in order to include $\sum_{i+1}^{k} \delta a_i 0\left(\frac{\nu}{L}\right)\delta_{x_{i+1}^-}$ in the second equation. Since the dimension of L^{ν}_{2m} is $4m$. we have to incorporate $2m$ other parameters. This is a technicality which we address later).

If we consider the homogeneous differential equation associated to this system, its resolvant varies continuously with the curve x in $W^{1,1}$. Under periodic boundary conditions, we might face a kernel of dimension 3. However, with the δs_i's and the δa_i's we have $2k$ independent variables.

Therefore, at any curve $x \in \Gamma_{2k}$, if we restrict to a subspace of dimension $2k-3$, we can solve this differential equation. The solution, furthermore, can be extended continuously to a $W^{1,1}$-neighborhood.

On this space of dimension $2k-3$ provided by the δs_i's and the δa_i's (after removing the kernel), we can ask whether $\int_0^1 \left(|b| - \frac{\nu}{L}\right)^+$ and $\int_0^1 a = \sum a_i$ are dependent.

This generically provides $2k-4$ conditions, hence in Γ_{2k} a stratified subset of dimension 4. If we assume $k_0 \geq 5$ this set will avoid the subset of curves having k_0 or more normals. Furthermore, outside of this set, there is a vector-field defined at infinity extending to a $W^{1,1}$ neighborhood and contolling $\int_0^1 \left(|b| - \frac{\nu}{L}\right)^+$ while decreasing $\int_0^1 a$.

We now display the other properties of this flow:

First, we have to take care of the Diracs $\sum \delta s_i \delta x_i^+$ and $\sum \delta a_i \delta_{x_{i+1}^-}$. We can regularize them, in this neighborhood into approximation of the identity $\delta^{\varepsilon_i}_{x_i^+}, \delta^{\varepsilon_i}_{x_{i+1}^-}, \varepsilon_i$ tending to zero as we approach infinity.

Our curves, besides those at infinity, are in $H^2(b \in H^1)$. When they are close $W^{1,1}$ to curves of Γ_{2k} or L^{ν}_{2m}, at each corner $x_i^{\pm}$, $|b|$ falls from values close to $|b|_\infty$ to values close to $\frac{\nu}{L}$. Of course, we might have additional sharp oscillations, but there are such drops.

We insert $\sum \delta s_i \delta^{\varepsilon_i}_{x_i^+}$ and $\sum \delta a_i \delta^{\varepsilon_i}_{x_{1+i}}$ in such drops i.e the approximation of identity have their support away from the set where $|b| = |b|_\infty$. We do not worry about the set where $|b| = \frac{\nu}{L}$ for this flow because the control of $\int_0^1 \left(|b| - \frac{\nu}{L}\right)^+$ is built already in its construction. ε_i is taken so small that, on an H^1-neighborhood in b, the approximation of identity have their support in such drops.

Coming back to (D), we then notice that, following this choice, when $|b| = |b|_\infty$, we have

$$\dot{\mu} = -a\eta\tau + b\eta\bar{\mu}_\xi$$

$$\overline{\lambda + \bar{\mu}\eta}^{\,\boldsymbol{\cdot}} = b\eta.$$

Thus, if s is the time along the deformation lines of this vector field,

$$\frac{\partial}{\partial s}\left(\frac{b}{a}\right) = o \text{ if } |b| \text{ is close to } |b|_\infty.$$

This shows that, besides $\int_0^1 \left(|b| - \frac{\nu}{L}\right)^+ \left(\text{or } \int_0^1 \left(|b| - 2\frac{\nu}{L}\right)^+\right), |b|_\infty$ is controlled on the flow-lines of this vector-field. Keeping a to be a constant is only a matter of reparametrization, which does not change this property.

We cannot expect, with this flow, to control $\int_0^1 \dot{b}^+$, since we are building up Diracs (or nearly so) along the v-branches.

However, we claim that we can control $\int_0^1 \dot{b}^2$. Indeed we have:

$$\frac{\partial}{\partial s}\int_0^1 \left(\frac{\dot{b}}{a}\right)^2 = -2\int_0^1 \frac{\ddot{b}}{a}\frac{\partial}{\partial s}\left(\frac{b}{a}\right).$$

Using (D), this yields, if we do not introduce the approximation of identity:

$$\frac{\partial}{\partial s}\int_0^1\left(\frac{\dot b}{a}\right)^2 = -2\sum_{i=1}^{k}\frac{\ddot b}{a^2}(x_i^+)\delta s_i + 2\sum_{i=1}^{k} b(x_{i+1}^-)\frac{\ddot b}{a^3}(x_{i+1}^-)\delta a_i.$$

This does not look very inviting. However, the above formula can be improved quite a lot: x_i^+ has to be thought as any point of a drop of b from a value near $|b|_\infty$ to some other value:

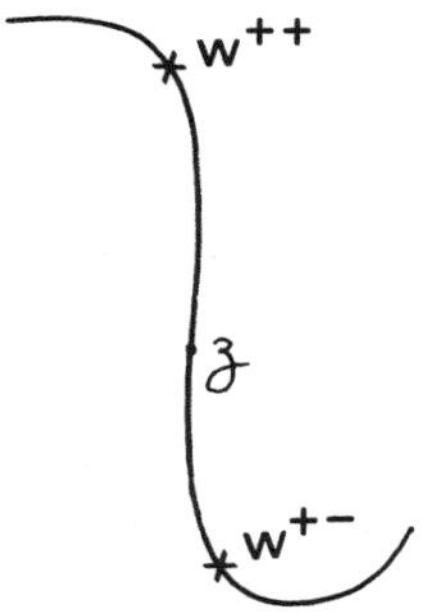

FIGURE A4.8

We can then take z to be any point between w^{++} and w^{+-}. Therefore, we can average our flow over all these values of $z = x_i^+$. The same procedure applies to x_{i+1}^-, with the provision that the averaging is completed with the weight $\frac{1}{b(z)}$ to get rid of the product $b\ddot b$. This yields, after averaging:

$$\frac{\partial}{\partial s}\int_0^1\left(\frac{\dot b}{a}\right)^2 = -2\sum_{i=1}^{k}\frac{\dot b(w_i^{+-}) - \dot b(w_i^{++})}{a^2\Delta_i^+}\delta s_i +$$
$$+2\sum_{i=1}^{k}\frac{\dot b(\tilde w_i^{+-}) - \dot b(\tilde w_i^{++})}{a^3}\frac{\delta a_i}{\Delta_i^-}$$

Δ_i^+ and Δ_i^- are lowerbounded and upperbounded in the H^1-neighborhood of a fixed b_0; they depend then only on b_0.

In order to control $\frac{\partial}{\partial s}\int_0^1\left(\frac{\dot b}{a}\right)^2$, let us analyze what happens along a drop for $|b|$:

When $|b|$ drops from a very large value to a value which is $0\left(\frac{\nu}{L}\right)$, this drop can either be very sharp i.e occurs over a period of time Δt less than $0\left(\frac{1}{|b|_\infty^{1000}}\right)$, a hypothesis which we can always assume for the flow defined with nearly Diracs, since we need it only at infinity; then, since the drop is at least 1, $|\dot{b}|$ assumes over the drop values larger than $|b|_\infty^{1000}$. $|\dot{b}|$ cannot stay over a period of time larger than $\frac{3}{|b|_\infty^{1000}}$, the drop would go over $2|b|_\infty$. Thus, there is a maximum for $|\dot{b}|$ within an interval of length $\frac{3}{|b|^{100}}_\infty$ of such values. All of this argument assumes $\dot{b}$ to be continuous, we will see how to proceed if $\dot{b}$ is not continuous.

If the drop occurs over a period of time larger than $\frac{1}{|b|_\infty^{1000}}$, a situation which might occur for the deconcentration flow (defined on a larger neighborhood of infinity, where the proximity is basically defined in graph), then $|\dot{b}|$ has to assume over such an interval values of the order of $|b|_\infty^{100000}$ at most at at least two times t_1, t_2 such that

$$|t_1 - t_2| \geq \frac{1}{4|b|_\infty^{1000}}.$$

Thus, when the drop is sharp, we can find families of values t_1, t_2, the first ones before the maximum, the second ones after the maximum such that

$$\dot{b}(t_1) = \dot{b}(t_2).$$

In fact, we can build then two intervals $[\underline{t}_1, \bar{t}_1]$ and $[\underline{t}_2, \bar{t}_2]$ such that, for every $t_1 \in [\underline{t}_1, \bar{t}_1]$, there exists t_2 in $[\underline{t}_2, \bar{t}_2]$ satisfying the above equation, and vice-versa. These intervals can be taken to be disjoint, one after the maximum of $|\dot{b}|$, the other before, as small as we wish. This includes the case where $|\dot{b}|$ is constant equal to its maximum on an interval.

When the drop is not so sharp, we can always find values t_1 and t_2 such that:

$$\begin{cases} |\dot{b}|(t_1) + |\dot{b}(t_2)| \leq 2|b|_\infty^{100000} \\ |t_1 - t_2| \geq \frac{1}{4|b|_\infty^{1000}}. \end{cases}$$

We will discuss this later.

For the flow involving near Diracs, coming back to $\frac{\partial}{\partial s} \int_0^1 \left(\frac{\dot{b}}{a}\right)^2$, using th fact that the drops can be assumed to be sharp, we can average letting w^{++} run into $[\underline{t}_1, \bar{t}_1], w^{+-}$ run into $[\underline{t}_2, \bar{t}_2]$, using the total weight $(\bar{t}_1 - \underline{t}_1) + (\bar{t}_2 - \underline{t}_2)$. The same procedure can be completed for $\tilde{w}^{+-}$ and $\tilde{w}^{++}$.

We then have, if $\dot{b}$ is continuous:

$$\frac{\partial}{\partial s} \int_0^1 \left(\frac{\dot{b}}{a}\right)^2 = o.$$

However, this condition is precisely fullfilled at b, using the fact that $\dot{b}$ is continuous. As it is, it is not possible to extend it to an H^1 weak-neighborhood. In order to do so, we notice that, if instead of $[\underline{t}_1,\bar{t}_1],[\underline{t}_2,\bar{t}_2],[\underline{t}'_1,\bar{t}'_1],\underline{t}'_2,\bar{t}'_2]$(for $\tilde{w}^{++},\tilde{w}^{+-}$), we would have used intervals $[\underline{\tilde{t}}_i,\bar{\bar{t}}_i],[\underline{\tilde{t}'}_i,\bar{\bar{t}}'_i]$, we would have derived.

$$\frac{\partial}{\partial s}\int_0^1\left(\frac{\dot b}{a}\right)^2=\frac{\dot b(\bar{\bar{t}}_1)-\dot b(\underline{\tilde{t}}_1)+\dot b(\bar{\bar{t}}_2)-\dot b(\underline{\tilde{t}}_2)}{\bar{\bar{t}}_1-\underline{\tilde{t}}_1+\bar{\bar{t}}_2-\underline{\tilde{t}}_2}+$$
$$+\frac{\dot b(\bar{\bar{t}}'_1)-\dot b(\underline{\tilde{t}'}_1)+\dot b(\bar{\bar{t}}'_2)-\dot b(\underline{\tilde{t}'}_2)}{\bar{\bar{t}}'_1-\underline{\tilde{t}'}_1+\bar{\bar{t}}'_2-\underline{\tilde{t}'}_2}.$$

At $\bar{t}_i,\underline{t}_i,\bar{t}'_i,\underline{t}'_i$, the right hand side vanishes.

Hence, averaging the flow over small interests around these values, we still can have:

$$\text{Averaged right hand side}\leq-\frac{\partial a}{\partial s}.$$

Now, after averaging, the right hand side is expressed in function of $\dot b(\underline{t}_i),\dot b(\bar{t}_i),\dot b(\bar{t}'_i),\dot b(\underline{t}'_i)$. The formula holds for every function $\tilde b$ instead of b. Therefore, we can complete an extension to an H^1 weak-neighborhood of b where we will have:

$$\frac{\partial}{\partial s}\int_0^1\left(\frac{\dot{\tilde b}}{\tilde a}\right)^2\leq-2\frac{\partial}{\partial s}.$$

We have a provision with this argument : namely, when we average, we could cross values where $\dot b=0$ or $|\dot b|=\frac{\nu}{L}$ or $\frac{2\nu}{L}$.

These values, we want to avoid in order to keep control of the number of zeros etc. However, this can be easily overcome: On all the intervals of averaging, $\dot b$ can be assumed to be non zero. Therefore, these values are isolated, at most 16 of them. We can then take smaller intervals of averaging and derive the same result without ever approaching these values.

This construction is completed when $\dot b$ is continuous. We need to modify it in order to transform the Diracs in ε_i-approximations of the identity. We can use functions which will have support of length $=\varepsilon_i$ around their peak (equal to $0\left(\frac{1}{\varepsilon_i}\right)$). Our former estimates extend then if ε_i is small enough. How small ε_i should be depends only on the modules of continuity of $\dot b$. In particular, we can also ask that every interval of averaging ($[\underline{t}_i,\bar{t}_i],[\underline{t}'_i,\bar{t}'_i]$, the other ones also) has length at most $2\varepsilon_i$ while these intervals are away at least by $2\varepsilon_i$ of the zeros of $\dot b$ etc.

This needs some justification:

Given two values, one before the maximum of $|\dot b|$ and the other one after, such that $\dot b(t_1)=\dot b(t_2)$, we can assume that t_1 is the last time before the maximum where $|\dot b|$ takes the value $|\dot b|(t_1)$ and that t_2 is the first time after the maximum with this property. Taking then $\varepsilon>o$ very small and $\theta>o$ given, we claim that, for $\varepsilon>o$ small enough, $|\dot b(t_1)|+\varepsilon$ will be achieved for the last time before the maximum in

$[t_1, t_1+\theta]$ and for the first time after in $[t_2-\theta, t_2]$. Indeed, otherwise, letting ε tend to zero, we acheive a condtradiction with the construction of t_1 and t_2.

θ is as small as we wish: thus, given t_1 and t_2, we can take it so small, also ε_i so small, that $[t_1-2\varepsilon_i, t_1+\theta+2\varepsilon_i], [t_2-\theta-2\varepsilon_i, t_2+2\varepsilon_i]$ keeps away from the zeros of b etc.

We can then construct our intervals $[\underline{t}_i, \bar{t}_i], [\underline{t}'_i, \bar{t}'_i]$ etc.

Within $[t_1, t_1+\theta], [t_2-\theta, t_2]$. We take $\theta \leq \frac{1}{2}\inf \varepsilon_i$ and our claim is fullfilled.

Summing up, when $\dot{b}$ is continuous, we have constructed a flow very close to the vector-field with Diracs which satisfies in an H^1 weak-neighborhood:

$$\frac{\partial}{\partial s}\int_0^1 \left(\frac{\dot{\bar{b}}}{\bar{a}}\right)^2 \leq -3\frac{\partial a}{\partial s}.$$

This flow does not touch the zeros of b, the values where $|b| = \frac{\nu}{L}$ or $\frac{2\nu}{L}$.

When $\dot{b}$ is not continuous, we use the small viscosity term in η, cb which yields in $\frac{\partial}{\partial s}\int_0^1 \left(\frac{\dot{b}}{a}\right)^2$ a term equal to $-c\int_0^1 \ddot{b}^2$. After averaging, all $\dot{b}(w_i^{+-}), \dot{b}(w_i^{++}$ etc. are $0(|\ddot{b}|_{L^2})$. Thus, if $\dot{b}$ is not continuous, $|\ddot{b}|_{L^2}$ is infinite and

$$\frac{\partial}{\partial s}\int_0^1 \left(\frac{\dot{b}}{a}\right)^2 \leq -c|\ddot{b}|_{L^2}^2 + 0(|\ddot{b}|_{L^2}) < o.$$

This property easily extends to an H^1 weak-neighborhood of b.

Here the averaging takes place over intervals as small as we wish. Therefore, the values where $b = o, |b| = \frac{\nu}{L}$ or $\frac{2\nu}{L}$ can be easily avoided.

c, in cb can be taken to tend to zero as we approach infinity i.e the curves of Γ_{2k} or the $\frac{\nu}{L}$-stretched curves. Our flow, by contruction, controls $|b|_\infty, \int \left(|b| - \frac{\nu}{L}\right)^+$, decreases a. We analyze its effect on $\int_0^1 \left(\frac{\dot{b}}{a}\right)^+$:

Before averaging, using (D), it is clear that

$$\frac{\partial}{\partial s}\int_0^1 \left(\frac{\dot{b}}{a}\right)^+ = \sum \delta s_i 0\left(\frac{1}{\varepsilon_i}\right) + \delta a_i 0\left(\frac{1}{\varepsilon_i}\right).$$

After averaging, the above formula still holds true.

Thus, clearly, if $\dot{b}$ is not continuous, we can take the H^1 weak-neighborhood of b so small that, for every $\tilde{b}$ in this neighborhood, we have:

$$\frac{\partial}{\partial s}\int_0^1 \left(\frac{\dot{\bar{b}}}{\bar{a}}\right)^2 \leq -\frac{9c}{10}\int_0^1 \ddot{\bar{b}}^2 + \frac{C}{c}\left(-\frac{\partial}{\partial s}\int_0^1 \dot{\bar{b}}^+ - \frac{\partial a}{\partial s} + C\right)\int_0^1 \left(\frac{\dot{\bar{b}}}{\bar{a}}\right)^2 + C.$$

This formula allows to convex-combine the global flow which we have built (which satisfies such a inequality) with this flow at infinity i.e to assume, in all our analysis, that the flow which we have at infinity is the addition of the cancellation/stretching flow with this new flow.

Of course, we do not know anymore that $-\frac{\partial}{\partial s}\int_0^1 \dot{b}^+ - \frac{\partial a}{\partial s} + C(1+\int_0^1 \dot{b}^+)$ is positive, which might impeed the derivation of the bound on $\int_0^1 \dot{b}^2$. However, we can complete this addition, near infinity, on a strip where c can be taken to be constant, very small, in function of $|b|_\infty$. On one side of the strip and a little bit in the strip, we have our usual flow. On the other side and a little bit in the strip we have our new flow. In between, a combination. Inside the strip, there cannot be explosions, since c is constant. On each side, also. A flow-line cannot, then, enter the inner strip, where the convex-combination actually takes place, and leave the outer strip an infinite number of times. The bound on $\int_0^1 \dot{b}^2$ follows using the above formula.

We need to prove that this inequality holds also true when $\dot{b}$ is continuous, in a small H^1 weak-neighborhood:

For this, we observe that, when $\dot{b}$ is continuous, the nearly Diracs, even after averaging have support in regions when $|\dot{b}|$ is larger than a given positive value $\theta = \theta(b) > o$.

Nearby b, in the H^1 weak-sense, maxima and minima might enter the support of the ε_i-approximations. If so, taking the H^1 weak-neighborhood very small, $\int_0^1 \ddot{\tilde{b}}^2$ must be extremely large. How large can be estimated as follows: Let J be one of the intervals of averaging. By construction, $|J|$ is at most $3\varepsilon_i$. We can cover an ε_i-neighborhood J by a finite number of intervals J_ℓ of length at most ε_i^2 and ask that, on the H^1 weak-neighborhood of b, we have:

$$\frac{1}{|J_\ell|}\Big|\int_{J_\ell} \dot{\overline{\tilde{b} - b}}\Big| < \frac{1}{2} \operatorname*{Min}_{J_\ell} |\dot{b}|.$$

Thus, $|\dot{b}|$ assume values on each J_ℓ larger than $\frac{\theta}{2}$. (ε_i small enough) $\frac{\partial}{\partial s}\int_0^1 \left(\frac{\dot{b}}{\tilde{a}}\right)^+$ could be non zero if a maximum or a minimum of $\tilde{b}$ was within an interval of length ε_i of an interval such as J. This follows from our construction. Then, we have with an appropriate J_ℓ:

$$\int_{J_\ell} |\ddot{\tilde{b}}| \geq \frac{\theta}{2}.$$

Thus,

$$\int_{J_\ell} |\ddot{\tilde{b}}|^2 \geq \frac{\theta}{2|J_\ell|} \geq \frac{\theta}{2\varepsilon_i^2}.$$

Thus,

$$\frac{c}{10}\int_0^1 |\ddot{\tilde{b}}|^2 \geq \frac{\theta}{2\varepsilon_i^2}$$

while

$$\frac{\partial}{\partial s}\int_0^1\left(\frac{\dot{\bar{b}}}{\bar{a}}\right)^+ = \sum \delta s_i 0\left(\frac{1}{\varepsilon_i}\right) + \sum \delta a_i 0\left(\frac{1}{\varepsilon_i}\right).$$

θ can be chosen uniform even as ε_i tends to zero. This allows, to derive again the inequality:

$$\frac{\partial}{\partial s}\int_0^1\left(\frac{\dot{\bar{b}}}{\bar{a}}\right)^2 \leq -\frac{9c}{10}\int_0^1 \ddot{\bar{b}}^2 + \frac{C}{c}\left(-\frac{\partial}{\partial s}\int_0^1 \dot{\bar{b}}^+ - \frac{\partial}{\partial s} + C\right)\int_0^1\left(\frac{\dot{\bar{b}}}{\bar{a}}\right)^2 + C$$

which now holds globally.

This shows that we can convex-combine this flow with the global flow which we have built in the book. The convex-combination has the same properties than the former flow: it controls $|b|_\infty, \int_0^1\left(|b| - \frac{\nu}{L}\right)^+$ etc. Furthermore, now at infinity, it extends with the flow with Diracs on $\bigcup_k \Gamma_{2k}$ or $\bigcup L^{\nu}_{2m}$ i.e the $\frac{\nu}{L}$-stretched curves. This allows us to control the number of characteristic jumps (i.e deconcentration jumps here) on a given deformation class.

We move to study the second flow, i.e the deconcentration flow, which we have defined above and prove also that, as long as we are in a given $W^{1,1}$ neighborhood of infinity, this flow is well defined and has all the required properties in order to be used in lieu of the cancellation / stretching flow when there is deconcentration, at least as long as we are near infinty.

In order to define analytically this flow we will defne η_0 as follows: We are considering a curve having a nearly v-jump, between two pieces when $|b| = 0\left(\frac{\nu}{L}\right)$. This curve is H^2

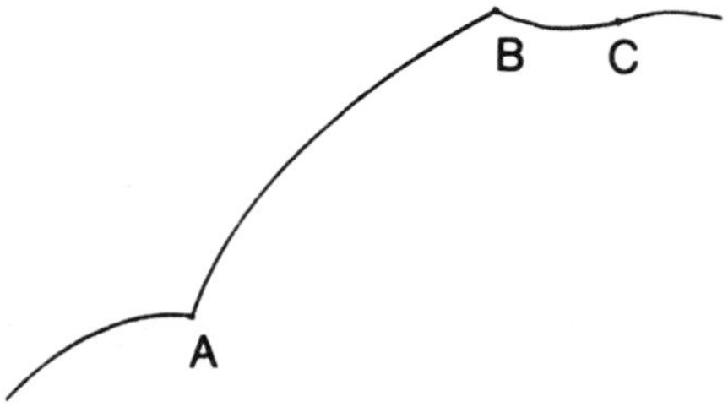

FIG. A4.9

Let A and B be near the edges of the nearly v-jump. Consider the vector $-\dot{x}(A)$, we use it as initial condition in the differential equation:

$$(E)\begin{cases} \dot{\eta}_o = \mu_o a - \lambda_o b \\ \dot{\mu}_o + a\eta_o\tau - b\eta_o\bar{\mu}_\xi = o \qquad \text{between } A \text{ and } C \\ \overline{\lambda_o + \mu\eta_o}^{\,\cdot} = b\eta_o \qquad (\lambda_o\xi + \mu_o v + \eta_o w)(A) = -\dot{x}(A). \end{cases}$$

We derive a certain function η_o on [A, C]. η_o and $\dot{\eta}_o$ vanish at A. η_o is essentially positive on [A, C] by our arguments above. Indeed, as we approach infinity, only in a $W^{1,1}$ sense, the piece between A and C becomes essentially tangent to v; (E) transforms into the transport equation along v, for which we have seen that

$$-\beta_{x_s}(D\phi_s(\xi_A)) = -\beta_{x_s}(D\phi_s(\dot{x}_A))$$

is non negative, positive on [A, B], thus on [A,C] by continuity.

η_o and $\dot{\eta}_o$ are zero at A but not at C. We therefore multiply η_o, near A, by a function ω going from o to 1 very rapidly: ω is C^∞, equal to zero before A and jumping to 1 in a time equal to ε:

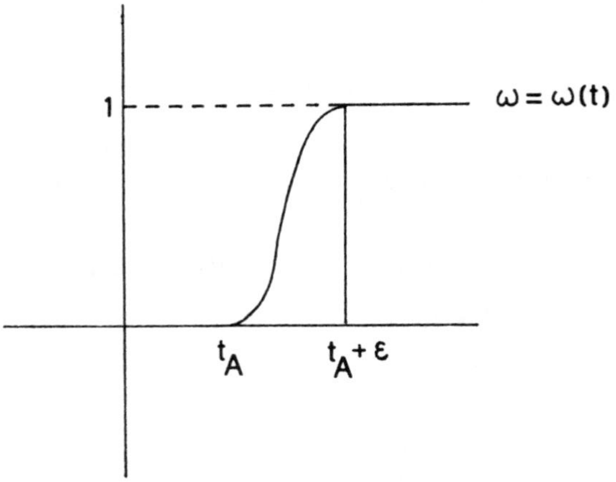

FIGURE A4.10

Near C, we will use a function $\psi_{\alpha,\beta}$ of b which is an ε-regularization of the function:

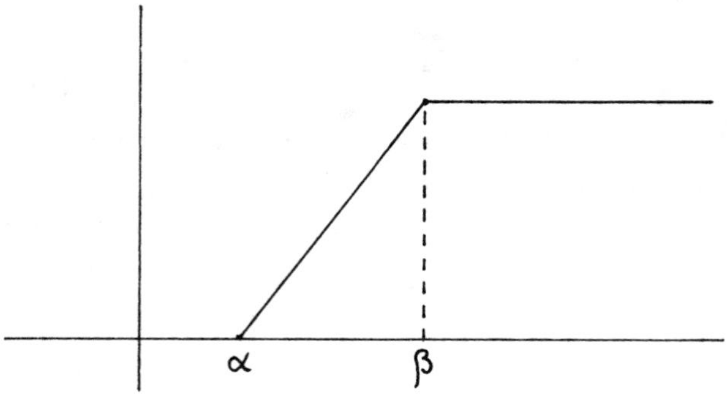

FIGURE A4.11

i.e we will choose values α, β appropriately near C, before C and associated times t_α, t_β such that:

$$\begin{cases} |b(t_\beta)| = \beta \qquad |b(t_\alpha)| = \alpha \\ t_\beta < t_\alpha. \end{cases}$$

We define N to be:

$$\begin{cases} N = o \text{ before } t_A \text{ and after } t_\alpha \\ N = \omega\eta_o \text{ on } [t_A, t_A + \varepsilon] \\ N = \eta_o \text{ on } [t_A + \varepsilon, t_\beta] \\ N = \psi_{\alpha,\beta}(|b|)\eta_o \text{ on } [t_\beta, t_\alpha]. \end{cases} \qquad \text{We write } N = \theta\eta_o.$$

We observe that η_o is positive on $[t_\beta, t_\alpha]$ since this interval is far from A and η_o has had to become and stay positive, away from zero.

As usual, to N is added cb. We study the properties of this flow which, when $\varepsilon = o$ and $t_\beta = t_\alpha$ is exactly the limit deconcentration flow which we envisioned.

We first notice that if, on $[t_A, t_A + \varepsilon]$ and on $[t_\beta, t_\alpha]|b|$ never achieves the value $|b|_\infty$, then

$$\frac{\partial}{\partial s}\frac{|b|_\infty}{a} \leq o.$$

Indeed, then the computation of $\left.\frac{\partial}{\partial s}\frac{b}{a}\right|_{|b|=|b|_\infty}$ has to be made outside $[t_A, t_A + \varepsilon]$

and $[t_\beta, t_\alpha]$. N is then equal to η_o and, using (E):

$$\frac{\partial}{\partial s} b\Big|_{|b|=|b|_\infty} = \frac{\overline{\dot{\eta_o + \lambda b}}}{a} + a\eta_o\tau - b\eta_o\bar{\mu}_\xi =$$

$$= \frac{\overline{\dot{\eta_o + \lambda_o b}}}{a} + a\eta_o\tau - b\eta_o\bar{\mu}_\xi + \frac{\overline{\dot{(\lambda - \lambda_o)b}}}{a} = \dot{\mu}_o + a\eta_o\tau - b\eta_o\bar{\mu}_\xi + \frac{\overline{\dot{\lambda - \lambda_o}}}{a} b =$$

$$= -\frac{\int_o^1 b\eta_o b}{a^2} \leq \frac{\frac{\partial a}{\partial s}}{a^2} b.$$

The claim follows.

Second, we compute:

$$\frac{\partial}{\partial s} \int \left(|b| - \frac{\nu}{L}\right)^+ = \int_{|b|>\frac{\nu}{L}} (\dot{\mu} + aN\tau - bN\bar{\mu}_\xi) \leq \sum_i \mu\Big|_{t_i}^{t_i+1} + (a+1)\int_A^C (1+|b|)|\eta_o|.$$

Using the $W^{1,1}$ proximity to infinity, it is easy to see that

$$\int_A^C (1+|b|)|\eta_o| \leq -C\frac{\partial a}{\partial s}$$

t_i are the various times where $|b| = \frac{\nu}{L}$.

$$\sum_i \mu\Big|_{t_i}^{t_i+1} = \sum_i \frac{\overline{\dot{\theta\eta_o}} + \lambda\frac{\nu}{L}}{a}\Big|_{t_i}^{t_i+1} = \sum_i \frac{\overline{\dot{\theta\eta_o}} - \bar{\mu}\theta\eta_o\frac{\nu}{L}}{a}\Big|_{t_i}^{t_i+1} +$$

$$+ \frac{(\lambda + \bar{\mu}N)b}{a}\Big|_{t_i}^{t_i+1} = \sum_i \frac{\dot{\theta}\eta_o + \theta(\dot{\eta}_o - \bar{\mu}\eta_o b)}{a}\Big|_{t_i}^{t_i+1} + \frac{\nu}{L}0\left(-\frac{\partial a}{\partial s}\right) =$$

$$= \sum_i \frac{\dot{\theta}\eta_o + \theta(\dot{\eta}_o + \lambda_o b)}{a}\Big|_{t_i}^{t_i+1} - \frac{\theta(\lambda_o + \bar{\mu}\eta_o)b}{a}\Big|_{t_i}^{t_i+1} + \frac{\nu}{L}0\left(-\frac{\partial a}{\partial s}\right).$$

Clearly, $\dot{\theta}\eta_o$ is zero if t_i or t_{i+1} is not in $[t_A, t_A + \varepsilon]$ or $[t_\beta, t_\alpha]$.

Since η_o is positive in $[t_\beta, t_\alpha]$ and $\psi_{\alpha,\beta}$ is increasing, $\dot{\theta}\eta_o(t_{i+1})$ is negative, $\dot{\theta}\eta_o(t_i)$ is positive if t_i or t_{i+1} is in $[t_\beta, t_\alpha]$. Thus, $\sum_i \frac{\dot{\theta}\eta_o}{a}\Big|_{t_i}^{t_i+1}$ is upperbounded by the contribution of the t_i, t_{i+1} in $[t_A, t_A + \varepsilon]$. On this interval, $\theta = \omega, \ddot{\theta} = 0\left(\frac{1}{\varepsilon^2}\right)$ and if ε can be chosen small enough, which we will assume and check, $\eta_o = 0(\varepsilon)$. Therefore,

$$\sum_i \frac{\dot{\theta}\eta_o}{a}\Big|_{t_i}^{t_i+1} \leq \frac{1}{\varepsilon} \times 0(\varepsilon) = 0(1).$$

For this flow, $-\frac{\partial a}{\partial s}$ is lowerbounded by a constant $\gamma > o$.
Hence

$$\sum_i \frac{\dot{\theta}\eta_o}{a}\Bigg|_{t_i}^{t_{i+1}} \leq -C' = \frac{\partial a}{\partial s}.$$

On the other hand, $\sum_i \frac{\theta(\dot{\eta}_o+\lambda_o b)}{a}\Bigg|_{t_i}^{t_{i+1}}$ can be upperbounded by $\sum_i \theta(t_i)\int_{t_i}^{t_{i+1}} \dot{\mu}_0 + \frac{\dot{\eta}_0+\lambda_0 b}{a}(t_{i+1})(\theta(t_{i+1})-\theta(t_i))$ which is equal to:

$$\sum_i \theta(t_i)\int_{t_i}^{t_{i+1}} (-a\eta_0\tau + b\eta_0\bar{\mu}_\xi) + \mu_0(t_{i+1})(\theta(t_{i+1})-\theta(t_i)) \leq$$

$$\leq (a+1)\int_A^C (1+|b|)|\eta_0| + \sum_i \mu_0(t_{i+1})(\theta(t_{i+1})-\theta(t_i)) \leq$$

$$\leq -C\frac{\partial a}{\partial s} + \sum_i \mu_0(t_{i+1})(\theta(t_{i+1})-\theta(t_i)).$$

Assume that $|b|(A) \leq 1$. Then $|\mu_0|$ is upperbounded by a constant C. Using the form of the function θ, it is easy then to see that

$$\sum_i \mu_0(t_{i+1})(\theta(t_{i+1})-\theta(t_i)) \leq 2C$$

because either $\theta = \omega$, which is increasing, or $\theta = 1$ and $\theta(t_{i+1}) - \theta(t_i) = 0$ or $\theta = \psi_{\beta,\alpha}(b)$ and $\theta(t_{i+1}) = \theta(t_i)$ again. Only, two terms might escape these estimates. They yield the upperbound of $2C$. Again, using the fact that $-\frac{\partial a}{\partial s}$ is lowerbounded by $\gamma > 0$ we derive that this term is upperbounded by $-C'\frac{\partial a}{\partial s}$.

Lastly, we consider $-\sum_i \theta\frac{(\lambda_o+\bar{\mu}\eta_o)}{a}b\Bigg|_{t_i}^{t_{i+1}}$ which can be upperbounded by:

$$-\sum_i \left(\theta(t_i)\frac{\nu}{L}\frac{\lambda_o+\bar{\mu}\eta_o}{a}\Bigg|_{t_i}^{t_{i+1}} + \frac{\nu}{L}\frac{\lambda_o+\bar{\mu}\eta_o}{a}(t_{i+1})\left(\theta(t_{i+1})-\theta(t_i)\right)\right) \leq$$

$$\leq -C\frac{\partial a}{\partial s} - \sum_i \frac{\nu}{L}\frac{\lambda_o+\bar{\mu}\eta_o}{a}(t_{i+1})\left(\theta(t_{i+1})-\theta(t_i)\right).$$

Clearly, $\lambda_o + \bar{\mu}\eta_o \leq 0\left(1-\frac{\partial a}{\partial s}\right)$. Thus, using the special form of θ and the fact that $\theta \leq 1$ we have:

$$-\sum_i \theta\frac{(\lambda_o+\bar{\mu}\eta_o)b}{a}\Bigg|_{t_i}^{t_{i+1}} \leq -C\frac{\partial a}{\partial s} + C.$$

But, we know, with this flow, that $-\frac{\partial a}{\partial s}$ is lowerbounded by a positive constant $\gamma > o$ Thus:

$$-\sum_i \theta \frac{(\lambda_o + \bar{\mu}\eta_o)b}{a}\Bigg|_{t_i}^{t_{i+1}} \leq C' \frac{\partial a}{\partial s}.$$

Summing up, we have:

$$\frac{\partial}{\partial s}\int (|b| - \frac{\nu}{L})^+ \leq -3C' \frac{\partial a}{\partial s}.$$

We now study

$$\frac{\partial}{\partial s}\int \frac{\dot{b}^+}{a}.$$

At a time t where $\dot{b}(t) = o$, we have:

$$\frac{\partial}{\partial s}\left(\frac{\dot{b}}{a}\right) = \frac{\overline{\theta\eta_o}^{\cdot\cdot} + \dot{\lambda} b}{a^2} + \theta\eta_o\tau - \theta\frac{b\eta_o}{a}\bar{\mu}_\xi + \frac{b\int_0^1 b\theta\eta_o}{a^2} =$$

$$= \frac{\ddot{\theta}\eta_o + 2\dot{\theta}\dot{\eta}_o}{a^2} + \theta\left(\frac{\ddot{\eta}_o}{a^2} + \eta_o\tau - \frac{b\eta_o}{a}\bar{\mu}_\xi\right) + \frac{b\int_0^1 b\theta\eta_o}{a^2} +$$

$$+ \frac{\dot{\overline{\lambda + \bar{\mu}\theta\eta_o}}}{a^2} b - \frac{\dot{\overline{\bar{\mu}\theta\eta_o b}}}{a^2}$$

Observe that $\dot{\overline{\lambda + \bar{\mu}\theta\eta_o}} + \int_0^1 b\theta\eta_o = b\theta\eta_o = \theta\dot{\overline{\lambda_o + \bar{\mu}\eta_o}}$.
Thus, using (E):

$$\frac{\partial}{\partial s}\frac{\dot{b}}{a} = \frac{\ddot{\theta}\eta_o + 2\dot{\theta}\dot{\eta}_o}{a^2} + \theta\left(\frac{\ddot{\eta}_o + \dot{\lambda}_o b}{a^2} + \eta_o\tau - \frac{b\eta_o}{a}\bar{\mu}_\xi\right) - \frac{\dot{\theta}\bar{\mu}\eta_o b}{a^2} =$$

$$= \frac{\ddot{\eta}_o + 2\dot{\theta}\dot{\eta}_o}{a^2} - \frac{\dot{\theta}\bar{\mu}\eta_o b}{a^2}.$$

If t does not belong to $[t_A, t_A + \varepsilon], \dot{\theta}$ is zero at such a t and $\ddot{\theta}$ is also zero unless $t \in [t_\beta, t_\alpha]$. Then, $\ddot{\theta}$ is negative at a maximum, positive at a minimum and η_o is positive, thus the contribution of $\ddot{\theta}\eta_o$ in $\frac{\partial}{\partial s}\int \frac{\dot{b}}{a}$ is negative for such values of t where $\dot{b}$ vanishes.

On the other hand, if $t \in [t_A, t_A + \varepsilon], \varepsilon$ small enough, since we are assuming $|\dot{b}(t_A)| \leq 1$, we have, if $\dot{b}$ is continuous:

$$\frac{\ddot{\theta}\eta_o + 2\dot{\theta}\dot{\eta}_o}{a^2} - \frac{\dot{\theta}\bar{\mu}_o b}{a^2} = 0(1).$$

Indeed, η_o and $\dot{\eta}_o$ vanishes at A and $\dot{\theta} = 0\left(\frac{1}{\varepsilon}\right), \ddot{\theta} = 0\left(\frac{1}{\varepsilon^2}\right)$. When $\dot{b}$ is continuous, $\dot{\eta}_o$ is then $0(\varepsilon)$ near A and $\eta_o = 0(\varepsilon^2)$), yielding our claim. Such an estimate can be extended to an H^1 weak-neighborhood of b easily. ($\dot{\eta}_o$ depends on b in an H^1-weak sense). This yields:

$$\frac{\partial}{\partial s}\int \frac{\dot{b}^+}{a} \leq 0(1) \leq -C'\frac{\partial a}{\partial s}.$$

if $\dot{b}$ is continuous. (Observe that $\ddot{\eta}_o$ is then bounded near A). This formula extend to an H^1 weak-neighborhood in b.

If $\dot{b}$ is not continuous, we still have:

$$\frac{\partial}{\partial s}\int \frac{\dot{b}^+}{a} \leq \frac{C}{\varepsilon}\int_A^{A+\varepsilon} |\dot{b}| + 0(1) \leq \frac{C}{\varepsilon}\int_A^{A+\varepsilon} |\dot{b}| - C'\frac{\partial a}{\partial s}.$$

Finally, we consider

$$\frac{\partial}{\partial s}\int_0^1 \left(\frac{\dot{b}}{A}\right)^2 = -2\int_0^1 \frac{\ddot{b}}{a^2}\left(\frac{\dot{\overline{\theta\eta_o + \lambda b}}}{a} + a\theta\eta_o\tau - b\eta_o\theta\bar{\mu}_\xi\right) + 2\frac{\int_0^1 \dot{b}^2 \frac{\partial a}{\partial s}}{a^3} =$$

$$= -2\int_0^1 \frac{\ddot{b}}{a^2}\left(\frac{\ddot{\theta}\eta_o + 2\dot{\theta}\dot{\eta}_o}{a} + \theta\left(\frac{\dot{\overline{\dot{\eta}_o + \lambda_o b}}}{a} + a\eta_o\tau - b\eta_o\bar{\mu}_\xi\right)\right) +$$

$$+ 2\frac{\int_0^1 \dot{b}^2 \frac{\partial a}{\partial s}}{a^3} - 2\int_0^1 \frac{\ddot{b}}{a^2}\frac{\dot{\overline{\lambda b}} - \theta\dot{\overline{\lambda_o b}}}{a}.$$

Observe that

$$\dot{\overline{\lambda b}} - \theta\dot{\overline{\lambda b}} = (\dot{\lambda} - \theta\dot{\lambda}_o)b + (\lambda - \theta\lambda_o)\dot{b} =$$

$$= \dot{\overline{\lambda + \bar{\mu}\theta\eta_o}} - \theta\dot{\overline{\lambda_o + \bar{\mu}\eta_o}} - \dot{\theta}\bar{\mu}\eta_o + (\lambda - \theta\lambda_o)\dot{b} =$$

$$= -\int_0^1 b\theta\eta_o - \dot{\theta}\bar{\mu}\eta_o + (\lambda - \theta\lambda_o)\dot{b}.$$

Observe also that, by (E):

$$\frac{\dot{\overline{\dot{\eta}_o + \lambda_o b}}}{a} + a\eta_o\tau - b\eta_o\bar{\mu}_\xi = o.$$

Thus,

$$\frac{\partial}{\partial s}\int_0^1 \left(\frac{\dot{b}}{a}\right)^2 = -2\int_0^1 \frac{\ddot{b}}{a^2}\left(\frac{\ddot{\theta}\eta_o + 2\dot{\theta}\dot{\eta}_o - \dot{\theta}\bar{\mu}\eta_o}{a}\right) - 2\int_0^1 \frac{\ddot{b}\dot{b}}{a^2}\frac{(\lambda - \theta\lambda_o)}{a} +$$

$$+2\frac{\int_0^1 \dot{b}^2 \frac{\partial a}{\partial s}}{a^3} = 2\int_0^1 \frac{\ddot{b}}{a^2}\left(\frac{\ddot{\theta}\eta_o + 2\dot{\theta}\dot{\eta} - \dot{\theta}\bar{\mu}\eta_o}{a}\right) -$$

$$-\int_0^1 \frac{\dot{b}^2}{a^2}\left(\frac{\int_0^1 b\theta\eta_o}{a} - \frac{\dot{\theta}(\lambda_o + \bar{\mu}\eta_o)}{a}\right) + \frac{2\int_0^1 \dot{b}^2 \frac{\partial a}{\partial s}}{a^3} =$$

$$= 2\int_0^1 \frac{\ddot{b}}{a^2}\frac{(\ddot{\theta}\eta_o + 2\dot{\theta}\dot{\eta}_o - \dot{\theta}\bar{\mu}\eta_o)}{a} + \frac{\int_0^1 \dot{b}^2 0\left(-\frac{\partial a}{\partial s}\right)}{a^3} - \int_0^1 \frac{\dot{b}^2\dot{\theta}(\lambda_o + \bar{\mu}\eta_o)}{a} =$$

$$2\int_0^1 \frac{\ddot{b}}{a^2}\frac{\ddot{\theta}\eta_o + 2\dot{\theta}\dot{\eta}_o\dot{\theta}\bar{\mu}\eta_o}{a} + \frac{1}{a^3}0\left(-\frac{\partial a}{\partial s}\right)\int_0^1 \dot{b}^2 + 2\int_0^1 \frac{\ddot{b}\dot{b}\theta(\lambda_o + \bar{\mu}\eta_o)}{a} +$$

$$+\int_0^1 \frac{\theta\dot{b}^2}{a} \times b\eta_o \leq \frac{c}{10}\int_0^1 \ddot{b}^2 + \frac{C}{c}\int_0^1 \dot{b}^2 + 2\int_0^1 \frac{\ddot{b}}{a^2}\frac{\ddot{\theta}\eta_o + 2\dot{\theta}\dot{\eta}_o - \dot{\theta}\bar{\mu}\eta_o}{a}.$$

Of course, we always add to our flow a small regularizing component i.e we use $N + cb$ and this yields a term $-c\int_0^1 \ddot{b}\frac{\ddot{\theta}\eta_o + 2\dot{\theta}\dot{\eta} - \dot{\theta}\bar{\mu}\eta_o}{a}$.

If $\dot{b}$ is continuous, $\ddot{\theta}\eta_o + 2\dot{\theta}\dot{\eta}_o - \dot{\theta}\bar{\mu}\eta_o = 0(1)$ on $[t_A, t_A + \varepsilon]$, for ε small enough. This extends to an H^1 weak-neighborhood of b, then and we thus have:

$$\int_{t_a}^{t_A+\varepsilon} \ddot{b}\frac{\ddot{\theta}\eta_o + 2\dot{\theta}\dot{\eta}_o - \dot{\theta}\bar{\mu}\eta_o}{a} = 0\left(\int_{t_A}^{t_A+\varepsilon} |\ddot{b}|\right) \leq \sqrt{\varepsilon}\int_0^1 \ddot{b}^2.$$

It suffices then to choose $\sqrt{\varepsilon} < \frac{c}{10}$ and this term is also absorbed. Observe that this construction is entirely compatible with the requirement $|b(t_A)| \leq 1$. Also, c for this flow, as well as for the deconcentration flow (before the introduction of the weights and the convex combination) has c away from zero. Thus, we can require $\sqrt{\varepsilon} < \frac{c}{10}$. We will see how to preserve suitable inequalities after convex-combination.

We are left with the contribution on $[t_\beta, t_\alpha]$.

Using the special form of the function $\psi_{\alpha\beta}$ and the same tricks as above, this reduces to:

$$\int \ddot{b}\ddot{\theta}\eta_o = -\left(\int \ddot{b}\psi''_{\alpha\beta}\dot{b}^2\eta_o + \int \ddot{b}^2\psi'_{\alpha\beta}\eta_o\right) \leq -\int \ddot{b}\psi''_{\alpha\beta}\dot{b}^2\eta_o$$

since η_o is positive on the support of $\psi'_{\alpha\beta}(|b|)$.

Indeed, we can absorb the contribution of $2\int_{t_\alpha}^{t_\beta} \ddot{b}\dot{\theta}(2\dot{\eta}_o - \bar{\mu}\eta_o)$ into $-\int \ddot{b}^2\psi'_{\alpha\beta}\eta_0$ as follows:

$$-\int \ddot{b}^2\psi'_{\alpha\beta}\eta_0 \text{ behaves as } -\frac{1}{\beta - \alpha}\int_{t_2}^{t_\beta} \ddot{b}^2.$$

On the other hand, since $\ddot{b}$ vanishes on $[t_\alpha, t_\beta]$ if the drop is sharp - if it is not, we can assume that $\beta - \alpha \geq \frac{1}{2}$ by a simple refinement of the argument about the behavior of $|\dot{b}|$ along drops; the argument becomes then easy - we have

$$|\dot{b}|(t) \leq \int_{t_\alpha}^{t_\beta} |\ddot{b}| \quad \forall t \in [t_\alpha, t_\beta].$$

Thus,

$$\Big|\int_{t_\alpha}^{t_\beta} \ddot{b}\dot{\theta}(2\dot{\eta}_0 - \bar{\mu}\eta_0)\Big| \leq \frac{C(|b|_\infty + 1)}{\beta - \alpha}\left(\int_{t_\alpha}^{t_\beta} |\ddot{b}|\right)^2 \leq$$
$$\leq \frac{C(1 + |b|_\infty)}{\beta - \alpha}\int_{t_\alpha}^{t_\beta} \ddot{b}^2 \times (t_\beta - t_\alpha).$$

Since $C(1 + |b|_\infty)(t_\beta - t_\alpha)$ can be asumed to be small, we have shown that this term is controlled. We return now to $\int \ddot{b}\psi''_{\alpha\beta}\dot{b}^2\eta_0$.

We can squeeze $\psi_{\alpha\beta}$ on an interval where $\dot{b}$ is non zero. This uses the drops which we must have in a $W^{1,1}$ neighborhood of infinity. On such an interval (assuming that $\dot{b}$ is continuous), $|\dot{b}|$ takes only once the value α and the value β. Clearly, (if b is positive on $[t_\beta, t_\alpha], \dot{b} < 0$).

$$\int \ddot{b}\psi''_{\alpha\beta}\dot{b}^2\eta_o = \frac{\ddot{b}\dot{b}(t_\beta)\eta_o(t_\beta) - \ddot{b}\dot{b}(t_\alpha)\eta_o(t_\alpha)}{\beta - \alpha}.$$

If the drop is very sharp, we can, by averaging over t_β and t_α, with weight $\frac{1}{\eta_o(t_\beta)}$ and $\frac{1}{\eta_o(t_\alpha)}$ respectively, derive a zero contribution of such a term; and this estimate, with zero replaced by 1, extends to an H^1 weak-neighborhood of b, making use of the viscosity term $-c\int_0^1 \ddot{b}^2$, which forces the strong H^1-convergence, or the estimate. Indeed, on any interval $A, |\int_A \dot{\tilde{b}}^2 - \dot{b}^2| \leq \frac{c}{100}\int_0^1 \ddot{b}^2 + \ddot{\tilde{b}}^2 + \frac{C}{c}|b - \tilde{b}|^2_{L^\infty}$. We can, as usual, avoid the values where $b = o$ or $|\dot{b}| = \frac{\nu}{L}, \frac{2\nu}{L}$.

Otherwise, we can use intervals of length $\frac{1}{|b|_\infty^{1000000}}$ between values where $|\dot{b}| \leq C(|b|_\infty)$. We can also assume that $\beta - \alpha \geq \frac{1}{2}$, otherwise we are brought back to sharp drops. By averaging again, we derive:

$$\int \ddot{b}\tilde{\psi}''_{\alpha\beta}\dot{b}^2\eta_o \leq C(|b|_\infty).(\tilde{\psi}_{\alpha\beta} \text{ stands for } \psi_{\alpha\beta} \text{ after averaging}).$$

Adding cb to N, we then have derived, if $\dot{b}$ was continuous:

$$\frac{\partial}{\partial s}\int_0^1 \left(\frac{\dot{b}}{a}\right)^2 \leq -\frac{9c}{10}\int_0^1 \ddot{b}^2 + \frac{C}{c}\int_0^1 \dot{b}^2 + C(|b|_\infty).$$

Since $\frac{\partial}{\partial s}\int_0^1 \frac{\dot{b}^+}{a} \leq -C'\frac{\partial a}{\partial s}$ under the same assumption, we have:

$$\frac{\partial}{\partial s}\int_0^1 \left(\frac{\dot{b}}{a}\right)^2 \leq \frac{C}{c}\left(1 - C'\frac{\partial a}{\partial s} - \frac{\partial \int_0^1 \dot{b}^+}{\partial s \quad a}\right)\int_0^1 \dot{b}^2 + C(|b|_\infty).$$

This formula convex-combines with the one we have for the cancellation / stretching flow. A common c away, from zero, can be taken for both of them. After convex-combining, we can multiply the convex-combination by a suitable function $\omega(x)$ which allows to keep the exit set unchanged and replace, on it and out of it, the former flow by the new one.

If b is not continuous, it is easy to absorb all terms in

$$\frac{\partial}{\partial s}\int_0^1 \left(\frac{\dot{b}}{a}\right)^2 \text{ into } -\frac{c}{10}\int_0^1 \ddot{b}^2,$$

this on an H^1 weak-neighborhood.

Then,

$$\frac{\partial}{\partial s}\int_0^1 \frac{\dot{b}^+}{a} \leq \frac{C}{\varepsilon}\int_A^{A+\varepsilon} |\dot{b}| - C'\frac{\partial a}{\partial s}$$

where ε is given, small. Taking a very small H^1 weak-neighborhood, we have:

$$\frac{\partial}{\partial s}\int_0^1 \frac{\dot{\tilde{b}}^+}{\tilde{a}} \leq \frac{c}{100\int_0^1 \dot{\tilde{b}}^2}\int_0^1 \ddot{\tilde{b}}^2 \text{ for ever } \tilde{b} \text{ in this neighborhood.}$$

Thus,

$$\frac{\partial}{\partial s}\int_0^1 \left(\frac{\dot{\tilde{b}}}{\tilde{a}}\right)^2 \leq \frac{C}{c}\left(1 - C'\frac{\partial a}{\partial s} - \frac{\partial\left(\frac{\int_0^1 \dot{\tilde{b}}^+}{\tilde{a}}\right)}{\partial s}\right)\int_0^1 \dot{\tilde{b}}^2 + C(|b|_\infty).$$

Furthermore, on a ball $B(o, R)$ of H^1, we use a finite number of functions θ, even after suitable regularization of $\psi_{\alpha\beta}$. This is due to the weak continuity of our construction.

The result follows. Our construction is complete.

It is quite clear that if a deconcentration jump is preceeded and followed by sizable ξ-pieces, the curve will decrease sizably. Using the flow at infinity, we can also manage to have on a topological class of dimension k_o at most k_o of such jumps.

Therefore, the curves which, flowing out of infinity, do not decrease sizably essentially remain at infinity.

The combination of the $Z_{\nu/L}/Z_{2\nu/L}$ flows used in I.4 and I.5 can be applied to all of them with the same estimates.

Locally, all curves will decrease: the ones which are $\frac{\nu}{L}$ or $2\nu/L$-stretched, their unstable manifolds which do not involve a modification of the v-branches and the other pieces of the unstable manifolds involving the deconcentration phenomenon: all of this scheme decreases together, untill reaching the curves of $\bigcup_k \Gamma_{2k}$ and their unstable manifolds.

Some more tedious details are certainly necessary to claim, up to the tiniest detail, the foreseen picture. But we have largely opened the road to it. We will write, after this lengthy paper, these tedious details some other time.

References

1. Bahri, A., *Pseudo-Orbits of Contact Forms*, Pitman Research Notes in Mathematics Series No. 173, Longman, London, 1988.
2. Bahri, A., *Critical Points at Infinity in Some Variational Problems*, Pitman Research Notes in Mathematics Series No. 182, Longman, London.
3. Bahri, A. and Coron, J.M., *Vers une Theoric des Points Critiques à l' Infini* (1984), Seminaire E.D.P. Ecole Polytechnique, France.
4. Bahri, A. and Coron, J.M., *On a nonlinear elliptic equation involving the critical Sobolev exponent: The effect of the topology of the domain*, Comm. Pure Appl. Math **41** (1988), 253–294.
5. Bahri, A., *Proof of the Yamabe Conjecture, Without the Positive Mass Theorem, for Locally Conformally Flat Manifolds, in Einstein Metrics and Yang-Mills Connections*, Proceedings of the 27th Taniguchi International Symposium (T. Mabuchi and S. Mukai, eds.).
6. Bahri, A and Brezis, H., *Nonlinear Elliptic Equations on Riemannian Manifolds with the Sobolev Critical Exponent*, Topics in Geometry Volume In Memory of Joseph D' Atri. (Simon Gindikin, ed.), Birkhaüser.
7. Bahri, A., *An Invariant for Yamabe-type Flows with Applications to the Scalar-Curvature Problem in Higher Dimension* (1996), Duke Mathematical Journal, Volume in Honor of Nash.
8. Abdelmoula, N. and Yãcoub, R., *Proof of the CR Yamabe conjecture via the critical points at infinity* (to appear).
9. Bahri, A. and Rabinowitz, P.H., *Periodic solutios of Hamiltonian systems of 3-body type*, Ann. Inst. Henri Poincaré **8, no. 6** (1991), 561–659.
10. Rabinowitz, P.H., *Periodic solutions of Hamiltonian systems*, Comm. Pure and Applied Math. **31** (1978), 157–184.
11. Weinstein, A., *Periodic orbits for convex Hamiltonian systems*, Ann. of Math **(2) 108** (1978 no. 3), 507–578.
12. Weinstein, A., *On the hypotheses of Rabinowitz' periodic orbit theorems*, J. Diff. Equ **133** (1979), 353–358.
13. Souriau, J.M., *Structure des Systèmes Dynamiques* (1970), Dunod, Paris.
14. Hofer, H. and Viterbo, C., *The Weinstein conjecture in cotangent bundles and related results*, Ann. Scuola. Norm. Sup. Pisa. Cl Sir. **(4) 15** (1988 no. 3), 411–445.
15. Floer, A., Hofer, H. and Viterbo, C., *The Weinstein conjecture in* $P \times \mathbb{C}^\ell$ *Math. Z.* **203** (1990 no. 3), 469–482.
16. Smale, S., *Regular curves on Riemannian manifolds*, Trans. Amer. Math. Soc. **87** (1958), 492–512.
17. Boothby, W., *On the integral curves of a linear differential form of maximum rank*, Math. Ann. **177** (1968), 1–104.
18. Schrödinger, E., *Memoirs sur la Mecanique Ondulatoire* (1933), Paris, Librairie Felix Alcan.

19. Heisenberg, W., *Les Principes Physiques de la Theorie des Quanta* (1932), Paris, Gauthier-Villars.
20. Bell, J.S., *Speakable and Unspeakable in Quantum Mechanics* (1993), Cambridge University Press.
21. Penrose, R., *The Emperor's New Mind* (1978), Penguin Books Oxford University Press.
22. Penrose, R., *Shadows of the Mind* (1989), Oxford University Press.
23. Conley, C.C. and Easton, R.W., *Isolated invariant sets isolating blocks*, Trans. Amer. Math. Soc. **188** (1971), 35–61.
24. Krasnosels'kii, M.A., *Topological Methods in the Theory of Nonlinear Integral Equations* (1964), Macmillan, New York.
25. Bahri, A. and Berestycki, H., *Forced vibrations of superquadratic Hamiltonian systems*, Acta Math **152** **3–4** (1984), 143–197.
26. Hofer, H., *Pseudo-holomorphic in symplectizations with applications to the three dimensional Weinstein conjecture*, Inventiones Math. **114** (1993), 515–563.